Chromatic Graph Theory

Textbooks in Mathematics
Series editors:
Al Boggess and Ken Rosen

CRYPTOGRAPHY: THEORY AND PRACTICE, FOURTH EDITION
Douglas R. Stinson and Maura B. Paterson

GRAPH THEORY AND ITS APPLICATIONS, THIRD EDITION
Jonathan L. Gross, Jay Yellen and Mark Anderson

COMPLEX VARIABLES: A PHYSICAL APPROACH WITH APPLICATIONS,
SECOND EDITION
Steven G. Krantz

GAME THEORY: A MODELING APPROACH
Richard Alan Gillman and David Housman

FORMAL METHODS IN COMPUTER SCIENCE
Jiacun Wang and William Tepfenhart

AN ELEMENTARY TRANSITION TO ABSTRACT MATHEMATICS
Gove Effinger and Gary L. Mullen

ORDINARY DIFFERENTIAL EQUATIONS: AN INTRODUCTION TO THE
FUNDAMENTALS, SECOND EDITION
Kenneth B. Howell

SPHERICAL GEOMETRY AND ITS APPLICATIONS
Marshall A. Whittlesey

COMPUTATIONAL PARTIAL DIFFERENTIAL PARTIAL EQUATIONS USING
MATLAB®, SECOND EDITION
Jichun Li and Yi-Tung Chen

AN INTRODUCTION TO MATHEMATICAL PROOFS
Nicholas A. Loehr

DIFFERENTIAL GEOMETRY WITH MANIFOLDS, SECOND EDITION
Stephen T. Lovett

MATHEMATICAL MODELING WITH EXCEL
Brian Albright and William P. Fox

THE SHAPE OF SPACE
Jeffrey R. Weeks

CHROMATIC GRAPH THEORY, SECOND EDITION
Gary Chartrand and Ping Zhang

https://www.crcpress.com/Textbooks-in-Mathematics/book-series/CANDHTEX-
BOOMT

Chromatic Graph Theory

Second Edition

Gary Chartrand and Ping Zhang

Western Michigan University

CRC Press
Taylor & Francis Group
Boca Raton London New York

CRC Press is an imprint of the
Taylor & Francis Group, an **informa** business

CRC Press
Taylor & Francis Group
6000 Broken Sound Parkway NW, Suite 300
Boca Raton, FL 33487-2742

First issued in paperback 2022

ISBN 13: 978-1-03-247510-3 (pbk)
ISBN 13: 978-1-138-34386-3 (hbk)

DOI: 10.1201/9780429438868

**Visit the Taylor & Francis Web site at
http://www.taylorandfrancis.com**

**and the CRC Press Web site at
http://www.crcpress.com**

To Ralph Stanton (1922-2010)

whose contributions to discrete mathematics and enthusiastic support of mathematicians are acknowledged and greatly appreciated by all who knew him; he will long be remembered.

Table of Contents

PREFACE TO THE SECOND EDITION

Graph coloring is one of the most popular areas of graph theory, likely due to its many fascinating problems and applications as well as the sheer beauty of the subject. Because of the numerous interesting problems that have been introduced in this area over many decades, there are many options available as to what might be included in a book on graph colorings. Surely, the topics that are considered fundamental and historically important in this area must be included. However, there are several topics on graph colorings that have appeared in recent years and which have attracted the attention and interest of many mathematicians but have not previously appeared in books. We decided to include a number of these as well.

In this second edition of *Chromatic Graph Theory*, we begin with a discussion of the origin of the Four Color Problem, which is the beginning of graph colorings. The primary goal of this book, however, is to introduce graph theory with a coloring theme and emphasis – to explore connections between major topics in graph theory and graph colorings and to look at graph colorings in a variety of ways. This book has been written with the intention of using it for one or more of the following purposes:

- for a course in graph theory with an emphasis on graph colorings, where this course could be either a beginning course in graph theory or a follow-up course to an elementary graph theory course;

- for a reading course on graph colorings;

- for a seminar on graph colorings;

- as a reference book for individuals interested in graph colorings.

To accomplish this, it has been our goal to write this book in an engaging, student-friendly style so that it contains carefully explained proofs and examples as well as many exercises of varying difficulty. Both the text and exercises contain problems that suggest research topics that can be explored further.

MATERIAL NEW TO THE SECOND EDITION

Among the material new to this second edition are the following:

(1) additional material on Ramsey theory, including topics dealing with bipartite, proper, and monochromatic Ramsey numbers and Gallai-Ramsey numbers;

(2) in addition to rainbow-connected graphs, there are discussions dealing with proper connected and rainbow Hamiltonian-connected graphs;

(3) rainbow disconnection in connected graphs;

(4) additional material that looks at domination through coloring;

(5) majestic, royal, and regal colorings, which are edge colorings that induce vertex coloring in a set theoretic manner;

(6) rainbow mean colorings, which are edge colorings that induce a vertex coloring defined in an arithmetic manner;

(7) zonal labelings of planar graphs and its connection with the Four Color Problem.

The second edition of *Chromatic Graph Theory* contains over 150 new exercises, some of which suggest new areas for research in graph colorings.

OUTLINE OF CONTENTS

The second edition of *Chromatic Graph Theory* consists of 18 chapters (Chapters 0–17). Chapter 0 sets the stage for the remainder of the book. This chapter describes the origin of graph colorings and, in a sense, how coloring contributed to graph theory becoming an area of mathematics. The origin of the Four Color Problem, some of its history, and attempts to solve the problem are discussed. This laid the groundwork for numerous types of colorings in graphs, many of which are presented in Chapters 6–17.

In order to have a better understanding of the material presented on graph coloring, it is important to be familiar with other fundamental areas of graph theory. To achieve the goal of having the book self-contained, Chapters 1-5 have been written to contain many of the fundamentals, concepts, and results of graph theory that lie outside of graph colorings.

⋆ Chapter 1 describes much of the basic terminology of graph theory, introducing the notation of the concepts that commonly appear in graph theory.

⋆ Much of graph theory deals with graphs that are connected. This is the primary topic of Chapter 2. The simplest type of connected graphs are the trees, also discussed in this chapter. Among the measures that describe how connected a graph may be, the two best known are connectivity and edge-connectivity, both of which are discussed in Chapter 2.

⋆ There are various ways that one may proceed about a connected graph. The two best known of these, each with an interesting origin and history, result in graphs referred to as Eulerian graphs and Hamiltonian graphs, named for the two famous mathematicians Leonhard Euler and Sir Willian Rowan Hamilton, each of whom played a role in the history of graph theory. These are the primary topics of Chapter 3.

⋆ Much of the interest and research in graph theory deals with whether a given graph contains a particular subgraph or possesses properties under which a graph may contain such a subgraph. One of these subgraphs deals with what is called a matching. A related topic concerns the problem of decomposing a graph into these types of subgraphs or other subgraphs of interest. These are the topics of Chapter 4.

⋆ Since the subject of graph coloring came from the Four Color Problem, which deals with coloring maps drawn on a sphere or in a plane, there has been an interest for many years on those graphs that can be drawn in the plane without any of its edges crossing. A discussion of these planar graphs is the primary topic of Chapter 5. If a graph is not planar, then there are more complex surfaces on which the graphs can be drawn or embedded. This is discussed in Chapter 5 as well.

⋆ The main topic of this book, namely graph colorings, is introduced in Chapter 6, where the subject of vertex coloring is described. As was the goal of solving the Four Color Problem, the major interest in a vertex coloring is one in which as few colors are used as possible so that no two neighboring vertices are colored the same. The basic terminology of vertex coloring is described here as well as a number of possibly unexpected applications of vertex colorings. If a subgraph of a given graph requires a certain number of colors, then at least that many colors are needed to color the vertices of the graph itself. This observation gave rise to the concept of perfect graphs, a topic also discussed in Chapter 6.

⋆ It is often extraordinarily difficult to determine the exact minimum number of colors needed to color the vertices of a graph so that no two neighboring vertices are colored the same. Because of this, there has been interest in knowing a number of colors, at most of which can be used to color the vertices of a graph. A number of such upper bounds are discussed in Chapter 7, the primary topic of this chapter.

⋆ One way of stating the Four Color Problem is that of determining whether 4 is the minimum number of colors needed to color the vertices of all planar graphs so that no two neighboring vertices are colored the same. During the 124-year period in which this problem was known but a solution unknown, other results, problems, and concepts came about in hopes of understanding graph coloring better. While the Four Color Problem was difficult, the Five Color Theorem was not. The number of ways to color the vertices of a graph with a specified number of colors was introduced and turned out to be a polynomial in the number of colors. These are the topics of Chapter 8.

⋆ The minimum number of colors needed to color the vertices of a graph G so that no two neighboring vertices are colored the same is the chromatic number of G. If the chromatic number of a graph G is k, then it's possible to partition the vertex set into k subsets, each subset containing vertices no two of which are neighboring. There are occasions when there is only one such partition. This is a topic discussed in Chapter 9. When there is a vertex coloring of a graph G, with k colors say, then each vertex of G can be assigned any one of these k colors with the only condition being that at the conclusion no two neighboring vertices are colored the same. There has been interest in vertex colorings where each vertex can be colored with a color in a prescribed proper subset or list of the k colors. Such list colorings are also discussed in this chapter. There are also situations where the vertices of a subgraph of a graph G is colored with k colors, bringing up the question as to whether this coloring can be extended to a vertex coloring of G itself – another topic discussed in Chapter 9.

⋆ Of the many attempts to solve the Four Color Problem and show that the vertices of every planar graph could be colored with four colors so that no two neighboring vertices are colored the same, one such attempt involved coloring the edges of certain planar graphs so that no two edges incident with a common vertex are colored the same. The number of edges incident with a vertex is its degree. While the minimum number of colors needed to color the edges of a graph is always at least the maximum degree, an important theorem states that it never exceeds this maximum degree by more than 1. Consequently, the minimum number of colors needed to color the edges of a graph is always one of two numbers. This is the primary topic of Chapter 10.

⋆ There is a theorem, called Ramsey's theorem, that implies that for every two graphs, say F and H, there are complete graphs (every two vertices are joined by an edge) such that if the edges of these complete graphs are colored red or blue in any manner whatsoever, then there results either a subgraph F all of whose edges are colored red or a subgraph H all of whose edges are colored blue. The minimum number of vertices in such a complete graph is the Ramsey number of F and H. This and related Ramsey numbers where complete graphs are replaced by other graphs are the primary topics of Chapter 11.

⋆ There have been numerous variations of Ramsey numbers introduced, many of which deal with properties required of the colors of given subgraphs. A number of these types of Ramsey numbers are described in Chapter 12.

⋆ The September 11, 2001 terrorist attacks have given rise to many concepts related to security of communication networks. Connected edge-colored graphs can be used to model and study the transfer of information. This has created an interest in edge colorings of connected graphs so that every two vertices are connected by a path whose edges satisfy some prescribed property and in the minimum number of colors that will accomplish this. One of these requires that the colors of the edges in one such path be distinct for each pair of vertices. This results in the concept of rainbow-connected graphs. Another concept deals with coloring the edges of a graph with as few colors as possible so that for each pair of vertices there always exists a set of edges no two of which are colored the same and whose removal results in a graph where the vertices are not connected. These are the major topics of Chapter 13.

⋆ Chapters 14 and 15 return to vertex colorings, each chapter discussing conditions to be satisfied by a coloring.

Chapter 14 deals with distance and lengths of paths connecting two vertices. Many of these concepts have come from graphs that model the Channel Assignment Problem in a certain way. Here the goal is to assign channels to transmitters located in some region so that clear reception of the transmitted signals results. In these cases, the channels are the colors and requirements on the channels result in conditions on the colors.

* A topic in graph theory that has gained increasing interest in recent decades is domination. Typically, a vertex in a graph dominates itself and each neighboring vertex and a dominating set is a set of vertices having the property that each vertex in the graph is dominated by at least one vertex in the set. The primary interest here has been determining the minimum number of vertices in such a dominating set. There are other sets of vertices satisfying other types of domination. In Chapter 15, it is shown that many of the best-known types of domination can be looked in terms of coloring the vertices of a graph with one of two colors.

* Chapter 16 deals with both edge colorings and vertex colorings of graphs. In each case, every edge of a graph is assigned a color which induces a color for each vertex in some manner. In the first instance, each edge is assigned a positive integer for its color and the vertex coloring is a set, namely the set of colors of the edges incident with the vertex. The main problem here is to minimize the number of colors needed for the edges so that the resulting vertex coloring has some prescribed property. This has been generalized to where each edge is assigned a nonempty subset of some set of colors so that each vertex is assigned either the union or the intersection of the colors of its incident edges and, again, the resulting vertex coloring satisfies some prescribed property. In the final edge coloring, each edge is assigned a positive integer for its color, resulting in a positive integer color for each vertex, where again, the vertex coloring satisfies some prescribed property.

* In the final Chapter 17, we close where we began, namely with the Four Color Problem. In this case, we describe a vertex labeling of planar graphs due to Cooroo Egan. Here, the problem is whether the vertices of a planar graph embedded in the plane can be labeled with the nonzero elements of the additive group $\mathbb{Z}_3$ in such a way that the sum of the labels of the vertices on the boundary of each zone (region) is a constant, namely 0. It is seen that there is a connection with this type of labeling, called a zonal labeling, with the Four Color Problem.

TO THE STUDENT

In the 1950s and 1960s, it was unusual for colleges and universities to offer courses in graph theory or even in discrete mathematics. However, as time went by, the importance and applicability of discrete mathematics has become increasingly clear. One of the most important areas within discrete mathematics is graph theory, a subject that is considered to have begun in 1736 when the famous Swiss mathematician Leonhard Euler solved the equally famous Königsberg Bridge Problem (which is discussed in Chapter 3). What Euler accomplished did not cause graph theory to become an area of mathematics soon afterward, however. Indeed, for the next 150-200 years, graphs primarily occurred indirectly in puzzles and other recreational mathematics. However, during the second half of the 19th century and especially beginning in 1936, when the first book on graph theory by the Hungarian mathematician Dénes König was published, graph theory began developing into a "theory".

As with much of mathematics, graph theory only grew significantly after World War II. In the case of graph theory, though, the early interest in this subject had much to do with a problem that occurred in 1852: The Four Color Problem. It was attempts to solve this problem during the next 124 years that was the impetus for graph coloring becoming a major area of graph theory. Indeed, this problem was the beginning of graph coloring becoming perhaps the most popular area within graph theory. In this second edition, Chapter 0 gives some historical background on the subject of graph coloring. The next five chapters provide some of the fundamental concepts and topics within graph theory that do not involve coloring. All of the remaining twelve chapters deal with graph coloring, including material that has become standard, followed by more recent material that mathematicians are investigating even now. Many of the topics discussed in these later chapters are still developing. Studying these topics gives one ideas on how some areas of research in graph theory came about and may suggest ideas for new contributions to graph theory.

TO THE INSTRUCTOR

If this book is being used as a textbook for a course or a reading course, then the way the material is to be covered depends on whether this is a first course in graph theory for the students. Regardless of the primary purpose, since this book is written with a graph coloring theme, it would be good for students to read Chapter 0 on their own to give them background on this subject.

Let's first consider the case where this is a first course in graph theory for the students. Chapters 1–5 cover some fundamental material on graph theory not dealing with graph colorings, but which is important in the study of various aspects of graph colorings. Chapter 6 covers vertex colorings. Consequently, it would be good to design a course to cover various parts of Chapters 1–6. Topics could then be chosen from the remaining chapters that are of interest to the instructor.

If the students have already had a beginning course on graph theory, then it would be good to quickly go through the material in Chapter 1 so that everyone is familiar with the terminology and notation being used throughout the book. Since the students already had a course in graph theory, it is likely that they have already encountered much of the material from Chapters 1-6. Nevertheless, it is good for students to review this material. The instructor can then design a course from the remaining chapters that not only interest the instructor but is likely to interest the students as well. Depending on what the purpose of the course is, topics can be chosen so that the students can be creative and ask questions of their own.

ACKNOWLEDGMENT

We thank Bob Ross (Senior Editor at CRC Press/Chapman & Hall, Taylor & Francis Group) for his constant encouragement and interest and for suggesting this writing project to us.

G.C. & P.Z.

List of Symbols

Chapter 0

The Origin of Graph Colorings

If the countries in a map of South America (see Figure 1) were to be colored in such a way that every two countries with a common boundary are colored differently, then this map could be colored using only four colors. Is this true of every map?

While it is not difficult to color a map of South America with four colors, it is not possible to color this map with less than four colors. In fact, every two of Brazil, Argentina, Bolivia, and Paraguay are neighboring countries and so four colors are required to color only these four countries.

It is probably clear why we might want two countries colored differently if they have a common boundary – so they can easily be distinguished as different countries in the map. It may not be clear, however, why we would think that four colors would be enough to color the countries of every map. After all, we can probably envision a complicated map having a large number of countries with some countries having several neighboring countries, so constructed that a great many colors might possibly be needed to color the entire map. Here we understand neighboring countries to mean two countries with a boundary *line* in common, not simply a single point in common.

While this problem may seem nothing more than a curiosity, it is precisely this problem that would prove to intrigue so many for so long and whose attempted solutions would contribute so significantly to the development of the area of mathematics known as Graph Theory and especially to the subject of graph colorings: Chromatic Graph Theory. This map coloring problem would eventually acquire a name that would become known throughout the mathematical world.

The Four Color Problem *Can the countries of every map be colored with four or fewer colors so that every two countries with a common boundary are colored differently?*

Figure 1: Map of South America

Many of the concepts, theorems, and problems of Graph Theory lie in the shadows of the Four Color Problem. Indeed ...

Graph Theory is an area of mathematics whose past is always present.

Since the maps we consider can be real or imagined, we can think of maps being divided into more general regions, rather than countries, states, provinces, or some other geographic entities.

So just how did the Four Color Problem come about? It turns out that this question has a rather well-documented answer. On 23 October 1852, a student, namely Frederick Guthrie (1833–1886), at University College London visited his mathematics professor, the famous Augustus De Morgan (1806–1871), to describe an apparent mathematical discovery of his older brother Francis. While coloring the counties of a map of England, Francis Guthrie (1831–1899) observed that he could color them with four colors, which led him to conjecture that no more than four colors would be needed to color the regions of any map.

The Four Color Conjecture *The regions of every map can be colored with four or fewer colors in such a way that every two regions sharing a common boundary are colored differently.*

Two years earlier, in 1850, Francis had earned a Bachelor of Arts degree from University College London and then a Bachelor of Laws degree in 1852. He would later become a mathematics professor himself at the University of Cape Town in South Africa. Francis developed a lifelong interest in botany and his extensive collection of flora from the Cape Peninsula would later be placed in the Guthrie Herbarium in the University of Cape Town Botany Department. Several rare species of flora are named for him.

Francis Guthrie attempted to prove the Four Color Conjecture and although he thought he may have been successful, he was not completely satisfied with his proof. Francis discussed his discovery with Frederick. With Francis's approval, Frederick mentioned the statement of this apparent theorem to Professor De Morgan, who expressed pleasure with it and believed it to be a new result. Evidently Frederick asked Professor De Morgan if he was aware of an argument that would establish the truth of the theorem.

This led De Morgan to write a letter to his friend, the famous Irish mathematician Sir William Rowan Hamilton (1805–1865), on 23 October 1852. These two mathematical giants had corresponded for years, although apparently had met only once. De Morgan wrote (in part):

My dear Hamilton:

A student of mine asked me to day to give him a reason for a fact which I did not know was a fact – and do not yet. He says that if a figure be any how divided and the compartments differently coloured so that figures with any portion of common boundary <u>lines</u> are differently coloured – four colours may be wanted but not more – the following is his case in which four <u>are</u> wanted.

 A B C D are
names of
colours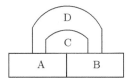

Query cannot a necessity for five or more be invented …

My pupil says he guessed it colouring a map of England …. The more I think of it the more evident it seems. If you retort with some very simple case which makes me out a stupid animal, I think I must do as the Sphynx did …

In De Morgan's letter to Hamilton, he refers to the "Sphynx" (or Sphinx). While the Sphinx is a male statue of a lion with the head of a human in ancient Egypt which guards the entrance to a temple, the Greek Sphinx is a female creature of bad luck who sat atop a rock posing the following riddle to all those who pass by:

> *What animal is that which in the morning goes on four feet, at noon on two, and in the evening upon three?*

Those who did not solve the riddle were killed. Only Oedipus (the title character in *Oedipus Rex* by Sophocles, a play about how people do not control their own destiny) answered the riddle correctly as "Man", who in childhood (the morning of life) creeps on hands and knees, in manhood (the noon of life) walks upright, and in old age (the evening of life) walks with the aid of a cane. Upon learning that her riddle had been solved, the Sphinx cast herself from the rock and perished, a fate De Morgan had envisioned for himself if his riddle (the Four Color Problem) had an easy and immediate solution.

In De Morgan's letter to Hamilton, De Morgan attempted to explain why the problem appeared to be difficult. He followed this explanation by writing:

> *But it is tricky work and I am not sure of all convolutions – What do you say? And has it, if true been noticed?*

Among Hamilton's numerous mathematical accomplishments was his remarkable work with quaternions. Hamilton's quaternions are a 4-dimensional system of numbers of the form $a + bi + cj + dk$, where $a, b, c, d \in \mathbb{R}$ and $i^2 = j^2 = k^2 = -1$. When $c = d = 0$, these numbers are the 2-dimensional system of complex numbers; while when $b = c = d = 0$, these numbers are simply real numbers. Although it is commonplace for binary operations in algebraic structures to be commutative, such is not the case for products of quaternions. For example, $i \cdot j = k$ but $j \cdot i = -k$. Since De Morgan had shown an interest in Hamilton's research on quaternions as well as other subjects Hamilton had studied, it is likely that De Morgan expected an enthusiastic reply to his letter to Hamilton. Such was not the case, however. Indeed, three days later, on 26 October 1852, Hamilton gave an unexpected response:

> *I am not likely to attempt your "quaternion" of colours very soon.*

Hamilton's response did nothing however to diminish De Morgan's interest in the Four Color Problem.

Since De Morgan's letter to Hamilton did not mention Frederick Guthrie by name, there may be reason to question whether Frederick was in fact the student to whom De Morgan was referring and that it was Frederick's older brother Francis who was the originator of the Four Color Problem.

In 1852 Frederick Guthrie was a teenager. He would go on to become a distinguished physics professor and founder of the Physical Society in London. An area that he studied was the science of thermionic emission – first reported by Frederick Guthrie in 1873. He discovered that a red-hot iron sphere with a positive charge would lose its charge. This effect was rediscovered by the famous American inventor Thomas Edison early in 1880. It was during 1880 (only six years before Frederick died) that Frederick wrote:

Some thirty years ago, when I was attending Professor De Morgan's class, my brother, Francis Guthrie, who had recently ceased to attend them (and who is now professor of mathematics at the South African University, Cape Town), showed me the fact that the greatest necessary number of colours to be used in colouring a map so as to avoid identity colour in lineally contiguous districts is four. I should not be justified, after this lapse of time, in trying to give his proof, but the critical diagram was as in the margin.

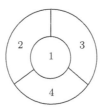

With my brother's permission I submitted the theorem to Professor De Morgan, who expressed himself very pleased with it; accepted it as new; and, as I am informed by those who subsequently attended his classes, was in the habit of acknowledging where he had got his information.

If I remember rightly, the proof which my brother gave did not seem altogether satisfactory to himself; but I must refer to him those interested in the subject. ....

The first statement in print of the Four Color Problem evidently occurred in an anonymous review written in the 14 April 1860 issue of the literary journal *Athenaeum*. Although the author of the review was not identified, De Morgan was quite clearly the writer. This review led to the Four Color Problem becoming known in the United States.

The Four Color Problem came to the attention of the American mathematician Charles Sanders Peirce (1839–1914), who found an example of a map drawn on a torus (a donut-shaped surface) that required six colors. (As we will see in Chapter 5, there is an example of a map drawn on a torus that requires seven colors.) Peirce expressed great interest in the Four Color Problem. In fact, he visited De Morgan in 1870, who by that time was experiencing poor health. Indeed, De Morgan died the following year. Not only had De Morgan made little progress towards a solution of the Four Color Problem at the time of his death, overall interest in this problem had faded. While Peirce continued to attempt to solve the problem, De Morgan's British acquaintances appeared to pay little attention to the problem – with at least one notable exception.

Arthur Cayley (1821–1895) graduated from Trinity College, Cambridge in 1842 and then received a fellowship from Cambridge, where he taught for four years. Afterwards, because of the limitations on his fellowship, he was required to choose a profession. He chose law, but only as a means to make money while he could continue to do mathematics. During 1849–1863, Cayley was a successful lawyer but published some 250 research papers during this period, including many for which he is well known. One of these was his pioneering paper on matrix algebra. Cayley was famous for his work on algebra, much of which was done with the British mathematician James Joseph Sylvester (1814–1897), a former student of De Morgan.

In 1863 Cayley was appointed a professor of mathematics at Cambridge. Two years later, the London Mathematical Society was founded at University College London and would serve as a model for the American Mathematical Society, founded in 1888. De Morgan became the first president of the London Mathematical Society, followed by Sylvester and then Cayley. During a meeting of the Society on 13 June 1878, Cayley raised a question about the Four Color Problem that brought renewed attention to the problem:

> *Has a solution been given of the statement that in colouring a map of a country, divided into counties, only four distinct colours are required, so that no two adjacent counties should be painted in the same colour?*

This question appeared in the Proceedings of the Society's meeting. In the April 1879 issue of the *Proceedings of the Royal Geographical Society*, Cayley reported:

> *I have not succeeded in obtaining a general proof; and it is worth while to explain wherein the difficulty consists.*

Cayley observed that if a map with a certain number of regions has been colored with four colors and a new map is obtained by adding a new region, then there is no guarantee that the new map can be colored with four colors – without first recoloring the original map. This showed that any attempted proof of the Four Color Conjecture using a proof by mathematical induction would not be straightforward. Another possible proof technique to try would be proof by contradiction. Applying this technique, we would assume that the Four Color Conjecture is false. This would mean that there are some maps that cannot be colored with four colors. Among the maps that require five or more colors are those with a smallest number of regions. Any one of these maps constitutes a minimum counterexample. If it could be shown that no minimum counterexample could exist, then this would establish the truth of the Four Color Conjecture.

For example, no minimum counterexample M could possibly contain a region R surrounded by three regions R_1, R_2, and R_3 as shown in Figure 2(a). In this case, we could shrink the region R to a point, producing a new map M' with one less region. The map M' can then be colored with four colors, only three of which are used to color R_1, R_2, and R_3 as in Figure 2(b). Returning to the original map M, we see that there is now an available color for R as shown in Figure 2(c), implying that M could be colored with four colors after all, thereby producing a contradiction. Certainly, if the map M contains a region surrounded by fewer than three regions, a contradiction can be obtained in the same manner.

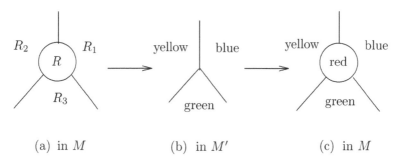

(a) in M (b) in M' (c) in M

Figure 2: A region surrounded by three regions in a map

Suppose, however, that the map M contained no region surrounded by three or fewer regions but did contain a region R surrounded by four regions, say R_1, R_2, R_3, R_4, as shown in Figure 3(a). If, once again, we shrink the region R to a point, producing a map M' with one less region, then we know that M' can be colored with four colors. If two or three colors are used to color R_1, R_2, R_3, R_4, then we can return to M and there is a color available for R. However, this technique does not work if the regions R_1, R_2, R_3, R_4 are colored with four distinct colors, as shown in Figure 3(b).

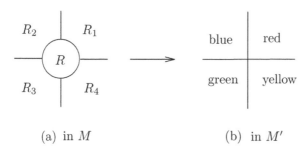

(a) in M (b) in M'

Figure 3: A region surrounded by four regions in a map

What we can do in this case, however, is to determine whether the map M' has a chain of regions, beginning at R_1 and ending at R_3, all of which are colored red or green. If no such chain exists, then the two colors of every red-green chain of regions beginning at R_1 can be interchanged. We can then return to the map M, where the color red is now available for R. That is, the map M can be colored with four colors, producing a contradiction. But what if a red-green chain of regions beginning at R_1 and ending at R_3 exists? (See Figure 4, where r, b, g, y denote the colors red, blue, green, yellow.) Then interchanging the colors red and green offers no benefit to us. However, in this case, there can be no blue-yellow chain of regions, beginning at R_2 and ending at R_4. Then the colors of every blue-yellow chain of regions beginning at R_2 can be interchanged. Returning to M, we see that the color blue is now available for R, which once again says that M can be colored with four colors and produces a contradiction.

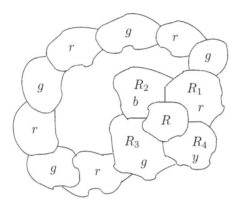

Figure 4: A red-green chain of regions from R_1 to R_3

It is possible to show (as we will see in Chapter 5) that a map may contain no region that is surrounded by four or fewer neighboring regions. Should this occur however, such a map must contain a region surrounded by exactly five neighboring regions.

We mentioned that James Joseph Sylvester worked with Arthur Cayley and served as the second president of the London Mathematical Society. Sylvester, a superb mathematician himself, was invited to join the mathematics faculty of the newly founded Johns Hopkins University in Baltimore, Maryland in 1875. Included among his attempts to inspire more research at the university was his founding in 1878 of the *American Journal of Mathematics*, of which he held the position of editor-in-chief. While the goal of the journal was to serve American mathematicians, foreign submissions were encouraged as well, including articles from Sylvester's friend Cayley.

Among those who studied under Arthur Cayley was Alfred Bray Kempe (1849–1922). Despite his great enthusiasm for mathematics, Kempe took up a career in the legal profession. Kempe was present at the meeting of the London Mathematical Society in which Cayley had inquired about the status of the Four Color Problem. Kempe worked on the problem and obtained a solution in 1879. Indeed, on 17 July 1879 a statement of Kempe's accomplishment appeared in the British journal *Nature*, with the complete proof published in Volume 2 of Sylvester's *American Journal of Mathematics*.

Kempe's approach for solving the Four Color Problem essentially followed the technique described earlier. His technique involved locating a region R in a map M such that R is surrounded by five or fewer neighboring regions and showing that for every coloring of M (minus the region R) with four colors, there is a coloring of the entire map M with four colors. Such an argument would show that M could not be a minimum counterexample. We saw how such a proof would proceed if R were surrounded by four or fewer neighboring regions. This included looking for chains of regions whose colors alternate between two colors and then interchanging these colors, if appropriate, to arrive at a coloring of the regions of M (minus R) with four colors so that the neighboring regions of R used at most three of these colors and thereby leaving a color available for R. In fact, these chains of regions became known as *Kempe chains*, for it was Kempe who originated this idea.

There was one case, however, that still needed to be resolved, namely the case where no region in the map was surrounded by four or fewer neighboring regions. As we noted, the map must then contain some region R surrounded by exactly five neighboring regions. At least three of the four colors must be used to color the five neighboring regions of R. If only three colors are used to color these five regions, then a color is available for R. Hence, we are left with the single situation in which all four colors are used to color the five neighboring regions surrounding R (see Figure 5), where once again r, b, g, y indicate the colors red, blue, green, yellow.

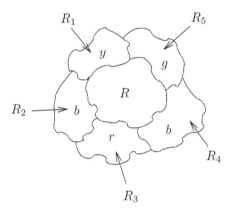

Figure 5: The final case in Kempe's solution of the Four Color Problem

Let's see how Kempe handled this final case. Among the regions adjacent to R, only the region R_1 is colored yellow. Consider all the regions of the map M that are colored either yellow or red and that, beginning at R_1, can be reached by an alternating sequence of neighboring yellow and red regions, that is, by a yellow-red Kempe chain. If the region R_3 (which is the neighboring region of R colored red) cannot be reached by a yellow-red Kempe chain, then the colors yellow and red can be interchanged for all regions in M that can be reached by a yellow-red Kempe chain beginning at R_1. This results in a coloring of all regions in M (except R) in which neighboring regions are colored differently and such that each neighboring region of R is colored red, blue, or green. We can then color R yellow to arrive at a 4-coloring of the entire map M. From this, we may assume that the region R_3 can be reached by a yellow-red Kempe chain beginning at R_1. (See Figure 6.)

Let's now look at the region R_5, which is colored green. We consider all regions of M colored green or red and that, beginning at R_5, can be reached by a green-red Kempe chain. If the region R_3 cannot be reached by a green-red Kempe chain that begins at R_5, then the colors green and red can be interchanged for all regions in M

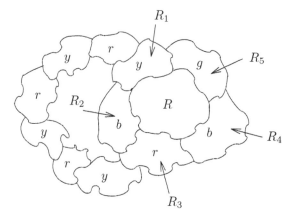

Figure 6: A yellow-red Kempe chain in the map M

that can be reached by a green-red Kempe chain beginning at R_5. Upon doing this, a 4-coloring of all regions in M (except R) is obtained, in which each neighboring region of R is colored red, blue, or yellow. We can then color R green to produce a 4-coloring of the entire map M. We may therefore assume that R_3 can be reached by a green-red Kempe chain that begins at R_5. (See Figure 7.)

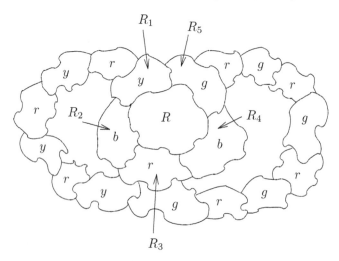

Figure 7: Yellow-red and green-red Kempe chains in the map M

Because there is a ring of regions consisting of R and a green-red Kempe chain, there cannot be a blue-yellow Kempe chain in M beginning at R_4 and ending at R_1. In addition, because there is a ring of regions consisting of R and a yellow-red Kempe chain, there is no blue-green Kempe chain in M beginning at R_2 and ending at R_5. Hence, we interchange the colors blue and yellow for all regions in M that can be reached by a blue-yellow Kempe chain beginning at R_4 and interchange the colors blue and green for all regions in M that can be reached by a blue-green Kempe chain beginning at R_2. Once these two color interchanges have been performed, each of the five neighboring regions of R is colored red, yellow, or green. Then R can be colored blue and a 4-coloring of the map M has been obtained, completing the proof.

As it turned out, the proof given by Kempe contained a fatal flaw, but one that would go unnoticed for a decade. Despite the fact that Kempe's attempted proof of the Four Color Problem was erroneous, he made a number of interesting observations in his article. He noticed that if a piece of tracing paper was placed over a map and a point was marked on the tracing paper over each region of the map and two points were joined by a line segment whenever the corresponding regions had a common boundary, then a diagram of a "linkage" was produced. Furthermore, the problem of determining whether the regions of the map can be colored with four colors so that neighboring regions are colored differently is the same problem as determining whether the points in the linkage can be colored with four colors so that every two points joined by a line segment are colored differently. (See Figure 8.)

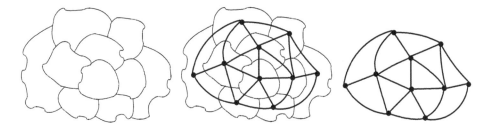

Figure 8: A map and corresponding planar graph

In 1878 Sylvester referred to a linkage as a graph and it is this terminology that became accepted. Later it became commonplace to refer to the points and lines of a linkage as the vertices and edges of the graph (with "vertex" being the singular of "vertices"). Since the graphs constructed from maps in this manner (referred to as the dual graph of the map) can themselves be drawn in the plane without two edges (line segments) intersecting, these graphs were called *planar graphs*. A planar graph that is actually drawn in the plane without any of its edges intersecting is called a *plane graph*. In terms of graphs, the Four Color Conjecture could then be restated.

The Four Color Conjecture *The vertices of every planar graph can be colored with four or fewer colors in such a way that every two vertices joined by an edge are colored differently.*

Indeed, the vast majority of this book will be devoted to coloring graphs (not coloring maps) and, in fact, to coloring graphs in general, not only planar graphs.

The colouring of abstract graphs is a generalization of the colouring of maps, and the study of the colouring of abstract graphs ... opens a new chapter in the combinatorial part of mathematics.

Gabriel Andrew Dirac (1951)

For the present, however, we continue our discussion in terms of coloring the regions of maps.

Kempe's proof of the theorem, which had become known as the Four Color Theorem, was accepted both within the United States and England. Arthur Cayley had accepted Kempe's argument as a valid proof. This led to Kempe being elected as a Fellow of the Royal Society in 1881.

The Four Color Theorem *The regions of every map can be colored with four or fewer colors so that every two adjacent regions are colored differently.*

Among the many individuals who had become interested in the Four Color Problem was Charles Lutwidge Dodgson (1832–1898), an Englishman with a keen interest in mathematics and puzzles. Dodgson was better known, however, under his pen-name Lewis Carroll and for his well-known books *Alice's Adventures in Wonderland* and *Through the Looking-Glass and What Alice Found There*.

Another well-known individual with mathematical interests, but whose primary occupation was not that of a mathematician, was Frederick Temple (1821–1902), Bishop of London and who would later become the Archbishop of Canterbury. Like Dodgson and others, Temple had a fondness for puzzles. Temple showed that it was impossible to have five mutually neighboring regions in any map and from this concluded that no map required five colors. Although Temple was correct about the non-existence of five mutually neighboring regions in a map, his conclusion that this provided a proof of the Four Color Conjecture was incorrect.

There was historical precedence about the non-existence of five mutually adjacent regions in any map. In 1840 the famous German mathematician August Möbius (1790–1868) reportedly stated the following problem, which was proposed to him by the philologist Benjamin Weiske (1748–1809).

Problem of Five Princes

There was once a king with five sons. In his will, he stated that after his death his kingdom should be divided into five regions in such a way that each region should have a common boundary with the other four. Can the terms of the will be satisfied?

As we noted, the conditions of the king's will cannot be met. This problem illustrates Möbius's interest in topology, a subject of which Möbius was one of the early pioneers. In a memoir written by Möbius and only discovered after his death, he discussed properties of one-sided surfaces, which became known as Möbius strips (even though it was determined that Johann Listing (1808–1882) had discovered these earlier).

In 1885 the German geometer Richard Baltzer (1818–1887) also lectured on the non-existence of five mutually adjacent regions. In the published version of his lecture, it was incorrectly stated that the Four Color Theorem followed from this. This error was repeated by other writers until the famous geometer Harold Scott MacDonald Coxeter (1907–2003) corrected the matter in 1959.

Mistakes concerning the Four Color Problem were not limited to mathematical errors however. Prior to establishing Francis Guthrie as the true and sole originator of the Four Color Problem, it was often stated in print that cartographers were aware that the regions of every map could be colored with four or less colors so that adjacent regions are colored differently. The well-known mathematical historian Kenneth O. May (1915–1977) investigated this claim and found no justification to it. He conducted a study of atlases in the Library of Congress and found no evidence of attempts to minimize the number of colors used in maps. Most maps used more than four colors and even when four colors were used, often less colors could have been used. There was never a mention of a "four color theorem".

Another mathematician of note around 1880 was Peter Guthrie Tait (1831–1901). In addition to being a scholar, he was a golf enthusiast. His son Frederick Guthrie Tait was a champion golfer and considered a national hero in Scotland. The first golf biography ever written was about Frederick Tait. Indeed, the Freddie Tait Golf Week is held every year in Kimberley, South Africa to commemorate his life as a golfer and soldier. He was killed during the Anglo-Boer War of 1899–1902.

Peter Guthrie Tait had heard of the Four Color Conjecture through Arthur Cayley and was aware of Kempe's solution. He felt that Kempe's solution of the Four Color Problem was overly long and gave several shorter solutions of the problem, all of which turned out to be incorrect. Despite this, one of his attempted proofs contained an interesting and useful idea. A type of map that is often encountered is a *cubic map*, in which there are exactly three boundary lines at each meeting point. In fact, every map M that has no region completely surrounded by another region can be converted into a cubic map M' by drawing a circle about each meeting point in M' and creating new meeting points and one new region (see Figure 9). If the map M' can be colored with four colors, then so can M.

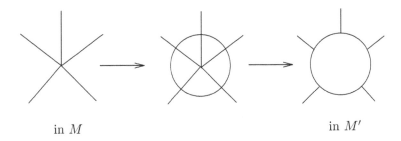

in M in M'

Figure 9: Converting a map into a cubic map

Tait's idea was to consider coloring the boundary lines of cubic maps. In fact, he stated as a lemma that:

The boundary lines of every cubic map can always be colored with three colors so that the three lines at each meeting point are colored differently.

Tait also mentioned that this lemma could be easily proved and showed how the lemma could be used to prove the Four Color Theorem. Although Tait was correct that this lemma could be used to prove the Four Color Theorem, he was incorrect when he said that the lemma could be easily proved. Indeed, as it turned out, this lemma is equivalent to the Four Color Theorem and, of course, is equally difficult to prove. (We will discuss Tait's coloring of the boundary lines of cubic maps in Chapter 10.)

The next important figure in the history of the Four Color Problem was Percy John Heawood (1861–1955), who spent the period 1887–1939 as a lecturer, professor, and vice-chancellor at Durham College in England. When Heawood was a student at Oxford University in 1880, one of his teachers was Professor Henry Smith who spoke often of the Four Color Problem. Heawood read Kempe's paper and it was he who discovered the serious error in the proof. In 1889 Heawood wrote a paper of his own, published in 1890, in which he presented the map shown in Figure 10.

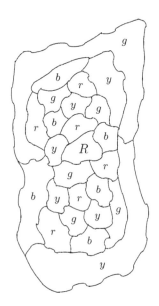

Figure 10: Heawood's counterexample to Kempe's proof

In the Heawood map, two of the five neighboring regions surrounding the uncolored region R are colored red; while for each of the colors blue, yellow, and green, there is exactly one neighboring region of R with that color. According to Kempe's argument, since blue is the color of the region that shares a boundary with R as well as with the two neighboring regions of R colored red, we are concerned with whether this map contains a blue-yellow Kempe chain between two neighboring regions of R as well as a blue-green Kempe chain between two neighboring regions of

R. It does. These Kempe chains are shown in Figures 11(a) and 11(b), respectively.

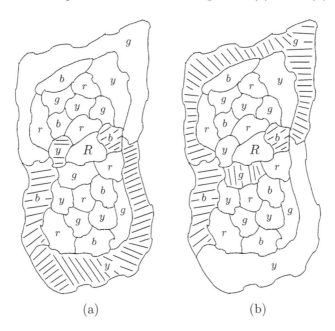

(a)   (b)

Figure 11: Blue-yellow and blue-green Kempe chains in the Heawood map

Because the Heawood map contains these two Kempe chains, it follows by Kempe's proof that this map does not contain a red-yellow Kempe chain between the two neighboring regions of *R* that are colored red and yellow *and* does not contain a red-green Kempe chain between the two neighboring regions of *R* that are colored red and green. This is, in fact, the case. Figure 12(a) indicates all regions that can be reached by a red-yellow Kempe chain beginning at the red region that borders *R* and that is not adjacent to the yellow region bordering *R*. Furthermore, Figure 12(b) indicates all regions that can be reached by a red-green Kempe chain beginning at the red region that borders *R* and that is not adjacent to the green region bordering *R*.

In the final step of Kempe's proof, the two colors within each Kempe chain are interchanged resulting in a coloring of the Heawood map with four colors. This double interchange of colors is shown in Figure 12(c). However, as Figure 12(c) shows, this results in neighboring regions with the same color. Consequently, Kempe's proof is unsuccessful when applied to the Heawood map, as colored in Figure 10. What Heawood had shown was that Kempe's method of proof was incorrect. That is, Heawood had discovered a counterexample to Kempe's technique, not to the Four Color Conjecture itself. Indeed, it is not particularly difficult to give a 4-coloring of the regions of the Heawood map so that every two neighboring regions are colored differently.

Other counterexamples to Kempe's proof were found after the publication of Heawood's 1890 paper, including a rather simple example (see Figure 13) given in

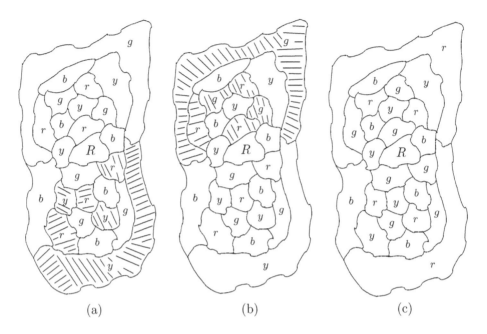

Figure 12: Steps in illustrating Kempe's technique

1921 by Alfred Errera (1886–1960), a student of Edmund Landau, well known for his work in analytic number theory and the distribution of primes.

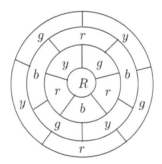

Figure 13: The Errera example

In addition to the counterexample to Kempe's proof, Heawood's paper contained several interesting results, observations, and comments. For example, although Kempe's attempted proof of the Four Color Theorem was incorrect, Heawood was able to use this approach to show that the regions of every map could be colored with five or fewer colors so that neighboring regions were colored differently (see Chapter 8).

Heawood also considered the problem of coloring maps that can be drawn on other surfaces. Maps that can be drawn in the plane are precisely those maps that can be drawn on the surface of a sphere. There are considerably more complex surfaces on which maps can be drawn, however. In particular, Heawood proved that the regions of every map drawn on the surface of a torus can be colored with seven or fewer colors and that there is, in fact, a map on the torus that requires seven colors (see Chapter 8). More generally, Heawood showed that the regions of every map drawn on a pretzel-shaped surface consisting of a sphere with k holes ($k > 0$) can be colored with $\left\lfloor \frac{7+\sqrt{1+48k}}{2} \right\rfloor$ colors. In addition, he stated that such maps requiring this number of colors exist. He never proved this latter statement, however. In fact, it would take another 78 years to verify this statement (see Chapter 8).

Thus, the origin of a curious problem by the young scholar Francis Guthrie was followed over a quarter of a century later by what was thought to be a solution to the problem by Alfred Bray Kempe. However, we were to learn from Percy John Heawood a decade later that the solution was erroneous, which returned the problem to its prior status. Well not quite – as these events proved to be stepping stones along the path to chromatic graph theory.

> *Is it five? Is it four?*
> *Heawood rephrased the query.*
> *Sending us back to before,*
> *But moving forward a theory.*

At the beginning of the 20th century, the Four Color Problem was still unsolved. Although possibly seen initially as a rather frivolous problem, not worthy of a serious mathematician's attention, it would become clear that the Four Color Problem was a very challenging mathematics problem. Many mathematicians, using a variety of approaches, would attack this problem during the 1900s. As noted, it was known that if the Four Color Conjecture could be verified for cubic maps, then the Four Color Conjecture would be true for all maps. Furthermore, every cubic map must contain a region surrounded by two, three, four, or five neighboring regions. These four kinds of configurations (arrangements of regions) were called *unavoidable* because every cubic map had to contain at least one of them. Thus, the arrangements of regions shown in Figure 14 make up an unavoidable set of configurations.

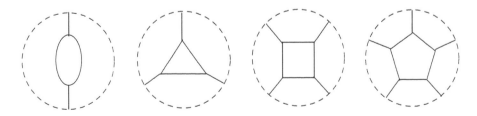

Figure 14: An unavoidable set of configurations in a cubic map

A region surrounded by k neighboring regions is called a *k-gon*. It is possible to show that any map that contains no k-gon where $k < 5$ must contain at least twelve pentagons (5-gons). In fact, there is a map containing exactly twelve regions, each of which is a pentagon. Such a map is shown in Figure 15, where one of the regions is the "exterior region". Since this map can be colored with four colors, any counterexample to the Four Color Conjecture must contain at least thirteen regions. Alfred Errera proved that no counterexample could consist only of pentagons and hexagons (6-gons).

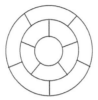

Figure 15: A cubic map with twelve pentagons

A *reducible configuration* is any configuration of regions that cannot occur in a minimum counterexample of the Four Color Conjecture. Many mathematicians who attempted to solve the Four Color Problem attempted to do so by trying to find an unavoidable set S of reducible configurations. Since S is unavoidable, this means that every cubic map must contain at least one configuration in S. Because each configuration in S is reducible, this means that it cannot occur in a minimum counterexample. Essentially then, a proof of the Four Color Conjecture by this approach would be a proof by minimum counterexample resulting in a number of cases (one case for each configuration in the unavoidable set S) where each case leads to a contradiction (that is, each configuration is shown to be reducible).

Since the only configuration in the unavoidable set shown in Figure 14 that could not be shown to be reducible was the pentagon, this suggested searching for more complex configurations that must also be part of an unavoidable set with the hope that these more complicated configurations could somehow be shown to be reducible. For example, in 1903 Paul Wernicke proved that every cubic map containing no k-gon where $k < 5$ must either contain two adjacent pentagons or two adjacent regions, one of which is a pentagon and the other a hexagon (see Chapter 5). That is, the troublesome case of a cubic map containing a pentagon could be eliminated and replaced by two different cases.

Finding new, large unavoidable sets of configurations was not a problem. Finding reducible configurations was. In 1913 the distinguished mathematician George David Birkhoff (1884–1944) published a paper called *The reducibility of maps* in which he considered rings of regions for which there were regions interior to as well as exterior to the ring. Since the map was a minimum counterexample, the ring together with the interior regions and the ring together with the exterior regions could both be colored with four colors. If two 4-colorings could be chosen so that they match along the ring, then there is a 4-coloring of the entire map. Since this can always be done if the ring consists of three regions, rings of three regions can never appear in a minimum counterexample. Birkhoff proved that rings of four regions also cannot appear in a minimum counterexample. In addition, he was successful in proving that rings of five regions cannot appear in a minimum counterexample either – unless the interior of the region consisted of a single region. This generalized Kempe's approach. While Kempe's approach to solving the Four Color Problem involved the removal of a single region from a map, Birkhoff's method allowed the removal of regions inside or outside some ring of regions. For example, a configuration that Birkhoff was able to prove was reducible consisted of a ring of six pentagons enclosing four pentagons. This became known as the *Birkhoff diamond* (see Figure 16).

Figure 16: The Birkhoff diamond (a reducible configuration)

Philip Franklin (1898–1965) wrote his doctoral dissertation in 1921 titled *The Four Color Problem* under the direction of Oswald Veblen (1880–1960). Veblen was the first professor at the Institute for Advanced Study at Princeton University. He was well known for his work in geometry and topology (called analysis situs at the time) as well as for his lucid writing. In his thesis, Franklin showed that if a cubic map does not contain a k-gon, where $k < 5$, then it must contain a pentagon adjacent to two other regions, each of which is a pentagon or a hexagon (see Chapter 5). This resulted in a larger unavoidable set of configurations.

In 1922 Franklin showed that every map with 25 or fewer regions could be colored with four or fewer colors. This number gradually worked its way up to 96 in a result established in 1975 by Jean Mayer, curiously a professor of French literature.

Favorable impressions of new areas of mathematics clearly did not occur quickly. Geometry of course had been a prominent area of study in mathematics for centuries. The origins of topology may only go back to 19th century however. In his 1927 survey paper about the Four Color Problem, Alfred Errera reported that some mathematicians referred to topology as the "geometry of drunkards". Graph theory belongs to the more general area of combinatorics. While combinatorial arguments can be found in all areas of mathematics, there was little recognition of combinatorics as a major area of mathematics until later in the 20th century, at which time topology was gaining in prominence. Indeed, John Henry Constantine Whitehead (1904–1960), one of the founders of homotopy theory in topology, reportedly said that "Combinatorics is the slums of topology." However, by the latter part of the 20th century, combinatorics had come into its own. The famous mathematician Israil Moiseevich Gelfand (1913–2009) stated (in 1990):

> *The older I get, the more I believe that at the bottom of most deep mathematical problems there is a combinatorial problem.*

Heinrich Heesch (1906–1995) was a German mathematician who was an assistant to Hermann Weyl, a gifted mathematician who was a colleague of Albert Einstein (1879–1955) and a student of the famous mathematician David Hilbert (1862–1943), whom he replaced as mathematics chair at the University of Göttingen. In 1900, Hilbert gave a lecture before the International Congress of Mathematicians in Paris in which he presented 23 extremely challenging problems. In 1935, Heesch solved one of these problems (Problem 18) dealing with tilings of the plane. One of Heesch's friends at Göttingen was Ernst Witt (1911–1991), who thought he had solved an even more famous problem: the Four Color Problem. Witt was anxious to show his proof to the famous German mathematician Richard Courant (1888–1972), who later moved to the United States and founded the Courant Institute of Mathematical Sciences. Since Courant was in the process of leaving Göttingen for Berlin, Heesch joined Witt to travel with Courant by train in order to describe the proof. However, Courant was not convinced and the disappointed young mathematicians returned to Göttingen. On their return trip, however, Heesch discovered an error in Witt's proof. Heesch too had become captivated by the Four Color Problem.

As Heesch studied this famous problem, he had become increasingly convinced that the problem could be solved by finding an unavoidable set of reducible configurations, even though such a set may very well be extremely large. He began lecturing on his ideas in the 1940s at the Universities of Hamburg and Kiel. A 1948 lecture at the University of Kiel was attended by the student Wolfgang Haken (born in 1928), who recalls Heesch saying that an unavoidable set of reducible configurations may contain as many as ten thousand members. Heesch discovered a method for creating many unavoidable sets of configurations. Since the method had an electrical flavor to it, electrical terms were chosen for the resulting terminology.

What Heesch did was to consider the dual planar graphs constructed from cubic maps. Thus, the configurations of regions in a cubic map became configurations of vertices in the resulting dual planar graph. These planar graphs themselves had regions, each necessarily a triangle (a 3-gon). Since the only cubic maps whose coloring was still in question were those in which every region was surrounded by five or more neighboring regions, five or more edges of the resulting planar graph met at each vertex of the graph. If k edges meet at a vertex, then the vertex is said to have *degree k*. Thus every vertex in each planar graph of interest had degree 5 or more. Heesch then assigned each vertex in the graph a "charge" of $6 - k$ if the degree of the vertex was k (see Chapter 5). The only vertices receiving a positive charge were therefore those of degree 5, which were given a charge of $+1$. The vertices of degree 6 had a charge of 0, those of degree 7 a charge of -1, and so on. It can be proved (see Chapter 5) that the sum of the charges of the vertices in such a planar graph is always positive (in fact exactly 12).

Heesch's plan consisted of establishing rules, called *discharging rules*, for moving a positive charge from one vertex to others in a manner that did not change the sum of the charges. The goal was to use these rules to create an unavoidable set of configurations by showing that if a minimum counterexample to the Four Color Conjecture contained none of these configurations, then the sum of the charges of its vertices was not 12.

Since Heesch's discharging method was successful in finding unavoidable sets, much of the early work in the 20th century on the Four Color Problem was focused on showing that certain configurations were reducible. Often showing that even one configuration was reducible became a monumental task. In the 1960s Heesch had streamlined Birkhoff's approach of establishing the reducibility of certain configurations. One of these techniques, called D-reduction, was sufficiently algorithmic in nature to allow this technique to be executed on a computer and, in fact, a computer program for implementing D-reducibility was written on the CDC 1604A computer by Karl Dürre, a graduate of Hanover.

Because of the large number of ways that the vertices on the ring of a configuration could be colored, the amount of computer time needed to analyze complex configurations became a major barrier to their work. Heesch was then able to develop a new method, called C-reducibility, where only some of the colorings of the ring vertices needed to be considered. Of course, one possible way to deal with the obstacles that Heesch and Dürre were facing was to find a more powerful computer on which to run Dürre's program.

While Haken had attended Heesch's talk at the University of Kiel on the Four Color Problem, the lectures that seemed to interest Haken the most were those on topology given by Karl Heinrich Weise in which he described three long-standing unsolved problems. One of these was the Poincaré Conjecture posed by the great mathematician and physicist Henri Poincaré in 1904 and which concerned the relationship of shapes, spaces, and surfaces. Another was the Four Color Problem and the third was a problem in knot theory. Haken decided to attempt to solve all three problems. Although his attempts to prove the Poincaré Conjecture failed, he was successful with the knot theory problem. A proof of the Poincaré Conjecture by the Russian mathematician Grigori Perelman was confirmed and reported in Trieste, Italy on 17 June 2006. For this accomplishment, he was awarded a Fields Medal (the mathematical equivalent of the Nobel Prize) on 22 August 2006. However, Perelman declined to attend the ceremony and did not accept the prize. As for the Four Color Problem, the story continues.

Haken's solution of the problem in knot theory led to his being invited to the University of Illinois as a visiting professor. After leaving the University of Illinois to spend some time at the Institute for Advanced Study in Princeton, Haken then returned to the University of Illinois to take a permanent position.

Heesch inquired, through Haken, about the possibility of using the new supercomputer at the University of Illinois (the ILLAC IV) but much time was still needed to complete its construction. The Head of the Department of Computer Science there suggested that Heesch contact Yoshio Shimamoto, Head of the Applied Mathematics Department at the Brookhaven Laboratory at the United States Atomic Energy Commission, which had access to the Stephen Cray-designed Control Data 6600, which was the fastest computer at that time.

Shimamoto himself had an interest in the Four Color Problem and had even thought of writing his own computer program to investigate the reducibility of configurations. Shimamoto arranged for Heesch and Dürre to visit Brookhaven in the late 1960s. Dürre was able to test many more configurations for reducibility. The configurations that were now known to be D-reducible still did not constitute an unavoidable set, however, and Heesch and Dürre returned to Germany. In August of 1970 Heesch visited Brookhaven again – this time with Haken visiting the following month. At the end of September, Shimamoto was able to show that if a certain configuration that he constructed (known as the *horseshoe configuration*) was D-reducible, then the Four Color Conjecture is true. Figure 17 shows the dual planar graph constructed from the horseshoe configuration. This was an amazing development. To make matters even more interesting, Heesch recognized the horseshoe configuration as one that had earlier been shown to be D-reducible. Because of the importance of knowing, with complete certainty, that this configuration was D-reducible, Shimamoto took the cautious approach of having a totally new computer program written to verify the D-reducibility of the horseshoe configuration.

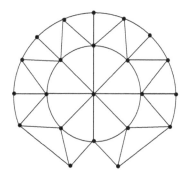

Figure 17: The Shimamoto horseshoe

Dürre was brought back from Germany because of the concern that the original verification of the horseshoe configuration being D-reducible might be incorrect. Also, the printout of the computer run of this was nowhere to be found. Finally, the new computer program was run and, after 26 hours, the program concluded that this configuration was *not* D-reducible. It was not only that this development was so very disappointing to Shimamoto but, despite the care he took, rumors had begun to circulate in October of 1971 that the Four Color Problem had been solved – using a computer!

Haken had carefully checked Shimamoto's mathematical reasoning and found it to be totally correct. Consequently, for a certain period, the only obstacle standing in the way of a proof of the Four Color Conjecture had been a computer. William T. Tutte (1917–2002) and Hassler Whitney (1907–1989), two of the great graph theorists at that time, had also studied Shimamoto's method of proof and found no flaw in his reasoning. Because this would have resulted in a far simpler proof of the Four Color Conjecture than could reasonably be expected, Tutte and Whitney concluded that the original computer result must be wrong. However, the involvement of Tutte and Whitney in the Four Color Problem resulted in a clarification of D-reducibility. Also because of their stature in the world of graph theory, there was even more interest in the problem.

It would not be hard to present the history of graph theory as an account of the struggle to prove the four color conjecture, or at least to find out why the problem is difficult.

William T. Tutte (1967)

In the April 1, 1975 issue of the magazine *Scientific American* the popular mathematics writer Martin Gardner (1914–2010) stunned the mathematical community (at least momentarily) when he wrote an article titled "Six Sensational Discoveries that Somehow Have Escaped Public Attention" that contained a map (see Figure 18) advertised as one that could *not* be colored with four colors. However, several individuals found that this map could in fact be colored with four colors, only to learn that Gardner had intended this article as an April Fool's joke.

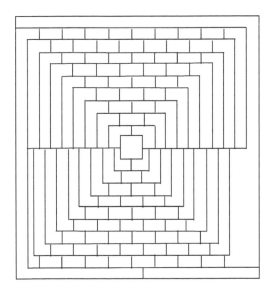

Figure 18: Martin Gardner's April Fool's Map

In the meantime, Haken had been losing faith in a computer-aided solution of the Four Color Problem despite the fact that he had a doctoral student at the University of Illinois whose research was related to the problem. One of the members of this student's thesis committee was Kenneth Appel (1932–2013). After completing his undergraduate degree at Queens College with a special interest in actuarial mathematics, Appel worked at an insurance company and shortly afterwards was drafted and began a period of military service. He then went to the University of Michigan for his graduate studies in mathematics. During the spring of 1956, the University of Michigan acquired an IBM 650 and the very first programming course offered at the university was taught by John W. Carr, III, one of the pioneers of computer education in the United States. Curiously, Carr's doctoral advisor at the Massachusetts Institute of Technology was Phillip Franklin, who, as we mentioned, wrote his dissertation on the Four Color Problem. Appel audited this programming course. Since the university did not offer summer financial support to Appel and Douglas Aircraft was recruiting computer programmers, he spent the summer of 1956 writing computer programs concerning the DC-8 jetliner, which was being designed at the time. Appel had become hooked on computers.

Kenneth Appel's area of research was mathematical logic. In fact, Appel asked Haken to give a talk at the logic seminar in the Department of Mathematics so he could better understand the thesis. In his talk, Haken included a discussion of the computer difficulties that had been encountered in his approach to solve the Four Color Problem and explained that he was finished with the problem for the present. Appel, however, with his knowledge of computer programming, convinced Haken that the two of them should "take a shot at it".

Together, Appel and Haken took a somewhat different approach. They devised an algorithm that tested for "reduction obstacles". The work of Appel and Haken was greatly aided by Appel's doctoral student John Koch who wrote a very efficient program that tested certain kinds of configurations for reducibility. Much of Appel and Haken's work involved refining Heesch's method for finding an unavoidable set of reducible configurations.

The partnership in the developing proof concerned the active involvement of a team of three, namely Appel, Haken, and a computer. As their work progressed, Appel and Haken needed ever-increasing amounts of time on a computer. Because of Appel's political skills, he was able to get time on the IBM 370-168 located in the university's administration building. Eventually, everything paid off. In June of 1976, Appel and Haken had constructed an unavoidable set of 1936 reducible configurations, which was later reduced to 1482. The proof was finally announced at the 1976 Summer Meeting of the American Mathematical Society and the Mathematical Association of America at the University of Toronto. Shortly afterwards, the University of Illinois employed the postmark

FOUR COLORS SUFFICE

on its outgoing mail.

In 1977 Frank Harary (1921–2005), editor-in-chief of the newly founded *Journal of Graph Theory*, asked William Tutte if he would contribute something for the first volume of the journal in connection with this announcement. Tutte responded with a short but pointed poem (employing his often-used pen-name Blanche Descartes) with the understated title *Some Recent Progress in Combinatorics*:

> *Wolfgang Haken*
> *Smote the Kraken*
> *One! Two! Three! Four!*
> *Quoth he: "The monster is no more".*

In the poem, Tutte likened the Four Color Problem to the legendary sea monster known as a kraken and proclaimed that Haken (along with Appel, of course) had slain this monster.

With so many mistaken beliefs that the Four Color Theorem had been proved during the preceding century, it was probably not surprising that the announced proof by Appel and Haken was met with skepticism by many. While the proof was received with enthusiasm by some, the reception was cool by others, even to the point of not being accepted by some that such an argument was a proof at all. It certainly didn't help matters that copying, typographical, and technical errors were found – even though corrected later. In 1977, the year following the announcement of the proof of the Four Color Theorem, Wolfgang Haken's son Armin, then a graduate student at the University of California at Berkeley, was asked to give a talk about the proof. He explained that

> the proof consisted of a rather short theoretical section, four hundred pages of detailed checklists showing that all relevant cases had been covered, and about 1800 computer runs totaling over a thousand hours of computer time.

He went on to say that the audience seemed split into two groups, largely by age and roughly at age 40. The older members of the audience questioned a proof that made such extensive use of computers, while the younger members questioned a proof that depended on hand-checking 400 pages of detail.

The proof of the Four Color Theorem initiated a great number of philosophical discussions as to whether such an argument was a proof and, in fact, what a proof is. Some believed that it was a requirement of a proof that it must be possible for a person to be able to read through the entire proof, even though it might be extraordinarily lengthy. Others argued that the nature of proof had changed over the years. Centuries ago a mathematician might have given a proof in a conversational style. As time went on, proofs had become more structured and were presented in a very logical manner. While some were concerned with the distinct possibility of computer error in a computer-aided proof, others countered this by saying that the literature is filled with incorrect proofs and misstatements since human error is always a possibility, perhaps even more likely. Furthermore, many proofs written by modern mathematicians, even though not computer-aided, were so long that it is likely that few, if any, had read through these proofs with care. Also, those who shorten proofs by omitting arguments of some claims within a proof may in fact be leaving out key elements of the proof, improving the opportunity for human error. Then there are those who stated that knowing the Four Color Theorem is true is not what is important. What is crucial is to know *why* only four colors are needed to color all maps. A computer-aided proof does not supply this information.

A second proof of the Four Color Theorem, using the same overall approach but a different discharging procedure, a different unavoidable set of reducible configurations, and more powerful proofs of reducibility was announced and described by Frank Allaire of Lakehead University, Canada in 1977, although the complete details were never published.

As Robin Thomas of the Georgia Institute of Technology reported, there appeared to be two major reasons for the lack of acceptance by some of the Appel-Haken proof: (1) part of the proof uses a computer and cannot be verified by hand; (2) the part that is supposed to be checked by hand is so complicated that no one may have independently checked it at all. For these reasons, in 1996, Neil Robertson,

Daniel P. Sanders, Paul Seymour, and Thomas constructed their own (computer-aided) proof of the Four Color Theorem. While Appel and Haken's unavoidable set of configurations consisted of 1482 graphs, this new proof had an unavoidable set of 633 graphs. In addition, while Appel and Haken used 487 discharging rules to construct their set of configurations, Robertson, Sanders, Seymour, and Thomas used only 32 discharging rules to construct their set of configurations. Thomas wrote:

> Appel and Haken's use of a computer 'may be a necessary evil', but the complication of the hand proof was more disturbing, particularly since the 4CT has a history of incorrect "proofs". So in 1993, mainly for our own peace of mind, we resolved to convince ourselves that the 4CT really was true.

The proof of the Four Color Theorem given by Robertson, Sanders, Seymour, and Thomas rested on the same idea as the Appel-Haken proof, however. These authors proved that none of the 633 configurations can be contained in a minimum counterexample to the Four Color Theorem and so each of these configurations is reducible.

As we noted, the Four Color Theorem could have been proved if any of the following could be shown to be true.

(1) The regions of every map can be colored with four or fewer colors so that neighboring regions are colored differently.

(2) The vertices of every planar graph can be colored with four or fewer colors so that every two vertices joined by an edge are colored differently.

(3) The edges of every cubic map can be colored with exactly three colors so that every three edges meeting at a vertex are colored differently.

Coloring the regions, vertices, and edges of maps and planar graphs, inspired by the desire to solve the Four Color Problem, has progressed far beyond this – to coloring more general graphs and even to reinterpreting what is meant by coloring.

For many decades, a coloring of (the vertices of) a graph G was always meant as an assignment of colors (elements of some set) to the vertices of G, one color to each vertex, so that adjacent vertices are colored differently. Of course, this is quite understandable as this came from the Four Color Problem where neighboring regions were required to be colored differently. In more recent decades, this interpretation of a coloring has changed – often dramatically. As we will see in the later chapters, there are occasions when we might not want any two vertices to be colored the same. And this is the case when coloring edges as well. Indeed, and opposite to this, we may want all vertices (or edges) to be colored the same. Of course, all this depends on what the goal of a coloring is. There are also occasions when a coloring (or the related concept of a labeling) of the vertices or edges of a graph itself gives rise to another coloring. It is the study of these topics into which we are about to venture.

Chapter 1

Introduction to Graphs

In the preceding chapter we were introduced to the famous map coloring problem known as the Four Color Problem. We saw that this problem can also be stated as a problem dealing with coloring the vertices of a certain class of graphs called *planar graphs* or as a problem dealing with coloring the edges of a certain subclass of planar graphs. This gives rise to coloring the vertices or coloring the edges of graphs in general. In order to provide the background needed to discuss this subject, we will describe, over the next five chapters, some of the fundamental concepts and theorems we will encounter in our investigation of graph colorings as well as some common terminology and notation in graph theory.

1.1   Fundamental Terminology

A **graph** G is a finite nonempty set V of objects called **vertices** (the singular is **vertex**) together with a set E of 2-element subsets of V called **edges**. Vertices are sometimes called **points** or **nodes**, while edges are sometimes referred to as **lines** or **links**. Each edge $\{u, v\}$ of G is commonly denoted by uv or vu. If $e = uv$, then the edge e is said to **join** u and v. The number of vertices in a graph G is the **order** of G and the number of edges is the **size** of G. We often use n for the order of a graph and m for its size. To indicate that a graph G has **vertex set** V and **edge set** E, we sometimes write $G = (V, E)$. To emphasize that V is the vertex set of a graph G, we often write V as $V(G)$. For the same reason, we also write E as $E(G)$. A graph of order 1 is called a **trivial graph** and so a **nontrivial graph** has two or more vertices. A graph of size 0 is an **empty graph** and so a **nonempty graph** has one or more edges.

Graphs are typically represented by diagrams in which each vertex is represented by a point or small circle (open or solid) and each edge is represented by a line segment or curve joining the corresponding small circles. A diagram that represents a graph G is referred to as the graph G itself and the small circles and lines representing the vertices and edges of G are themselves referred to as the vertices and edges of G.

27

Figure 1.1 shows a graph G with vertex set $V = \{t, u, v, w, x, y, z\}$ and edge set $E = \{tu, ty, uv, uw, vw, vy, wx, wz, yz\}$. Thus, the order of this graph G is 7 and its size is 9. In this drawing of G, the edges tu and vw intersect. This has no significance. In particular, the point of intersection of these two edges is not a vertex of G.

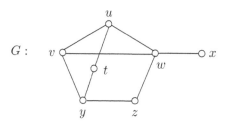

Figure 1.1: A graph

If uv is an edge of G, then u and v are **adjacent vertices**. Two adjacent vertices are referred to as **neighbors** of each other. The set of neighbors of a vertex v is called the **open neighborhood** of v (or simply the **neighborhood** of v) and is denoted by $N(v)$. The set $N[v] = N(v) \cup \{v\}$ is called the **closed neighborhood** of v. If uv and vw are distinct edges in G, then uv and vw are **adjacent edges**. The vertex u and the edge uv are said to be **incident** with each other. Similarly, v and uv are incident.

For the graph G of Figure 1.1, the vertices u and w are therefore adjacent in G, while the vertices u and x are not adjacent. The edges uv and uw are adjacent in G, while the edges vy and wz are not adjacent. The vertex v is incident with the edge vw but is not incident with the edge wz.

For nonempty disjoint sets A and B of vertices of G, we denote by $[A, B]$ the set of edges of G joining a vertex of A and a vertex of B. For the sets $A = \{u, v, y\}$ and $B = \{w, z\}$ in the graph G of Figure 1.1, $[A, B] = \{uw, vw, yz\}$.

The **degree of a vertex** v in a graph G is the number of vertices in G that are adjacent to v. Thus, the degree of a vertex v is the number of the vertices in its neighborhood $N(v)$. Equivalently, the degree of v is the number of edges of G incident with v. The degree of a vertex v is denoted by $\deg_G v$ or, more simply, by $\deg v$ if the graph G under discussion is clear. A vertex of degree 0 is referred to as an **isolated vertex** and a vertex of degree 1 is an **end-vertex** or a **leaf**. An edge incident with an end-vertex is called a **pendant edge**. The largest degree among the vertices of G is called the **maximum degree** of G is denoted by $\Delta(G)$. The **minimum degree** of G is denoted by $\delta(G)$. Thus, if v is a vertex of a graph G of order n, then

$$0 \leq \delta(G) \leq \deg v \leq \Delta(G) \leq n - 1.$$

For the graph G of Figure 1.1,

$$\deg x = 1, \deg t = \deg z = 2, \deg u = \deg v = \deg y = 3, \text{ and } \deg w = 4.$$

Thus, $\delta(G) = 1$ and $\Delta(G) = 4$.

A well-known theorem in graph theory deals with the sum of the degrees of the vertices of a graph. This theorem was indirectly observed by the great Swiss mathematician Leonhard Euler in a 1736 paper [80] that is now considered the first paper ever written on graph theory – even though graphs were never mentioned in the paper. It is often referred to as the First Theorem of Graph Theory. (Some have called this theorem the **Handshaking Lemma**, although Euler never used this name.)

Theorem 1.1 **(The First Theorem of Graph Theory)** *If G is a graph of size m, then*

$$\sum_{v \in V(G)} \deg v = 2m.$$

Proof. When summing the degrees of the vertices of G, each edge of G is counted twice, once for each of its two incident vertices. ∎

The sum of the degrees of the vertices of the graph G of Figure 1.1 is 18, which is twice the size 9 of G, as is guaranteed by Theorem 1.1.

A vertex v in a graph G is **even** or **odd**, according to whether its degree in G is even or odd. Thus, the graph G of Figure 1.1 has three even vertices and four odd vertices. While a graph can have either an even or odd number of even vertices, this is not the case for odd vertices.

Corollary 1.2 *Every graph has an even number of odd vertices.*

Proof. Suppose that G is a graph of size m. By Theorem 1.1,

$$\sum_{v \in V(G)} \deg v = 2m,$$

which is, of course, an even number. Since the sum of the degrees of the even vertices of G is even, the sum of the degrees of the odd vertices of G must be even as well, implying that G has an even number of odd vertices. ∎

A graph H is said to be a **subgraph** of a graph G if $V(H) \subseteq V(G)$ and $E(H) \subseteq E(G)$. If $V(H) = V(G)$, then H is a **spanning subgraph** of G. If H is a subgraph of a graph G and either $V(H)$ is a proper subset of $V(G)$ or $E(H)$ is a proper subset of $E(G)$, then H is a **proper subgraph** of G. For a nonempty subset S of $V(G)$, the **subgraph $G[S]$ of G induced by** S has S as its vertex set and two vertices u and v in S are adjacent in $G[S]$ if and only if u and v are adjacent in G. (The subgraph of G induced by S is also denoted by $\langle S \rangle_G$ or simply by $\langle S \rangle$ when the graph G is understood.) A subgraph H of a graph G is called an **induced subgraph** if there is a nonempty subset S of $V(G)$ such that $H = G[S]$. Thus, $G[V(G)] = G$. For a nonempty set X of edges of a graph G, the **subgraph $G[X]$ induced by** X has X as its edge set and a vertex v belongs to $G[X]$ if v is incident with at least one edge in X. A subgraph H of G is **edge-induced** if there is a nonempty subset X of $E(G)$ such that $H = G[X]$. Thus, $G[E(G)] = G$ if and only if G has no isolated vertices.

Figure 1.2 shows six graphs, namely G and the graphs H_i for $i = 1, 2, \ldots, 5$. All six of these graphs are proper subgraphs of G, except G itself and H_1. Since G is a subgraph of itself, it is not a proper subgraph of G. The graph H_1 contains the edge uz, which G does not and so H_1 is not even a subgraph of G. The graph H_3 is a spanning subgraph of G since $V(H_3) = V(G)$. Since $xy \in E(G)$ but $xy \notin E(H_4)$, the subgraph H_4 is not an induced subgraph of G. On the other hand, the subgraphs H_2 and H_5 are both induced subgraphs of G. Indeed, for $S_1 = \{v, x, y, z\}$ and $S_2 = \{u, v, y, z\}$, $H_2 = G[S_1]$ and $H_5 = G[S_2]$. The subgraph H_4 of G is edge-induced; in fact, $H_4 = G[X]$, where $X = \{uw, wx, wy, xz, yz\}$.

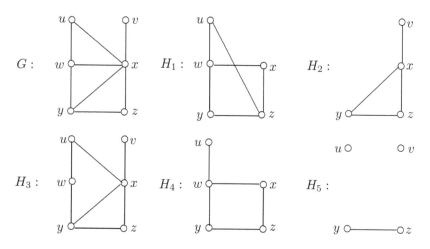

Figure 1.2: Graphs and subgraphs

For a vertex v and an edge e in a nonempty graph $G = (V, E)$, the subgraph $G - v$, obtained by deleting v from G, is the induced subgraph $G[V - \{v\}]$ of G and the subgraph $G - e$, obtained by deleting e from G, is the spanning subgraph of G with edge set $E - \{e\}$. More generally, for a proper subset U of V, the graph $G - U$ is the induced subgraph $G[V - U]$ of G. For a subset X of E, the graph $G - X$ is the spanning subgraph of G with edge set $E - X$. If u and v are distinct nonadjacent vertices of G, then $G + uv$ is the graph with $V(G + uv) = V(G)$ and $E(G + uv) = E(G) \cup \{uv\}$. Thus, G is a spanning subgraph of $G + uv$. For the graph G of Figure 1.3, the set $U = \{t, x\}$ of vertices, and the set $X = \{tw, ux, vx\}$ of edges, the subgraphs $G - u$, $G - wx$, $G - U$, and $G - X$ of G are also shown in that figure, as is the graph $G + uv$.

1.2   Connected Graphs

There are several types of sequences of vertices in a graph as well as subgraphs of a graph that can be used to describe ways in which one can move about within the graph. For two (not necessarily distinct) vertices u and v in a graph G, a $u - v$ **walk** W in G is a sequence of vertices in G, beginning at u and ending at v such that

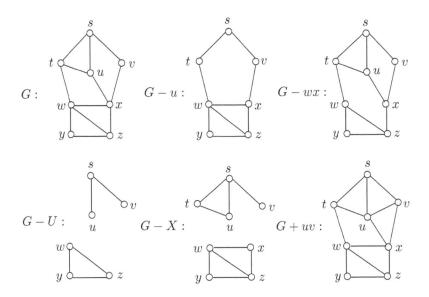

Figure 1.3: Deleting vertices and edges from and adding edges to a graph

consecutive vertices in W are adjacent in G. Such a walk W in G can be expressed as

$$W = (u = v_0, v_1, \ldots, v_k = v), \qquad (1.1)$$

where $v_i v_{i+1} \in E(G)$ for $0 \le i \le k - 1$. (The walk W is also commonly denoted by $W : u = v_0, v_1, \ldots, v_k = v$.) Non-consecutive vertices in W need not be distinct. The walk W is said to contain each vertex v_i $(0 \le i \le k)$ and each edge $v_i v_{i+1}$ $(0 \le i \le k - 1)$. The walk W can therefore be thought of as beginning at the vertex $u = v_0$, proceeding along the edge $v_0 v_1$ to the vertex v_1, then along the edge $v_1 v_2$ to the vertex v_2, and so forth, until finally arriving at the vertex $v = v_k$. The number of edges encountered in W (including multiplicities) is the **length** of W. Hence, the length of the walk W in (1.1) is k. In the graph G of Figure 1.4,

$$W_1 = (x, w, y, w, v, u, w) \qquad (1.2)$$

is an $x - w$ walk of length 6. This walk encounters the vertex w three times and the edge wy twice.

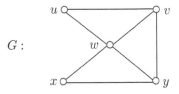

Figure 1.4: Walks in a graph

A walk whose initial and terminal vertices are distinct is an **open walk**; otherwise, it is a **closed walk**. Thus, the walk W_1 in (1.2) in the graph G of Figure 1.4 is an open walk. It is possible for a walk to consist of a single vertex, in which case it is a **trivial walk**. A trivial walk is therefore a closed walk.

A walk in a graph G in which no edge is repeated is a **trail** in G. For example, in the graph G of Figure 1.4, $T = (u, v, y, w, v)$ is a $u - v$ trail of length 4. While no edge of T is repeated, the vertex v is repeated, which is allowed. On the other hand, a walk in a graph G in which no vertex is repeated is called a **path**. Every nontrivial path is necessarily an open walk. Thus, $P' = (u, v, w, y)$ is a $u - y$ path of length 3 in the graph G of Figure 1.4. Many proofs in graph theory make use of $u - v$ walks or $u - v$ paths of minimum length (or of maximum length) for some pair u, v of vertices of a graph. The proof of the following theorem illustrates this.

Theorem 1.3 *If a graph G contains a $u - v$ walk, then G contains a $u - v$ path.*

Proof.   Among all $u - v$ walks in G, let

$$P = (u = u_0, u_1, \ldots, u_k = v)$$

be a $u - v$ walk of minimum length. Thus, the length of P is k. We claim that P is a $u - v$ path. Assume, to the contrary, that this is not the case. Then some vertex of G must be repeated in P, say $u_i = u_j$ for some i and j with $0 \le i < j \le k$. If we then delete the vertices $u_{i+1}, u_{i+2}, \ldots, u_j$ from P, we arrive at the $u - v$ walk

$$(u = u_0, u_1, \ldots, u_{i-1}, u_i = u_j, u_{j+1}, \ldots, u_k = v)$$

whose length is less than k, which is impossible.   ∎

A nontrivial closed walk in a graph G in which no edge is repeated is a **circuit** in G. For example,

$$C = (u, w, x, y, w, v, u)$$

is a $u - u$ circuit in the graph G of Figure 1.4. In addition to the required repetition of u in this circuit, w is repeated as well. This is acceptable provided no edge is repeated. A circuit

$$C = (v = v_0, v_1, \ldots, v_k = v),$$

$k \ge 2$, for which the vertices v_i, $0 \le i \le k-1$, are distinct is a **cycle** in G. Therefore,

$$C' = (u, v, y, x, w, u)$$

is a $u - u$ cycle of length 5 in the graph G of Figure 1.4. A cycle of length $k \ge 3$ is called a k-**cycle**. A 3-cycle is also referred to as a **triangle**. A cycle of even length is an **even cycle**, while a cycle of odd length is an **odd cycle**.

There are also subgraphs of a graph referred to as paths and cycles. A subgraph P of a graph G is a **path** in G if the vertices of P can be labeled as $v_1, v_2, \cdots, v_k$ so that its edges are $v_1 v_2, v_2 v_3, \cdots, v_{k-1} v_k$. A subgraph C of G is a **cycle** in G if the vertices of C can be labeled as $v_1, v_2, \cdots, v_k$ $(k \ge 3)$ so that its edges are

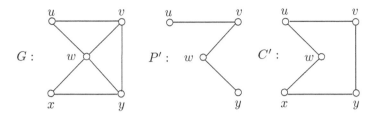

Figure 1.5: A path and cycle in a graph

$v_1v_2, v_2v_3, \cdots, v_{k-1}v_k, v_kv_1$. Consequently, paths and cycles have two interpretations in graphs – as sequences of vertices and as subgraphs. This is the case with trails and circuits as well. The path P' and the cycle C' described earlier in the graph G of Figure 1.4 correspond to the subgraphs shown in Figure 1.5.

Two vertices u and v in a graph G are **connected** if G contains a $u - v$ path. The graph G itself is **connected** if every two vertices of G are connected. By Theorem 1.3, a graph G is connected if G contains a $u - v$ walk for every two vertices u and v of G. A graph G that is not connected is a **disconnected graph**. The graph F of Figure 1.6 is connected since F contains a $u - v$ path (and a $u - v$ walk) for every two vertices u and v in F. On the other hand, the graph H is disconnected since, for example, H contains no $y_4 - y_5$ path.

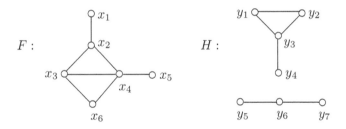

Figure 1.6: A connected graph and disconnected graph

A connected subgraph H of a graph G is a **component** of G if H is not a proper subgraph of a connected subgraph of G. The number of components in a graph G is denoted by $k(G)$. Thus, G is connected if and only if $k(G) = 1$. For the sets $S_1 = \{y_1, y_2, y_3, y_4\}$ and $S_2 = \{y_5, y_6, y_7\}$ of vertices of the graph H of Figure 1.6, the induced subgraphs $H[S_1]$ and $H[S_2]$ are (the only) components of H. Therefore, $k(H) = 2$.

As would be expected, if the degrees of the vertices of a graph G are large enough, then G is connected. The following theorem gives such a result. For a nontrivial graph G that is not complete, let

$$\sigma_2(G) = \min\{\deg u + \deg v : \ uv \notin E(G)\}.$$

Theorem 1.4 *If G is a graph of order $n \geq 2$ such that $\sigma_2(G) \geq n - 1$, then G is connected.*

The following corollary is then a consequence of Theorem 1.4.

Corollary 1.5 *If G is a graph of order $n \geq 2$ such that $\delta(G) \geq (n-1)/2$, then G is connected.*

Another corollary of Theorem 1.4 gives a sufficient condition for a graph to be connected in terms of its size.

Corollary 1.6 *If G is a graph of order $n \geq 2$ and size $m > \binom{n-1}{2}$, then G is connected.*

1.3　Distance in Graphs

If u and v are distinct vertices in a connected graph G, then there is a $u - v$ path in G. In fact, there may very well be several $u - v$ paths in G, possibly of varying lengths. This information can be used to provide a measure of how close u and v are to each other or how far from each other they are. The most common definition of distance between two vertices in a connected graph is the following.

The **distance** $d(u, v)$ from a vertex u to a vertex v in a connected graph G is the minimum of the lengths of the $u - v$ paths in G. A $u - v$ path of length $d(u, v)$ is called a $u - v$ **geodesic**. In the graph G of Figure 1.7, the path $P = (v_1, v_5, v_6, v_{10})$ is a $v_1 - v_{10}$ geodesic and so $d(v_1, v_{10}) = 3$. Furthermore,

$$d(v_1, v_1) = 0, \ d(v_1, v_2) = 1, \ d(v_1, v_6) = 2, \ d(v_1, v_7) = 3, \text{ and } d(v_1, v_8) = 4.$$

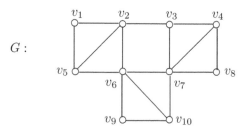

Figure 1.7: Distances in a graph

The distance d defined above satisfies each of the following properties in a connected graph G:

(1) $d(u, v) \geq 0$ for every two vertices u and v of G;

(2) $d(u, v) = 0$ if and only if $u = v$;

(3) $d(u, v) = d(v, u)$ for all $u, v \in V(G)$ (the **symmetric property**);

(4) $d(u, w) \leq d(u, v) + d(v, w)$ for all $u, v, w \in V(G)$ (the **triangle inequality**).

Since d satisfies the four properties (1)–(4), d is a **metric** on $V(G)$ and $(V(G), d)$ is a **metric space**. Since d is symmetric, we can speak of the distance between two vertices u and v rather than the distance from u to v.

The **eccentricity** $e(v)$ of a vertex v in a connected graph G is the distance between v and a vertex farthest from v in G. The **diameter** diam(G) of G is the greatest eccentricity among the vertices of G, while the **radius** rad(G) is the smallest eccentricity among the vertices of G. The diameter of G is also the greatest distance between any two vertices of G. A vertex v with $e(v) = \text{rad}(G)$ is called a **central vertex** of G and a vertex v with $e(v) = \text{diam}(G)$ is called a **peripheral vertex** of G. Two vertices u and v of G with $d(u, v) = \text{diam}(G)$ are **antipodal vertices** of G. Necessarily, if u and v are antipodal vertices in G, then each of u and v is a peripheral vertex. For the graph G of Figure 1.7,

$$e(v_6) = 2, \ e(v_2) = e(v_3) = e(v_4) = e(v_5) = e(v_7) = e(v_9) = e(v_{10}) = 3,$$
$$e(v_1) = e(v_8) = 4$$

and so diam$(G) = 4$ and rad$(G) = 2$. In particular, v_6 is the only central vertex of G and v_1 and v_8 are the only peripheral vertices of G. Since $d(v_1, v_8) = 4 = \text{diam}(G)$, it follows that v_1 and v_8 are antipodal vertices of G. It is certainly not always the case that diam$(G) = 2\,\text{rad}(G)$ as, for example, diam$(P_4) = 3$ and rad$(P_4) = 2$. Indeed, the following can be said about the radius and diameter of a connected graph.

Theorem 1.7 *For every nontrivial connected graph G,*

$$\text{rad}(G) \le \text{diam}(G) \le 2\,\text{rad}(G).$$

Proof. The inequality rad$(G) \le \text{diam}(G)$ is immediate from the definitions. Let u and w be two vertices such that $d(u, w) = \text{diam}(G)$ and let v be a central vertex of G. Therefore, $e(v) = \text{rad}(G)$. By the triangle inequality, diam$(G) = d(u, w) \le d(u, v) + d(v, w) \le 2e(v) = 2\,\text{rad}(G)$. ∎

The subgraph induced by the central vertices of a connected graph G is the **center** of G and is denoted by Cen(G). If every vertex of G is a central vertex, then Cen$(G) = G$ and G is **self-centered**. The subgraph induced by the peripheral vertices of a connected graph G is the **periphery** of G and is denoted by Per(G).

For the graph G of Figure 1.7, the center of G consists of the isolated vertex v_6 and the periphery consists of the two isolated vertices v_1 and v_8. The graph H of Figure 1.8 has radius 2 and diameter 3. Therefore, every vertex of H is either a central vertex or a peripheral vertex. Indeed, the center of H is the triangle induced by the three "exterior" vertices of H, while the periphery of H is the 6-cycle induced by the six "interior" vertices of H.

In an observation first made by Stephen Hedetniemi (see [31]), there is no restriction of which graphs can be the center of some graph.

Theorem 1.8 *Every graph is the center of some graph.*

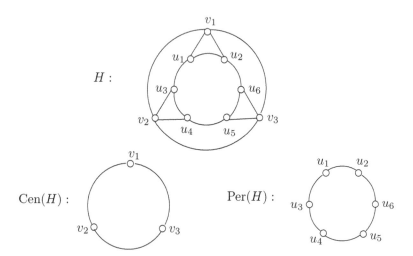

Figure 1.8: The center and periphery of a graph

Proof. Let G be a graph. We construct a graph H from G by first adding two new vertices u and v to G and joining them to every vertex of G but not to each other, and then adding two other vertices u_1 and v_1, where we join u_1 to u and join v_1 to v. Since $e(u_1) = e(v_1) = 4$, $e(u) = e(v) = 3$, and $e(x) = 2$ for every vertex x in G, it follows that $V(G)$ is the set of central vertices of H and so $\text{Cen}(H) = H[V(G)] = G$.   ∎

While every graph can be the center of some graph, Halina Bielak and Maciej Syslo [21] showed that only certain graphs can be the periphery of a graph.

Theorem 1.9 *A nontrivial graph G is the periphery of some graph if and only if every vertex of G has eccentricity 1 or no vertex of G has eccentricity 1.*

Proof. If every vertex of G has eccentricity 1, then G is complete and $\text{Per}(G) = G$; while if no vertex of G has eccentricity 1, then let F be the graph obtained from G by adding a new vertex w and joining w to each vertex of G. Since $e_F(w) = 1$ and $e_F(x) = 2$ for every vertex x of G, it follows that every vertex of G is a peripheral vertex of F and so $\text{Per}(F) = F[V(G)] = G$.

For the converse, let G be a graph that contains some vertices of eccentricity 1 and some vertices whose eccentricity is not 1 and suppose that there exists a graph H such that $\text{Per}(H) = G$. Necessarily, G is a proper induced connected subgraph of H. Thus, $\text{diam}(H) = k \geq 2$. Furthermore, $e_H(v) = k \geq 2$ for each $v \in V(G)$ and $e_H(v) < k$ for $v \in V(H) - V(G)$. Let u be a vertex of G such that $e_G(u) = 1$ and let w be a vertex of H such that $d(u, w) = e_H(u) = k \geq 2$. Since w is not adjacent to u, it follows that $w \notin V(G)$. On the other hand, $d(u, w) = k$ and so $e_H(w) = k$. This implies that w is a peripheral vertex of H and so $w \in V(G)$, which is impossible.   ∎

The distance d defined above on the vertex set of a connected graph G is not the only metric that can be defined on $V(G)$. The **detour distance** $D(u, v)$ from a vertex u to a vertex v in G is the length of a *longest* $u - v$ path in G. Thus, $D(u, u) = 0$ and if $u \neq v$, then $1 \leq D(u, v) \leq n - 1$. A $u - v$ path of length $D(u, v)$ is called a $u - v$ **detour**. If $D(u, v) = n - 1$, then G contains a spanning $u - v$ path. For the graph G of Figure 1.9,

$$D(w, x) = 1, \; D(u, w) = 3, \; D(t, x) = 4, \text{ and } D(u, t) = 6.$$

The $u - t$ path $P = (u, z, x, w, y, v, t)$ is a spanning $u - t$ detour in G.

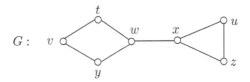

Figure 1.9: Detour distance graphs

Theorem 1.10 *Detour distance is a metric on the vertex set of a connected graph.*

Proof. Let G be a connected graph. The detour distance D on $V(G)$ certainly satisfies properties (1)-(3) of a metric. Hence, only the triangle inequality (property (4)) needs to be verified.

Let u, v, and w be any three vertices of G. Let P be a $u - w$ detour in G. Suppose first that v lies on P. Let P' be the $u - v$ subpath of P and let P'' be the $v - w$ subpath of P. Since the length $\ell(P')$ of P' is at most $D(u, v)$ and the length $\ell(P'')$ of P'' is at most $D(v, w)$, it follows that $D(u, v) + D(v, w) \geq D(u, w)$.

Thus, we may assume that v does not lie on P. Let Q be a path of minimum length from v to a vertex of P. Suppose that Q is a $v - x$ path. Thus, x is the only vertex of Q that lies on P. Let Q' be the $u - x$ subpath of P and let Q'' be the $x - w$ subpath of P. Since

$$D(u, v) \geq \ell(Q') + \ell(Q) \text{ and } D(v, w) \geq \ell(Q) + \ell(Q''),$$

it follows that $D(u, v) + D(v, w) > D(u, w)$.

In either case, $D(u, w) \geq D(u, v) + D(v, w)$ and the triangle inequality is satisfied. ∎

1.4 Isomorphic Graphs

Two graphs G and H are **isomorphic** (have the same structure) if there exists a bijective function $\phi : V(G) \to V(H)$ such that two vertices u and v are adjacent in G if and only if $\phi(u)$ and $\phi(v)$ are adjacent in H. The function ϕ is then called an **isomorphism**. If G and H are isomorphic, we write $G \cong H$. If there is no such function ϕ as described above, then G and H are **non-isomorphic graphs** and so $G \not\cong H$.

The graphs G and H in Figure 1.10 are isomorphic; in fact, the function ϕ : $V(G) \to V(H)$ defined by

$$\phi(u_1) = v_4,\ \phi(u_2) = v_2,\ \phi(u_3) = v_6,\ \phi(u_4) = v_1,$$
$$\phi(u_5) = v_5,\ \phi(u_6) = v_3,\ \phi(u_7) = v_7$$

is an isomorphism.

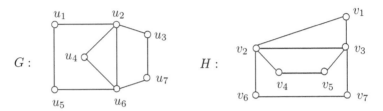

Figure 1.10: Isomorphic graphs

The graphs F_1 and F_2 in Figure 1.11 are not isomorphic; for if there were an isomorphism $\phi : V(F_1) \to V(F_2)$, then the four pairwise adjacent vertices s_4, s_5, s_7, and s_8 of F_1 must map into four pairwise adjacent vertices of F_2. Since F_2 does not contain four such vertices, there is no such function ϕ and so $F_1 \not\cong F_2$.

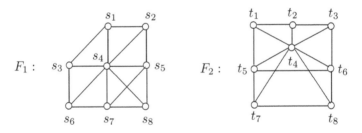

Figure 1.11: Non-isomorphic graphs

The following theorem is a consequence of the definition of isomorphism.

Theorem 1.11   *If two graphs G and H are isomorphic, then they have the same order and the same size, and the degrees of the vertices of G are the same as the degrees of the vertices of H.*

From Theorem 1.11, it follows that if G and H are two graphs such that (1) the orders of G and H are different, or (2) the sizes of G and H are different, or (3) the degrees of the vertices of G and those of the vertices of H are different, then G and H are non-isomorphic. The conditions described in Theorem 1.11 are strictly necessary for two graphs to be isomorphic – they are not sufficient. Indeed, the graphs F_1 and F_2 of Figure 1.11 have the same order, the same size, and the degrees of the vertices of F_1 and F_2 are the same; yet F_1 and F_2 are not isomorphic.

There are many necessary conditions for two graphs to be isomorphic in addition to those presented in Theorem 1.11. All of the (non-isomorphic) graphs of order 4 or less are shown in Figure 1.12.

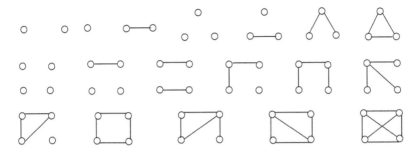

Figure 1.12: The (non-isomorphic) graphs of order 4 or less

1.5 Common Graphs and Graph Operations

There are certain graphs that are encountered so frequently that there is special notation reserved for them. We will see many of these in this section.

The graph that is itself a **cycle** of order $n \geq 3$ is denoted by C_n and the graph that is a **path** of order n is denoted by P_n. Thus, C_n is a graph of order n and size n, while P_n is a graph of order n and size $n - 1$. Some cycles and paths of small order are shown in Figure 1.13.

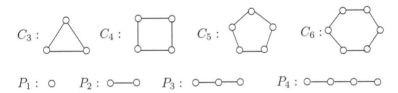

Figure 1.13: Cycles and paths

A graph is **complete** if every two distinct vertices in the graph are adjacent. The complete graph of order n is denoted by K_n. Therefore, K_n is a graph of order n and size $\binom{n}{2} = \frac{n(n-1)}{2}$. The complete graphs K_n, $1 \leq n \leq 5$, are shown in Figure 1.14.

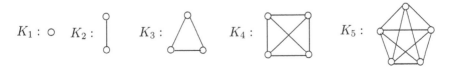

Figure 1.14: Complete graphs

Every vertex of C_n has degree 2, while every vertex of K_n has degree $n - 1$. If all of the vertices of a graph G have the same degree, then G is **a regular graph**. If every vertex of G has degree r, then G is r-**regular**. Hence, C_n is 2-regular and K_n is $(n - 1)$-regular.

Opposite to regular graphs are nontrivial graphs in which no two vertices have the same degree, sometimes called **irregular graphs**. Despite the following result, the term "irregular" has been applied to graphs in a variety of ways.

Theorem 1.12 *No nontrivial graph is irregular.*

Proof. Suppose that there exists an irregular graph G of order $n \geq 2$, where $V(G) = \{v_1, v_2, \ldots, v_n\}$. Since the degrees of the vertices of G are distinct, we may assume that

$$0 \leq \deg v_1 < \deg v_2 < \cdots < \deg v_n \leq n - 1.$$

This, however, implies that $\deg v_i = i - 1$ for all i ($1 \leq i \leq n$). Since $\deg v_1 = 0$, v_1 is not adjacent to v_n; and since $\deg v_n = n - 1$, it follows that v_n is adjacent to v_1. This is impossible.   ∎

A 3-regular graph is also called a **cubic graph**. The complete graph K_4 is a cubic graph. The best-known cubic graph (indeed, one of the best-known graphs) is the **Petersen graph**. Three different drawings of the Petersen graph are shown in Figure 1.15. Thus, the Petersen graph is a cubic graph of order 10. It contains no triangles or 4-cycles but it does have 5-cycles. We will encounter this graph often in the future.

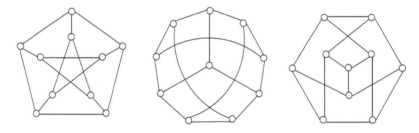

Figure 1.15: Three drawings of the Petersen graph

A graph without triangles is called **triangle-free**. The **girth** of a graph G with cycles is the length of a smallest cycle in G. The **circumference** $\mathrm{cir}(G)$ of a graph G with cycles is the length of a longest cycle in G. Thus, the girth of the Petersen graph is 5 and its circumference is 9. Obviously, the Petersen graph is triangle-free. We now consider an important class of triangle-free graphs.

A nontrivial graph G is a **bipartite graph** if it is possible to partition $V(G)$ into two subsets U and W, called **partite sets** in this context, such that every edge of G joins a vertex of U and a vertex of W. Figure 1.16 shows a bipartite graph G and a graph F that is not bipartite. That F contains odd cycles is essential in the observation that F is not bipartite, as the following characterization indicates.

Theorem 1.13 *A nontrivial graph G is a bipartite graph if and only if G contains no odd cycles.*

Proof. Suppose first that G is bipartite. Then $V(G)$ can be partitioned into partite sets U and W (and so every edge of G joins a vertex of U and a vertex of W). Let

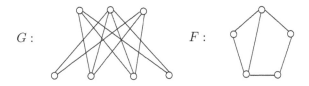

Figure 1.16: A bipartite graph and a graph that is not bipartite

$C = (v_1, v_2, \ldots, v_k, v_1)$ be a k-cycle of G. We may assume that $v_1 \in U$. Thus, $v_2 \in W$, $v_3 \in U$, and so forth. In particular, $v_i \in U$ for every odd integer i with $1 \le i \le k$ and $v_j \in W$ for every even integer j with $2 \le j \le k$. Since $v_1 \in U$, it follows that $v_k \in W$ and so k is even.

For the converse, let G be a nontrivial graph containing no odd cycles. If G is empty, then G is clearly bipartite. Hence, it suffices to show that every nontrivial component of G is bipartite and so we may assume that G itself is connected. Let u be a vertex of G and let

$$U = \{x \in V(G): \ d(u,x) \text{ is even}\}$$
$$W = \{x \in V(G): \ d(u,x) \text{ is odd}\},$$

where $u \in U$, say. We show that G is bipartite with partite sets U and W. It remains to show that no two vertices of U are adjacent and no two vertices of W are adjacent. Suppose that W contains two adjacent vertices w_1 and w_2. Let P_1 be a $u - w_1$ geodesic and P_2 a $u - w_2$ geodesic. Let z be the last vertex that P_1 and P_2 have in common (possibly $z = u$). Then the length of the $z - w_1$ subpath P_1' of P_1 and the length of the $z - w_2$ subpath P_2' of P_2 are of the same parity. Thus, the paths P_1' and P_2' together with the edge $w_1 w_2$ produce an odd cycle. This is a contradiction. The argument that no two vertices of U are adjacent is similar. ∎

A bipartite graph having partite sets U and W is a **complete bipartite graph** if every vertex of U is adjacent to every vertex of W. If the partite sets U and W of a complete bipartite graph contain s and t vertices, then this graph is denoted by $K_{s,t}$ or $K_{t,s}$. The graph $K_{1,t}$ is called a **star**. The graph $K_{s,t}$ has order $s + t$ and size st. In particular, the r-regular complete bipartite graph $K_{r,r}$ has order $n = 2r$ and size $m = r^2$. Therefore, $\delta(K_{r,r}) = n/2$. Of course, $K_{r,r}$ is triangle-free. On the other hand, every graph G of order $n \ge 3$ with $\delta(G) > n/2$ contains a triangle (see Exercise 32).

The following result gives a necessary and sufficient condition for a connected bipartite graph to be a complete bipartite graph.

Theorem 1.14 *Let G be a connected bipartite graph. Then G is a complete bipartite graph if and only if G does not contain P_4 as an induced subgraph.*

Proof. Suppose that the partite sets of G are U and W. Assume, first that G is a complete bipartite graph. Let $P = (v_1, v_2, v_3, v_4)$ be a path of order 4 in G. Then one of v_1 and v_4 belongs to U and the other to W. Since G is a complete bipartite graph, v_1 is adjacent to v_4 in G and so P is not an induced subgraph.

For the converse, suppose that G does not contain P_4 as an induced subgraph and that G is not a complete bipartite graph. Then there is a vertex $u \in U$ and a vertex $w \in W$ that are not adjacent. Since G is connected, $d(u,w) = k$ for some odd integer $k \geq 3$. Let $P = (u = u_0, u_1, \ldots, u_k = w)$ be a shortest $u - w$ path in G. Then $(u = u_0, u_1, u_2, u_3)$ is an induced P_4, producing a contradiction. ∎

More generally, for an integer $k \geq 2$ and positive integers $n_1, n_2, \ldots, n_k$, a **complete multipartite graph** (or **complete k-partite graph**) $K_{n_1, n_2, \ldots, n_k}$ is that graph G whose vertex set can be partitioned into k subsets $V_1, V_2, \ldots, V_k$ (also called **partite sets**) with $|V_i| = n_i$ for $1 \leq i \leq k$ such that $uv \in E(G)$ if $u \in V_i$ and $v \in V_j$, where $1 \leq i, j \leq k$ and $i \neq j$. The four graphs in Figure 1.17 are complete multipartite graphs, where $K_{2,3}$ and $K_{1,5}$ are complete bipartite graphs, the latter of which is a star.

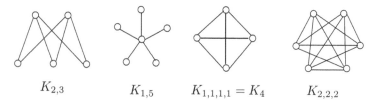

$K_{2,3}$ $K_{1,5}$ $K_{1,1,1,1} = K_4$ $K_{2,2,2}$

Figure 1.17: Complete multipartite graphs

There are many ways of producing a new graph from one or more given graphs. The **complement** $\overline{G}$ of a graph G is that graph whose vertex set is $V(G)$ and where uv is an edge of $\overline{G}$ if and only if uv is not an edge of G. Observe that if G is a graph of order n and size m, then $\overline{G}$ is a graph of order n and size $\binom{n}{2} - m$. Furthermore, if G is isomorphic to $\overline{G}$, then G is said to be **self-complementary**. Both P_4 and C_5 are self-complementary graphs.

For two (vertex-disjoint) graphs G and H, the **union** $G + H$ of G and H is the (disconnected) graph with

$$V(G + H) = V(G) \cup V(H) \text{ and } E(G + H) = E(G) \cup E(H).$$

If G and H are both isomorphic to a graph F, then we write $G + H$ as $2F$, that is, $F + F = 2F$. The **join** $G \vee H$ of two vertex-disjoint graphs G and H has $V(G \vee H) = V(G) \cup V(H)$ and

$$E(G \vee H) = E(G) \cup E(H) \cup \{uv : u \in V(G), v \in V(H)\}.$$

Therefore, $\overline{K}_s \vee \overline{K}_t = K_{s,t}$ for positive integers s and t. Also, for a graph G of order n, the graph $F = G \vee K_1$ is the graph of order $n + 1$ obtained by adding a new vertex v to G and joining v to each vertex of G. In this case, $\deg_F v = n$ and $\deg_F x = \deg_G x + 1$ for each $x \in V(G)$. For example, $C_n \vee K_1$ is called the **wheel** of order $n + 1$ and is denoted by W_n. (See Figure 1.18.)

Recall that a graph G is called irregular if no two vertices of G have the same degree. By Theorem 1.12 no nontrivial graph is irregular. A graph G is **nearly irregular** if G contains only two vertices of the same degree.

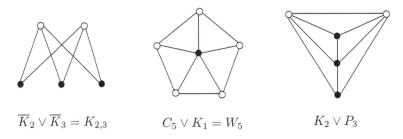

$$\overline{K}_2 \vee \overline{K}_3 = K_{2,3} \qquad\qquad C_5 \vee K_1 = W_5 \qquad\qquad K_2 \vee P_3$$

Figure 1.18: Joins of two graphs

Theorem 1.15 *For every integer $n \geq 2$, there is exactly one connected nearly irregular graph of order n.*

Proof. For each $n \geq 2$, we define a connected graph F_n of order n recursively as follows. For $n = 2$, let $F_2 = K_2$; while for $n \geq 3$, let $F_n = \overline{F_{n-1}} \vee K_1$. Then F_n is nearly irregular for each $n \geq 2$. It remains to show that F_n is the only connected nearly irregular graph of order n. We verify this by induction on n. Since K_2 is the only such graph of order 2, the basis step of the induction is established. Assume that there is a unique connected nearly irregular graph H of order n. Let G be a connected nearly irregular graph of order $n + 1$. Then the degrees of G must be $1, 2, \ldots, n$, where one of these degrees is repeated. Necessarily, G does not contain two vertices of degree n, for otherwise, G has no vertex of degree 1. Then $G = F \vee K_1$ for some graph F of order n. Necessarily, F is a nearly irregular graph. Since G does not contain two vertices of degree n, it follows that F does not contain a vertex of degree $n - 1$. This, however, implies that F contains an isolated vertex and so F is disconnected. Thus, $\overline{F}$ is a connected nearly irregular graph of order n. By the induction hypothesis, $\overline{F} = H$. Thus, $F = \overline{H}$ and so $G = \overline{H} \vee K_1$ is the only connected nearly irregular graph of order $n + 1$. ∎

For $k \geq 2$ mutually vertex-disjoint graphs $G_1, G_2, \ldots, G_k$, the **union**

$$G = G_1 + G_2 + \cdots + G_k$$

of these k graphs is defined by $V(G) = \bigcup_{i=1}^{k} V(G_i)$ and $E(G) = \bigcup_{i=1}^{k} E(G_i)$; while the **join** $H = G_1 \vee G_2 \vee \cdots \vee G_k$ of these k graphs is defined by $V(H) = \bigcup_{i=1}^{k} V(G_i)$ and

$$E(H) = \bigcup_{i=1}^{k} E(G_i) \cup \{v_i v_j : v_i \in V(G_i), v_j \in V(G_j), i \neq j\}.$$

For example, if $G_i = 2K_1 = \overline{K}_2$ for $i = 1, 2, 3$, then

$$G_1 + G_2 + G_3 = 2K_1 + 2K_1 + 2K_1 = 6K_1 = \overline{K}_6 \text{ and}$$
$$G_1 \vee G_2 \vee G_3 = 2K_1 \vee 2K_1 \vee 2K_1 = K_{2,2,2}.$$

The **Cartesian product** $G \,\square\, H$ of two graphs G and H has vertex set

$$V(G \,\square\, H) = V(G) \times V(H)$$

and two distinct vertices (u, v) and (x, y) of $G \, \Box \, H$ are adjacent if either

$$(1) \ u = x \text{ and } vy \in E(H) \text{ or } (2) \ v = y \text{ and } ux \in E(G).$$

(Sometimes the Cartesian product of G and H is denoted by $G \times H$.) For the graphs $G = P_3$ and $H = P_3$ of Figure 1.19, the graphs $G + H = 2P_3$, $G \vee H = P_3 \vee P_3$, and $G \, \Box \, H = P_3 \, \Box \, P_3$ are also shown in Figure 1.19. A graph $P_s \, \Box \, P_t$ is referred to as a **grid**.

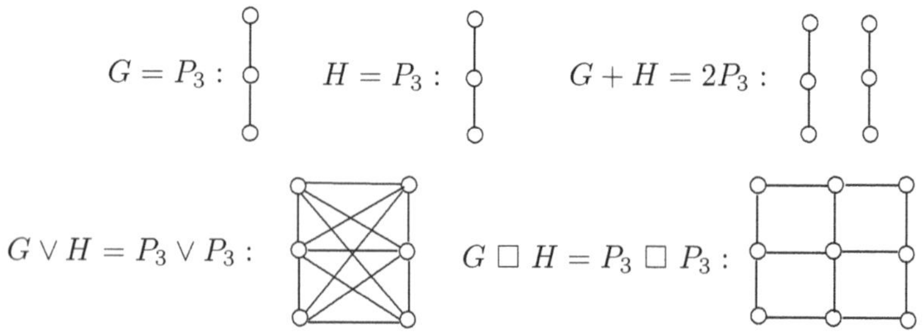

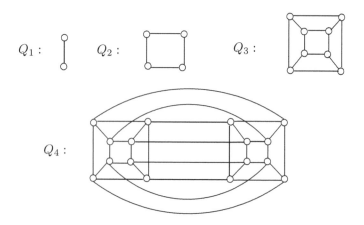

Figure 1.19: Graph operations

The well-known class of graphs Q_n called n-**cubes** or **hypercubes** is defined recursively as a Cartesian product. The graph Q_1 is the graph K_2, while $Q_2 = K_2 \, \Box \, K_2 = C_4$. In general, for $n \geq 2$, $Q_n = Q_{n-1} \, \Box \, K_2$. The hypercubes Q_n, $1 \leq n \leq 4$ are shown in Figure 1.20.

Figure 1.20: Hypercubes

The **line graph** $L(G)$ of a nonempty graph G is that graph whose vertex set is $E(G)$ and two vertices e and f of $L(G)$ are adjacent if and only if e and f are adjacent edges in G. For the graph G of Figure 1.21, its line graph $H = L(G)$ is shown in Figure 1.21 as well. A graph H is called **a line graph** if $H = L(G)$ for some graph G. Obviously, the graph H of Figure 1.21 is a line graph since

$H = L(G)$ for the graph G of Figure 1.21. The graph F of Figure 1.21 is not a line graph however. (See Exercise 36.)

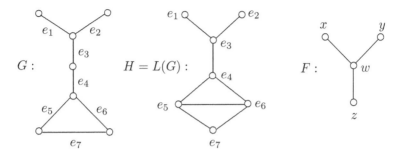

Figure 1.21: Line graphs and non-line graphs

1.6 Multigraphs and Digraphs

There will be occasions when it is useful to consider structures that are not exactly those represented by graphs. A **multigraph** M is a nonempty set of vertices, every two of which are joined by a finite number of edges. Two or more edges that join the same pair of distinct vertices are called **parallel edges**. An edge joining a vertex to itself is called a **loop**. Structures that permit both parallel edges and loops (including parallel loops) are called **pseudographs**. There are authors who refer to multigraphs or pseudographs as graphs and those who refer to what we call graphs as **simple graphs**. Consequently, when reading any material written on graph theory, it is essential that there is a clear understanding of the use of the term *graph*. According to the terminology introduced here then, every multigraph is a pseudograph and every graph is both a multigraph and a pseudograph. In Figure 1.22, M_1 and M_4 are multigraphs, while M_2 and M_3 are pseudographs. Of course, M_1 and M_4 are also pseudographs, while M_4 is the only graph in Figure 1.22.

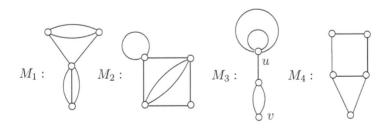

Figure 1.22: Multigraphs and pseudographs

For a vertex v in a multigraph G, the **degree** $\deg v$ of v in G is the number of edges of G incident with v. In a pseudograph, there is a contribution of 2 for each loop at v. For the pseudograph M_3 of Figure 1.22, $\deg u = 5$ and $\deg v = 2$.

When describing walks in multigraphs or in pseudographs, it is often necessary to list edges in a sequence as well as vertices in order to know which edges are used in the walk. For example, $W = (u, e_1, u, v, e_6, w, e_6, v, e_7, w)$ is a $u - w$ walk in the pseudograph G of Figure 1.23.

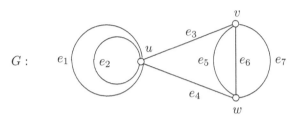

Figure 1.23: Walks in a pseudograph

A **digraph** (or **directed graph**) D is a finite nonempty set V of vertices and a set E of ordered pairs of distinct vertices. The elements of E are called **directed edges** or **arcs**. The digraph D with vertex set $V = \{u, v, w, x\}$ and arc set $E = \{(u, v), (v, u), (u, w), (w, v), (w, x)\}$ is shown in Figure 1.24. This digraph has order 4 and size 5.

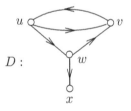

Figure 1.24: A digraph

If $a = (u, v)$ is an arc of a digraph D, then a is said to **join** u and v (and a is **incident from** u and **incident to** v). Furthermore, u is **adjacent to** v and v is **adjacent from** u. For a vertex v in a digraph D, the **outdegree** od v of v is the number of vertices of D to which v is adjacent, while the **indegree** id v of v is the number of vertices of D from which v is adjacent. The **degree** deg v of a vertex v is defined by deg $v = $ od $v + $ id v. The directed graph version of Theorem 1.1 is stated below.

Theorem 1.16 (The First Theorem of Digraph Theory) *If D is a digraph of size m, then*

$$\sum_{v \in V(D)} \text{od} \, v = \sum_{v \in V(D)} \text{id} \, v = m.$$

In a **multidigraph**, parallel arcs are permitted. If, for each pair u, v of distinct vertices in a digraph D, at most one of (u, v) and (v, u) is a directed edge, then D is called an **oriented graph**. Thus, an oriented graph D is obtained by assigning a direction to each edge of some graph G. In this case, the digraph D is also called an **orientation** of G. The **underlying graph** of a digraph D is the graph obtained from D by replacing each arc (u, v) or a pair (u, v), (v, u) of arcs by the edge uv. An orientation of a complete graph is a **tournament**. The digraph D_1 of Figure 1.25 is an oriented graph; it is an orientation of the graph G_1. The digraph D_2 is not an oriented graph, while D_3 is a tournament.

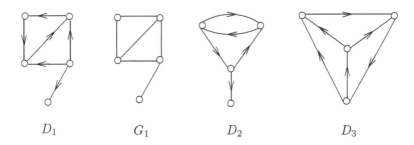

$$D_1 \qquad G_1 \qquad D_2 \qquad D_3$$

Figure 1.25: Orientations of graphs

A sequence $W = (u = u_0, u_1, \ldots, u_k = v)$ of vertices of a digraph D such that (u_i, u_{i+1}) is an arc of D for all i $(1 \leq i \leq k-1)$ is called a **(directed)** $u-v$ **walk** in D. The number of occurrences of arcs in a walk is the **length** of the walk. A walk in which no arc is repeated is a **(directed) trail**; while a walk in which no vertex is repeated is a **(directed) path**. A $u-v$ walk is **closed** if $u = v$ and is **open** if $u \neq v$. A closed trail of length at least 2 is a **(directed) circuit**; a closed walk of length at least 2 in which no vertex is repeated except for the initial and terminal vertices is a **(directed) cycle**. A digraph is **acyclic** if it contains no directed cycles. For example, the orientation D of C_4 in Figure 1.26 is an acyclic digraph. On the other hand, none of D_1, D_2, and D_3 in Figure 1.25 is acyclic.

$D :$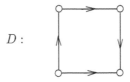

Figure 1.26: An acyclic digraph

A digraph D is **connected** if the underlying graph of D is connected. A digraph D is **strong** (or **strongly connected**) if D contains both a $u-v$ path and a $v-u$ path for every pair u, v of distinct vertices of D. For example, the digraphs D_1 and D_2 of Figure 1.25 are connected but not strong, while the digraph D_3 of Figure 1.25 is strong.

Exercises for Chapter 1

1. Give an example of a graph of order $3n \geq 15$ containing at least n vertices of degree $2n - 1$ and at most one vertex of degree 1 such that every vertex of degree $k > 1$ is adjacent to at least one vertex of degree less than k.

2. Suppose that G is a graph of order $2r + 1 \geq 5$ such that every vertex of G has degree $r + 1$ or degree $r + 2$. Prove that either G contains at least $r + 2$ vertices of degree $r + 1$ or G contains at least $r + 1$ vertices of degree $r + 2$.

3. Let S be a finite set of positive integers whose largest element is n. Prove that there exists a graph G of order $n + 1$ such that (1) $\deg u \in S$ for every vertex u of G and (2) for every $d \in S$, there exists a vertex v in G such that $\deg v = d$.

4. Let $d_1, d_2, \ldots, d_n$ be a sequence of nonnegative integers such that $d_1 \geq d_2 \geq \cdots \geq d_n$, where $d_1 \geq 1$. Prove that there exists a graph G of order n with $V(G) = \{v_1, v_2, \ldots, v_n\}$ such that $\deg v_i = d_i$ for all i, $1 \leq i \leq n$, if and only if there exists a graph H of order $n - 1$ with $V(H) = \{u_1, u_2, \ldots, u_{n-1}\}$ such that

$$\deg u_i = \begin{cases} d_{i+1} - 1 & \text{if } 1 \leq i \leq d_1 \\ d_{i+1} & \text{if } d_1 + 1 \leq i \leq n - 1. \end{cases}$$

5. Let G be a graph of order 3 or more. Prove that G is connected if and only if G contains two distinct vertices u and v such that $G - u$ and $G - v$ are connected.

6. (a) A graph G of order 6 and unknown size m has vertex set $V(G) = \{v_1, v_2, \ldots, v_6\}$. For $i = 1, 2, \ldots, 6$, the subgraph $G_i = G - v_i$ has size m_i. It is known that $m_1 = 4$, $m_2 = m_3 = 5$ and $m_4 = m_5 = m_6 = 6$. What is m?

   (b) The problem in (a) should suggest a theorem to you. State and prove this theorem.

7. Prove that if G is a connected graph of order $n \geq 2$, then the vertices of G can be listed as $v_1, v_2, \ldots, v_n$ such that each vertex v_i ($2 \leq i \leq n$) is adjacent to some vertex in the set $\{v_1, v_2, \ldots, v_{i-1}\}$.

8. Let k and n be integers with $2 \leq k < n$ and let G be a graph of order n. Prove that if every vertex of G has degree exceeding $(n - k)/k$, then G has fewer than k components.

9. Prove Theorem 1.4: *If G is a graph of order $n \geq 2$ such that $\sigma_2(G) \geq n - 1$, then G is connected.*

10. Prove Corollary 1.6: *If G is a graph of order $n \geq 2$ and size $m > \binom{n-1}{2}$, then G is connected.*

11. Prove that if G is a graph of order $n \geq 4$ and size $m > n^2/4$, then G contains an odd cycle.

12. Let G be a connected graph of order $n \geq 3$. Suppose that each vertex of G is colored with one of the colors red, blue, and green such that for each color, there exists at least one vertex of G assigned that color.

    (a) Show that G contains two adjacent vertices that are colored differently.

    (b) Show that, regardless of how large n may be, G may not contain two adjacent vertices that are colored the same.

    (c) Show that G has a path containing at least one vertex of each of these three colors.

    (d) The question in (c) should suggest another question to you. Ask and answer such a question.

13. Let G be a graph with $\delta(G) = \delta$. Prove each of the following.

    (a) The graph G contains a path of length δ.

    (b) If $\delta \geq 2$, then G contains a cycle of length at least $\delta + 1$.

14. For vertices u and v in a connected graph G, let $d(u,v)$ be the length of a shortest $u - v$ path in G. Prove that d satisfies the triangle inequality.

15. Let G be a connected graph and let u and v be distinct vertices of G. Recall that a $u - v$ path P is a $u - v$ geodesic if the length of P is $d(u,v)$. On the other hand, a $u - v$ path P' is a $u - v$ detour if the length of P' is $D(u,v)$. If P is a $u - v$ geodesic and x and y are non-consecutive vertices on P, then x and y are nonadjacent. If P' is a $u - v$ detour and x and y are non-consecutive vertices on P', then x and y may be adjacent. For two vertices u and v of G, the *induced distance* $\rho(u,v)$ *between u and v* is the maximum length of a $u - v$ path P'' such that every two non-consecutive vertices on P'' are nonadjacent. Thus, $G[V(P'')] = P''$. The $u - v$ path P'' is called a *maximum induced $u - v$ path* if the length of P'' is $\rho(u,v)$. Therefore, $d(u,v) \leq \rho(u,v) \leq D(u,v)$.

    (a) Give an example of a connected graph G and two vertices u and v such that $d(u,v) < \rho(u,v) < D(u,v)$.

    (b) Prove or disprove: The induced distance is a metric on the vertex set of a connected graph.

16. Prove or disprove: If u and v are peripheral vertices in a connected graph G, then u and v are antipodal vertices of G.

17. Prove that for every two positive integers r and d such that $r \leq d \leq 2r$, there exists a connected graph G such that $\mathrm{rad}(G) = r$ and $\mathrm{diam}(G) = d$.

18. Prove that if G is a graph of order $n \geq 5$ containing three distinct vertices $u, v,$ and w such that $d(u,v) = d(u,w) = n - 2$, then G is connected.

19. Let G be a connected graph of order n. For a vertex v of G and an integer k with $1 \leq k \leq n - 1$, let $d_k(v)$ be the number of vertices at distance k from v.

(a) What is $d_1(v)$?

(b) Show that $\sum_{v \in V(G)} d_k(v)$ is even for every integer k with $1 \le k \le n-1$.

(c) What is the value of $\sum_{v \in V(G)} \left(\sum_{k=1}^{n-1} d_k(v) \right)$?

20. Let G be a connected graph and let $u \in V(G)$. Prove or disprove: If $v \in V(G)$ such that $d(u, v) = e(u)$, then v is a peripheral vertex of G.

21. Give an example of connected graphs G and H of order 3 or more such that (1) $D(u, v) = d(u, v)$ for every two vertices u and v of G and (2) $D(u, v) \ne d(u, v)$ for every two vertices u and v of H.

22. Let G be a connected graph containing a vertex v. Prove that if u and w are any two adjacent vertices of G, then $|d(u, v) - d(v, w)| \le 1$.

23. Prove Theorem 1.11: *If two graphs G and H are isomorphic, then they have the same order and the same size, and the degrees of the vertices of G are the same as the degrees of the vertices of H.*

24. Let G and H be isomorphic graphs. Prove the following.

    (a) If G contains a k-cycle for some integer $k \ge 3$, then so does H.

    (b) If G contains a path of length k, then H contains a path of length k.

    (c) The graph G is bipartite if and only if H is bipartite.

    (d) The graph G is connected if and only if H is connected.

25. Let G and H be two graphs, where S is the set of vertices of degree r in G and T is the set of vertices of degree r in H.

    (a) Prove that if G and H are isomorphic, then $G[S]$ and $H[T]$ are isomorphic.

    (b) Give an example of two (non-isomorphic) graphs G and H having the same order and same size, where the degrees of the vertices of G are the same as the degrees of the vertices of H but where the statement in (a) is false for some r.

26. How many (non-isomorphic) graphs of order 5 are there?

27. For an integer $n \ge 3$, each edge of K_n is colored red, blue, or yellow. The spanning subgraphs of K_n whose edges are all red, blue, or yellow are denoted by G_r, G_b, and G_y.

    (a) For $n = 4$, does there exist a coloring of the edges of K_n such that every two of G_r, G_b, and G_y are isomorphic?

    (b) Repeat (a) for $n = 5$ and $n = 6$.

28. Let G be a (not necessarily connected) graph of size $m \geq 3$. Each edge of G is assigned one of the two colors red and blue or one of the three colors red, blue and green. The subgraphs of G induced by the edges colored red, blue or green is denoted by G_r, G_b or G_g, respectively. Give an example of a graph G of size 6 such that

   (a) there is an assignment of two colors red and blue to G for which $G_r \cong G_b$ and there is an assignment of three colors to G for which $G_r \cong G_b \cong G_g$.

   (b) there is no assignment of two colors red and blue to G for which $G_r \cong G_b$ and there is no assignment of three colors to G for which $G_r \cong G_b \cong G_g$.

   (c) there is an assignment of two colors red and blue to G for which $G_r \cong G_b$ but there is no assignment of three colors to G for which $G_r \cong G_b \cong G_g$.

   (d) there is no assignment of two colors red and blue to G for which $G_r \cong G_b$ but there is an assignment of three colors to G for which $G_r \cong G_b \cong G_g$.

29. Prove that if G is a disconnected graph, then $\overline{G}$ is connected and $\text{diam}(\overline{G}) \leq 2$.

30. We showed in Theorem 1.15 that for every integer $n \geq 2$, there is exactly one connected graph F_n of order n containing exactly two vertices of the same degree. What is this degree? What are the degrees of the vertices of $\overline{F}_n$?

31. Prove that if G is an r-regular bipartite graph, $r \geq 1$, with partite sets U and V, then $|U| = |V|$.

32. Prove that if G is any graph of order $n \geq 3$ with $\delta(G) > n/2$, then G contains a triangle.

33. Let r and n be integers with $0 \leq r \leq n - 1$. Prove that there exists an r-regular graph of order n if and only if at least one of r and n is even.

34. Let $k \geq 2$ be an integer. Prove that if G is a graph of order $n \geq k+1$ and size $m \geq (k-1)(n-k-1) + \binom{k+1}{2}$, then G contains a subgraph having minimum degree k.

35. Suppose that a graph G and its complement $\overline{G}$ are both connected graphs of order $n \geq 5$.

   (a) Prove that if the diameter of G is at least 3, then the diameter of its complement is at most 3.

   (b) What diameters are possible for self-complementary graphs with at least three vertices?

   (c) If the diameter of G is 2, then what is the smallest and largest diameter of its complement $\overline{G}$, expressed in terms of n?

36. Show that the graph $K_{1,3}$ is not a line graph.

37. Show that there exist two non-isomorphic connected graphs G_1 and G_2 such that $L(G_1) = L(G_2)$.

38. (a) Prove that if G is a nonempty connected graph, then $L(G)$ is connected.

(b) Show that there are connected graphs F and H of diameter 3 such that $\overline{L(F)}$ is disconnected, while $\overline{L(H)}$ is connected.

(c) Prove or disprove: If G is a connected graph of diameter $d \geq 4$, then $\overline{L(G)}$ is connected.

39. Let G be a graph of order n and size m such that $n = 4k+3$ for some positive integer k. Suppose that the complement $\overline{G}$ of G has size $\overline{m}$. Prove that either $m > \frac{1}{2}\binom{n}{2}$ or $\overline{m} > \frac{1}{2}\binom{n}{2}$.

40. Let G be a graph of order 3 or more. Prove that if for each $S \subseteq V(G)$ with $|S| \geq 3$, the size of $G[S]$ is at least the size of $\overline{G}[S]$, then G is connected.

41. Let G be a disconnected graph of order $n \geq 6$ having three components. Prove that $\Delta(\overline{G}) \geq \frac{2n+3}{3}$.

42. Prove that every graph has an acyclic orientation.

43. (a) Show that every connected graph has an orientation that is not strong.

(b) Show that there are connected graphs where no orientation is strong.

44. Let G be a connected graph of order $n \geq 3$. Prove that there is an orientation of G in which no directed path has length 2 if and only if G is bipartite.

45. Let u and v be two vertices in a tournament T. Prove that if u and v do not lie on a common cycle, then od $u \neq$ od v.

46. Let T be a tournament of order 10. Suppose that the outdegree of each vertex of T is 2 or more. Determine the maximum number of vertices in T whose outdegree can be exactly 2.

47. Let T be a tournament of order $n \geq 10$. Suppose that T contains two vertices u and v such that when the directed edge joining u and v is removed, the resulting digraph D does not contain a directed $u - v$ path or a directed $v - u$ path. Show that $\text{od}_D u = \text{od}_D v$.

48. Let T be a tournament with $V(T) = \{v_1, v_2, \ldots, v_n\}$. We know that

$$\sum_{i=1}^{n} \text{od } v_i = \sum_{i=1}^{n} \text{id } v_i.$$

(a) Prove that $\sum_{i=1}^{n}(\text{od } v_i)^2 = \sum_{i=1}^{n}(\text{id } v_i)^2$.

(b) Prove or disprove: $\sum_{i=1}^{n}(\text{od } v_i)^3 = \sum_{i=1}^{n}(\text{id } v_i)^3$.

49. Prove that if T is a tournament of order $4r$ with $r \geq 1$, where $2r$ vertices of T have outdegree $2r$ and the other $2r$ vertices have outdegree $2r - 1$, then T is strong.

Chapter 2

Trees and Connectivity

Although the property of a graph G being connected depends only on whether G contains a $u - v$ path for every pair u, v of vertices of G, there are varying degrees of connectedness that a graph may possess. Some of the best-known measures of connectedness are discussed in this chapter.

2.1   Cut-Vertices, Bridges, and Blocks

There are some graphs that are so slightly connected that they can be disconnected by the removal of a single vertex or a single edge.

Let v be a vertex and e an edge of a graph G. If $G - v$ has more components than G, then v is a **cut-vertex** of G; while if $G - e$ has more components than G, then e is a **bridge** of G. In particular, if v is a cut-vertex of a connected graph G, then $G - v$ is disconnected; and if e is a bridge of a connected graph G, then $G - e$ is disconnected – necessarily a graph with exactly two components. While K_2 is a connected graph of order 2 containing a bridge but no cut-vertices, every connected graph of order 3 or more that contains bridges also contains cut-vertices (see Exercise 1).

For the graph G of Figure 2.1, only u, w, and y are cut-vertices while only uv, wy, and yz, are bridges. The subgraphs $G - u$ and $G - wy$ are shown in Figure 2.1. For $n \geq 2$, the path P_n of order n has exactly $n - 2$ cut-vertices. Indeed, this graph shows that the following theorem cannot be improved.

Theorem 2.1 *Every nontrivial connected graph contains at least two vertices that are not cut-vertices.*

Proof. Let G be a nontrivial connected graph and let P be a longest path in G. Suppose that P is a $u - v$ path. We show that u and v are not cut-vertices. Assume, to the contrary, that u is a cut-vertex of G. Then $G - u$ is disconnected and so contains two or more components. Let w be the vertex adjacent to u on P and let P' be the $w - v$ subpath of P. Necessarily, P' belongs to a component, say G_1, of $G - u$. Let G_2 be another component of $G - u$. Then G_2 contains some vertex x

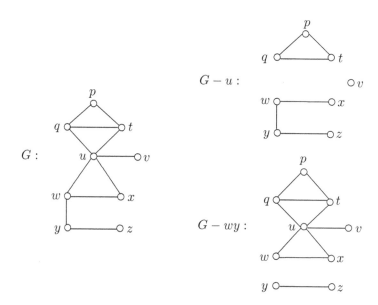

Figure 2.1: Cut-vertices and bridges in graphs

that is adjacent to u. This produces an $x - v$ path that is longer than P, which is impossible. Similarly, v is not a cut-vertex of G.   ∎

The following theorems provide characterizations of cut-vertices and bridges in a graph (see Exercises 2 and 3).

Theorem 2.2   *A vertex v in a graph G is a cut-vertex of G if and only if there are two vertices u and w distinct from v such that v lies on every $u - w$ path in G.*

Theorem 2.3   *An edge e in a graph G is a bridge of G if and only if e lies on no cycle in G.*

Often we are interested in nontrivial connected graphs that contain no cut-vertices. A nontrivial connected graph having no cut-vertices is called **nonseparable**. In particular, the cycles C_n, $n \geq 3$, and the complete graphs K_n, $n \geq 2$, are nonseparable graphs.

Although the complete graph K_2 is the only nonseparable graph of order less than 3, each nonseparable graph of order 3 or more has an interesting property.

Theorem 2.4   *Every two distinct vertices in a nonseparable graph G of order 3 or more lie on a common cycle of G.*

Proof.   Since G is a nonseparable graph of order 3 or more, it follows, as we noted earlier, that G contains neither bridges nor cut-vertices. Assume, to the contrary, that there are pairs of vertices of G that do not lie on a common cycle. Among all such pairs, let u, v be a pair for which $d(u, v)$ is minimum. Now $d(u, v) \neq 1$, for otherwise $uv \in E(G)$. Since uv is not a bridge, it follows from Theorem 2.3 that uv lies on a cycle of G. Therefore, $d(u, v) = k \geq 2$.

Let $P = (u = v_0, v_1, \ldots, v_{k-1}, v_k = v)$ be a $u - v$ geodesic in G. Since $d(u, v_{k-1}) = k - 1 < k$, there is a cycle C containing u and v_{k-1}. By assumption, v is not on C. Since v_{k-1} is not a cut-vertex of G and u and v are distinct from v_{k-1}, it follows from Theorem 2.2 that there is a $v - u$ path Q that does not contain v_{k-1}. Since u is on C, there is a first vertex x of Q that is on C. Let Q' be the $v - x$ subpath of Q and let P' be a $v_{k-1} - x$ path on C that contains u. (If $x \neq u$, then the path P' is unique.) However, the cycle C' produced by proceeding from v to its neighbor v_{k-1}, along P' to x, and then along Q' to v contains both u and v, a contradiction. ∎

Corollary 2.5 *For every two distinct vertices u and v in a nonseparable graph G of order 3 or more, there are two distinct $u - v$ paths in G having only u and v in common.*

A nonseparable subgraph B of a nontrivial connected graph G is a **block** of G if B is not a proper subgraph of any nonseparable subgraph of G. Every two distinct blocks of G have at most one vertex in common; and if they have a vertex in common, then this vertex is a cut-vertex of G. A block of G containing exactly one cut-vertex of G is called an **end-block** of G. A graph G and its five blocks B_i, $1 \leq i \leq 5$, are shown in Figure 2.2. The end-blocks of G are B_1, B_2, and B_5. A connected graph with cut-vertices must contain two or more end-blocks (see Exercise 6).

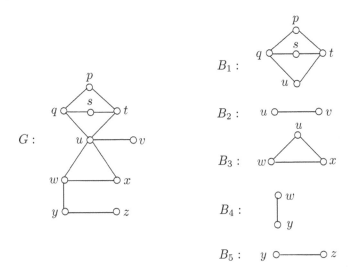

Figure 2.2: The blocks of a graph

Theorem 2.6 *Every connected graph containing cut-vertices contains at least two end-blocks.*

If a graph G has components $G_1, G_2, \ldots, G_k$ and a nonempty connected graph H has blocks $B_1, B_2, \ldots, B_\ell$, then $\{V(G_1),\ V(G_2),\ \ldots,\ V(G_k)\}$ is a partition of $V(G)$ and $\{E(B_1),\ E(B_2),\ \ldots,\ E(B_\ell)\}$ is a partition of $E(H)$.

For a cut-vertex v of a connected graph G, suppose that the disconnected graph $G - v$ has k components $G_1, G_2, \ldots, G_k$ $(k \geq 2)$. The induced subgraphs

$$B_i = G[V(G_i) \cup \{v\}]$$

are connected and referred to as the **branches** of G at v. If a subgraph G_i contains no cut-vertices of G, then the branch B_i is a block of G, in fact, an end-block of G.

A connected graph G containing three cut-vertices u, v, and w, three bridges uv, ux, and vy, and six blocks is shown in Figure 2.3. Four of these blocks are end-blocks. The graph G has four branches at v, all of which are shown in Figure 2.3. Two of the four branches at v are end-blocks of G.

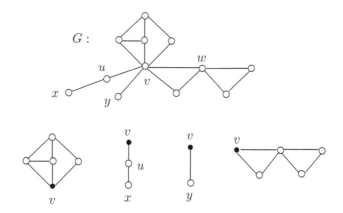

Figure 2.3: The four branches of a graph G at a cut-vertex v

2.2   Trees

According to Theorem 2.1, it is impossible for every vertex of a connected graph G to be a cut-vertex. It is possible, however, for every edge of G to be a bridge. By Theorem 2.3, this can only occur if G has no cycles. This brings us to one of the most studied and best-known classes of graphs.

A connected graph without cycles is a **tree**. All of the graphs T_1, T_2, and T_3 of Figure 2.4 are trees. Also, all paths and stars are trees. There are other well-known classes of trees. A tree containing exactly two vertices that are not leaves (which are necessarily adjacent) is called a **double star**. Thus, a double star is a tree of diameter 3. A tree T of order 3 or more is a **caterpillar** if the removal of its leaves produces a path. Thus, every path and star (of order at least 3) and every double star is a caterpillar. The trees T_2 and T_3 in Figure 2.4 are caterpillars, while T_1 is not a caterpillar. The tree T_2 is also a double star.

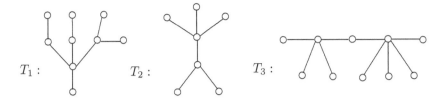

Figure 2.4: Trees

Trees can be characterized as those graphs in which every two vertices are connected by a single path.

Theorem 2.7 *A graph G is a tree if and only if every two vertices of G are connected by a unique path.*

Proof. First, suppose that G is a tree and that u and v are two vertices of G. Since G is connected, G contains at least one $u - v$ path. On the other hand, if G were to contain at least two $u - v$ paths, then G would contain a cycle, which is impossible. Therefore, G contains exactly one $u - v$ path.

Conversely, let G be a graph in which every two vertices are connected by a unique path. Certainly then, G is connected. If G were to contain a cycle C, then every two vertices on C would be connected by two paths. Thus, G contains no cycle and G is a tree. ∎

While every vertex of degree 2 or more in a tree is a cut-vertex, the vertices of degree 1 (the leaves) are not. These observations provide a corollary of Theorem 2.1.

Corollary 2.8 *Every nontrivial tree contains at least two leaves.*

For a cut-vertex v of T, there are $\deg v$ branches of T at v. In the tree T of Figure 2.5, the four branches of T at v are shown in that figure.

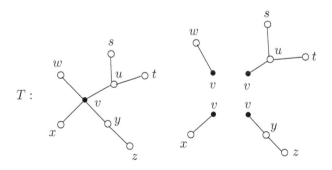

Figure 2.5: The branches of a tree at a vertex

In the tree T of Figure 2.5 and in each of the trees T_1, T_2, and T_3 of Figure 2.4, the size of the tree is one less than its order. With the aid of Corollary 2.8, this can be verified in general. Observe that if v is a leaf in a nontrivial tree T, then $T - v$ is also a tree of order one less than that of T.

Theorem 2.9 *If T is a tree of order n and size m, then $m = n - 1$.*

Proof.   We proceed by induction on the order of a tree. There is only one tree of order 1, namely K_1, and it has no edges. Thus, the basis step of the induction is established. Assume that the size of every tree of order $n - 1$ where $n \geq 2$ is $n - 2$ and let T be a tree of order n and size m. By Corollary 2.8, T has at least two leaves. Let v be one of them. As we observed, $T - v$ is a tree of order $n - 1$. By the induction hypothesis, the size of $T - v$ is $n - 2$. Thus, $m = (n - 2) + 1 = n - 1$. ∎

A graph without cycles is a **forest**. Thus, each tree is a forest and every component of a forest is a tree. All of the graphs F_1, F_2, and F_3 in Figure 2.6 are forests but none are trees.

Figure 2.6: Forests

The following is an immediate corollary of Theorem 2.9.

Corollary 2.10 *The size of a forest of order n having k components is $n - k$.*

By Theorem 2.9, if G is a graph of order n and size m such that G is connected and has no cycles (that is, G is a tree), then $m = n - 1$. It is easy to see that the converse of this statement is not true. However, if we were to add to the hypothesis of the converse either of the two defining properties of a tree, then the converse would be true.

Theorem 2.11 *Let G be a graph of order n and size m. If G has no cycles and $m = n - 1$, then G is a tree.*

Proof.   It remains only to show that G is connected. Suppose that the components of G are $G_1, G_2, \ldots, G_k$, where $k \geq 1$. Let n_i be the order of G_i ($1 \leq i \leq k$) and m_i the size of G_i. Since each graph G_i is a tree, it follows by Theorem 2.9 that $m_i = n_i - 1$ and by Corollary 2.10 that $m = n - k$. Hence

$$n - 1 = m = \sum_{i=1}^{k} m_i = \sum_{i=1}^{k} (n_i - 1) = n - k.$$

Thus, $k = 1$ and so G is connected. Therefore, G is a tree.   ∎

Theorem 2.12 *Let G be a graph of order n and size m. If G is connected and $m = n - 1$, then G is a tree.*

Proof. Assume, to the contrary, that there exists some connected graph of order n and size $m = n - 1$ that is not a tree. Necessarily then, G contains one or more cycles. By successively deleting an edge from a cycle in each resulting subgraph, a tree of order n and size less than $n - 1$ is obtained. This contradicts Theorem 2.9. ∎

Combining Theorems 2.9, 2.11, and 2.12, we have the following.

Corollary 2.13 *Let G be a graph of order n and size m. If G satisfies any two of the following three properties, then G is a tree:*

(1) *G is connected,*   (2) *G has no cycles,*   (3) *$m = n - 1$.*

A tree that is a spanning subgraph of a connected graph G is a **spanning tree** of G. If G is a connected graph of order n and size m, then $m \geq n - 1$. If T is a spanning tree of G, then the size of T is $n - 1$. Hence, $m - (n - 1) = m - n + 1$ edges must be deleted from G to obtain T. The number $m - n + 1$ is referred to as the **cycle rank** of G. Since $m - n + 1 \geq 0$, the cycle rank of a connected graph is a nonnegative integer. A graph with cycle rank 0 is therefore a tree.

If G is a graph of order n and size m having cycle rank 1, then $n - m + 1 = 1$ and so $n = m$. Such a graph is therefore a connected graph with exactly one cycle. These graphs are often called **unicyclic graphs**. All of the graphs in Figure 2.7 are unicyclic graphs.

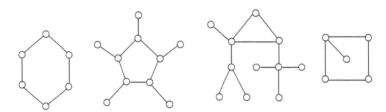

Figure 2.7: Unicyclic graphs

Suppose that a tree T of order $n \geq 3$, size m, and maximum degree $\Delta(T) = \Delta$ has n_i vertices of degree i ($1 \leq i \leq \Delta$). Then

$$\sum_{v \in V(T)} \deg v = \sum_{i=1}^{\Delta} i n_i = 2m = 2n - 2 = 2 \sum_{i=1}^{\Delta} n_i - 2. \tag{2.1}$$

Solving (2.1) for n_1, we have the following.

Theorem 2.14 *Let T be a tree of order $n \geq 3$ having maximum degree Δ and containing n_i vertices of degree i ($1 \leq i \leq \Delta$). Then*

$$n_1 = 2 + n_3 + 2n_4 + \cdots + (\Delta - 2)n_\Delta.$$

2.3   Connectivity and Edge-Connectivity

Each tree of order 3 or more contains at least one vertex whose removal results in a disconnected graph. In fact, every vertex in a tree that is not a leaf has this property. Furthermore, the removal of any edge in a tree results in a disconnected graph (with exactly two components). On the other hand, no vertex or edge in a nonseparable graph of order 3 or more has this property. Hence, in this sense, nonseparable graphs possess a greater degree of connectedness than trees. We now look at the two most common measures of connectedness of graphs. In the process of doing this, we will encounter some of most famous theorems in graph theory.

A **vertex-cut** of a graph G is a set S of vertices of G such that $G - S$ is disconnected. A vertex-cut of minimum cardinality in G is called a **minimum vertex-cut** of G and this cardinality is called the **vertex-connectivity** (or, more simply, the **connectivity**) of G and is denoted by $\kappa(G)$. (The symbol κ is the Greek letter *kappa*.)

Let S be a minimum vertex-cut of a (noncomplete) connected graph G and let $G_1, G_2, \ldots, G_k$ $(k \geq 2)$ be the components of $G - S$. Then the subgraphs $B_i = G[V(G_i) \cup S]$ are called the **branches** of G at S or the **S-branches** of G. For the minimum vertex-cut $S = \{u, v\}$ of the graph G of Figure 2.8, the three S-branches of G are also shown in that figure.

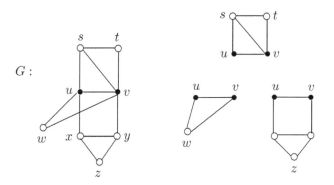

Figure 2.8: The branches of a graph at $S = \{u, v\}$

Complete graphs do not contain vertex-cuts. Indeed, the removal of any proper subset of vertices from a complete graph results in a smaller complete graph. The connectivity of the complete graph of order n is defined as $n - 1$, that is, $\kappa(K_n) = n - 1$. In general then, the **connectivity** $\kappa(G)$ of a graph G is the smallest number of vertices whose removal from G results in either a disconnected graph or a trivial graph. Therefore, for every graph G of order n,

$$0 \leq \kappa(G) \leq n - 1.$$

Thus, a graph G has connectivity 0 if and only if either $G = K_1$ or G is disconnected; a graph G has connectivity 1 if and only if $G = K_2$ or G is a connected graph with cut-vertices; and a graph G has connectivity 2 or more if and only if G is a nonseparable graph of order 3 or more.

Often it is more useful to know that a given graph G cannot be disconnected by the removal of a certain number of vertices rather than to know the actual connectivity of G. A graph G is k-**connected**, $k \geq 1$, if $\kappa(G) \geq k$. That is, G is k-connected if the removal of fewer than k vertices from G results in neither a disconnected nor a trivial graph. The 1-connected graphs are then the nontrivial connected graphs, while the 2-connected graphs are the nonseparable graphs of order 3 or more.

How connected a graph G is can be measured not only in terms of the number of vertices that need to be deleted from G to arrive at a disconnected or trivial graph but in terms of the number of edges that must be deleted from G to produce a disconnected or trivial graph.

An **edge-cut** of a graph G is a subset X of $E(G)$ such that $G - X$ is disconnected. An edge-cut of minimum cardinality in G is a **minimum edge-cut** and this cardinality is the **edge-connectivity** of G, which is denoted by $\lambda(G)$. (The symbol λ is the Greek letter *lambda*.) The trivial graph K_1 does not contain an edge-cut but we define $\lambda(K_1) = 0$. Therefore, $\lambda(G)$ is the minimum number of edges whose removal from G results in a disconnected or trivial graph. Thus,

$$0 \leq \lambda(G) \leq n - 1$$

for every graph G of order n. A graph G is k-**edge-connected**, $k \geq 1$, if $\lambda(G) \geq k$. That is, G is k-edge-connected if the removal of fewer than k edges from G results in neither a disconnected graph nor a trivial graph. Thus, a 1-edge-connected graph is a nontrivial connected graph and a 2-edge-connected graph is a nontrivial connected bridgeless graph.

For the graph G of Figure 2.9, $\kappa(G) = 2$ and $\lambda(G) = 3$. Both $\{u, v_1\}$ and $\{u, v_2\}$ are minimum vertex-cuts, while $\{e_1, e_2, e_3\}$ is a minimum edge-cut.

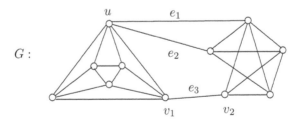

Figure 2.9: Connectivity and edge-connectivity

The edge-connectivity of every complete graph is given in the next theorem.

Theorem 2.15 *For every positive integer n,*

$$\lambda(K_n) = n - 1.$$

Proof. Since the edge-connectivity of K_1 is defined to be 0, we may assume that $n \geq 2$. If the $n - 1$ edges incident with any vertex of K_n are removed from K_n, then a disconnected graph results. Thus, $\lambda(K_n) \leq n-1$. Now let X be a minimum edge-cut of K_n. Then $|X| = \lambda(K_n)$ and $G - X$ consists of two components, say G_1 and G_2. Suppose that G_1 has order k. Then G_2 has order $n - k$. Thus, $|X| = k(n - k)$. Since $k \geq 1$ and $n - k \geq 1$, it follows that $(k - 1)(n - k - 1) \geq 0$ and so

$$(k - 1)(n - k - 1) = k(n - k) - (n - 1) \geq 0,$$

which implies that

$$\lambda(K_n) = |X| = k(n - k) \geq n - 1.$$

Therefore, $\lambda(K_n) = n - 1$.   ∎

The connectivity, edge-connectivity, and minimum degree of a graph satisfy inequalities, first observed by Hassler Whitney [201].

Theorem 2.16 *For every graph G,*

$$\kappa(G) \leq \lambda(G) \leq \delta(G).$$

Proof.   Let G be a graph of order n. If G is disconnected, then $\kappa(G) = \lambda(G) = 0$; while if G is complete, then $\kappa(G) = \lambda(G) = \delta(G) = n - 1$. Thus, the desired inequalities hold in these two cases. Hence, we may assume that G is a connected graph that is not complete.

Since G is not complete, $\delta(G) \leq n - 2$. Let v be a vertex of G such that $\deg v = \delta(G)$. If the edges incident with v are deleted from G, then a disconnected graph is produced. Hence, $\lambda(G) \leq \delta(G) \leq n - 2$.

It remains to show that $\kappa(G) \leq \lambda(G)$. Let X be a minimum edge-cut of G. Then $|X| = \lambda(G) \leq n - 2$. Necessarily, $G - X$ consists of two components, say G_1 and G_2. Suppose that the order of G_1 is k. Then the order of G_2 is $n - k$, where $k \geq 1$ and $n - k \geq 1$. Also, every edge in X joins a vertex of G_1 and a vertex of G_2. We consider two cases.

Case 1.   Every vertex of G_1 is adjacent to every vertex of G_2. Then $|X| = k(n - k)$. Since $k - 1 \geq 0$ and $n - k - 1 \geq 0$, it follows that

$$(k - 1)(n - k - 1) = k(n - k) - (n - 1) \geq 0$$

and so

$$\lambda(G) = |X| = k(n - k) \geq n - 1.$$

This, however, contradicts $\lambda(G) \leq n - 2$ and so Case 1 cannot occur.

Case 2.   There exist a vertex u in G_1 and a vertex v in G_2 such that $uv \notin E(G)$. We now define a set U of vertices of G. Let $e \in X$. If e is incident with u, say $e = uv'$, then the vertex v' is placed in the set U. If e is not incident with u, say $e = u'v'$ where u' is in G_1, then the vertex u' is placed in U. Hence, for every edge $e \in X$, one of its two incident vertices belongs to U but $u, v \notin U$. Thus, $|U| \leq |X|$ and U is a vertex-cut. Therefore,

$$\kappa(G) \le |U| \le |X| = \lambda(G),$$

as desired. ∎

We observed that $\kappa(G) = 2$ and $\lambda(G) = 3$ for the graph G of Figure 2.9. Since $\delta(G) = 4$, this graph shows that the two inequalities stated in Theorem 2.16 can be strict. The first of these inequalities cannot be strict for cubic graphs, however.

Theorem 2.17 *For every cubic graph G,*

$$\kappa(G) = \lambda(G).$$

Proof. For a cubic graph G, it follows that $\kappa(G) = \lambda(G) = 0$ if and only if G is disconnected. If $\kappa(G) = 3$, then $\lambda(G) = 3$ by Theorem 2.16. So two cases remain, namely $\kappa(G) = 1$ or $\kappa(G) = 2$. Let U be a minimum vertex-cut of G. Then $|U| = 1$ or $|U| = 2$. So $G - U$ is disconnected. Let G_1 and G_2 be two components of $G - U$. Since G is cubic, for each $u \in U$, at least one of G_1 and G_2 contains exactly one neighbor of u.

Case 1. $\kappa(G) = |U| = 1$. Thus, U consists of a cut-vertex u of G. Since some component of $G - U$ contains exactly one neighbor w of u, the edge uw is a bridge of G and so $\lambda(G) = \kappa(G) = 1$.

Case 2. $\kappa(G) = |U| = 2$. Let $U = \{u, v\}$. Assume that each of u and v has exactly one neighbor, say u' and v', respectively, in the same component of $G - U$. (This is the case that holds if $uv \in E(G)$.) Then $X = \{uu', vv'\}$ is an edge-cut of G and $\lambda(G) = \kappa(G) = 2$. (See Figure 2.10(a) for the situation when u and v are not adjacent.)

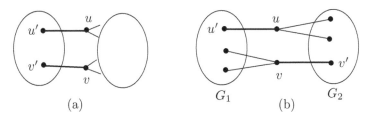

Figure 2.10: A step in the proof of Case 2

Hence, we may assume that u has one neighbor u' in G_1 and two neighbors in G_2; while v has two neighbors in G_1 and one neighbor v' in G_2 (see Figure 2.10(b)). Therefore, $uv \notin E(G)$ and $X = \{uu', vv'\}$ is an edge-cut of G; so $\lambda(G) = \kappa(G) = 2$. ∎

2.4 Menger's Theorem

For two nonadjacent vertices u and v in a graph G, a $u - v$ **separating set** is a set $S \subseteq V(G) - \{u, v\}$ such that u and v lie in different components of $G - S$. A $u - v$ separating set of minimum cardinality is called a **minimum $u - v$ separating set**.

For two distinct vertices u and v in a graph G, a collection of $u - v$ paths is **internally disjoint** if every two paths in the collection have only u and v in common. Internally disjoint $u - v$ paths and $u - v$ separating sets are linked according to one of the best-known theorems in graph theory, due to Karl Menger [149]. This 1927 "min-max" theorem received increased recognition when Menger mentioned his theorem to Dénes König [137] who, as a result, added a chapter to his 1936 book *Theorie der endlichen und unendliehen Graphen*, which was the first book written on graph theory.

Theorem 2.18 (**Menger's Theorem**) *Let u and v be nonadjacent vertices in a graph G. The minimum number of vertices in a $u - v$ separating set equals the maximum number of internally disjoint $u - v$ paths in G.*

Proof.   We proceed by induction on the size of graphs. The theorem is certainly true for every empty graph. Assume that the theorem holds for all graphs of size less than m, where $m \geq 1$, and let G be a graph of size m. Moreover, let u and v be two nonadjacent vertices of G. If u and v belong to different components of G, then the result follows. So we may assume that u and v belong to the same component of G. Suppose that a minimum $u - v$ separating set consists of $k \geq 1$ vertices. Then G contains at most k internally disjoint $u - v$ paths. We show, in fact, that G contains k internally disjoint $u - v$ paths. Since this is obviously true if $k = 1$, we may assume that $k \geq 2$. We now consider three cases.

Case 1. Some minimum $u - v$ separating set X in G contains a vertex x that is adjacent to both u and v. Then $X - \{x\}$ is a minimum $u - v$ separating set in $G - x$ consisting of $k - 1$ vertices. Since the size of $G - x$ is less than m, it follows by the induction hypothesis that $G - x$ contains $k - 1$ internally disjoint $u - v$ paths. These paths together with the path $P = (u, x, v)$ produce k internally disjoint $u - v$ paths in G.

Case 2. For every minimum $u - v$ separating set S in G, either every vertex in S is adjacent to u and not to v or every vertex of S is adjacent to v and not to u. Necessarily then, $d(u, v) \geq 3$. Let $P = (u, x, y, \ldots, v)$ be a $u - v$ geodesic in G, where $e = xy$. Every minimum $u - v$ separating set in $G - e$ contains at least $k - 1$ vertices. We show, in fact, that every minimum $u - v$ separating set in $G - e$ contains k vertices.

Suppose that there is some minimum $u - v$ separating set Z in $G - e$ with $k - 1$ vertices, say $Z = \{z_1, z_2, \ldots, z_{k-1}\}$. Then $Z \cup \{x\}$ is a $u - v$ separating set in G and therefore a minimum $u - v$ separating set in G. Since x is adjacent to u (and not to v), it follows that every vertex z_i $(1 \leq i \leq k - 1)$ is also adjacent to u and not adjacent to v.

Since $Z \cup \{y\}$ is also a minimum $u - v$ separating set in G and each vertex z_i $(1 \leq i \leq k - 1)$ is adjacent to u but not to v, it follows that y is adjacent to u. This, however, contradicts the assumption that P is a $u - v$ geodesic. Thus, k is the minimum number of vertices in a $u - v$ separating set in $G - e$. Since the size of $G - e$ is less than m, it follows by the induction hypothesis that there are k internally disjoint $u - v$ paths in $G - e$ and in G as well.

Case 3. *There exists a minimum $u - v$ separating set W in G in which no vertex is adjacent to both u and v and containing at least one vertex not adjacent to u and at least one vertex not adjacent to v.* Let $W = \{w_1, w_2, \ldots, w_k\}$. Let G_u be the subgraph of G consisting of, for each i with $1 \leq i \leq k$, all $u - w_i$ paths in G in which $w_i \in W$ is the only vertex of the path belonging to W. Let G'_u be the graph constructed from G_u by adding a new vertex v' and joining v' to each vertex w_i for $1 \leq i \leq k$. The graphs G_v and G'_v are defined similarly.

Since W contains a vertex that is not adjacent to u and a vertex that is not adjacent to v, the sizes of both G'_u and G'_v are less than m. So G'_u contains k internally disjoint $u - v'$ paths A_i $(1 \leq i \leq k)$, where A_i contains w_i. Also, G'_v contains k internally disjoint $u' - v$ paths B_i $(1 \leq i \leq k)$, where B_i contains w_i. Let A'_i be the $u - w_i$ subpath of A_i and let B'_i be the $w_i - v$ subpath of B_i for $1 \leq i \leq k$. The k paths constructed from A'_i and B'_i for each i $(1 \leq i \leq k)$ are internally disjoint $u - v$ paths in G. ■

If G is a k-connected graph $(k \geq 1)$ and v is a vertex of G, then $G - v$ is $(k-1)$-connected. In fact, if $e = uv$ is an edge of G, then $G - e$ is also $(k-1)$-connected (see Exercise 32). With the aid of Menger's theorem, a useful characterization of k-connected graphs, due to Hassler Whitney [201], can be proved. Since nonseparable graphs of order 3 or more are 2-connected, this gives a generalization of Theorem 2.4.

Theorem 2.19 (Whitney's Theorem) *A nontrivial graph G is k-connected for some integer $k \geq 2$ if and only if for each pair u, v of distinct vertices of G, there are at least k internally disjoint $u - v$ paths in G.*

Proof. First, suppose that G is a k-connected graph, where $k \geq 2$, and let u and v be two distinct vertices of G. Assume first that u and v are not adjacent. Let U be a minimum $u - v$ separating set. Then

$$k \leq \kappa(G) \leq |U|.$$

By Menger's theorem, G contains at least k internally disjoint $u - v$ paths.

Next, assume that u and v are adjacent, where $e = uv$. As observed earlier, $G - e$ is $(k-1)$-connected. Let W be a minimum $u - v$ separating set in $G - e$. So,

$$k - 1 \leq \kappa(G - e) \leq |W|.$$

By Menger's theorem, $G - e$ contains at least $k - 1$ internally disjoint $u - v$ paths, implying that G contains at least k internally disjoint $u - v$ paths.

For the converse, assume that G contains at least k internally disjoint $u - v$ paths for every pair u, v of distinct vertices of G. If G is complete, then $G = K_n$, where $n \geq k + 1$, and so $\kappa(G) = n - 1 \geq k$. Hence, G is k-connected. Thus, we may assume that G is not complete.

Let U be a minimum vertex-cut of G. Then $|U| = \kappa(G)$. Let x and y be vertices in distinct components of $G - U$. Thus, U is an $x - y$ separating set of G. Since there are at least k internally disjoint $x - y$ paths in G, it follows by Menger's theorem that

$$k \leq |U| = \kappa(G),$$

and so G is k-connected.   ∎

The following two results are consequences of Theorem 2.19.

Corollary 2.20   *Let G be a k-connected graph, $k \geq 1$, and let S be any set of k vertices of G. If a graph H is obtained from G by adding a new vertex and joining this vertex to the vertices of S, then H is also k-connected.*

Corollary 2.21   *If G is a k-connected graph, $k \geq 2$, and $u, v_1, v_2, \ldots, v_t$ are $t+1$ distinct vertices of G, where $2 \leq t \leq k$, then G contains a $u - v_i$ path for each i ($1 \leq i \leq t$), every two paths of which have only u in common.*

By Theorem 2.4, every two vertices in a 2-connected graph lie on a common cycle of the graph. Gabriel Dirac [68] generalized this to k-connected graphs.

Theorem 2.22   *If G is a k-connected graph, $k \geq 2$, then every k vertices of G lie on a common cycle of G.*

Proof.   Let $S = \{v_1, v_2, \ldots, v_k\}$ be a set of k vertices of G. Among all cycles in G, let C be one containing a maximum number ℓ of vertices of S. Then $\ell \leq k$. If $\ell = k$, then the result follows, so we may assume that $\ell < k$. Since G is k-connected, G is 2-connected and so by Theorem 2.4, $\ell \geq 2$. We may further assume that $v_1, v_2, \ldots, v_\ell$ lie on C. Let u be a vertex of S that does not lie on C. We consider two cases.

Case 1. The cycle C contains exactly ℓ vertices, say $C = (v_1, v_2, \ldots, v_\ell, v_1)$. By Corollary 2.21, G contains a $u - v_i$ path P_i for each i with $1 \leq i \leq \ell$ such that every two of the paths $P_1, P_2, \ldots, P_\ell$ have only u in common. Replacing the edge $v_1 v_2$ on C by P_1 and P_2 produces a cycle containing at least $\ell + 1$ vertices of S. This is a contradiction.

Case 2. The cycle C contains at least $\ell + 1$ vertices. Let v_0 be a vertex on C that does not belong to S. Since $2 < \ell + 1 \leq k$, it follows by Corollary 2.21 that G contains a $u - v_i$ path P_i for each i with $0 \leq i \leq \ell$ such that every two of the paths $P_0, P_1, \ldots, P_\ell$ have only u in common. For each i ($0 \leq i \leq \ell$), let u_i be the first vertex of P_i that belongs to C and let P_i' be the $u - u_i$ subpath of P_i. Suppose that the vertices u_i ($0 \leq i \leq \ell$) are encountered in the order $u_0, u_1, \ldots, u_\ell$ as we proceed about C in some direction. For some i with $0 \leq i \leq \ell$ and $u_{\ell+1} = u_0$, there is a $u_i - u_{i+1}$ path P on C, none of whose internal vertices belong to S. Replacing P on C by P_i' and P_{i+1}' produces a cycle containing at least $\ell + 1$ vertices of S. Again, this is a contradiction.   ∎

There are analogues to Theorem 2.18 (Menger's theorem) and Theorem 2.19 (Whitney's theorem) in terms of edge-cuts.

Theorem 2.23   *For distinct vertices u and v in a graph G, the minimum cardinality of a set X of edges of G such that u and v lie in distinct components of $G - X$ equals the maximum number of pairwise edge-disjoint $u - v$ paths in G.*

Theorem 2.24 *A nontrivial graph G is k-edge-connected if and only if G contains k pairwise edge-disjoint $u - v$ paths for each pair u, v of distinct vertices of G.*

Exercises for Chapter 2

1. Prove that if G is a connected graph of order 3 or more, then every bridge of G is incident with a cut-vertex of G.

2. Prove Theorem 2.2: *A vertex v in a graph G is a cut-vertex if and only if there are two vertices u and w distinct from v such that v lies on every $u - w$ path in G.*

3. Prove Theorem 2.3: *An edge e in a graph G is a bridge if and only if e lies on no cycle in G.*

4. (a) For each integer $n \geq 4$, give an example of a connected graph G_n of order n that has no bridge but for every edge e of G_n, every edge of $G_n - e$ is a bridge.

   (b) Prove for each integer $n \geq 4$ that the connected graph of order n given in (a) is the only possible example.

5. Prove Corollary 2.5: *For every two distinct vertices u and v in a nonseparable graph G of order 3 or more, there are two distinct $u - v$ paths in G having only u and v in common.*

6. Prove Theorem 2.6: *Every connected graph containing cut-vertices contains at least two end-blocks.*

7. Prove or disprove: If B is a block of order 3 or more in a connected graph G, then there is a cycle in B that contains all the vertices of B.

8. Prove that if G is a graph of order $n \geq 3$ such that $\deg u + \deg v \geq n$ for every pair u, v of nonadjacent vertices in G, then G is nonseparable.

9. Let u be a cut-vertex in a connected graph G and let v be a vertex of G such that $d(u, v) = k \geq 1$. Show that G contains a vertex w such that $d(v, w) > k$.

10. Let G be a connected graph of order n and size m such that $V(G) = \{v_1, v_2, \cdots, v_n\}$. Let $b(v_i)$ be the number of blocks to which v_i belongs.

    (a) Show that $\sum_{i=1}^{n} b(v_i) \leq 2m$.

    (b) Show that $\sum_{i=1}^{n} b(v_i) = 2m$ if and only if G is a tree.

11. Determine all trees T such that $\overline{T}$ is also a tree.

12. Let T be a tree of order k. Prove that if G is a graph with $\delta(G) \geq k - 1$, then T is isomorphic to some subgraph of G.

13. Prove Corollary 2.10: *The size of a forest of order n having k components is $n - k$.*

14. Prove Corollary 2.13: *Let G be a graph of order n and size m. If G satisfies any two of the following three properties, then G is a tree:* (1) *G is connected,* (2) *G has no cycles,* (3) *$m = n - 1$.*

15. Let $s : d_1, d_2, \cdots, d_n$ be a non-increasing sequence of $n \geq 2$ positive integers, that is, $d_1 \geq d_2 \geq \cdots \geq d_n \geq 1$. Prove that the sequence s is the degree sequence of a tree of order n if and only if $\sum d_i = 2(n-1)$.

16. Prove in two different ways (using the hints in (a) and (b) below) that there is no tree T with maximum degree $\Delta = \Delta(T) \geq 3$ having the property that every vertex of degree $k \geq 2$ in T is adjacent to at most $k - 2$ leaves of T.

    (a) **Hint 1**. Let P be a longest path in T.

    (b) **Hint 2**. Let n_i denote the number of vertices of degree i in T for $i = 1, 2, \cdots, \Delta$, where $\sum_{i=1}^{\Delta} n_i = n$. First prove the following lemma.

      Lemma $\sum_{i=1}^{\Delta}(2 - i)n_i = 2$.

      Assign to every vertex v of T a "charge" of $2 - \deg v$. Then for each leaf u of T, move the charge of u to its neighbor in T.

17. A tree T of order n is known to satisfy the following properties: (1) $95 < n < 100$, (2) the degree of every vertex of T is either 1, 3, or 5, and (3) T has twice as many vertices of degree 3 as that of degree 5. What is n?

18. Let T be a tree of order n with degree sequence $d_1 \geq d_2 \geq \ldots \geq d_n$. Prove that $d_i \leq \lceil \frac{n-1}{i} \rceil$ for each integer i with $1 \leq i \leq n$. [Hint: (1) Use a proof by contradiction and (2) note that if $\deg v > r$, where $v \in V(T)$ and r is an integer, then $\deg v \geq r + 1$.]

19. (a) Let G be a graph of even order $n \geq 6$, every vertex of which has degree 3 or 4. If G contains two spanning trees T_1 and T_2 such that $\{E(T_1), E(T_2)\}$ is a partition of $E(G)$, then how many vertices of degree 4 must G contain?

    (b) Show that there exists an even integer $n \geq 6$ and a connected graph G of order n such that

    (1) every vertex of G has degree 3 or 4,

    (2) G contains the number of vertices of degree 4 determined in (a), and

    (3) G does not contain two spanning trees T_1 and T_2 for which $\{E(T_1), E(T_2)\}$ is a partition of $E(G)$.

20. Prove that if G is a k-connected graph, then $G \vee K_1$ is $(k + 1)$-connected.

21. Let G be a noncomplete graph of order n and connectivity k such that $\deg v \geq (n + 2k - 2)/3$ for every vertex v of G. Show that if S is a minimum vertex-cut of G, then $G - S$ has exactly two components.

22. Let G be a noncomplete graph with $\kappa(G) = k \geq 1$. Prove that for every minimum vertex-cut S of G, each vertex of S is adjacent to one or more vertices in each component of $G - S$.

23. Prove that a nontrivial graph G is k-edge-connected if and only if there exists no nonempty proper subset W of $V(G)$ such that the number of edges joining W and $V(G) - W$ is less than k.

24. Prove that if G is a connected graph of diameter 2, then $\lambda(G) = \delta(G)$.

25. What is the minimum size of a k-connected graph of order n?

26. Prove that if G is a 2-connected graph of order 4 or more such that each vertex of G is colored with one of the four colors red, blue, green, and yellow and each color is assigned to at least one vertex of G, then there exists a path containing at least one vertex of each of the four colors.

27. Prove Corollary 2.20: *Let G be a k-connected graph, $k \geq 1$, and let S be any set of k vertices of G. If a graph H is obtained from G by adding a new vertex and joining this vertex to the vertices of S, then H is also k-connected.*

28. Prove Corollary 2.21: *If G is a k-connected graph, $k \geq 2$, and $u, v_1, v_2, \ldots, v_t$ are $t + 1$ distinct vertices of G, where $2 \leq t \leq k$, then G contains a $u - v_i$ path for each i $(1 \leq i \leq t)$, every two paths of which have only u in common.*

29. Prove that the converse of Theorem 2.4 is true but the converse of Theorem 2.22 is not true in general.

30. Prove that a graph G of order $n \geq 2k$ is k-connected if and only if for every two disjoint sets V_1 and V_2 of k distinct vertices each, there exist k pairwise disjoint paths connecting V_1 and V_2.

31. Prove that a graph G of order $n \geq k + 1 \geq 3$ is k-connected if and only if for each set S of k distinct vertices of G and for each two-vertex subset T of S, there is a cycle of G that contains both vertices of T but no vertices of $S - T$.

32. Prove that if G is a k-connected graph, $k \geq 2$, then $G - e$ is a $(k-1)$-connected graph for every edge e of G.

33. (a) Show that for every positive integer k, there exists a connected graph G and a non-cut-vertex u of G such that $\mathrm{rad}(G - u) = \mathrm{rad}(G) + k$.

    (b) Prove for every nontrivial connected graph G and every non-cut-vertex v of G that $\mathrm{rad}(G - v) \geq \mathrm{rad}(G) - 1$.

    (c) Let G be a nontrivial connected graph with $\mathrm{rad}(G) = r$. Among all connected induced subgraphs of G having radius r, let H be one of minimum order. Prove that $\mathrm{rad}(H - v) = r - 1$ for every non-cut-vertex v of H.

34. (a) Show that the complete graphs K_4 and K_6 contain pairwise edge-disjoint spanning paths such that every edge of these complete graphs lies on one of these paths.

    (b) Show that the complete graph K_5 does not have the property possessed by K_4 and K_6 described in (a).

35. Let G be a graph of order $n \geq 6$ and $\kappa(G) = k$, where $1 \leq k \leq n - 3$. Then there exists a set (namely a vertex-cut) S such that $|S| = k$ and $G - S$ is disconnected. Prove that for every integer j with $k+1 \leq j \leq n-2$ that there is a set S' with $|S'| = j$ such that $G - S'$ is disconnected.

36. For two nonadjacent vertices u and v in a connected graph G, let $\mu(u, v)$ denote the number of vertices in a minimum $u - v$ separating set and define

$$\mu^-(G) = \min_{uv \notin E(G)} \{\mu(u,v)\} \text{ and } \mu^+(G) = \max_{uv \notin E(G)} \{\mu(u,v)\}.$$

    (a) What familiar concept is $\mu^-(G)$?

    (b) Show for every two positive integers a and b with $a \leq b$ that there exists a connected graph G such that $\mu^-(G) = a$ and $\mu^+(G) = b$.

37. Prove that if G is a connected graph such that $\delta(G) \geq 2$ and $\Delta(G) \geq 3$, then G contains two distinct cycles C and C' such that $V(C) \neq V(C')$.

Chapter 3

Eulerian and Hamiltonian Graphs

A solution to a famous puzzle called the Königsberg Bridge Problem appeared in a 1736 paper [80] of Leonhard Euler titled "Solutio problematis ad geometriam situs pertenentis" [The solution of a problem related to the geometry of position]. As the title of the paper suggests, Euler was aware that he was dealing with a different kind of geometry, namely one in which position was the relevant feature, not distance. Indeed, this gave rise to a new subject which for many years was known as *Analysis Situs* – the Analysis of Position. In the 19th century, this subject became Topology, a word that first appeared in print in an 1847 paper titled "Vorstudien zur Topologie" [Preliminary Studies of Topology] written by Johann Listing, even though he had already used the word Topologie for ten years in correspondence. Many of Listing's topological ideas were due to Carl Friedrich Gauss, who never published any papers in topology. Euler's 1736 paper is also considered the beginning of graph theory. Over a century later, in 1857, William Rowan Hamilton introduced a game called the *Icosian*. The Königsberg Bridge Problem and Hamilton's game would give rise to two concepts in graph theory named after Euler and Hamilton. These concepts are the main subjects of the current chapter.

3.1   Eulerian Graphs

Early in the 18th century, the East Prussian city of Königsberg (now called Kaliningrad) occupied both banks of the River Pregel and the island of Kneiphof, lying in the river at a point where it branches into two parts. There were seven bridges that spanned the various sections of the river. (See Figure 3.1.) A popular puzzle, called the **Königsberg Bridge Problem**, asked whether there was a route that crossed each of these bridges exactly once. Although such a route was long thought to be impossible, the first mathematical verification of this was presented by the famed mathematician Leonhard Euler at the Petersburg Academy on 26 August

1735. Euler's proof was contained in a paper [80] that appeared in the 1736 volume of the proceedings of the Petersburg Academy (the *Commentarii*). Euler's paper consisted of 21 paragraphs, beginning with the following two paragraphs (translated into English), the first of which, as it turned out, contains the elements of the new mathematical area of graph theory:

1 *In addition to that branch of geometry which is concerned with magnitudes, and which has always received the greatest attention, there is another branch, previously almost unknown, which Leibniz first mentioned, calling it the geometry of position. This branch is concerned only with the determination of position and its properties; it does not involve measurements, nor calculations made with them. It has not yet been satisfactorily determined what kind of problems are relevant to this geometry of position, or what methods should be used in solving them. Hence, when a problem was recently mentioned, which seemed geometrical but was so constructed that it did not require the measurement of distances, nor did calculation help at all, I had no doubt that it was concerned with the geometry of position – especially as its solution involved only position, and no calculation was of any use. I have therefore decided to give here the method which I have found for solving this kind of problem, as an example of the geometry of position.*

2 *The problem, which I am told is widely known, is as follows: in Königsberg in Prussia, there is an island A, called the Kneiphof; the river which surrounds it is divided into two branches, as can be seen in Fig. 3.1, and these branches are crossed by seven bridges, a, b, c, d, e, f, and g. Concerning these bridges, it was asked whether anyone could arrange a route in such a way that he would cross each bridge once and only once. I was told that some people asserted that this was impossible, while others were in doubt: but no one would actually assert that it could be done. From this, I have formulated the general problem: whatever be the arrangement and division of the river into branches, and however many bridges there be, can one find out whether or not it is possible to cross each bridge exactly once?*

Euler then proceeded to describe what would be required if, in fact, there existed a route in which every one of the seven Königsberg bridges was crossed exactly once. Euler observed that such a route could be represented as a sequence of letters, each term of the sequence chosen from A, B, C, or D. Two consecutive letters in the sequence would indicate that at some point in the route, the traveler had reached the land area of the first letter and had then crossed a bridge that led him to the land area represented by the second letter. Since there are seven bridges, the sequence must consist of eight terms.

Since there are five bridges leading into (or out of) land area A (the island Kneiphof), each occurrence of the letter A must indicate that either the route had

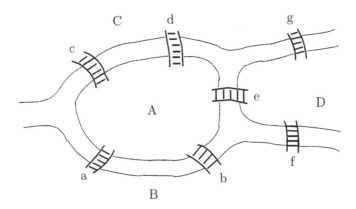

Figure 3.1: The bridges of Königsberg

started at A, had ended at A, or had progressed to and then exited A. Necessarily
then, the term A must appear three times in the sequence. Since three bridges enter
or exit B (as well as C and D), each of B, C, and D must appear twice. However, this
implies that such a sequence must contain nine terms, not eight, which produces a
contradiction and shows that there is no route in Königsberg that crosses each of
its seven bridges exactly once.

As Euler mentioned in paragraph 2 of his paper, he "formulated the general
problem". In order to describe and present a solution to the general problem, we
turn to the modern-day approach in which both the Königsberg Bridge Problem
and its generalization are described in terms of graphs.

Let G be a nontrivial connected graph. A circuit C of G that contains every
edge of G (necessarily exactly once) is an **Eulerian circuit**, while an open trail
that contains every edge of G is an **Eulerian trail**. (Some refer to an Eulerian
circuit as an **Euler tour**.) These terms are defined in exactly the same way if G
is a nontrivial connected multigraph. In fact, the map of Königsberg in Figure 3.1
can be represented by the multigraph shown in Figure 3.2. Then the Königsberg
Bridge Problem can be reformulated as: Does the multigraph shown in Figure 3.2
contain either an Eulerian circuit or an Eulerian trail? As Euler showed (although
not using this terminology, of course, nor even graphs), the answer to this question
is *no*.

A connected graph G is called **Eulerian** if G contains an Eulerian circuit. The
following characterization of Eulerian graphs is attributed to Euler.

Theorem 3.1 *A nontrivial connected graph G is Eulerian if and only if every
vertex of G has even degree.*

Proof.   Assume first that G is an Eulerian graph. Then G contains an Eulerian
circuit C. Let v be a vertex of G. Suppose first that v is not the initial vertex of
C (and thus, not the terminal vertex of C either). Since each occurrence of v in C
indicates that v is both entered and exited on C and produces a contribution of 2
to the degree of v, the degree of v is even. Next, suppose that v is the initial and

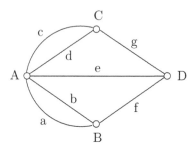

Figure 3.2: The multigraph of Königsberg

terminal vertex of C. As the initial vertex of C, this represents a contribution of 1 to the degree of v. There is also a contribution of 1 to the degree of v because v is the terminal vertex of C as well. Every other occurrence of v on C again represents a contribution of 2 to the degree of v. Here too v has even degree.

We now turn to the converse. Let G be a nontrivial connected graph in which every vertex is even. Let u be a vertex of G. First, we show that G contains a $u-u$ circuit. Construct a trail T beginning at u that contains a maximum number of edges of G. We claim that T is, in fact, a circuit; for suppose that T is a $u-v$ trail, where $u \neq v$. Then there is an odd number of edges incident with v and belonging to T. Since the degree of v in G is even, there is at least one edge incident with v that does not belong to T. Suppose that vw is such an edge. However then, T followed by w produces a trail T' with initial vertex u containing more edges than T, which is impossible. Thus, T is a circuit with initial and terminal vertex u. We now denote T by C.

If C is an Eulerian circuit of G, then the proof is complete. Hence, we may assume that C does not contain all edges of G. Since G is connected, there is a vertex x on C that is incident with an edge that does not belong to C. Let $H = G - E(C)$. Since every vertex on C is incident with an even number of edges on C, it follows that every vertex of H is even. Let H' be the component of H containing x. Consequently, every vertex of H' has positive even degree. By the same argument as before, H' contains a circuit C' with initial and terminal vertex x. By inserting C' at some occurrence of x in C, a $u-u$ circuit C'' in G is produced having more edges than C. This is a contradiction.   ■

With the aid of Theorem 3.1, connected graphs possessing an Eulerian trail can be characterized.

Corollary 3.2 *A connected graph G contains an Eulerian trail if and only if exactly two vertices of G have odd degree. Furthermore, each Eulerian trail of G begins at one of these odd vertices and ends at the other.*

Theorem 3.1 and Corollary 3.2 hold for connected multigraphs as well as for connected graphs.

In paragraph 20 (the next-to-last paragraph) of Euler's paper, Euler actually wrote (again an English translation):

20 *So whatever arrangement may be proposed, one can easily deter-mine whether or not a journey can be made, crossing each bridge once, by the following rules:*

> *If there are more than two areas to which an odd number of bridges lead, then such a journey is impossible.*
>
> *If, however, the number of bridges is odd for exactly two areas, then the journey is possible if it starts in either of these areas.*
>
> *If, finally, there are no areas to which an odd number of bridges lead, then the required journey can be accomplished from any starting point.*

With these rules, the given problem can also be solved.

Euler ended his paper by writing the following:

21 *When it has been determined that such a journey can be made, one still has to find how it should be arranged. For this I use the following rule: let those pairs of bridges which lead from one area to another be mentally removed, thereby considerably reducing the number of bridges; it is then an easy task to construct the required route across the remaining bridges, and the bridges which have been removed will not significantly alter the route found, as will become clear after a little thought. I do not therefore think it worthwhile to give any further details concerning the finding of the routes.*

In Euler's paper therefore, he actually only verified that every vertex being even is a necessary condition for a connected graph to be Eulerian and that exactly two vertices being odd is a necessary condition for a connected graph to contain an Eulerian trail. He did not show that these are sufficient conditions. The first proof of this would not be published for another 137 years, in an 1873 paper authored by Carl Hierholzer [127]. Carl Hierholzer was born in 1840, received his Ph.D. in 1870, and died in 1871. Thus, his paper was published two years after his death. He had told colleagues of what he had done but died before he could write a paper containing this work. His colleagues wrote the paper on his behalf and had it published for him.

The concepts introduced for graphs and multigraphs have analogues for digraphs and multidigraphs as well. An **Eulerian circuit** in a connected digraph (or mul-tidigraph) D is a directed circuit that contains every arc of D; while an **Eulerian trail** in D is an open directed trail that contains every arc of D. A connected digraph that contains an Eulerian circuit is an **Eulerian digraph**.

The following two theorems characterize those connected digraphs that contain Eulerian circuits or Eulerian trails.

Theorem 3.3 *Let D be a nontrivial connected digraph. Then D is Eulerian if and only if $\operatorname{od} v = \operatorname{id} v$ for every vertex v of D.*

Theorem 3.4 *Let D be a nontrivial connected digraph.   Then D contains an Eulerian trail if and only if D contains two vertices u and v such that*

$$\text{od } u = \text{id } u + 1 \quad and \quad \text{id } v = \text{od } v + 1,$$

while od $w =$ id w *for all other vertices w of D. Furthermore, each Eulerian trail of D begins at u and ends at v.*

Thus, the digraph D_1 of Figure 3.3 contains an Eulerian circuit, D_2 contains an Eulerian $u - v$ trail, and D_3 contains neither an Eulerian circuit nor an Eulerian trail.

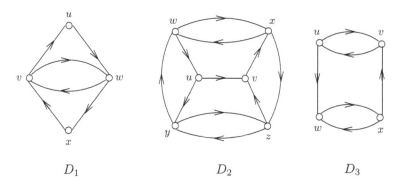

$$D_1 \qquad\qquad\qquad D_2 \qquad\qquad\qquad D_3$$

Figure 3.3: Eulerian circuits and trails in digraphs

3.2   de Bruijn Digraphs

In 1951 Nicolaas Govert de Bruijn (1918–2012) and Tatyana Pavlovna van Aardenne-Ehrenfest (1905–1984) and in 1941 Cedric Austen Bardell Smith (1917–2002) and William Thomas Tutte (1917–2002) independently discovered a result in [191] and [190], respectively, that gives the number of distinct Eulerian circuits in an Eulerian digraph. This theorem is commonly called the **BEST Theorem** after the initials of the two pairs of coauthors (de **B**ruijn, van Aardenne-**E**hrenfest, **S**mith, **T**utte). Nicolaas de Bruijn was a Dutch mathematician who was a faculty member at the Eindhoven University of Technology. The parents of Tatyana van Aardenne-Ehrenfest were both scientists with her father Paul Ehrenfest a distinguished theoretical physicist. The famous physicist Albert Einstein was a friend of the family. Although van Aardenne-Ehrenfest received a Ph.D. from Leiden University and did a great deal of research, she never held a faculty position. Smith was known for his research in genetic statistics. Tutte was one of the great graph theorists and his name will be encountered again and often.

For a digraph D of order n with $V(D) = \{v_1, v_2, \ldots, v_n\}$, the **adjacency matrix** $A = [a_{ij}]$ of D is the $n \times n$ matrix where $a_{ij} = 1$ if v_i is adjacent to v_j and $a_{ij} = 0$ otherwise, while the **outdegree matrix** $B = [b_{ij}]$ is the $n \times n$ matrix with $b_{ii} = $ od v_i and $b_{ij} = 0$ if $i \neq j$. The matrix M is defined by $M = B - A$. For $1 \leq i, j \leq n$, the (i, j)-cofactor of M is $(-1)^{i+j} \cdot \det(M_{ij})$, where M_{ij} is the $(n-1) \times (n-1)$ submatrix of M obtained by deleting row i and column j of M and $\det(M_{ij})$ is the determinant of M_{ij}. It is known that the values of the cofactors of each such matrix M is a constant. (See the discussion by Frank Harary [114], for example.)

Theorem 3.5 (**The BEST Theorem**) *Let D be an Eulerian digraph of order n with $V(D) = \{v_1, v_2, \ldots, v_n\}$, where od $v_i = $ id $v_i = d_i$ and c is the common cofactor of the matrix M. Then the number of distinct Eulerian circuits in D is*

$$c \cdot \prod_{i=1}^{n}(d_i - 1)!.$$

For the Eulerian digraph D of Figure 3.4, the adjacency matrix A, the outdegree matrix B, and the matrix M are shown as well. Since the common cofactor of M is 2, the number of distinct Eulerian circuits in D is $2 \cdot \prod_{i=1}^{n}(d_i - 1)! = 2$. These two Eulerian circuits are

$$(v_1, v_3, v_2, v_4, v_3, v_4, v_1, v_1) \text{ and } (v_1, v_3, v_4, v_3, v_2, v_4, v_1, v_1).$$

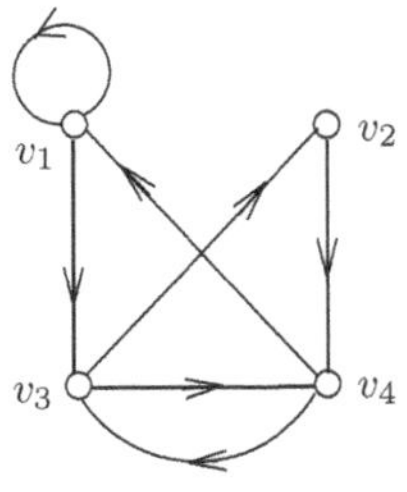

$$A = \begin{bmatrix} 1 & 0 & 1 & 0 \\ 0 & 0 & 0 & 1 \\ 0 & 1 & 0 & 1 \\ 1 & 0 & 1 & 0 \end{bmatrix} \quad B = \begin{bmatrix} 2 & 0 & 0 & 0 \\ 0 & 1 & 0 & 0 \\ 0 & 0 & 2 & 0 \\ 0 & 0 & 0 & 2 \end{bmatrix}$$

$$M = B - A = \begin{bmatrix} 1 & 0 & -1 & 0 \\ 0 & 1 & 0 & -1 \\ 0 & -1 & 2 & -1 \\ -1 & 0 & -1 & 2 \end{bmatrix}$$

Figure 3.4: Eulerian circuits in an Eulerian digraph

Although de Bruijn was an innovative researcher in many areas of mathematics, including studying models of the human brain, he is perhaps best known for a sequence and digraph that bear his name. Let A be a set consisting of $k \geq 2$ elements. For a positive integer n, an n-**word** over A is a sequence of length n whose terms belong to A. There are therefore k^n distinct n-words over A. A **de Bruijn sequence** is a sequence $a_0 a_1 \cdots a_{N-1}$ of elements of A having length $N = k^n$ such that for each n-word w over A, there is a unique integer i with $0 \leq i \leq N-1$ such that $w = a_i a_{i+1} \cdots a_{i+n-1}$ where addition in the subscripts is performed modulo N.

For example, if $k = 3$ and $n = 2$ (so $A = \{0, 1, 2\}$), then $N = k^n = 3^2$ and the nine distinct 2-words over A are

$$00, \ 01, \ 11, \ 10, \ 02, \ 22, \ 21, \ 12, \ 20.$$

In fact, 001102212 is a de Bruijn sequence in this case.

During 1944–1946, de Bruijn worked as a mathematician in Eindhoven at the Philips Research Laboratory. In early 1946, there was a conjecture made at the laboratory by the telecommunications engineer Kees Posthumus that the number of distinct de Bruijn sequences (of course, not being called by that name then) for $k = 2$ and an arbitrary n was $2^{2^{n-1}-n}$, which had been verified by Posthumus for $1 \leq n \leq 5$. During a single weekend, de Bruijn was able to verify the conjecture when $n = 6$, which led de Bruijn to construct a complete proof of the general conjecture of Posthumus.

As it turned out, the sequences known as de Bruijn sequences had been discussed in a 1934 paper by M. H. Martin [147] and, in fact, had been counted by C. Flye Sainte-Marie [85] in 1894. As was typically the case in the 19th century and before, mathematical papers were often not written in a particular mathematical manner. Indeed, combinatorics and graph theory had not yet blossomed into fully accepted areas of mathematics.

While it may not be difficult to construct a de Bruijn sequence for small values of k and n, this is not the case when k or n is large. However, de Bruijn sequences can be constructed with the aid of a digraph (actually a **pseudodigraph** since it contains directed loops).

For integers $k, n \geq 2$, the **de Bruijn digraph** $B(k, n)$ is that pseudodigraph of order k^{n-1} whose vertex set is the set of $(n-1)$-words over $A = \{0, 1, \cdots, k-1\}$ and size k^n whose arc set consists of all n-words over A, where the arc $a_1 a_2 \cdots a_n$ is the ordered pair $(a_1 a_2 \cdots a_{n-1},\ a_2 a_3 \cdots a_n)$ of vertices. Since the vertex $a_1 a_2 \cdots a_{n-1}$ is adjacent to the vertex $a_2 a_3 \cdots a_n$, we need only label the arc from $a_1 a_2 \cdots a_{n-1}$ to $a_2 a_3 \cdots a_n$ by a_n to indicate that the initial term a_1 is removed from $a_1 a_2 \cdots a_{n-1}$ and a_n is added as the final term to produce $a_2 a_3 \cdots a_n$. The de Bruijn digraph $B(3, 2)$ is shown in Figure 3.5.

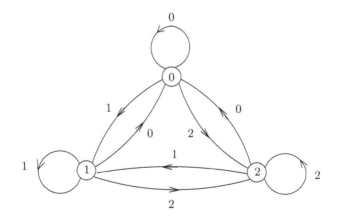

Figure 3.5: The de Bruijn digraph $B(3, 2)$

Since the outdegree and indegree of every vertex of $B(3, 2)$ are equal, it follows that $B(3, 2)$ is Eulerian. One Eulerian circuit of $B(3, 2)$ is $(0, 0, 1, 1, 0, 2, 2, 1, 2, 0)$, which results in the de Bruijn sequence 001102212.

Because the de Bruijn digraph $B(k,n)$ is connected and the outdegree and indegree of every vertex of $B(k,n)$ is k, we have the following consequence of Theorem 3.3.

Theorem 3.6 *For every two integers $k, n \geq 2$, the de Bruijn digraph $B(k,n)$ is Eulerian.*

The de Bruijn digraph $B(2,4)$ is shown in Figure 3.6.

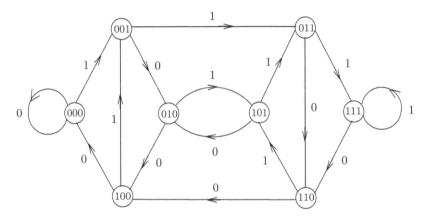

Figure 3.6: The de Bruijn digraph $B(2,4)$

3.3 Hamiltonian Graphs

William Rowan Hamilton (1805–1865) was gifted even as a child and his numerous interests and talents ranged from languages (having mastered many by age 10) to mathematics to physics. (Hamilton is mentioned in Chapter 0.) In 1832 he predicted that a ray of light passing through a biaxial crystal would be refracted into the shape of a cone. When this was experimentally confirmed, it was considered a major discovery and led to his being knighted in 1835, thereby becoming *Sir* William Rowan Hamilton. Even today, Hamilton is regarded as one of the leading mathematicians and physicists of the 19th century.

Although Hamilton's accomplishments were many, one of his best known in mathematics was his creation in 1843 of a new algebraic system called *quaternions*. This system dealt with a set of "numbers" of the form $a + bi + cj + dk$, where $a, b, c, d \in \mathbb{R}$, subject to certain arithmetic rules. Hamilton's discovery of quaternions was inspired by his search for an algebraic system that provided an interpretation of 3-dimensional space, much like the algebraic system of complex numbers $a + bi$ provided an interpretation of the 2-dimensional plane. This geometric interpretation of the complex numbers had only been introduced early in the 19th century. Hamilton's development of complex numbers as ordered pairs of real numbers was described in an 1837 essay he wrote. He concluded this essay by mentioning that he hoped he would soon be able to publish something similar on the algebra of ordered triples. Hamilton worked on this problem for years, obtaining success in 1843 only after he decided to relinquish the commutative property. In particular, in the quaternions, $ij = k$ and $ji = -k$; so, $ij \neq ji$. Also, Hamilton moved from a desired 3-dimensional system to the 4-dimensional system of quaternions $a + bi + cj + dk$. Shortly after Hamilton's discovery, physicists saw that the vector portion $b\mathbf{i} + c\mathbf{j} + d\mathbf{k}$ in Hamilton's quaternions could be used to represent 3-dimensional space. The algebraic system of 3-dimensional vectors retained the noncommutativity of quaternions, as can be seen in vector cross-product and matrix multiplication.

In 1856 Hamilton developed another example of a noncommutative algebraic system in a game he called the *Icosian Game*, initially exhibited by Hamilton at a meeting of the British Association in Dublin. The Icosian Game (the prefix *icos* is from the Greek for *twenty*) consisted of a board on which were placed twenty holes and some lines between certain pairs of holes. The diagram for this game is shown in Figure 3.7, where the holes are designated by the twenty consonants of the English alphabet.

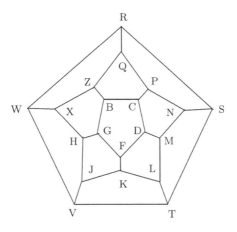

Figure 3.7: Hamilton's Icosian Game

Hamilton later sold the rights to his game for 25 pounds to Jaques and Son of London (now known as Jaques of London), a game manufacturer founded in 1795, especially well known as a dealer in chess sets. (The company introduced table tennis to the world in 1891, which took on the name ping pong in 1902.) The preface to the instruction pamphlet for the Icosian game, prepared by Hamilton for marketing the game in 1859, read as follows:

> *In this new Game (invented by Sir WILLIAM ROWAN HAMILTON, LL.D., & c., of Dublin, and by him named Icosian from a Greek word signifying 'twenty') a player is to place the whole or part of a set of twenty numbered pieces or men upon the points or in the holes of a board,*

represented by the diagram above drawn, in such a manner as always to proceed along the lines of the figure, and also to fulfill certain other conditions, which may in various ways be assigned by another player. Ingenuity and skill may thus be exercised in proposing as well as in resolving problems of the game. For example, the first of the two players may place the first five pieces in any five consecutive holes, and then require the second player to place the remaining fifteen men consecutively in such a manner that the succession may be cyclical, that is, so that No. 20 may be adjacent to No. 1; and it is always possible to answer any question of this kind. Thus, if B C D F G be the five given initial points, it is allowed to complete the succession by following the alphabetical order of the twenty consonants, as suggested by the diagram itself; but after placing the piece No. 6 in hole H, as above, it is also allowed (by the supposed conditions) to put No. 7 in X instead of J, and then to conclude with the succession, W R S T V J K L M N P Q Z. Other Examples of Icosian Problems, with solutions of some of them, will be found in the following page.

Of course, the diagram of Hamilton's Icosian game shown in Figure 3.7 can be immediately interpreted as a graph (see Figure 3.8), where the lines in the diagram have become the edges of the graph and the holes have become the vertices. Indeed, the graph of Figure 3.8 can be considered as the graph of the geometric solid called the **dodecahedron** (where the prefix *dodec* is from the Greek for *twelve*, pertaining to the twelve faces of the solid and the twelve regions determined by the graph, including the *exterior region*). This subject will be discussed in more detail in Chapter 5.

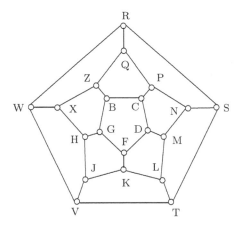

Figure 3.8: The graph of the dodecahedron

The problems proposed by Hamilton in his Icosian game gave rise to concepts in graph theory, which eventually became a popular subject of study by mathematicians. Let G be a graph. A path in G that contains every vertex of G is called a **Hamiltonian path** of G, while a cycle in G that contains every vertex of G is called a **Hamiltonian cycle** of G. A graph that contains a Hamiltonian cycle is itself called **Hamiltonian**. Certainly, the order of every Hamiltonian graph is at least 3 and every Hamiltonian graph contains a Hamiltonian path. On the other hand, a graph with a Hamiltonian path need not be Hamiltonian. The graph G_1 of Figure 3.9 is Hamiltonian and therefore contains both a Hamiltonian cycle and a Hamiltonian path. The graph G_2 contains a Hamiltonian path but is not Hamiltonian; while G_3 contains neither a Hamiltonian cycle nor a Hamiltonian path.

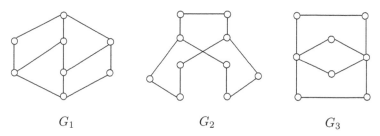

Figure 3.9: Hamiltonian paths and cycles in graphs

As implied by Hamilton's remarks, the graph of the dodecahedron in Figure 3.8 is Hamiltonian. Indeed, Hamilton's statement implies that this graph has a much stronger property. If one begins with any path of order 5 in the graph of Figure 3.8, then the path can be extended to a Hamiltonian cycle. As Hamilton stated, the path (B, C, D, F, G) can be extended to each of the Hamiltonian cycles

$$(B, C, D, F, G, H, J, K, L, M, N, P, Q, R, S, T, V, W, X, Z, B)$$

and

$$(B, C, D, F, G, H, X, W, R, S, T, V, J, K, L, M, N, P, Q, Z, B).$$

Hamilton proposed a number of additional problems in his Icosian Game such as showing the existence of three initial points for which the board cannot be covered "noncyclically", that is, there is a path of order 3 in the graph of the dodecahedron that can be extended to a Hamiltonian path but which cannot in turn be extended to a Hamiltonian cycle. Another problem of Hamilton was to find a path of order 3 such that whenever it is extendable to a Hamiltonian path, it is necessarily extendable to a Hamiltonian cycle.

In 1855 the Reverend Thomas Penyngton Kirkman (1840–1892) studied such questions as whether it is possible to visit all corners (vertices) of a polyhedron exactly once by moving along edges of the polyhedron and returning to the starting vertex. He observed that this could be done for some polyhedra but not all. While Kirkman had studied *Hamiltonian cycles* on general polyhedra and had preceded Hamilton's work on the dodecahedron by several months, it is Hamilton's name that became associated with spanning cycles of graphs, not Kirkman's. Quite possibly these cycles should have been named for Kirkman. But perhaps it was Hamilton's fame, his work on quaternions and physics, and that his questions dealing with spanning cycles on the graph of the dodecahedron were deeper and more varied that led to Hamilton's name being forever linked with these cycles.

Since the concepts of a circuit that contains every edge of a graph and a cycle that contains every vertex sound so similar and since there is a simple and useful characterization of graphs that are Eulerian, one might very well anticipate the existence of such a characterization for graphs that are Hamiltonian. However, no such theorem has ever been discovered. On the other hand, it is much more likely that a graph is Hamiltonian if the degrees of its vertices are large. It wasn't until 1952 that a general theorem by Gabriel Andrew Dirac on Hamiltonian graphs appeared, giving a sufficient condition for a graph to be Hamiltonian. However, in 1960 a more general theorem, due to Oystein Ore [156], would be discovered and lead to a host of other sufficient conditions.

Theorem 3.7 *Let G be a graph of order $n \geq 3$. If $\deg u + \deg v \geq n$ for each pair u, v of nonadjacent vertices of G, then G is Hamiltonian.*

Proof. Suppose that the statement is false. Then for some integer $n \geq 3$, there exists a graph H of order n such that $\deg u + \deg v \geq n$ for each pair u, v of nonadjacent vertices of H but yet H is not Hamiltonian. Add as many edges as possible between pairs of nonadjacent vertices of H so that the resulting graph G is still not Hamiltonian. Hence, G is a maximal non-Hamiltonian graph. Certainly, G is not a complete graph. Furthermore, $\deg_G u + \deg_G v \geq n$ for every pair u, v of nonadjacent vertices of G.

If the edge xy were to be added between two nonadjacent vertices x and y of G, then necessarily $G + xy$ is Hamiltonian and so $G + xy$ contains a Hamiltonian cycle C. Since C must contain the edge xy, the graph G contains a Hamiltonian $x - y$ path $(x = v_1, v_2, \ldots, v_n = y)$. If $xv_i \in E(G)$, where $2 \leq i \leq n - 1$, then $yv_{i-1} \notin E(G)$; for otherwise,

$$C' = (x, v_2, \ldots, v_{i-1}, y, v_{n-1}, v_{n-2}, \ldots, v_i, x)$$

is a Hamiltonian cycle of G, which is impossible. Hence, for each vertex of G adjacent to x, there is a vertex of $V(G) - \{y\}$ not adjacent to y, which implies that

$$\deg_G y \leq (n - 1) - \deg_G x$$

and so $\deg_G x + \deg_G y \leq n - 1$, producing a contradiction.   ∎

The aforementioned 1952 paper of Dirac [66] contained the following sufficient condition for a graph to be Hamiltonian, which is a consequence of Theorem 3.7.

Corollary 3.8 *If G is a graph of order $n \geq 3$ such that $\deg v \geq n/2$ for each vertex v of G, then G is Hamiltonian.*

With the aid of Theorem 3.7, a sufficient condition for a graph to have a Hamiltonian path can also be given.

Corollary 3.9 *Let G be a graph of order $n \geq 2$. If $\deg u + \deg v \geq n - 1$ for each pair u, v of nonadjacent vertices of G, then G contains a Hamiltonian path.*

Proof. Let $H = G \vee K_1$, where w is the vertex of H that does not belong to G. Then H has order $n + 1$ and $\deg u + \deg v \geq n + 1$ for each pair u, v of nonadjacent vertices of H. Then H is Hamiltonian by Theorem 3.7. Let C be a Hamiltonian cycle of H. Deleting w from C produces a Hamiltonian path in G.   ∎

J. Adrian Bondy and Vašek Chvátal [25] observed that the proof of Ore's theorem (Theorem 3.7) neither uses nor needs the full strength of the requirement that the degree sum of each pair of nonadjacent vertices is at least the order of the graph being considered. Initially, Bondy and Chvátal observed the following result.

Theorem 3.10 *Let u and v be nonadjacent vertices in a graph G of order n such that $\deg u + \deg v \geq n$. Then $G + uv$ is Hamiltonian if and only if G is Hamiltonian.*

Proof. Certainly if G is Hamiltonian, then $G + uv$ is Hamiltonian. For the converse, suppose that $G + uv$ is Hamiltonian but G is not. Then every Hamiltonian cycle in $G + uv$ contains the edge uv, implying that G contains a Hamiltonian $u - v$ path. We can now proceed exactly as in the proof of Theorem 3.7 to produce a contradiction.   ∎

The preceding result inspired a definition. Let G be a graph of order n. The **closure** $CL(G)$ of G is the graph obtained from G by recursively joining pairs of nonadjacent vertices whose degree sum is at least n (in the resulting graph at each stage) until no such pair remains. A graph G and its closure are shown in Figure 3.10.

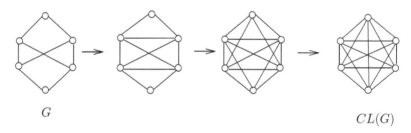

$$G \qquad\qquad\qquad\qquad\qquad\qquad\qquad\qquad\qquad CL(G)$$

Figure 3.10: Constructing the closure of a graph

First, it is known that this concept is well-defined, that is, the same graph is obtained regardless of the order in which edges are added.

Theorem 3.11 *Let G be a graph of order n. If G_1 and G_2 are graphs obtained by recursively joining pairs of nonadjacent vertices of G and each succeeding graph whose degree sum is at least n until no such pair remains, then $G_1 = G_2$.*

Proof. Suppose that G_1 is obtained by adding the edges $e_1, e_2, \ldots, e_r$ to G in the given order and G_2 is obtained from G by adding the edges $f_1, f_2, \ldots, f_s$ in the given order. Assume, to the contrary, that $G_1 \neq G_2$. Then $E(G_1) \neq E(G_2)$. Thus, we may assume that there is a first edge $e_i = xy$ in the sequence $e_1, e_2, \ldots, e_r$ that does not belong to G_2. Let $H = G + \{e_1, e_2, \ldots, e_{i-1}\}$. Then H is a subgraph of G_2.

Since $\deg_H x + \deg_H y \geq n$, it follows that $\deg_{G_2} x + \deg_{G_2} y \geq n$, which produces a contradiction since x and y are not adjacent in G_2. ∎

Repeated applications of Theorem 3.10 give us the following result.

Theorem 3.12 *A graph is Hamiltonian if and only if its closure is Hamiltonian.*

While we have described two sufficient conditions for a graph to be Hamiltonian, there is also a useful necessary condition. Recall that $k(H)$ denotes the number of components in a graph H.

Theorem 3.13 *If G is a Hamiltonian graph, then*

$$k(G - S) \leq |S|$$

for every nonempty proper subset S of $V(G)$.

Proof. Let S be a nonempty proper subset of $V(G)$. If $G - S$ is connected, then certainly, $k(G - S) \leq |S|$. Hence, we may assume that $k(G - S) = k \geq 2$ and that $G_1, G_2, \ldots, G_k$ are the components of $G - S$. Let $C = (v_1, v_2, \ldots, v_n, v_1)$ be a Hamiltonian cycle of G. Without loss of generality, we may assume that $v_1 \in V(G_1)$. For $1 \leq j \leq k$, let v_{i_j} be the last vertex of C that belongs to G_j. Necessarily then, $v_{i_j+1} \in S$ for $1 \leq j \leq k$ and so $|S| \geq k = k(G - S)$. ∎

Because Theorem 3.13 presents a necessary condition for a graph to be Hamiltonian, it is most useful in its contrapositive formulation:

> *If there exists a nonempty proper subset S of the vertex set $V(G)$ of a graph G such that $k(G - S) > |S|$, then G is not Hamiltonian.*

Certainly every Hamiltonian graph is connected. As a consequence of Theorem 3.13, no graph with a cut-vertex is Hamiltonian. The graph G of Figure 3.11 is not Hamiltonian either for if we let $S = \{w, x\}$, then $k(G - S) = 3$ and so $k(G - S) > |S|$.

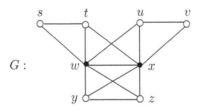

Figure 3.11: A non-Hamiltonian graph

The famous Petersen graph is not Hamiltonian but Theorem 3.13 cannot be used to verify this. Recall that the girth of the Petersen graph (the length of a smallest cycle) is 5.

Theorem 3.14 *The Petersen graph is not Hamiltonian.*

Proof.   Assume, to the contrary, that the Petersen graph is Hamiltonian. Then P has a Hamiltonian cycle

$$C = (v_1, v_2, \ldots, v_{10}, v_1).$$

Since P is cubic, v_1 is adjacent to exactly one of the vertices $v_3, v_4, \ldots, v_9$. However, since P contains neither a 3-cycle nor a 4-cycle, v_1 is adjacent to exactly one of v_5, v_6, and v_7. Because of the symmetry of v_5 and v_7, we may assume that v_1 is adjacent to either v_5 or v_6.

Case 1. v_1 *is adjacent to* v_5 *in* P. Then v_{10} is adjacent to exactly one of v_4, v_5, and v_6, which results in a 4-cycle, a 3-cycle, or a 4-cycle, respectively, each of which is impossible.

Case 2. v_1 *is adjacent to* v_6 *in* P. Again, v_{10} is adjacent to exactly one of v_4, v_5, and v_6. Since P does not contain a 3-cycle or a 4-cycle, v_{10} must be adjacent to v_4. However, then this returns us to Case 1, where v_1 and v_5 are replaced by v_{10} and v_4, respectively.   ∎

There is a much-studied class of Hamiltonian graphs in which Hamiltonian paths play a key role. A graph G is **Hamiltonian-connected** if for every pair u, v of vertices of G, there is a Hamiltonian $u-v$ path in G. Necessarily, every Hamiltonian-connected graph of order 3 or more is Hamiltonian but the converse is not true. The cubic graph $G_1 = C_3 \,\square\, K_2$ of Figure 3.12 is Hamiltonian-connected, while the cubic graph $G_2 = C_4 \,\square\, K_2 = Q_3$ is not Hamiltonian-connected. The graph G_2 contains no Hamiltonian $u - v$ path, for example. (See Exercise 20.)

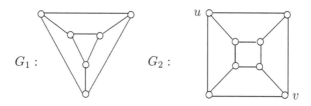

Figure 3.12: Hamiltonian-connected and non-Hamiltonian-connected graphs

There is a sufficient condition for a graph to be Hamiltonian-connected that is similar in statement to the sufficient condition for a graph to contain a Hamiltonian cycle presented in Theorem 3.7. The following theorem is also due to Oystein Ore [158] and provides a sufficient condition for a graph to be Hamiltonian-connected.

Theorem 3.15   *Let G be a graph of order $n \geq 4$. If $\deg u + \deg v \geq n+1$ for every pair u, v of nonadjacent vertices of G, then G is Hamiltonian-connected.*

Proof.   Suppose that the theorem is false. Then there exist two nonadjacent vertices u and v of G such that G does not contain a Hamiltonian $u - v$ path. On the other hand, G contains a Hamiltonian cycle C by Theorem 3.7. Let $C =$

$(v_1, v_2, \ldots, v_n, v_1)$, where $u = v_n$ and $v = v_j$ for some $j \notin \{1, n-1, n\}$. The vertex v_1 is not adjacent to v_{j+1}, for otherwise

$$u = v_n, v_{n-1}, \ldots, v_{j+1}, v_1, v_2, \ldots, v_j = v$$

is a Hamiltonian $u - v$ path in G.

If $v_1 v_i \in E(G)$, $2 \le i \le j$, then $v_{j+1} v_{i-1} \notin E(G)$; for otherwise,

$$u = v_n, v_{n-1}, \ldots, v_{j+1}, v_{i-1}, v_{i-2}, \ldots, v_1, v_i, v_{i+1}, \ldots, v_j = v$$

is a Hamiltonian $u - v$ path in G. Also, if $v_1 v_i \in E(G)$, $j + 2 \le i \le n - 1$, then $v_{j+1} v_{i+1} \notin E(G)$; for otherwise

$$u = v_n, v_{n-1}, \ldots, v_{i+1}, v_{j+1}, v_{j+2}, \ldots, v_i, v_1, v_2, \ldots, v_j = v$$

is a Hamiltonian $u - v$ path in G.

Since $v_1 v_{j+1} \notin E(G)$, there are $\deg v_1 - 1$ vertices in the set

$$\{v_2, v_3, \ldots, v_j\} \cup \{v_{j+2}, v_{j+3}, \ldots, v_{n-1}\}$$

that are adjacent to v_1. For each of these vertices, there is a vertex in the set

$$\{v_1, v_2, \ldots, v_{j-1}\} \cup \{v_{j+3}, v_{j+4}, \ldots, v_n\}$$

that is not adjacent to v_{j+1}. This implies that

$$\deg v_{j+1} \le 2 + [(n-3) - (\deg v_1 - 1)]$$

or that $\deg v_1 + \deg v_{j+1} \le n$, which is a contradiction. ∎

Exercises for Chapter 3

1. In Euler's solution to the Königsberg Bridge Problem, he observed that if there was a route that crossed each bridge exactly once, then this route could be represented by a sequence of eight letters, each of which is one of the four land regions A, B, C, and D shown in Figure 3.1. Show that it is impossible for any of these letters to appear only among the middle six terms of the sequence. What conclusion can be made from this observation?

2. Let G be a nontrivial connected graph. Prove that G is Eulerian if and only if $E(G)$ can be partitioned into subsets E_i, $1 \le i \le k$, where the subgraph $G[E_i]$ induced by each set E_i is a cycle.

3. Let G be a connected graph containing $2k$ odd vertices, where $k \ge 1$. Prove that $E(G)$ can be partitioned into subsets E_i, $1 \le i \le k$, where each subgraph $G[E_i]$ induced by E_i is an open trail, at most one of which has odd length.

4. The complement $\overline{G}$ of a connected 6-regular graph G of order n is Eulerian. The edge set $E(G)$ of G is partitioned into subsets E_i, $1 \le i \le k$, where the subgraph $G[E_i]$ induced by each set E_i is a cycle. Prove that there is an odd number of the sets E_i for which $|E_i|$ is odd.

5. Prove or disprove: There exists a strong digraph with an Eulerian trail.

6. Prove that an Eulerian graph G has even size if and only if G has an even number of vertices v such that $\deg v \equiv 2 \pmod 4$.

7. Let G be an Eulerian graph of order $n \geq 4$. Prove that G contains at least three vertices all of which have the same degree.

8. Prove or disprove: There exist two connected graphs G and H both of order at least 3 and neither of which is Eulerian such that $G \vee H$ is Eulerian.

9. (a) Find an Eulerian circuit in the de Bruijn digraph $B(2, 4)$ shown in Figure 3.6.

   (b) Use the information in (a) to construct the corresponding de Bruijn sequence.

   (c) Locate the 4-words 1010, 0101, 1001, and 0110 in the de Bruijn sequence in (b).

10. (a) Draw the de Bruijn digraph $B(3, 3)$.

    (b) Construct an Eulerian circuit in $B(3, 3)$.

    (c) Use your answer in (b) to construct the corresponding de Bruijn sequence.

11. Determine the order, size, and the outdegree and indegree of every vertex of the de Bruijn digraph that can be used to construct a de Bruijn sequence that will give all 5-words, each term of which is an element of $A = \{0, 1, 2, 3\}$.

12. Prove that if T is a tree of order at least 4 that is not a star, then $\overline{T}$ contains a Hamiltonian path.

13. Let $S = \{1, 2, 3, 4\}$ and for $1 \leq i \leq 4$, let S_i denote the set of i-element subsets of S. Let $G = (V, E)$ be a graph with $V = S_2 \cup S_3$ such that A is adjacent to B in G if $|A \cap B| = 2$. Prove or disprove each of the following.

    (a) G is Eulerian.

    (b) G has an Eulerian trail.

    (c) G is Hamiltonian.

14. (a) Give an example of a graph G containing a Hamiltonian path for which $k(G - S) > |S|$ for some nonempty proper subset S of $V(G)$.

    (b) State and prove a result analogous to Theorem 3.13 that gives a necessary condition for a graph to contain a Hamiltonian path.

15. Let $K_{s,t}$ be the complete bipartite graph where $2 \leq s \leq t$. Prove that $K_{s,t}$ is Hamiltonian if and only if $s = t$.

16. Suppose that it is possible to assign every vertex of a graph G of odd order $n \geq 3$ either the color red or the color blue in such a way that every red vertex is adjacent only to blue vertices and every blue vertex is adjacent only to red vertices. Show that G is not Hamiltonian.

17. (a) Prove that if G is a graph of order 101 and $\delta(G) = 51$, then every vertex of G lies on a cycle of length 27.

    (b) State and prove a generalization of (a).

18. Give an alternative proof of Theorem 3.15:

    > Let G be a graph of order $n \geq 4$. If $\deg u + \deg v \geq n + 1$ for every pair u, v of nonadjacent vertices of G, then G is Hamiltonian-connected.

    by first observing that $G - v$ is Hamiltonian for every vertex v of G.

19. Prove that every Hamiltonian-connected graph of order 4 or more is 3-connected.

20. Prove that no bipartite graph of order 3 or more is Hamiltonian-connected.

21. Let G be a graph of order $n \geq 7$ such that $\delta(G) \geq \frac{n+5}{2}$. By Theorem 3.15, G is Hamiltonian-connected and also Hamiltonian. Prove that G contains both a Hamiltonian cycle C and a Hamiltonian path P such that every edge e of G belongs to at most one of C and P.

22. Let G be a graph of even order $n \geq 4$ such that $\delta(G) \geq \frac{n+2}{2}$. Prove that G contains two Hamiltonian cycles C and C' such that $|E(C) \cap E(C')| \leq \frac{n}{2}$.

23. A graph G of order $n \geq 3$ is k-**leaf connected** for an integer k with $2 \leq k \leq n - 1$ if for every set S of k vertices, there exists a spanning tree T of G such that S is the set of end-vertices (leaves) of T. Therefore, 2-leaf connected and Hamiltonian-connected are the same concept.

    (a) By Theorem 3.15, if G is a graph of order $n \geq 4$ such that $\deg u + \deg v \geq n + 1$ for every pair u, v of nonadjacent vertices of G, then G is 2-leaf connected. Show that G need not be 3-leaf connected.

    (b) Prove that if G is a graph of order $n \geq 5$ such that $\deg u + \deg v \geq n + 2$ for every pair u, v of nonadjacent vertices of G, then G is 3-leaf connected.

24. Give an example of a Hamiltonian graph G and a path P of order 2 in G that cannot be extended to a Hamiltonian cycle of G.

25. Let G be a graph of order $n \geq 3$ and let k be an integer such that $1 \leq k \leq n-1$. Prove that if $\deg v \geq (n + k - 1)/2$ for every vertex v of G, then every path of order k in G can be extended to a Hamiltonian cycle of G.

26. In Hamilton's Icosian game, he stated that every path of order 5 in the graph of the dodecahedron can be extended to a Hamiltonian cycle in this graph. A graph G of order $n \geq 3$ is **Hamiltonian extendable** if every path of G can be extended to a Hamiltonian cycle of G.

   (a) Show that K_n, C_n, and $K_{\frac{n}{2},\frac{n}{2}}$ (n even) are Hamiltonian extendable.

   Now let G be a Hamiltonian extendable graph of order $n \geq 3$ with a Hamiltonian cycle $C = (v_1, v_2, \ldots, v_n, v_1)$.

   (b) Show that if $v_a v_b \in E(G)$, then $v_{a+1} v_{b+1} \in E(G)$.

   (c) Show that G is r-regular for some integer $r \geq 2$.

   (d) Show that if G is an r-regular Hamiltonian extendable graph where $r > n/2$, then $G = K_n$.

   (e) Show that if G is an r-regular Hamiltonian extendable graph where $r < n/2$, then $G = C_n$.

   (f) Determine all Hamiltonian extendable graphs of a given order $n \geq 3$.

   [Hint for (d)-(f): Let G be a Hamiltonian extendable graph of order $n \geq 3$. (i) If there is a vertex that is adjacent to two consecutive vertices of the Hamiltonian cycle C, then $G = K_n$. (ii) If there is a vertex that is not adjacent to two consecutive vertices of the Hamiltonian cycle C, then $G = C_n$. If (i) and (ii) don't occur, then $G = K_{\frac{n}{2},\frac{n}{2}}$. Thus, $G \in \{K_n, C_n, K_{\frac{n}{2},\frac{n}{2}}\}$.]

Chapter 4

Matchings and Factorization

There are numerous problems concerning graphs that contain sets of edges or sets of vertices possessing a property of particular interest. Often we are interested in partitioning the edge set or vertex set of a graph into subsets so that each subset possesses this property. In this chapter, we describe some of the best-known examples of such sets and the subgraphs they induce.

4.1   Matchings and Independence

Suppose that A and B are finite nonempty sets with $|A| = s$ and $|B| = t$, say

$$A = \{a_1, a_2, \ldots, a_s\} \text{ and } B = \{b_1, b_2, \ldots, b_t\}.$$

Does there exist a one-to-one function from A to B? Clearly, this question cannot be answered without more information. Surely such a function exists if and only if $s \leq t$. But what if the image of each element a_i ($1 \leq i \leq s$) must be selected from some nonempty list $L(a_i)$ of elements of B? Then this question still cannot be answered without knowing more about the lists $L(a_i)$.

Let's consider an example of this. If $A = \{a_1, a_2, a_3, a_4\}$ and $B = \{b_1, b_2, b_3, b_4\}$, where

$$L(a_1) = \{b_1, b_2\},\ L(a_2) = \{b_1, b_4\},\ L(a_3) = \{b_2, b_3\},\ L(a_4) = \{b_3, b_4\},$$

then the function $f : A \to B$ defined by

$$f(a_1) = b_1,\ f(a_2) = b_4,\ f(a_3) = b_2,\ f(a_4) = b_3 \tag{4.1}$$

is one-to-one. This situation can be modeled by a bipartite graph G with partite sets A and B, where $a_i b_j \in E(G)$ if $b_j \in L(a_i)$. (See Figure 4.1.) To show that there is a one-to-one function f from A to B, it is sufficient to show that G contains a set of pairwise nonadjacent edges such that each vertex of A is incident with one of these edges. The edges corresponding to the function f defined in (4.1) are shown in bold in Figure 4.1.

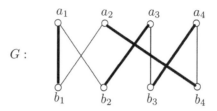

Figure 4.1: A bipartite graph modeling a function problem

We consider an additional example. Suppose that $A = \{a_1, a_2, \ldots, a_5\}$ and $B = \{b_1, b_2, \ldots, b_6\}$, where

$$L(a_1) = \{b_3, b_5\}, \; L(a_2) = \{b_1, b_2, b_3, b_4, b_6\},$$
$$L(a_3) = \{b_1, b_2, b_4, b_5, b_6\}, \; L(a_4) = \{b_3, b_5\}, \; L(a_5) = \{b_3, b_5\}.$$

Again we ask whether there is a one-to-one function from A to B. This situation can also be modeled by a bipartite graph, namely the graph G shown in Figure 4.2. In this case, however, notice that the image of each of the elements a_1, a_4, and a_5 of A must be chosen from the same two elements of B (namely b_3 and b_5). Consequently, there is no way to construct a function from A to B such that every two elements of $\{a_1, a_4, a_5\}$ have distinct images and so no one-to-one function from A to B exists.

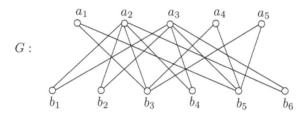

Figure 4.2: A bipartite graph modeling a function problem

The problem mentioned above pertains to an important concept in graph theory that is encountered in a variety of circumstances. In a graph G, a set M of edges, no two edges of which are adjacent, is called a **matching**.

Although matchings are of interest in all graphs, they are of particular interest in bipartite graphs. For example, we saw that the bipartite graph of Figure 4.1 contains a matching of size 4 (and so there is a one-to-one function from A to B under the given restrictions) but that there is no matching of size 5 in the bipartite graph of Figure 4.2 (and so there is no one-to-one function from A to B under the given restrictions).

Let G be a bipartite graph with partite sets U and W, where $|U| \leq |W|$. A matching in G is therefore a set $M = \{e_1, e_2, \ldots, e_k\}$ of edges, where $e_i = u_i w_i$ for $1 \leq i \leq k$ such that $u_1, u_2, \ldots, u_k$ are k distinct vertices of U and $w_1, w_2, \ldots, w_k$ are k distinct vertices of W. In this case, M **matches** the set $\{u_1, u_2, \ldots, u_k\}$ to the set $\{w_1, w_2, \ldots, w_k\}$. Necessarily, for any matching of k edges, $k \leq |U|$. If $|U| = k$, then U is said to be **matched** to a subset of W.

For a bipartite graph G with partite sets U and W and for $S \subseteq U$, let $N(S)$ be the set of all vertices in W having a neighbor in S. The condition that

$$|N(S)| \geq |S| \text{ for all } S \subseteq U$$

is referred to as **Hall's condition**. This condition is named for Philip Hall (1904–1982), a British group theorist. The following 1935 theorem of Hall [112] shows that this condition provides a necessary and sufficient condition for one partite set of a bipartite graph to be matched to a subset of the other.

Theorem 4.1 *Let G be a bipartite graph with partite sets U and W, where $|U| \leq |W|$. Then U can be matched to a subset of W if and only if Hall's condition is satisfied.*

Proof. If Hall's condition is not satisfied, then there is some subset S of U such that $|S| > |N(S)|$. Since S cannot be matched to a subset of W, it follows that U cannot be matched to a subset of W.

The converse is verified by the Strong Principle of Mathematical Induction. We proceed by induction on the cardinality of U. Suppose first that Hall's condition is satisfied and $|U| = 1$. Since $|N(U)| \geq |U| = 1$, there is a vertex in W adjacent to the vertex in U and so U can be matched to a subset of W. Assume, for an integer $k \geq 2$, that if G_1 is any bipartite graph with partite sets U_1 and W_1, where

$$|U_1| \leq |W_1| \text{ and } 1 \leq |U_1| < k,$$

that satisfies Hall's condition, then U_1 can be matched to a subset of W_1. Let G be a bipartite graph with partite sets U and W, where $k = |U| \leq |W|$, such that Hall's condition is satisfied. We show that U can be matched to a subset of W. We consider two cases.

Case 1. For every subset S of U such that $1 \leq |S| < |U|$, it follows that $|N(S)| > |S|$. Let $u \in U$. By assumption, u is adjacent to two or more vertices of W. Let w be a vertex adjacent to u. Now let H be the bipartite subgraph of G with partite sets $U - \{u\}$ and $W - \{w\}$. For each subset S of $U - \{u\}$, $|N(S)| \geq |S|$ in H. By the induction hypothesis, $U - \{u\}$ can be matched to a subset of $W - \{w\}$. This matching together with the edge uw shows that U can be matched to a subset of W.

Case 2. There exists a proper subset X of U such that $|N(X)| = |X|$. Let F be the bipartite subgraph of G with partite sets X and $N(X)$. Since Hall's condition is satisfied in F, it follows by the induction hypothesis that X can be matched to a subset of $N(X)$. Indeed, since $|N(X)| = |X|$, the set X can be matched to $N(X)$. Let M' be such a matching.

Next, consider the bipartite subgraph H of G with partite sets $U - X$ and $W - N(X)$. Let S be a subset of $U - X$ and let

$$S' = N(S) \cap (W - N(X)).$$

We show that $|S| \leq |S'|$. By assumption, $|N(X \cup S)| \geq |X \cup S|$. Hence,

$$|N(X)| + |S'| = |N(X \cup S)| \geq |X \cup S| = |X| + |S|.$$

Since $|N(X)| = |X|$, it follows that $|S'| \geq |S|$. Thus, Hall's condition is satisfied in H and so there is a matching M'' from $U - X$ to $W - N(X)$. Therefore, $M' \cup M''$ is a matching from U to W in G.   ■

There are certain kinds of matchings in graphs (bipartite or not) which will be of special interest. A matching M in a graph G is a

(1) **maximum matching** of G if G contains no matching with more than $|M|$ edges;

(2) **maximal matching** of G if M is not a proper subset of any other matching in G;

(3) **perfect matching** of G if every vertex of G is incident with some edge in M.

If M is a perfect matching in G, then G has order $n = 2k$ for some positive integer k and $|M| = k$. Thus, only a graph of even order can have a perfect matching. Furthermore, every perfect matching is a maximum matching and every maximum matching is a maximal matching, but neither converse is true. For the graph $G = P_6$ of Figure 4.3, the matching $M = \{v_1v_2, v_3v_4, v_5v_6\}$ is both a perfect and maximum matching, while $M' = \{v_2v_3, v_5v_6\}$ is maximal matching that is not a maximum matching.

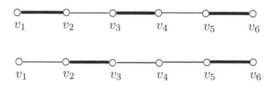

Figure 4.3: Maximum and maximal matchings in a graph

If G is a bipartite graph with partite sets U and W where $|U| \leq |W|$ and Hall's condition is satisfied, then by Theorem 4.1, G contains a matching of size $|U|$, which is a maximum matching. If $|U| = |W|$, then such a matching is a perfect matching in G. The following result is another consequence of Theorem 4.1.

Theorem 4.2 *Every r-regular bipartite graph $(r \geq 1)$ has a perfect matching.*

Proof.   Let G be an r-regular bipartite graph with partite sets U and W. Then $|U| = |W|$. Let S be a nonempty subset of U. Suppose that a total of k edges join the vertices of S and the vertices of $N(S)$. Thus, $k = r|S|$. Since there are $r|N(S)|$ edges incident with the vertices of $N(S)$, it follows that $k \leq r|N(S)|$ or $r|S| \leq r|N(S)|$. Therefore, $|S| \leq |N(S)|$ and Hall's condition is satisfied in G. By Theorem 4.1, G has a perfect matching.   ■

There is a popular reformulation of Theorem 4.2 that is referred to as the **Marriage Theorem**.

Theorem 4.3 (**The Marriage Theorem**) *In a group of r women and r men, r marriages between acquainted couples possible if and only if for each integer k with $1 \le k \le r$, each subset of k women is acquainted with at least k men.*

Another theorem closely related to Hall's theorem and the Marriage Theorem involves the concept of systems of distinct representatives. A collection of finite nonempty sets $S_1, S_2, \ldots, S_n$ has a **system of distinct representatives** if there exist n distinct elements $x_1, x_2, \ldots, x_n$ such that $x_i \in S_i$ for $1 \le i \le n$.

Theorem 4.4 *A collection $\{S_1, S_2, \ldots, S_n\}$ of finite nonempty sets has a system of distinct representatives if and only if for each integer k with $1 \le k \le n$, the union of any k of these sets contains at least k elements.*

Proof. Assume first that $\{S_1, S_2, \ldots, S_n\}$ has a system of distinct representatives. Then, for each integer k with $1 \le k \le n$, the union of any k of these sets contains at least k elements.

For the converse, suppose that $\{S_1, S_2, \ldots, S_n\}$ is a collection of n sets such that for each integer k with $1 \le k \le n$, the union of any k of these sets contains at least k elements. We now consider the bipartite graph G with partite sets

$$U = \{S_1, S_2, \ldots, S_n\} \text{ and } W = S_1 \cup S_2 \cup \cdots \cup S_n$$

such that a vertex S_i $(1 \le i \le n)$ in U is adjacent to a vertex w in W if $w \in S_i$. Let X be a subset of U with $|X| = k$, where $1 \le k \le n$. Since the union of any k sets in U contains at least k elements, it follows that $|N(X)| \ge |X|$. Thus, G satisfies Hall's condition. By Theorem 4.1, G contains a matching from U to a subset of W. This matching pairs off the sets $S_1, S_2, \ldots, S_n$ with n distinct elements in $S_1 \cup S_2 \cup \ldots \cup S_n$, producing a system of distinct representatives for $\{S_1, S_2, \ldots, S_n\}$. ∎

Let G be a bipartite graph with partite sets U and W where $|U| = |W|$. By Theorem 4.1, G has a perfect matching if and only if for every subset S of U, the inequality $|N(S)| \ge |S|$ holds. This, of course, says that G has a perfect matching if and only if Hall's condition is satisfied in G. In 1954 William Tutte [189] established a necessary and sufficient condition for graphs in general to have a perfect matching. A component of a graph is **odd** or **even** according to whether its order is odd or even. The number of odd components in a graph G is denoted by $k_o(G)$.

Theorem 4.5 *A nontrivial graph G contains a perfect matching if and only if $k_o(G - S) \le |S|$ for every proper subset S of $V(G)$.*

Proof. First, suppose that G contains a perfect matching M. Let S be a proper subset of $V(G)$. If $G-S$ has no odd components, then $k_o(G-S) = 0$ and $k_o(G-S) \le |S|$. Thus, we may assume that $k_o(G - S) = k \ge 1$. Let $G_1, G_2, \ldots, G_k$ be the odd components of $G - S$. (There may be some even components of $G - S$ as well.) For each component G_i of $G - S$, there is at least one edge of M joining a vertex of G_i and a vertex of S. Thus, $k_o(G - S) \le |S|$.

We now verify the converse. Let G be a graph such that $k_o(G - S) \le |S|$ for every proper subset S of $V(G)$. In particular, $k_o(G - \emptyset) \le |\emptyset| = 0$, implying that every component of G is even and so G itself has even order. We now show that G has a perfect matching by employing induction on the (even) order of G. Since K_2 is the only graph of order 2 having no odd components and K_2 has a perfect matching, the base case of the induction is verified.

For a given even integer $n \geq 4$, assume that all graphs H of even order less than n and satisfying $k_o(H - S) \leq |S|$ for every proper subset S of $V(H)$ contain a perfect matching. Now let G be a graph of order n satisfying $k_o(G - S) \leq |S|$ for every proper subset S of $V(G)$. As we saw above, every component of G has even order. We show that G has a perfect matching.

For a vertex v of G that is not a cut-vertex (see Theorem 2.1) and $R = \{v\}$, it follows that $k_o(G - R) = |R| = 1$. Hence, there are nonempty proper subsets T of $V(G)$ for which $k_o(G - T) = |T|$. Among all such sets T, let S be one of maximum cardinality. Suppose that $k_o(G - S) = |S| = k \geq 1$ and let $G_1, G_2, \ldots, G_k$ be the odd components of $G - S$.

We claim that $k(G - S) = k$, that is, $G_1, G_2, \ldots, G_k$ are the *only* components of $G - S$. Assume, to the contrary, that $G - S$ has an even component G_0. Let v_0 be a vertex of G_0 that is not a cut-vertex of G_0. Let $S_0 = S \cup \{v_0\}$. Since G_0 has even order,

$$k_o(G - S_0) \geq k_o(G - S) + 1 = k + 1.$$

On the other hand, $k_o(G - S_0) \leq |S_0| = k + 1$. Therefore,

$$k_o(G - S_0) = |S_0| = k + 1,$$

which is impossible. Thus, as claimed, $G_1, G_2, \ldots, G_k$ are the only components of $G - S$.

For each integer i ($1 \leq i \leq k$), let S_i denote the set of those vertices in S adjacent to at least one vertex of G_i. Since G has only even components, each set S_i is nonempty.

We claim, for each integer ℓ with $1 \leq \ell \leq k$, that the union of any ℓ of the sets $S_1, S_2, \ldots, S_k$ contains at least ℓ vertices. Assume, to the contrary, that this is not the case. Then there is an integer j such that the union S' of j of the sets $S_1, S_2, \ldots, S_k$ has fewer than j elements. Suppose that $S_1, S_2, \ldots, S_j$ have this property. Thus,

$$S' = S_1 \cup S_2 \cup \cdots \cup S_j \text{ and } |S'| < j.$$

Then $G_1, G_2, \ldots, G_j$ are at least some of the components of $G - S'$ and so

$$k_o(G - S') \geq j > |S'|,$$

which contradicts the hypothesis. Thus, as claimed, for each integer ℓ with $1 \leq \ell \leq k$, the union of any ℓ of the sets $S_1, S_2, \ldots, S_k$ contains at least ℓ vertices.

By Theorem 4.4, there is a set $\{v_1, v_2, \ldots, v_k\}$ of k distinct vertices of S such that $v_i \in S_i$ for $1 \leq i \leq k$. Since every component G_i of $G - S$ contains a vertex u_i such that $u_i v_i$ is an edge of G, it follows that $\{u_i v_i : 1 \leq i \leq k\}$ is a matching of G.

We now show that for each nontrivial component G_i of $G - S$ ($1 \leq i \leq k$), the graph $G_i - u_i$ contains a perfect matching. Let W be a proper subset of $V(G_i - u_i)$. We claim that

$$k_o(G_i - u_i - W) \le |W|.$$

Assume, to the contrary, that $k_o(G_i - u_i - W) > |W|$. Since G_i has odd order, $G_i - u_i$ has even order and so $k_o(G_i - u_i - W)$ and $|W|$ are both even or both odd. Hence,

$$k_o(G_i - u_i - W) \ge |W| + 2.$$

Let $X = S \cup W \cup \{u_i\}$. Then

$$
\begin{aligned}
|X| &= |S| + |W| + 1 = |S| + (|W| + 2) - 1 \\
&\le k_o(G - S) + k_o(G_i - u_i - W) - 1 \\
&= k_o(G - X) \le |X|,
\end{aligned}
$$

which implies that $k_o(G - X) = |X|$ and contradicts the defining property of S. Thus, as claimed, $k_o(G_i - u_i - W) \le |W|$.

Therefore, by the induction hypothesis, for each nontrivial component G_i of $G - S$ ($1 \le i \le k$), the graph $G_i - u_i$ has a perfect matching. The collection of perfect matchings of $G_i - u_i$ for all nontrivial graphs G_i of $G - S$ together with the edges in $\{u_i v_i : 1 \le i \le k\}$ produce a perfect matching of G. ∎

Clearly, every 1-regular graph contains a perfect matching, while only the 2-regular graphs containing no odd cycles have a perfect matching. As expected, determining whether a cubic graph has a perfect matching is more challenging. One of the best-known theorems in this connection is due to Julius Petersen [159] who showed that every bridgeless cubic graph contains a perfect matching.

Theorem 4.6 *Every bridgeless cubic graph contains a perfect matching.*

Proof. Let G be a bridgeless cubic graph, and let S be a proper subset of $V(G)$ with $|S| = k$. We show that $k_o(G - S) \le |S|$. This is true if $G - S$ has no odd components; so we assume that $G - S$ has $\ell \ge 1$ odd components, say $G_1, G_2, \ldots, G_\ell$.

Let E_i ($1 \le i \le \ell$) denote the set of edges joining the vertices of G_i and the vertices of S. Since G is cubic, every vertex of G_i has degree 3 in G. Because the sum of the degrees in G of the vertices of G_i is odd and the sum of the degrees in G_i of the vertices of G_i is even, it follows that $|E_i|$ is odd. Because G is bridgeless, $|E_i| \ne 1$ and so $|E_i| \ge 3$ for $1 \le i \le \ell$. This implies that there are at least 3ℓ edges joining the vertices of $G - S$ and the vertices of S. Since $|S| = k$, at most $3k$ edges join the vertices of $G - S$ and the vertices of S. Thus,

$$3k_o(G - S) = 3\ell \le 3k = 3|S|$$

and so $k_o(G - S) \le |S|$. By Theorem 4.10, G has a perfect matching. ∎

In fact, Petersen showed that a cubic graph with at most two bridges contains a perfect matching (see Exercise 2). This result cannot be extended further, however, since the graph G of Figure 4.4 is cubic and contains three bridges but no perfect matching since $k_o(G - v) = 3 > 1 = |\{v\}|$.

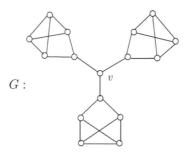

G :

Figure 4.4: A cubic graph with no perfect matching

A set M of edges in a graph G is **independent** if no two edges in M are adjacent. Therefore, M is an independent set of edges of G if and only if M is a matching in G. The maximum number of edges in an independent set of edges of G is called the **edge independence number** of G and is denoted by $\alpha'(G)$. Therefore, if M is an independent set of edges in G such that $|M| = \alpha'(G)$, then M is a maximum matching in G. If G has order n, then $\alpha'(G) \leq n/2$ and $\alpha'(G) = n/2$ if and only if G has a perfect matching.

An independent set M of edges of G is a **maximal independent set** if M is a maximal matching in G. Thus, M is not a proper subset of any independent set of edges of G. The **lower edge independence number** $\alpha'_o(G)$ of G is the minimum cardinality of a maximal independent set of edges (or maximal matching) in G. For the graph $G = P_6$ of Figure 4.3, $\alpha(G) = 3$ and $\alpha'_o(G) = 2$. Clearly, $\alpha'_o(G) \leq \alpha'(G)$ for every graph G. In fact, we have the following result.

Theorem 4.7 *For every nonempty graph G,*

$$\alpha'_o(G) \leq \alpha'(G) \leq 2\alpha'_o(G).$$

Proof.   As we noted, the inequality $\alpha'_o(G) \leq \alpha'(G)$ follows from the definitions of the edge independence number and lower edge independence number. Suppose that $\alpha'_o(G) = k$. Thus, G contains a maximal matching $M = \{e_1, e_2, \ldots, e_k\}$, where, say $e_i = u_i v_i$ $(1 \leq i \leq k)$. Thus, every edge of G is incident with at least one vertex in the set $W = \{u_1, u_2, \ldots, u_k, v_1, v_2, \ldots, v_k\}$. Now let M' be an independent set of edges such that $|M'| = \alpha'(G)$. Since each vertex of W is incident with at most one edge in M', it follows that $|M'| \leq 2k$, that is, $\alpha'(G) \leq 2\alpha'_o(G)$.   ∎

Independence in graph theory applies to sets of vertices as well as to sets of edges. A set U of vertices in a graph G is **independent** if no two vertices in U are adjacent. (Some refer to an independent set of vertices as a **stable set**.) The maximum number of vertices in an independent set of vertices of G is called the **vertex independence number**, or more simply, the **independence number** of G and is denoted by $\alpha(G)$. At the other extreme are sets of vertices that induce complete subgraphs in G. A complete subgraph of G is also called a **clique** of G. A clique of order k is a **k-clique**. The maximum order of a clique of G is called the **clique number** of G and is denoted by $\omega(G)$. Thus, $\alpha(G) = \omega(\overline{G})$ for every graph G. Vašek Chvátal and Paul Erdős [55] showed that if G is a graph of order at least 3 whose independence number doesn't exceed its connectivity, then G is Hamiltonian.

Theorem 4.8 *Let G be a graph of order at least 3. If $\kappa(G) \geq \alpha(G)$, then G is Hamiltonian.*

Proof. If $\alpha(G) = 1$, then G is complete and therefore Hamiltonian. Hence, we may assume that $\alpha(G) = k \geq 2$. Since $\kappa(G) \geq 2$, it follows that G is 2-connected and so G contains cycles by Theorem 2.19. Let C be a longest cycle in G. By Theorem 2.19, C contains at least k vertices. We show that C is a Hamiltonian cycle. Assume, to the contrary, that C is not a Hamiltonian cycle. Then there is some vertex w of G that does not lie on C. Since G is k-connected, it follows, with the aid of Corollary 2.21, that G contains k paths $P_1, P_2, \cdots, P_k$ such that P_i is a $w - v_i$ path where v_i is the only vertex of P_i on C and such that the paths are pairwise-disjoint except for w.

In a cyclic ordering of the vertices of C, let u_i be the vertex that follows v_i on C for each i $(1 \leq i \leq k)$. No vertex u_i is adjacent to w, for otherwise, replacing the edge $v_i u_i$ by P_i and $w u_i$ produces a cycle whose length exceeds that of C. Let $S = \{w, u_1, u_2, \ldots, u_k\}$. Since $|S| = k + 1 > \alpha(G)$ and $w u_i \notin E(G)$ for each i $(1 \leq i \leq k)$, there are distinct integers r and s such that $1 \leq r, s \leq k$ and $u_r u_s \in E(G)$. Replacing the edges $u_r v_r$ and $u_s v_s$ by the edge $u_r u_s$ and the paths P_r and P_s produces a cycle that is longer than C. This is a contradiction. $\blacksquare$

4.2 Factorization and Decomposition

We are often interested in collections of spanning subgraphs of a given graph G such that every edge of G belongs to exactly one of these subgraphs. By a **factor** of a graph G, we mean a spanning subgraph of G. A k-regular factor is called a **k-factor**. Thus, the edge set of a 1-factor in a graph G is a perfect matching in G. So, a graph G has a 1-factor if and only if G has a perfect matching. Therefore, the theorems stated earlier concerning perfect matchings can be restated in terms of 1-factors. First, we restate Theorem 4.2 in terms of 1-factors.

Theorem 4.9 *Every r-regular bipartite graph $(r \geq 1)$ has a 1-factor.*

Theorem 4.5 (restated here as Theorem 4.10) is often referred to as **Tutte's 1-factor theorem**.

Theorem 4.10 *A graph G contains a 1-factor if and only if $k_o(G - S) \leq |S|$ for every proper subset S of $V(G)$.*

The following is a restatement of Petersen's theorem (Theorem 4.6) in terms of 1-factors.

Theorem 4.11 *Every bridgeless cubic graph contains a 1-factor.*

A **factorization** $\mathcal{F}$ of a graph G is a collection of factors of G such that every edge of G belongs to exactly one factor in $\mathcal{F}$. Therefore, if each factor in $\mathcal{F}$ is nonempty, then the edge sets of the factors produce a partition of $E(G)$.

The factorizations of a graph in which we are primarily interested are those in which the factors are isomorphic (resulting in an **isomorphic factorization**) and where each factor has some specified property. Figure 4.5 shows an isomorphic factorization of K_6 into three factors, each isomorphic to the same double star.

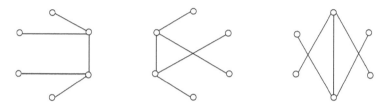

Figure 4.5: An isomorphic factorization of K_6

Figure 4.6 shows another isomorphic factorization $\mathcal{F} = \{F_1, F_2, F_3\}$ of K_6 into the same double star shown in Figure 4.5 but in this case the factors F_i ($i = 2, 3$) can be obtained from the factor F_1 by a clockwise rotation of F_1 through an angle of $2\pi(i-1)/3$ radians. For this reason, $\mathcal{F}$ is called a *cyclic factorization* of K_6. An isomorphic factorization of a graph G into k copies of a graph H is a **cyclic factorization** if H is drawn in an appropriate manner so that rotating H through an appropriate angle $k - 1$ times produces this isomorphic factorization.

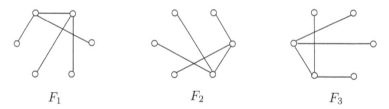

$$F_1 \qquad\qquad F_2 \qquad\qquad F_3$$

Figure 4.6: A cyclic factorization of K_6

Probably the most-studied factorizations are those in which each factor is a k-factor for a fixed positive integer k, especially when $k = 1$. A k-**factorization** of a graph G is a factorization of G into k-factors. A graph G is k-**factorable** if there exists a k-factorization of G. If G is a k-factorable graph, then G is r-regular for an integer r that is a multiple of k. Furthermore, if k is odd, then G has even order. In particular, every 1-factorable graph is a regular graph of even order.

The major problem here is that of determining which regular graphs are 1-factorable. Certainly, every 1-regular graph is 1-factorable. Also, a 2-regular graph G is 1-factorable if and only if G contains a 1-factor, that is, every component of G is an even cycle. Determining which cubic graphs are 1-factorable is considerably more complicated. For example, the cubic graphs K_4, $K_{3,3}$, and Q_3 are 1-factorable. We saw that the cubic graph G of Figure 4.4 does not contain a perfect matching.

Hence, this graph does not contain a 1-factor and so is not 1-factorable either.

The well-known Petersen graph P (see Figure 4.7) clearly has a 1-factor. Indeed, $M = \{u_i v_i : 1 \leq i \leq 5\}$ is a perfect matching. Thus, P can be factored into a 1-factor and a 2-factor. Because P contains no triangles or 4-cycles, each 2-factor in P is either a Hamiltonian cycle or two 5-cycles. By Theorem 3.14, the Petersen graph is not Hamiltonian. Thus, the 2-factor consists of two 5-cycles. Since the 2-factor is not 1-factorable, P is not 1-factorable.

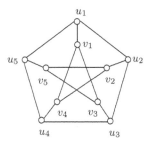

Figure 4.7: The Petersen graph: a cubic graph
with a 1-factor that is not 1-factorable

We now consider two classes of 1-factorable graphs. We saw in Theorem 4.9 that every r-regular bipartite graph, $r \geq 1$, contains a 1-factor. Even more can be said.

Theorem 4.12 *Every r-regular bipartite graph, $r \geq 1$, is 1-factorable.*

Proof. We proceed by induction on r. The result is obvious if $r = 1$. Assume that every $(r-1)$-regular bipartite graph is 1-factorable, where $r-1 \geq 1$ and let G be an r-regular bipartite graph. By Theorem 4.9, G contains a 1-factor F_1. Then $G - E(F_1)$ is an $(r-1)$-regular bipartite graph. By the induction hypothesis, $G - E(F_1)$ can be factored into $r - 1$ 1-factors, say $F_2, F_3, \ldots, F_r$. Then $\{F_1, F_2, \ldots, F_r\}$ is a 1-factorization of G. ∎

Theorem 4.13 *For each positive integer k, the complete graph K_{2k} is 1-factorable.*

Proof. Since K_2 is trivially 1-factorable, we assume that $k \geq 2$. Let $G = K_{2k}$, where $V(G) = \{v_0, v_1, v_2, \ldots, v_{2k-1}\}$. Place the vertices $v_1, v_2, \ldots, v_{2k-1}$ cyclically about a regular $(2k - 1)$-gon and place v_0 in the center of the $(2k - 1)$-gon. Join every two vertices of G by a straight line segment. For $1 \leq i \leq 2k - 1$, let the edge set of the factor F_i of G consist of the edge $v_0 v_i$ together with all edges of G that are perpendicular to $v_0 v_i$. Then F_i is a 1-factor of G for $1 \leq i \leq 2k - 1$ and $\{F_1, F_2, \ldots, F_{2k-1}\}$ is a 1-factorization of G. ∎

The proof of Theorem 4.13 provides a cyclic 1-factorization of K_{2k} where the 1-factor F_i, $2 \leq i \leq 2k - 1$, is obtained by rotating F_1 through an angle of $\frac{2\pi(i-1)}{2k-1}$ radians. This is illustrated in Figure 4.8 where a 1-factorization of K_6 into five 1-factors $F_1, F_2, \ldots, F_5$ is shown.

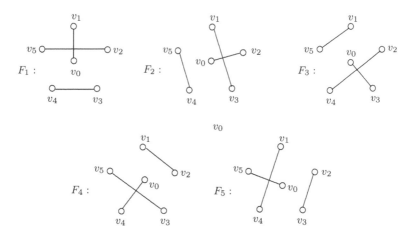

Figure 4.8: A 1-factorization of K_6

We saw in Corollary 3.8 (a result by Gabriel Dirac) that if G is a graph of order $n \geq 3$ such that $\deg v \geq n/2$ for every vertex v of G, then G is Hamiltonian. Consequently, if G is an r-regular graph of even order $n \geq 4$ such that $r \geq n/2$, then G contains a Hamiltonian cycle C. Since C is an even cycle, C can be factored into two 1-factors. If there exists a 1-factorization of $G - E(C)$, then clearly G is 1-factorable. This is certainly the case if $r = 3$. For which values of r and n this can be done is not known; however, there is a conjecture in this connection.

The 1-Factorization Conjecture If G is an r-regular graph of even order n such that $r \geq n/2$, then G is 1-factorable.

While the origin of this conjecture is unclear, its first mention in print appears to be in a 1986 paper of Amanda G. Chetwynd and Anthony J. W. Hilton [51]. Even though it is not known whether the 1-Factorization Conjecture is true for all r-regular graphs of even order $n \geq 4$ and $r \geq n/2$, Béla Csaba, Daniela Kühn, Allan Lo, Deryk Osthus, and Andrew Treglown [63] showed in 2014 that the conjecture is true for graphs of sufficiently large even order.

Theorem 4.14 *There exists an even integer n_0 such that the 1-Factorization Conjecture is true for all even integers $n \geq n_0$.*

If, in fact, the 1-Factorization Conjecture is true for all even integers $n \geq 4$, then it cannot be improved. To see this, let H_1 and H_2 be two graphs with $H_i = K_k$, where $k \geq 5$ and k is odd. Let $u_1 v_1 \in E(H_1)$ and $u_2 v_2 \in E(H_2)$. Furthermore, let

$$H = H_1 + H_2 \text{ and } G = H - u_1 v_1 - u_2 v_2 + u_1 u_2 + v_1 v_2.$$

Then G is a $(k-1)$-regular graph of order $2k$. We claim that G is not 1-factorable, for suppose that it is. Let $\mathcal{F}$ be a 1-factorization into $k - 1$ (≥ 4) 1-factors. Then there exists a 1-factor $F \in \mathcal{F}$ containing neither $u_1 u_2$ nor $v_1 v_2$. Since H_1 and H_2

have odd orders, some edge of F must join a vertex of H_1 and a vertex of H_2. This is impossible.

Thus, it is not true that if G is an r-regular graph of even order n such that $r \geq \frac{n}{2} - 1$, then G is 1-factorable. Hence, if the 1-Factorization Conjecture is true for all even integers $n \geq 4$, then the resulting theorem is sharp.

An obvious necessary condition for a graph G to be 2-factorable is that G is $2k$-regular for some positive integer k. Julius Petersen [159] showed that this condition is sufficient as well as necessary.

Theorem 4.15 *A graph G is 2-factorable if and only if G is $2k$-regular for some positive integer k.*

Proof. Since every 2-factorable graph is necessarily regular of positive even degree, it only remains to verify the converse.

Let G be a $2k$-regular graph, where $k \geq 1$, and let $V(G) = \{v_1, v_2, \ldots, v_n\}$. We may assume that G is connected. Hence, G is Eulerian and therefore contains an Eulerian circuit C. Necessarily, each vertex of G appears in C a total of k times.

We construct a bipartite graph H with partite sets

$$U = \{u_1, u_2, \ldots, u_n\} \text{ and } W = \{w_1, w_2, \ldots, w_n\}$$

such that u_i is adjacent to w_j ($1 \leq i, j \leq n$) if v_j immediately follows v_i on C. Thus, the graph H is k-regular. By Theorem 4.12, H is 1-factorable. Let $\{H_1, H_2, \ldots, H_k\}$ be a 1-factorization of H.

For each 1-factor H_ℓ in H for $1 \leq \ell \leq k$, we define a permutation α_ℓ on the set $\{1, 2, \ldots, n\}$ by

$$\alpha_\ell(i) = j \text{ if } u_i w_j \in E(H_\ell).$$

The permutation α_ℓ is now expressed as a product of disjoint permutation cycles. There is no permutation cycle of length 1 in this product, for otherwise $\alpha_\ell(i) = i$ for some integer i and so $u_i w_i \in E(H_\ell)$. This would imply that $v_i v_i \in E(G)$, which is impossible. Also, there is no permutation cycle of length 2 in this product, for otherwise, $\alpha_\ell(i) = j$ and $\alpha_\ell(j) = i$ for some integers i and j. This would imply that $u_i w_j, u_j w_i \in E(H_\ell)$ and so the edge $v_i v_j$ is repeated on the circuit. Thus, the length of every permutation cycle in the product of α_ℓ is at least 3.

Each permutation cycle in α_ℓ gives rise to a cycle in G, and the product of disjoint permutation cycles in α_ℓ produces a collection of mutually disjoint cycles in G that contain all vertices of G, that is, a 2-factor F_ℓ of G. Since the 1-factors H_ℓ in H are mutually edge-disjoint, the resulting 2-factors F_ℓ in G are mutually edge-disjoint. Therefore, the 1-factors $H_1, H_2, \ldots, H_k$ of H produce a 2-factorization $\{F_1, F_2, \ldots, F_k\}$ of G. ∎

As a consequence of Theorem 4.15, every complete graph of odd order at least 3 is 2-factorable. In fact, more can be said. A graph G is **Hamiltonian factorable** if there exists a factorization $\mathcal{F}$ of G such that each factor in $\mathcal{F}$ is a Hamiltonian cycle of G.

Theorem 4.16 *For every positive integer k, the complete graph K_{2k+1} is Hamiltonian factorable.*

Proof. We construct $G = K_{2k+1}$ on the vertex set $V(G) = \{v_0, v_1, \ldots, v_{2k}\}$ by first placing the vertices $v_1, v_2, \ldots, v_{2k}$ cyclically about a regular $2k$-gon and placing v_0 at some convenient position in the $2k$-gon. Let F_1 be the Hamiltonian cycle with edges

$$v_0 v_1, \; v_1 v_2, \; v_2 v_{2k}, \; v_{2k} v_3, \; v_3 v_{2k-1}, \; \ldots, \; v_{k+2} v_{k+1}, \; v_{k+1} v_0.$$

For $2 \le i \le k$, let F_i be the Hamiltonian cycle with edges $v_0 v_i$ and $v_{i+k} v_0$ and where all other edges are obtained from the edges of F_1 that are not incident with v_0 by rotating them clockwise through an angle of $2\pi(i-1)/k$ radians. Since $\{E(F_1), E(F_2), \ldots, E(F_k)\}$ is a partition of $E(G)$, the graph G is Hamiltonian factorable. ∎

The proof of Theorem 4.16 is illustrated in Figure 4.9.

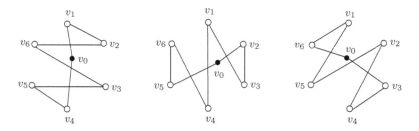

Figure 4.9: A Hamiltonian factorization of K_7

In every factorization $\mathcal{F} = \{F_1, F_2, \ldots, F_k\}$ of a graph G, (1) each factor F_i $(1 \le i \le k)$ is a spanning subgraph of G and (2) every edge of G belongs to exactly one factor in $\mathcal{F}$. There are other collections of subgraphs of a graph G where only condition (2) is a requirement. In particular, a **decomposition** of a graph G is a collection $\mathcal{D} = \{G_1, G_2, \ldots, G_k\}$ of subgraphs of G such that (1) no subgraph G_i $(1 \le i \le k)$ has isolated vertices and (2) every edge of G belongs to exactly one subgraph in $\mathcal{D}$. If such a decomposition exists, then G is said to be **decomposable** into the subgraphs $G_1, G_2, \ldots, G_k$. If each $G_i \cong H$ for some graph H, then G is H-**decomposable** and the resulting isomorphic decomposition is an H-**decomposition** of G.

For example, the complete graph K_7 is K_3-decomposable. One way to see this is to let $v_1, v_2, \ldots, v_7$ be the seven vertices of a regular 7-gon and join each pair of vertices by a straight line segment (resulting in K_7). Consider the triangle with vertices v_1, v_2, and v_4, which we denote by G_1 (see Figure 4.10). By rotating G_1 clockwise about the center of the 7-gon through an angle of $2\pi/7$ radians, another triangle G_2 is produced. Doing this five more times produces not only a K_3-decomposition of K_7 but a **cyclic** K_3-decomposition of K_7.

One of the best-known conjectures on decompositions is due to Gerhard Ringel [165].

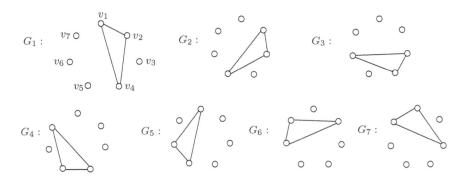

Figure 4.10: A cyclic K_3-decomposition of K_7

Ringel's Conjecture *For every tree T of size m, the complete graph K_{2m+1} is T-decomposable.*

There is, in fact, a stronger conjecture due jointly to Gerhard Ringel and Anton Kotzig.

The Ringel-Kotzig Conjecture *For every tree T of size m, the complete graph K_{2m+1} is cyclically T-decomposable.*

Exercises for Chapter 4

1. Prove or disprove: Let G be a bipartite graph with partite sets U and W such that $|U| \leq |W|$. If U can be matched to a subset of W, then for every nonempty subset S of U, the set S can be matched to a subset of $N(S)$.

2. Prove that every cubic graph with at most two bridges contains a perfect matching.

3. A connected bipartite graph G has partite sets U and W, where $|U| = |W| = k \geq 2$. Prove that if every two vertices of U have distinct degrees in G, then G contains a perfect matching.

4. (a) Let G be a bipartite graph with partite sets U and W where $|W| = 2|U| = 2n$. State a theorem analogous to Hall's Theorem that gives a necessary and sufficient condition for G to possess n pairwise vertex-disjoint copies of P_3.

   (b) Prove the theorem stated in (a).

5. Figure 4.4 shows a cubic graph with three bridges containing no perfect matching. Show, however, that for every integer $k \geq 3$ there exists a connected cubic graph with k bridges that contains a perfect matching.

6. Prove that every tree has at most one perfect matching.

7. For a graph G, the **maximal independent graph** $MI(G)$ of G has the set of maximal independent sets of the vertices of G as its vertex set and two vertices U and W are adjacent if $U \cap W \neq \emptyset$.

   (a) For each positive integer n, give an example of a graph G such that $MI(G) = K_n$.

   (b) Give an example of a graph G such that $MI(G) = K_{1,3}$.

8. For a nonempty graph G and its line graph $L(G)$, express $\alpha(L(G))$ and $\omega(L(G))$ in terms of some parameter or parameters involving G.

9. Determine the largest positive integer n such that there is a graph G of order n such that $\alpha(G) = \alpha(\overline{G}) = 2$.

10. Show that there exists a connected graph G whose vertex set $V(G)$ can be partitioned into three independent sets but no fewer and where each vertex of G can be colored red, blue, or green producing two partitions $\{V_1, V_2, V_3\}$ and $\{V_1', V_2', V_3'\}$ of $V(G)$ into independent sets such that

    (1) $|V_i| = |V_j'| \geq 2$ for all i and j with $1 \leq i, j \leq 3$,

    (2) every two vertices of V_i are colored the same for each i $(1 \leq i \leq 3)$, and

    (3) no two vertices of V_i' are colored the same for each i $(1 \leq i \leq 3)$.

11. Figure 4.11 shows a 6-regular graph G of order 14.

    (a) Is G 1-factorable?

    (b) Is G 2-factorable?

    (c) Is G Hamiltonian-factorable?

    (d) Is G K_3-decomposable?

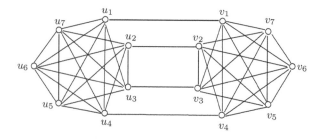

Figure 4.11: A 6-regular graph of order 14 in Exercise 11

12. Show that the Petersen graph does not contain two edge-disjoint perfect matchings.

13. Give a cyclic factorization of the Petersen graph into

    (a) three factors,

(b) five factors.

14. Give an example of an isomorphic cyclic factorization of K_7 into three factors that is not a Hamiltonian factorization.

15. Give an example of a connected 4-regular graph that is not Hamiltonian factorable.

16. Give an example of an isomorphic factorization of $K_{2,2,2}$ into three factors.

17. For the 4-regular graph $G = K_{2,2,2}$ in Figure 4.12, the circuit $C = (v_1, v_2, v_3, v_1, v_4, v_5, v_6, v_3, v_5, v_2, v_4, v_6, v_1)$ of G is Eulerian.

   (a) Construct the bipartite graph H described in the proof of Theorem 4.15.

   (b) Show that $E_1 = \{u_1w_4, u_2w_3, u_3w_5, u_4w_6, u_5w_2, u_6w_1\}$ is the edge set of a 1-factor H_1 of H.

   (c) Use E_1 in (b) to construct the corresponding 2-factorization of H described in the proof of Theorem 4.15.

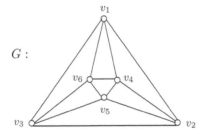

G :

Figure 4.12: The graph $G = K_{2,2,2}$ in Exercise 17

18. Prove that for every positive integer k, the complete graph K_{2k} can be factored into $k - 1$ Hamiltonian cycles and a 1-factor.

19. Show that there exists a tree T of size 5 for which the Petersen graph is cyclically T-decomposable.

20. Show that K_7 is cyclically $K_{1,3}$-decomposable.

21. Show that it is possible to color each edge of $K_{4,4}$ with one of four colors such that there are two 1-factorizations $\mathcal{F}$ and $\mathcal{F}'$ of $K_{4,4}$ for which every two edges of each factor in $\mathcal{F}$ are colored the same and no two edges of each factor in $\mathcal{F}'$ are colored the same.

22. Show that it is possible to color each edge of $K_{5,5}$ with one of five colors such that there are two 1-factorizations $\mathcal{F}$ and $\mathcal{F}'$ of $K_{5,5}$ for which every two edges of each factor in $\mathcal{F}$ are colored the same and no two edges of each factor in $\mathcal{F}'$ are colored the same.

23. For each integer $k \geq 2$, give an example of a connected graph G_k of size k^2 for which it is possible to color each edge of G_k with one of k colors, say $1, 2, \ldots, k$, such that there are two isomorphic factorizations $\mathcal{F}$ and $\mathcal{F}'$ of G_k where every two edges of each factor in $\mathcal{F}$ are colored the same and no two edges of each factor in $\mathcal{F}'$ are colored the same.

Chapter 5

Graph Embeddings

When considering a graph G, a diagram of G is often drawn (in the plane). Sometimes no edges cross in a drawing, while on other occasions some pairs of edges may cross. Even if some pairs of edges cross in a diagram of G, there may very well be other drawings of G in which no edges cross. On the other hand, it may be impossible to draw G without some of its edges crossing. Even in this case, there is a variety of other surfaces on which we may attempt to draw G so that none of its edges cross. This is the subject of the current chapter.

5.1   Planar Graphs and the Euler Identity

A **polyhedron** is a 3-dimensional object whose boundary consists of polygonal plane surfaces. These surfaces are typically called the **faces** of the polyhedron. The boundary of a face consists of the edges and vertices of the polygon. In this setting, the total number of faces in the polyhedron is commonly denoted by F, the total number of edges in the polyhedron by E, and the total number of vertices by V. The best-known polyhedra are the so-called **Platonic solids**: the **tetrahedron**, **cube (hexahedron)**, **octahedron**, **dodecahedron**, and **icosahedron**. These are shown in Figure 5.1, together with the values of V, E, and F for these polyhedra.

During the 18th century, many letters (over 160) were exchanged between Leonhard Euler (who, as we saw in Chapter 3, essentially introduced graph theory to the world when he solved and then generalized the Königsberg Bridge Problem) and Christian Goldbach (well known for stating the conjecture that every even integer greater than or equal to 4 can be expressed as the sum of two primes). In a letter that Euler wrote to Goldbach on 14 November 1750, he stated a relationship that existed among the numbers V, E, and F for a polyhedron and which would later become known as:

The Euler Polyhedral Formula *If a polyhedron has V vertices, E edges, and F faces, then*

$$V - E + F = 2.$$

109

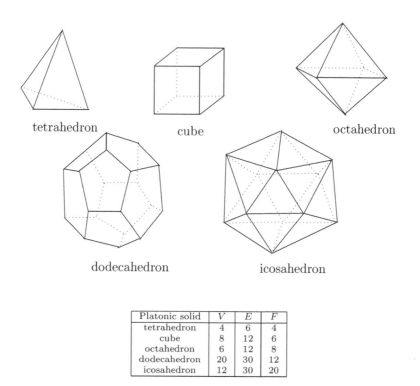

tetrahedron   cube   octahedron

dodecahedron   icosahedron

Platonic solid	V	E	F
tetrahedron	4	6	4
cube	8	12	6
octahedron	6	12	8
dodecahedron	20	30	12
icosahedron	12	30	20

Figure 5.1: The five Platonic solids

That Euler was evidently the first mathematician to observe this formula (which is actually an identity rather than a formula) may be somewhat surprising in light of the fact that Archimedes (287 BC – 212 BC) and René Descartes (1596 – 1650) both studied polyhedra long before Euler. A possible explanation as to why others had overlooked this identity might be due to the fact that geometry had primarily been a study of distances.

The Euler Polyhedral Formula appeared in print two years later (in 1752) in two papers by Euler. In the first of these two papers, Euler stated that he had been unable to prove the formula. However, in the second paper, he presented a proof by dissecting polyhedra into tetrahedra. Although his proof was clever, he nonetheless made some missteps. The first generally accepted proof of this identity was obtained by the French mathematician Adrien-Marie Legendre.

By applying a stereographic projection of a polyhedron onto the plane, a map is produced. Each map can be converted into a graph by inserting a vertex at each meeting point of the map (which is actually a vertex of the polyhedron). This is illustrated in Figure 5.2 for the cube.

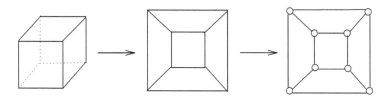

Figure 5.2: From a polyhedron to a map to a graph

The graphs obtained from the five Platonic solids are shown in Figure 5.3. These graphs have a property in which we will be especially interested: No two edges cross (intersect each other) in the graph.

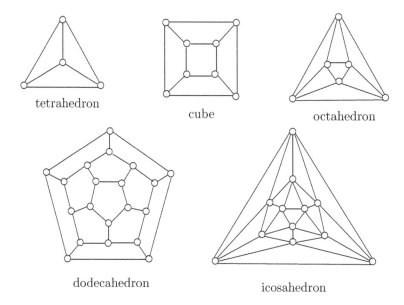

tetrahedron

cube

octahedron

dodecahedron

icosahedron

Figure 5.3: The graphs of the five Platonic solids

A graph G is called a **planar graph** if G can be drawn in the plane without any two of its edges crossing; otherwise, G is **nonplanar**. Such a drawing is also called an **embedding** of G in the plane. In this case, the embedding is a **planar embedding**. A graph G that is already drawn in the plane in this manner is a **plane graph**. Certainly then, every plane graph is planar and every planar graph can be drawn as a plane graph. In particular, all five graphs of the Platonic solids are planar.

When considering a plane graph G of a polyhedron, the faces of the polyhedron become the regions of G, one of which is the **exterior region** of G. On the other hand, a planar graph need not be the graph of any polyhedron. The plane graph H of Figure 5.4 is not the graph of any polyhedron. This graph has five regions, denoted by R_1, R_2, R_3, R_4, and R_5, where R_5 is the exterior region.

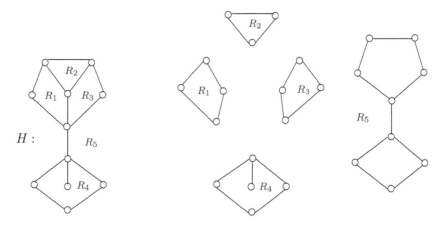

Figure 5.4: The boundaries of the regions of a plane graph

For a region R of a plane graph G, the vertices and edges incident with R form a subgraph of G called the **boundary** of R. Every edge of G that lies on a cycle belongs to the boundary of two regions of G, while every bridge of G belongs to the boundary of a single region. In Figure 5.4, the boundaries of the five regions of H are shown as well.

The five graphs G_1, G_2, G_3, G_4, and G_5 shown in Figure 5.5 are all planar, although G_1 and G_3 are not plane graphs. The graph G_1 can be drawn as G_2, while G_3 can be drawn as G_4. In fact, G_1 (and G_2) is the graph of the tetrahedron. For each graph, its order n, its size m, and the number r of regions are shown as well.

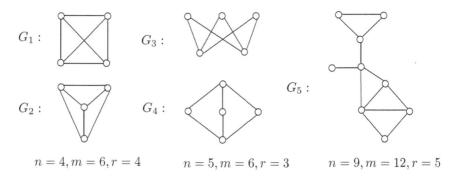

$$n = 4, m = 6, r = 4 \qquad n = 5, m = 6, r = 3 \qquad n = 9, m = 12, r = 5$$

Figure 5.5: Planar graphs

Observe that $n - m + r = 2$ for each graph of Figure 5.5. Of course, this is not surprising for G_2 since this is the graph of a polyhedron (the tetrahedron) and $n = V$, $m = E$, and $r = F$. In fact, this identity holds for every connected plane graph.

Theorem 5.1 (**The Euler Identity**) *For every connected plane graph of order n, size m, and having r regions,*

$$n - m + r = 2.$$

Proof. We proceed by induction on the size m of a connected plane graph. There is only one connected graph of size 0, namely K_1. In this case, $n = 1$, $m = 0$, and $r = 1$. Since $n - m + r = 2$, the base case of the induction holds.

Assume for a positive integer m that if H is a connected plane graph of order n' and size m', where $m' < m$ such that there are r' regions, then $n' - m' + r' = 2$. Let G be a connected plane graph of order n and size m with r regions. We consider two cases.

Case 1. G is a tree. In this case, $m = n - 1$ and $r = 1$. Thus, $n - m + r = n - (n - 1) + 1 = 2$, producing the desired result.

Case 2. G is not a tree. Since G is connected and is not a tree, G contains an edge e that is not a bridge. In G, the edge e is on the boundaries of two regions. So in $G - e$ these two regions merge into a single region. Since $G - e$ has order n, size $m - 1$, and $r - 1$ regions and $m - 1 < m$, it follows by the induction hypothesis that $n - (m - 1) + (r - 1) = 2$ and so $n - m + r = 2$. ∎

The Euler Polyhedron Formula is therefore a special case of Theorem 5.1. Recall that Euler struggled with the verification of $V - E + F = 2$, while the more general result was not all that difficult to prove. Of course, Euler did not have the luxury of a developed graph theory at his disposal.

If G is a plane graph of order 4 or more, then the boundary of every region of G must contain at least three edges. This observation is helpful in showing that the size of a planar graph cannot be too large in terms of its order.

Theorem 5.2 *If G is a planar graph of order $n \geq 3$ and size m, then*

$$m \leq 3n - 6.$$

Proof. Since the size of every graph of order 3 cannot exceed 3, the inequality holds for $n = 3$. So we may assume that $n \geq 4$. Furthermore, we may assume that the planar graphs under consideration are connected, for otherwise edges can be added to produce a connected graph. Suppose that G is a connected planar graph of order $n \geq 4$ and size m and that there is a given planar embedding of G, resulting in r regions. By the Euler Identity, $n - m + r = 2$. Let $R_1, R_2, \ldots, R_r$ be the regions of G and suppose that we denote the number of edges on the boundary of R_i $(1 \leq i \leq r)$ by m_i. Then $m_i \geq 3$. Since each edge of G is on the boundary of at most two regions of G, it follows that

$$3r \leq \sum_{i=1}^{r} m_i \leq 2m.$$

Hence,

$$6 = 3n - 3m + 3r \leq 3n - 3m + 2m = 3n - m$$

and so $m \le 3n - 6$.   ∎

By expressing Theorem 5.2 in its contrapositive form, it follows that:

> If G is a graph of order $n \ge 5$ and size m such that $m > 3n - 6$, then G is nonplanar.

This provides us with a large class of nonplanar graphs.

Corollary 5.3 *Every complete graph K_n of order $n \ge 5$ is nonplanar.*

Proof. Since $n \ge 5$, it follows that $(n-3)(n-4) > 0$ and so $n^2 - 7n + 12 > 0$. Hence, $n^2 - n > 6n - 12$, which implies that $\binom{n}{2} = \frac{n(n-1)}{2} > 3n - 6$ and so the size of K_n exceeds $3n - 6$. By Theorem 5.2, K_n is nonplanar.   ∎

Since it is evident that any graph containing a nonplanar subgraph is itself nonplanar, once we know that K_5 is nonplanar, we can conclude that K_n is nonplanar for every integer $n \ge 5$. Of course, K_n is planar for $1 \le n \le 4$. Another corollary of Theorem 5.2 provides us with a useful property of planar graphs.

Corollary 5.4 *Every planar graph contains a vertex of degree 5 or less.*

Proof. The result is obvious for planar graphs of order 6 or less. Let G be a graph of order n and size m, all of whose vertices have degree 6 or more. Then $n \ge 7$ and

$$2m = \sum_{v \in V(G)} \deg v \ge 6n$$

and so $m \ge 3n$. By Theorem 5.2, G is nonplanar.   ∎

We will soon see that both K_5 and $K_{3,3}$ are important nonplanar graphs. Since $K_{3,3}$ has order $n = 6$ and size $m = 9$ but $m < 3n - 6$, Theorem 5.2 cannot be used to establish the nonplanarity of $K_{3,3}$. However, we can use the fact that $K_{3,3}$ is bipartite to verify this property.

Corollary 5.5 *The graph $K_{3,3}$ is nonplanar.*

Proof. Suppose that $K_{3,3}$ is planar. Let there be given a planar embedding of $K_{3,3}$, resulting in r regions. Thus, by the Euler Identity, $n - m + r = 6 - 9 + r = 2$ and so $r = 5$. Let $R_1, R_2, \ldots, R_5$ be the five regions, and let m_i be the number of edges on the boundary of R_i ($1 \le i \le 5$). Since $K_{3,3}$ is bipartite, it follows that $K_{3,3}$ has no triangle and so $m_i \ge 4$ for $1 \le i \le 5$. Since every edge of $K_{3,3}$ lies on the boundary of a cycle, every edge of $K_{3,3}$ belongs to the boundary of two regions. Thus,

$$20 = 4r \le \sum_{i=1}^{5} m_i = 2m = 18,$$

which is impossible.   ∎

A planar graph G is **maximal planar** if the addition to G of any edge joining two nonadjacent vertices of G results in a nonplanar graph. Necessarily then, if a maximal planar graph G of order $n \geq 3$ and size m is embedded in the plane, then the boundary of every region of G is a triangle and so $3r = 2m$. It then follows by the proof of Theorem 5.2 that $m = 3n - 6$. All of the graphs shown in Figure 5.6 are maximal planar. A graph G is **nearly maximal planar** if there exists a planar embedding of G such that the boundary of every region of G is a cycle, at most one of which is not a triangle. For example, the wheels $W_n = C_n + K_1$ $(n \geq 3)$ are nearly maximal planar.

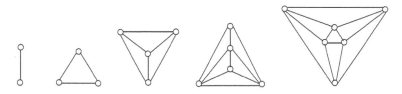

Figure 5.6: Maximal planar graphs

We now derive some results concerning the degrees of the vertices of a maximal planar graph.

Theorem 5.6 *If G is a maximal planar graph of order 4 or more, then the degree of every vertex of G is at least 3.*

Proof. Let G be a maximal planar graph of order $n \geq 4$ and size m and let v be a vertex of G. Since $m = 3n - 6$, it follows that $G - v$ has order $n - 1$ and size $m - \deg v$. Since $G - v$ is planar and $n - 1 \geq 3$,

$$m - \deg v \leq 3(n - 1) - 6$$

and so $m - \deg v = 3n - 6 - \deg v \leq 3n - 9$. Thus, $\deg v \geq 3$. ∎

In 1904 Paul August Ludwig Wernicke was awarded a Ph.D. from the University of Göttingen in Germany under the supervision of the famed geometer Hermann Minkowski. (Three years later, Dénes König, who wrote the first book [137] on graph theory, published in 1936, would receive a Ph.D. from the same university and have the same supervisor. Thus, Wernicke and König were "academic brothers".) By Corollary 5.4 and Theorem 5.6, every maximal planar graph of order 4 or more contains a vertex of degree 3, 4, or 5. In the very same year that he received his Ph.D., Wernicke [200] proved that every planar graph that didn't have a vertex of a degree less than 5 must contain a vertex of degree 5 that is adjacent either to a vertex of degree 5 or to a vertex of degree 6. In the case of maximal planar graphs, Wernicke's result states the following.

Theorem 5.7 *If G is a maximal planar graph of order 4 or more, then G contains at least one of the following: (1) a vertex of degree 3, (2) a vertex of degree 4, (3) two adjacent vertices of degree 5, (4) two adjacent vertices, one of which has degree 5 and the other has degree 6.*

Proof.   Assume, to the contrary, that there exists a maximal planar graph G of order $n \geq 4$ and size m containing none of (1)–(4). By Corollary 5.4, $\delta(G) = 5$. Let G be embedded in the plane, resulting in r regions. Then

$$n - m + r = 2.$$

Suppose that G has n_i vertices of degree i for $5 \leq i \leq \Delta(G) = \Delta$. Then

$$\sum_{i=5}^{\Delta} n_i = n \ \text{ and } \ \sum_{i=5}^{\Delta} in_i = 2m = 3r.$$

We now compute the number of regions that contain either a vertex of degree 5 or a vertex of degree 6 on its boundary. Since the boundary of every region is a triangle, it follows, by assumption, that no region has two vertices of degree 5 on its boundary or a vertex of degree 5 and a vertex of degree 6 on its boundary. On the other hand, the boundary of a region could contain two or perhaps three vertices of degree 6. Each vertex of degree 5 lies on the boundaries of five regions and every vertex of degree 6 lies on the boundaries of six regions. Furthermore, every region containing a vertex of degree 6 on its boundary can contain as many as three vertices of degree 6. Therefore, G has $5n_5$ regions whose boundary contains a vertex of degree 5 and at least $6n_6/3 = 2n_6$ regions whose boundary contains at least one vertex of degree 6. Thus,

$$
\begin{aligned}
r \ &\geq \ 5n_5 + 2n_6 \geq 5n_5 + 2n_6 - n_7 - 4n_8 - \cdots - (20 - 3\Delta)n_\Delta \\
&= \ \sum_{i=5}^{\Delta}(20 - 3i)n_i = 20n - 3\sum_{i=5}^{\Delta} in_i = 20(m - r + 2) - 3(2m) \\
&= \ (20m - 20r + 40) - 9r = (30r - 20r + 40) - 9r \\
&= \ r + 40,
\end{aligned}
$$

which is a contradiction.   ∎

The following result gives a relationship among the degrees of the vertices in a maximal planar graph of order at least 4.

Theorem 5.8   *Let G be a maximal planar graph of order $n \geq 4$ and size m containing n_i vertices of degree i for $3 \leq i \leq \Delta = \Delta(G)$. Then*

$$3n_3 + 2n_4 + n_5 = 12 + n_7 + 2n_8 + \cdots + (\Delta - 6)n_\Delta.$$

Proof.   Since $m = 3n - 6$, it follows that $2m = 6n - 12$. Therefore,

$$\sum_{i=3}^{\Delta} in_i = \sum_{i=3}^{\Delta} 6n_i - 12$$

and so

$$\sum_{i=3}^{\Delta}(6 - i)n_i = 12. \tag{5.1}$$

Hence, $3n_3 + 2n_4 + n_5 = 12 + n_7 + 2n_8 + \cdots + (\Delta - 6)n_\Delta$. ∎

Heinrich Heesch (discussed in Chapter 0) introduced the idea of assigning what is called a "charge" to each vertex of a planar graph as well as discharging rules which indicate how charges are to be redistributed among the vertices. In a maximal planar graph G, every vertex v of G is assigned a charge of $6 - \deg v$. In particular, every vertex of degree 5 receives a charge of $+1$, every vertex of degree 6 receives a charge of 0, and every vertex of degree 7 or more receives a negative charge. By appropriately redistributing positive charges, some useful results can often be obtained. According to Equation (5.1) in the proof of Theorem 5.8, the sum of the charges of the vertices of a maximal planar graph of order 4 or more is 12.

Theorem 5.9 *If G is a maximal planar graph of order $n \geq 4$, size m, and maximum degree $\Delta(G) = \Delta$ such that G has n_i vertices of degree i for $3 \leq i \leq \Delta$, then*

$$\sum_{i=3}^{\Delta}(6 - i)n_i = 12.$$

We now use the discharging method to give an alternative proof of Theorem 5.7.

Theorem 5.10 *If G is a maximal planar graph of order 4 or more, then G contains at least one of the following: (1) a vertex of degree 3, (2) a vertex of degree 4, (3) two adjacent vertices of degree 5, (4) two adjacent vertices, one of which has degree 5 and the other has degree 6.*

Proof. Assume, to the contrary, that there exists a maximal planar graph G of order $n \geq 4$, where there are n_i vertices of degree i for $3 \leq i \leq \Delta = \Delta(G)$ such that G contains none of (1)–(4). Thus, $\delta(G) = 5$. To each vertex v of G, we assign the charge $6 - \deg v$. Hence, each vertex of degree 5 receives a charge of $+1$, each vertex of degree 6 receives no charge, and each vertex of degree 7 or more receives a negative charge. By Theorem 5.9, the sum of the charges of the vertices of G is

$$\sum_{i=3}^{\Delta}(6 - i)n_i = 12.$$

Let there be given a planar embedding of G. For each vertex v of degree 5 in G, redistribute its charge of $+1$ by moving a charge of $\frac{1}{5}$ to each of its five neighbors, resulting in v now having a charge of 0. Hence, the sum of the charges of the vertices of G remains 12. By (3) and (4), no vertex of degree 5 or 6 will have its charges increased. Consider a vertex u with $\deg u = k \geq 7$. Thus, u received an initial charge of $6 - k$. Because no adjacent neighbors of u in the embedding can have degree 5, the vertex u can receive an added charge of $+\frac{1}{5}$ from at most $k/2$ of its neighbors. After the redistribution of charges, the new charge of u is at most

$$6 - k + \frac{k}{2} \cdot \frac{1}{5} = 6 - \frac{9k}{10} < 0.$$

Hence, no vertex of G now has a positive charge. This is impossible since the sum of the charges of the vertices of G is 12. ∎

Another result concerning maximal planar graphs that can be proved with the aid of the discharging method (see Exercise 6) is due to Philip Franklin [87].

Theorem 5.11 *If G is a maximal planar graph of order 4 or more, then G contains at least one of the following: (1) a vertex of degree 3, (2) a vertex of degree 4, (3) a vertex of degree 5 that is adjacent to two vertices, each of which has degree 5 or 6.*

5.2   Hamiltonian Planar Graphs

In the previous section, we saw several necessary conditions for a connected graph to be planar and several necessary conditions for a graph of order at least 4 to be maximal planar. In this section, we will be introduced to one result, namely, a necessary condition for a planar graph to be Hamiltonian.

Let G be a Hamiltonian planar graph of order n and let there be given a planar embedding of G with Hamiltonian cycle C. Any edge of G not lying on C is then a **chord** of G. Every chord and every region of G then either lies interior to C or exterior to C. For $i = 3, 4, \ldots, n$, let r_i denote the number of regions interior to C whose boundary contains exactly i edges and let r_i' denote the number of regions exterior to C whose boundary contains exactly i edges.

The plane graph G of Figure 5.7 of order 12 is Hamiltonian. With respect to the Hamiltonian cycle $C = (v_1, v_2, \ldots, v_{12}, v_1)$, we have

$$r_3 = r_3' = 1, \ r_4 = 3, \ r_4' = 2, \ r_5 = r_7' = 1,$$

while $r_i = 0$ for $6 \leq i \leq 12$ and $r_i' = 0$ for $i = 5, 6$, and $8 \leq i \leq 12$.

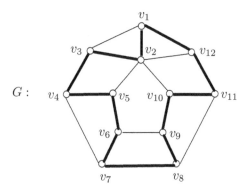

Figure 5.7: A Hamiltonian planar graph

In 1968 a necessary condition for a planar graph to be Hamiltonian was discovered by the Latvian mathematician Emanuels Ja. Grinberg [103].

Theorem 5.12 *For a plane graph G of order n with Hamiltonian cycle C,*

$$\sum_{i=3}^{n} (i-2)(r_i - r_i') = 0.$$

Proof. Suppose that c chords of G lie interior to C. Then $c + 1$ regions of G lie interior to C. Therefore,

$$\sum_{i=3}^{n} r_i = c + 1 \text{ and so } c = \sum_{i=3}^{n} r_i - 1.$$

Let N denote the result obtained by summing over all regions interior to C the number of edges on the boundary of each such region. Then each edge on C is counted once and each chord interior to C is counted twice, that is,

$$N = \sum_{i=3}^{n} i r_i = n + 2c.$$

Therefore,

$$\sum_{i=3}^{n} i r_i = n + 2c = n + 2\sum_{i=3}^{n} r_i - 2$$

and so

$$\sum_{i=3}^{n} (i - 2) r_i = n - 2.$$

Similarly,

$$\sum_{i=3}^{n} (i - 2) r_i' = n - 2,$$

giving the desired result $\sum_{i=3}^{n}(i - 2)(r_i - r_i') = 0$. $\blacksquare$

Since Theorem 5.12 gives a necessary condition for a planar graph to be Hamiltonian, this theorem also provides a sufficient condition for a planar graph to be non-Hamiltonian. We see how Grinberg's theorem can be used to show that the plane graph of Figure 5.8 is not Hamiltonian. This graph is called the **Tutte graph** (after William Tutte) and has a great deal of historical interest. We will encounter this graph again in Chapter 10.

Suppose that the Tutte graph G is Hamiltonian. Then G has a Hamiltonian cycle C. Necessarily, C must contain exactly two of the three edges e, f_1, and f_2, say f_1 and either e of f_2. Similarly, C must contain exactly two edges of the three edges e', f_2, and f_3. Since we may assume that C contains f_2, we may further assume that e is not on C. Consequently, R_1 and R_2 lie interior to C.

Let G_1 denote the component of $G - \{e, f_1, f_2\}$ containing w. Thus, G_1 contains a Hamiltonian $v_1 - v_2$ path P'. Therefore, $G_2 = G_1 + v_1 v_2$ is Hamiltonian and contains a Hamiltonian cycle C' consisting of P' and $v_1 v_2$. Applying Grinberg's theorem to G_2 with respect to C', we obtain

$$1(r_3 - r_3') + 2(r_4 - r_4') + 3(r_5 - r_5') + 6(r_8 - r_8') = 0. \tag{5.2}$$

Since $v_1 v_2$ is on C' and the exterior region of G_2 lies exterior to C', it follows that

$$r_3 - r_3' = 1 - 0 = 1 \quad \text{and} \quad r_8 - r_8' = 0 - 1 = -1.$$

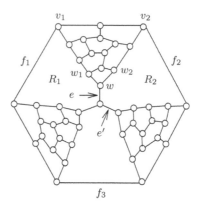

Figure 5.8: The Tutte graph

Therefore, from (5.2), we have

$$2(r_4 - r_4') + 3(r_5 - r_5') = 5.$$

Necessarily, both ww_1 and ww_2 are edges of C' and so $r_4 \geq 1$, implying that either

$$r_4 - r_4' = 1 - 1 = 0 \text{ or } r_4 - r_4' = 2 - 0 = 2.$$

If $r_4 - r_4' = 0$, then $3(r_5 - r_5') = 5$, which is impossible. On the other hand, if $r_4 - r_4' = 2$, then $3(r_5 - r_5') = 1$, which is also impossible. Hence, G is not Hamiltonian.

5.3   Planarity versus Nonplanarity

In the preceding two sections, we have discussed several properties of planar graphs. However, a fundamental question remains. For a given graph G, how does one determine whether G is planar or nonplanar? Of course, if G can be drawn in the plane without any of its edges crossing, then G is planar. On the other hand, if G cannot be drawn in the plane without edges crossing, then G is nonplanar. However, it may very well be difficult to see how to draw a graph G in the plane without edges crossing or to know that such a drawing is impossible. We saw from Theorem 5.2 that if G has order $n \geq 3$ and size m where $m > 3n - 6$, then G is nonplanar. Also, as a consequence of Theorem 5.2, we saw in Corollary 5.4 that if G contains no vertex of degree less than 6, then G is nonplanar.

Any graph that is a subgraph of a planar graph must surely be planar. Equivalently, every graph containing a nonplanar subgraph must itself be nonplanar. Thus, to show that a disconnected graph G is planar it suffices to show that each component of G is planar. Hence, when considering planarity, we may restrict our attention to connected graphs. Since a connected graph G with cut-vertices is planar if and only if each block of G is planar, it is sufficient to concentrate on 2-connected graphs only.

According to Corollaries 5.3 and 5.5, the graphs K_5 and $K_{3,3}$ are nonplanar. Hence, if a graph G should contain a subgraph that is isomorphic to either K_5 or $K_{3,3}$, then G is nonplanar. For the maximal planar graph G of order 5 and size 9 shown in Figure 5.9 (that is, G is obtained by deleting one edge from K_5), we consider the graph $F = G \times K_3$, shown in Figure 5.9. Thus, F consists of three copies of G, denoted by G_1, G_2, and G_3, where $u_1u_2 \notin E(G_1)$, $v_1v_2 \notin E(G_2)$, and $w_1w_2 \notin E(G_3)$. To make it easier to draw G, the nine edges of each graph G_i $(1 \le i \le 3)$ are not drawn. The graph F has order 15 and size $m = 42$. Since $m = 42 > 39 = 3n - 6$, it follows that F is nonplanar. Furthermore, it can be shown that no subgraph of F is isomorphic to either K_5 or $K_{3,3}$. Thus, despite the fact that F contains no subgraph isomorphic to either K_5 or $K_{3,3}$, the graph F is nonplanar. Consequently, there must exist some other explanation as to why this graph is nonplanar.

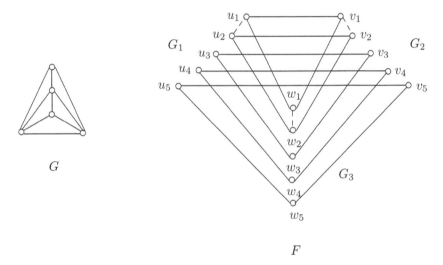

Figure 5.9: The graph $F = G \times K_3$

A graph H is a **subdivision** of a graph G if either $H = G$ or H can be obtained from G by inserting vertices of degree 2 into the edges of G. Thus, for the graph G of Figure 5.10, all of the graphs H_1, H_2, and H_3 are subdivisions of G. Indeed, H_3 is a subdivision of H_2.

Certainly, a subdivision H of a graph G is planar if and only if G is planar. Therefore, K_5 and $K_{3,3}$ are nonplanar as is any subdivision of K_5 or $K_{3,3}$. This provides a necessary condition for a graph to be planar.

Theorem 5.13 *A graph G is planar only if G contains no subgraph that is a subdivision of K_5 or $K_{3,3}$.*

The remarkable feature about this necessary condition for a graph to be planar is that the condition is also sufficient. The first published proof of this fact occurred in 1930. This theorem is due to the well-known Polish topologist Kazimierz

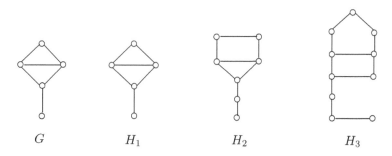

Figure 5.10: Subdivisions of a graph

Kuratowski (1896–1980), who first announced this theorem in 1929. The title of Kuratowski's paper is "Sur le problème des courbes gauches en topologie" [On the problem of skew curves in topology], which suggests, and rightly so, that the setting of his theorem was in topology – not graph theory. Nonplanar graphs were sometimes called *skew graphs* during that period. The publication date of Kuratowski's paper was critical to having the theorem credited to him, for, as it turned out, later in 1930 two American mathematicians Orrin Frink and Paul Althaus Smith submitted a paper containing a proof of this theorem as well but withdrew it after they became aware that Kuratowski's proof had preceded theirs, although just barely. They did publish a one-sentence announcement [88] of what they had accomplished in the *Bulletin of the American Mathematical Society* and, as the title of their note indicates ("Irreducible non-planar graphs"), the setting for their proof was graph theoretical in nature.

It is believed by some that a proof of this theorem may have been discovered somewhat earlier by the Russian topologist Lev Semenovich Pontryagin (1908–1988), who was blind his entire adult life. Because the first proof of this theorem may have occurred in Pontryagin's unpublished notes, this result is sometimes referred to as the Pontryagin-Kuratowski theorem in Russia and elsewhere. However, since the possible proof of this theorem by Pontryagin did not satisfy the established practice of appearing in print in an accepted refereed journal, the theorem is generally recognized as Kuratowski's theorem. We now present a proof of this famous theorem [138].

Theorem 5.14 (**Kuratowski's Theorem**) *A graph G is planar if and only if G contains no subgraph that is a subdivision of K_5 or $K_{3,3}$.*

Proof.   We have already noted the necessity of this condition for a graph to be planar. Hence, it remains to verify its sufficiency, namely that every graph containing no subgraph which is a subdivision of K_5 or $K_{3,3}$ is planar. Suppose that this statement is false. Then there is a nonplanar 2-connected graph G of minimum size containing no subgraph that is a subdivision of K_5 or $K_{3,3}$.

First, we show that G is necessarily 3-connected. Suppose that this is not the case. Then G contains a minimum vertex-cut consisting of two vertices x and y. Since G has no cut-vertices, it follows that each of x and y is adjacent to one or more vertices in each component of $G - \{x, y\}$. Let F_1 be one component of $G - \{x, y\}$ and let F_2 be the union of the remaining components of $G - \{x, y\}$. Furthermore, let

$$G_i = G[V(F_i) \cup \{x, y\}] \text{ for } i = 1, 2.$$

We consider two cases, depending on whether x and y are adjacent. Suppose first that x and y are adjacent. We claim, in this case, that at least one of G_1 and G_2 is nonplanar. If both G_1 and G_2 are planar, then there exist planar embeddings of these two graphs in which xy is on the boundary of the exterior region in each embedding. This, however, implies that G itself is planar, which is impossible. Thus, G_1, say, is nonplanar. Since G_1 is a subgraph of G, it follows that G_1 contains no subgraph that is a subdivision of K_5 or $K_{3,3}$. However, the size of G_1 is less than the size of G, which contradicts the defining property of G. Hence, x and y must be nonadjacent.

Let f be the edge obtained by joining x and y, and let $H_i = G_i + f$ for $i = 1, 2$. If H_1 and H_2 are both planar, then, as above, there is a planar embedding of $G + f$ and of G as well. Since this is impossible, at least one of H_1 and H_2 is nonplanar, say H_1. Because the size of H_1 is less than the size of G, the graph H_1 contains a subgraph F that is a subdivision of K_5 or $K_{3,3}$. Since G_1 contains no such subgraph, it follows that $f \in E(F)$. Let P be an $x - y$ path in F_2. By replacing f in F by P, we obtain a subgraph of G that is a subdivision of K_5 or $K_{3,3}$. This produces a contradiction. Hence, as claimed, G is 3-connected.

To summarize then, G is a nonplanar graph of minimum size containing no subgraph that is a subdivision of K_5 or $K_{3,3}$ and, as we just saw, G is 3-connected. Let $e = uv$ be an edge of G. Then $H = G - e$ is planar. Let there be given a planar embedding of H. Since G is 3-connected, H is 2-connected. By Theorem 2.4, there are cycles in H containing both u and v. Among all such cycles, let

$$C = (u = v_0, v_1, \ldots, v_\ell = v, \ldots, v_k = u)$$

be one for which the number of regions interior to C is maximum.

It is convenient to define two subgraphs of H. The *exterior subgraph* of H is the subgraph induced by those edges lying exterior to C and the *interior subgraph* of H is the subgraph induced by those edges lying interior to C. Both subgraphs exist for otherwise the edge e could be added either to the exterior or interior subgraph of H so that the resulting graph (namely G) is planar.

No two distinct vertices of $\{v_0, v_1, \ldots, v_\ell\}$ or of $\{v_\ell, v_{\ell+1}, \ldots, v_k\}$ are connected by a path in the exterior subgraph of H, for otherwise there is a cycle in H containing u and v and having more regions interior to it than C has. Since G is nonplanar, there must be a $v_s - v_t$ path P in the exterior subgraph of H, where $0 < s < \ell < t < k$, such that only v_s and v_t belong to C. (See Figure 5.11.) Necessarily, no vertex of P different from v_s and v_t is adjacent to a vertex of C or is even connected to a vertex of C by a path, all of whose edges belong to the exterior subgraph of H.

Let S be the set of vertices on C different from v_s and v_t, that is,

$$S = V(C) - \{v_s, v_t\},$$

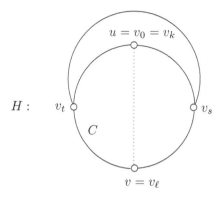

Figure 5.11: A step in the proof of Theorem 5.14

and let H_1 be the component of $H - S$ that contains P. By the defining property of C, the subgraph H_1 cannot be moved to the interior of C in a plane manner. This fact together with the fact that $G = H + e$ is nonplanar implies that the interior subgraph of H must contain one of the following:

(1) A $v_a - v_b$ path with $0 < a < s$ and $\ell < b < t$ such that only v_a and v_b belong to C. (See Figure 5.12(a).)

(2) A vertex w not on C that is connected to C by three internally disjoint paths such that the terminal vertex of one such path P' is one of v_0, v_s, v_ℓ and v_t. If, for example, the terminal vertex of P' is v_0, then the terminal vertices of the other two paths are v_a and v_b, where $s \leq a < \ell$ and $\ell < b \leq t$ where not both $a = s$ and $b = t$ occur. (See Figure 5.12(b).) If the terminal vertex of P' is one of v_s, v_ℓ and v_t, then there are corresponding bounds for a and b for the terminal vertices of the other two paths.

(3) A vertex w not on C that is connected to C by three internally disjoint paths P_1, P_2, and P_3 such that the terminal vertices of these paths are three of the four vertices v_0, v_s, v_ℓ and v_t, say v_0, v_ℓ and v_s, respectively, together with a $v_c - v_t$ path P_4 ($v_c \neq v_0, v_\ell, w$), where v_c is on P_1 or P_2, and P_4 is disjoint from P_1, P_2, and C except for v_c and v_t. (See Figure 5.12(c).) The remaining choices for P_1, P_2, and P_3 produce three analogous cases.

(4) A vertex w not on C that is connected to v_0, v_s, v_ℓ and v_t by four internally disjoint paths. (See Figure 5.12(d).)

In the first three cases, there is a subgraph of G that is a subdivision of $K_{3,3}$; while in the fourth case, there is a subgraph of G, which is a subdivision of K_5. This is a contradiction.   ∎

As a consequence of Kuratowski's theorem, the 4-regular graph G shown in Figure 5.13(a) is nonplanar since G contains the subgraph H in Figure 5.13(b), which is a subdivision of $K_{3,3}$.

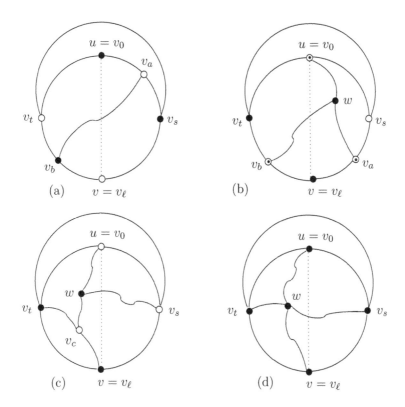

Figure 5.12: Situations (1)-(4) in the proof of Theorem 5.14

There is another characterization of planar graphs closely related to that given in Kuratowski's theorem. Before presenting this theorem, we have some additional terminology to introduce. For an edge $e = uv$ of a graph G, the graph G' obtained from G by **contracting the edge** e is that graph produced by identifying the adjacent vertices u and v, where $u = v$ and where this vertex is denoted by either u or v, say v in this case, and the edge set

$$E(G') = \{xy : xy \in E(G), x,y \in V(G) - \{u,v\}\} \cup$$
$$\{vx : ux \in E(G) \text{ or } vx \in E(G), x \in V(G) - \{u,v\}\}.$$

Thus, $V(G') - \{u = v\} = V(G) - \{u,v\}$ and $|V(G')| = |V(G)| - 1$. For the graph G of Figure 5.14, G' is obtained by contracting the edge uv in G and where G'' is obtained by contracting the edge wy in G'.

When dealing with edge contractions, it is often the case that we begin with a graph G, contract an edge in G to obtain a graph G', contract some edge in G' to obtain another graph G'', and so on, until finally arriving at a graph H. Any such graph H can be obtained in a different and perhaps simpler manner. In particular, H can be obtained from G by a succession of edge contractions if and only if the vertex set of H is the set of elements in a partition $\{V_1, V_2, \ldots, V_k\}$ of $V(G)$ where

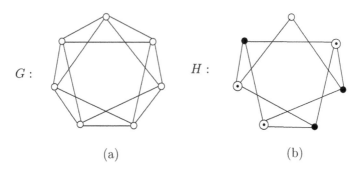

G : H :

(a)   (b)

Figure 5.13: A nonplanar graph

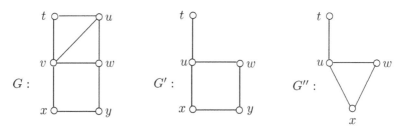

Figure 5.14: Contracting an edge

each induced subgraph $G[V_i]$ is connected and V_i is adjacent to V_j $(i \neq j)$ if and only if some vertex in V_i is adjacent to some vertex in V_j in G. For example, in the graph G of Figure 5.14, if we were to let

$$V_1 = \{t\}, \ V_2 = \{u, v\}, \ V_3 = \{x\}, \text{ and } V_4 = \{w, y\},$$

then the resulting graph H is shown in Figure 5.15. This is the graph G'' of Figure 5.14 obtained by successively contracting the edge uv in G and then the edge wy in G'.

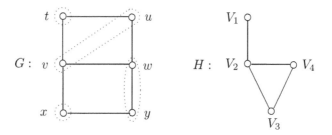

Figure 5.15: Edge contractions

A graph H is called a **minor** of a graph G if either $H = G$ or (a graph isomorphic to) H can be obtained from G by a succession of edge contractions, edge deletions, or vertex deletions (in any order). Equivalently, H is a minor of G if $H = G$ or

H can be obtained from a subgraph of G by a succession of edge contractions. In particular, the graph H of Figure 5.15 is a minor of the graph G of that figure. Consequently, a graph G is a minor of itself. If H is a minor of G such that $H \neq G$, then H is called a **proper minor** of G.

Consider next the graph G_1 of Figure 5.16, where

$$V_1 = \{t_1, t_2\}, \; V_2 = \{u_1, u_2, u_3, u_4\},$$
$$V_3 = \{v_1\}, \; V_4 = \{w_1, w_2, w_3\},$$
$$V_5 = \{x_1, x_2\}, \; V_6 = \{y_1\}, \text{ and } V_7 = \{z_1\}.$$

Then the graph H_1 of Figure 5.16 can be obtained from G_1 by successive edge contractions. Thus, H_1 is a minor of G_1. By deleting the edge $V_2 V_6$ and the vertices V_6 and V_7 from H_1 (or equivalently, deleting V_6 and V_7 from H_1), we see that K_5 is also a minor of G_1.

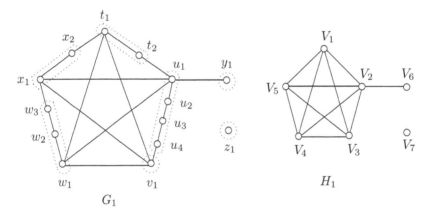

Figure 5.16: Minors of graphs

The example in Figure 5.16 serves to illustrate the following.

Theorem 5.15 *If a graph G is a subdivision of a graph H, then H is a minor of G.*

The following is therefore an immediate consequence of Theorem 5.15.

Theorem 5.16 *If G is a nonplanar graph, then K_5 or $K_{3,3}$ is a minor of G.*

The German mathematician Klaus Wagner (1910–2000) showed [197] that the converse of Theorem 5.16 is true only a year after obtaining his Ph.D. from Universität zu Köln (the University of Cologne), thereby giving another characterization of planar graphs.

Theorem 5.17 (**Wagner's Theorem**) *A graph G is planar if and only if neither K_5 nor $K_{3,3}$ is a minor of G.*

Proof. It was mentioned in Theorem 5.16 that if a graph G is nonplanar, then either K_5 or $K_{3,3}$ is a minor of G. It remains therefore only to verify the converse.

Let G be a graph having K_5 or $K_{3,3}$ as a minor. We show that G is nonplanar. We consider two cases, according to whether $K_{3,3}$ or K_5 is a minor of G.

Case 1. $H = K_{3,3}$ is a minor of G. The graph H can be obtained by first deleting edges and vertices of G (if necessary), obtaining a connected graph G', and then by a succession of edge contractions in G'. We show, in this case, that G' contains a subgraph that is a subdivision of $K_{3,3}$ and therefore that G', and G as well, is nonplanar.

Since $H = K_{3,3}$, the vertices of H can be denoted by the sets U_i and W_i $(1 \le i \le 3)$, where $\{U_1, U_2, U_3\}$ and $\{W_1, W_2, W_3\}$ are the partite sets of H. Since H is obtained from G' by a succession of edge contractions, the subgraphs

$$F_i = G'[U_i] \text{ and } H_i = G'[W_i]$$

are connected. Since $U_i W_j \in E(H)$ for $1 \le i, j \le 3$, there is a vertex $u_{i,j} \in U_i$ that is adjacent in H to a vertex $w_{i,j} \in W_j$. Among the vertices $u_{i,1}, u_{i,2}, u_{i,3}$ in U_i two or possibly all three may represent the same vertex. If $u_{i,1} = u_{i,2} = u_{i,3}$, then set $u_{i,j} = u_i$; if two of $u_{i,1}, u_{i,2}, u_{i,3}$ are the same, say $u_{i,1} = u_{i,2}$, then set $u_{i,1} = u_i$; if $u_{i,1}, u_{i,2}$ and $u_{i,3}$ are distinct, then set u_i to be a vertex in U_i that is connected to $u_{i,1}, u_{i,2}$ and $u_{i,3}$ by internally disjoint paths in F_i. (Possibly $u_i = u_{i,j}$ for some j.) We proceed in the same manner to obtain vertices $w_i \in W_i$ for $1 \le i \le 3$. The subgraph of G induced by the nine edges $u_i w_j$ together with the edge sets of all of the previously mentioned paths in F_i and H_j $(1 \le i, j \le 3)$ is a subdivision of $K_{3,3}$.

Case 2. $H = K_5$ is a minor of G. Then H can be obtained by first deleting edges and vertices of G (if necessary), obtaining a connected graph G', and then by a succession of edge contradictions in G'. We show in this case that either G' contains a subgraph that is a subdivision of K_5 or G' contains a subgraph that is a subdivision of $K_{3,3}$.

We may denote the vertices of H by the sets V_i $(1 \le i \le 5)$, where $G_i = G'[V_i]$ is a connected subgraph of G' and each subgraph G_i contains a vertex that is adjacent to G_j for each pair i, j of distinct integers where $1 \le i, j \le 5$. For $1 \le i \le 5$, let $v_{i,j}$ be a vertex of G_i that is adjacent to a vertex of G_j, where $1 \le j \le 5$ and $j \ne i$.

For a fixed integer i with $1 \le i \le 5$, if the vertices $v_{i,j}$ $(i \ne j)$ represent the same vertex, then denote this vertex by v_i. If three of the four vertices $v_{i,j}$ are the same, then we also denote this vertex by v_i. If two of the vertices $v_{i,j}$ are the same, the other two are distinct, and there exist internally disjoint paths from the coinciding vertices to the other two vertices, then we denote the two coinciding vertices by v_i. If the vertices $v_{i,j}$ are distinct and G_i contains a vertex from which there are four internally disjoint paths (one of which may be trivial) to the vertices $v_{i,j}$, then denote this vertex by v_i. Hence, there are several instances in which we have defined a vertex v_i. Should v_i be defined for all i $(1 \le i \le 5)$, then G' (and therefore G as well) contains a subgraph that is a subdivision of K_5 and so G is nonplanar.

We may assume then that for one or more integers i $(1 \le i \le 5)$, the vertex v_i has not been defined. For each such i, there exist distinct vertices u_i and w_i, each of which is connected to two of the vertices $v_{i,j}$ by internally disjoint (possibly trivial)

paths, while u_i and w_i are connected by a path none of whose internal vertices
are the vertices $v_{i,j}$ and where every two of the five paths have only u_i or w_i in
common. If two of the vertices $v_{i,j}$ coincide, then we denote this vertex by u_i. If
the remaining two vertices $v_{i,j}$ should also coincide, then we denote this vertex by
w_i. We may assume that $i = 1$, that u_1 is connected to $v_{1,2}$ and $v_{1,3}$, and that w_1 is
connected to $v_{1,4}$ and $v_{1,5}$, as described above. Denote the edge set of these paths
by E_1.

We now consider G_2. If $v_{2,1} = v_{2,4} = v_{2,5}$, then let w_2 be this vertex and set
$E_2 = \emptyset$; otherwise, there is a vertex w_2 of G_2 (which may coincide with $v_{2,1}$, $v_{2,4}$, or
$v_{2,5}$) connected by internally disjoint (possibly trivial) paths to the distinct vertices
in $\{v_{2,1}, v_{2,4}, v_{2,5}\}$. We then let E_2 denote the edge set of these paths. Similarly,
the vertices w_3, u_2, and u_3 and the sets E_3, E_4, and E_5 are defined with the aid
of the sets $\{v_{3,1}, v_{3,4}, v_{3,5}\}$, $\{v_{4,1}, v_{4,4}, v_{4,5}\}$, and $\{v_{5,1}, v_{5,2}, v_{5,3}\}$, respectively. The
subgraph of G' induced by the union of the sets E_i and the edges $v_{i,j}v_{j,i}$ contains
a subdivision of $K_{3,3}$ with partite sets $\{u_1, u_2, u_3\}$ and $\{w_1, w_2, w_3\}$. Thus, G is
nonplanar. ∎

In the proof of Wagner's theorem, it was shown that if K_5 is a minor of a graph
G, then G contains a subdivision of K_5 or a subdivision of $K_{3,3}$. In other words,
we were unable to show that G necessarily contains a subdivision of K_5. There is
good reason for this, which is illustrated in the next example.

The Petersen graph P is a graph of order $n = 10$ and size $m = 15$. Since
$m < 3n - 6$, no conclusion can be drawn from Theorem 5.2 regarding the planarity
or nonplanarity of P. Nevertheless, the Petersen graph is, in fact, nonplanar. Theorems 5.14 and 5.17 give two ways to establish this fact. Figures 5.17(a) and 5.17(b)
show P drawn in two ways. Since $P - x$ (shown in Figure 5.17(c)) is a subdivision
of $K_{3,3}$, the Petersen graph is nonplanar. The partition $\{V_1, V_2, \ldots, V_5\}$ of $V(P)$
shown in Figure 5.17(d), where $V_i = \{u_i, v_i\}$, $1 \le i \le 5$, shows that K_5 in Figure 5.17(d) is a minor of P and is therefore nonplanar. Since P is a cubic graph,
there is no subgraph of P that is subdivision of K_5, however.

5.4   Outerplanar Graphs

A graph G is **outerplanar** if there exists a planar embedding of G so that every
vertex of G lies on the boundary of the exterior region. If there is a planar embedding
of G so that every vertex of G lies on the boundary of a single region of G, then G
is outerplanar since there is always a planar embedding of G when this region is the
exterior region. The following two results provide characterizations of outerplanar
graphs.

Theorem 5.18   *A graph G is outerplanar if and only if $G vee K_1$ is planar.*

Proof. Let G be an outerplanar graph and suppose that G is embedded in the
plane such that every vertex of G lies on the boundary of the exterior region. Then
a vertex v can be placed in the exterior region of G and joined to all vertices of G
in such a way that a planar embedding of $G \vee K_1$ results. Thus, $G \vee K_1$ is planar.

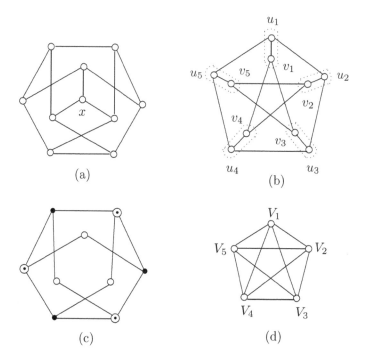

Figure 5.17: Showing that the Petersen graph is nonplanar

For the converse, assume that G is a graph such that $G \vee K_1$ is planar. Hence, $G \vee K_1$ contains a vertex v that is adjacent to every vertex of G. Let there be a planar embedding of $G \vee K_1$. Upon deleting the vertex v, we arrive at a planar embedding of G in which all vertices of G lie on the boundary of the same region. Thus, G is outerplanar. ∎

The following characterization of outerplanar graphs is analogous to the characterization of planar graphs stated in Kuratowski's theorem (Theorem 5.14).

Theorem 5.19 *A graph G is outerplanar if and only if G contains no subgraph that is a subdivision of K_4 or $K_{2,3}$.*

Proof.   Suppose first that there exists some outerplanar graph G that contains a subgraph H that is a subdivision of K_4 or $K_{2,3}$. By Theorem 5.18, $G \vee K_1$ is planar. Since $H \vee K_1$ is a subdivision of K_5 or contains a subdivision $K_{3,3}$, it follows that $G \vee K_1$ contains a subgraph that is a subdivision of K_5 or contains a subdivision $K_{3,3}$ and so is nonplanar. This produces a contradiction.

For the converse, assume, to the contrary, that there exists a graph G that is not outerplanar but contains no subgraph that is a subdivision of K_4 or $K_{2,3}$. By Theorem 5.18, $G \vee K_1$ is not planar, but $G \vee K_1$ contains a subgraph that is a subdivision of K_5 or $K_{3,3}$. This contradicts Theorem 5.14. ∎

An outerplanar graph G is **maximal outerplanar** if the addition to G of any edge joining two nonadjacent vertices of G results in a graph that is not outerplanar. Necessarily then, there is a planar embedding of a maximal outerplanar graph G of order at least 3, the boundary of whose exterior region of is a Hamiltonian cycle of G. We now describe some other facts about outerplanar graphs.

Theorem 5.20 *Every nontrivial outerplanar graph contains at least two vertices of degree 2 or less.*

Proof. Let G be a nontrivial outerplanar graph. The result is obvious if the order of G is 4 or less, so we may assume that the order of G is at least 5. Add edges to G, if necessary, to obtain a maximal outerplanar graph. Thus, the boundary of the exterior region of G is a Hamiltonian cycle of G. Among the chords of C, let uv be one such that uv and a $u - v$ path on C produce a cycle containing a minimum number of interior regions. Necessarily, this minimum is 1. Then the degree of the remaining vertex y on the boundary of this region is 2. There is such a chord wx of C on the other $u - v$ path of C, producing another vertex z of degree 2. In G, the degrees of y and z are therefore 2 or less. ∎

Theorem 5.21 *The size of every outerplanar graph of order $n \geq 2$ is at most $2n - 3$.*

Proof. We proceed by induction on n. The result clearly holds for $n = 2$. Assume that the size of every outerplanar graph of order k, where $k \geq 2$, is at most $2k - 3$ and let G be a outerplanar graph of order $k + 1$. We show that the size of G is at most $2(k + 1) - 3 = 2k - 1$. By Theorem 5.20, G contains a vertex v of degree at most 2. Then $G - v$ is an outerplanar graph of order k. By the induction hypothesis, the size of $G - v$ is at most $2k - 3$. Hence, the size of G is at most

$$2k - 3 + \deg v \leq 2k - 3 + 2 = 2k - 1. \qquad ∎$$

In view of Theorem 5.21, an outerplanar graph of order $n \geq 2$ is maximal outerplanar if and only if its size is $2n - 3$.

5.5 Embedding Graphs on Surfaces

We have seen that a graph G is planar if G can be drawn in the plane in such a way that no two edges cross and that such a drawing is called an *embedding* of G in the plane or a planar embedding. We have also remarked that a graph G can be embedded in the plane if and only if G can be embedded on (the surface of) a sphere.

Of course, not all graphs are planar. Indeed, Kuratowski's theorem (Theorem 5.14) and Wagner's theorem (Theorem 5.17) describe conditions (involving the two nonplanar graphs K_5 and $K_{3,3}$) under which G can be embedded in the plane. Graphs that are not embeddable in the plane (or on a sphere) may be embeddable on other surfaces, however. Another common surface on which a graph may be embedded is the **torus**, a doughnut-shaped surface (see Figure 5.18(a)). Two different

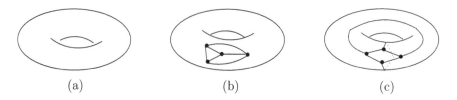

(a)   (b)   (c)

Figure 5.18: Embedding K_4 on a torus

embeddings of the (planar) graph K_4 on a torus are shown in Figures 5.18(b) and 5.18(c).

While it is easy to see that every planar graph can be embedded on a torus, some nonplanar graphs can be embedded on a torus as well. For example, embeddings of K_5 and $K_{3,3}$ on a torus are shown in Figures 5.19(a) and 5.19(b).

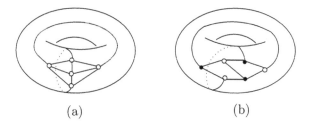

(a)   (b)

Figure 5.19: Embedding K_5 and $K_{3,3}$ on a torus

Another way to represent a torus and to visualize an embedding of a graph on a torus is to begin with a rectangular piece of (flexible) material, as in Figure 5.20, and first make a cylinder from it by identifying sides a and c, which are the same after the identification occurs. Sides b and d are then circles. These circles are then identified to produce a torus.

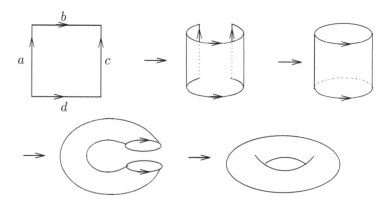

Figure 5.20: Constructing a torus

After seeing how a torus can be constructed from a rectangle, it follows that the points labeled A in the rectangle in Figure 5.21(a) represent the same point on the torus. This is also true of the points labeled B and the points labeled C. Figures 5.21(b) and 5.21(c) show embeddings of K_5 and $K_{3,3}$ on the torus. There are five regions in the embedding of K_5 on the torus shown in Figure 5.21(b) as R is a single region in this embedding. Moreover, there are three regions in the embedding of $K_{3,3}$ on the torus shown in Figure 5.21(c) as R' is a single region.

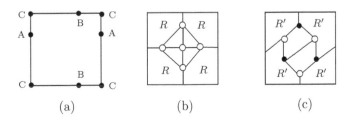

(a)   (b)   (c)

Figure 5.21: Embedding K_5 and $K_{3,3}$ on a torus

Another way to represent a torus and an embedding of a graph on a torus is to begin with a sphere, insert two holes in its surface (as in Figure 5.22(a)), and attach a handle on the sphere, where the ends of the handle are placed over the two holes (as in Figure 5.22(b)). An embedding of K_5 on the torus constructed in this manner is shown in Figure 5.22(c).

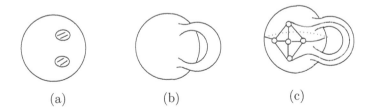

(a)   (b)   (c)

Figure 5.22: Embedding K_5 on a torus

While a torus is a sphere with a handle, a sphere with k handles, $k \geq 0$, is called a **surface of genus** k and is denoted by S_k. Thus, S_0 is a sphere and S_1 is a torus. The surfaces S_k are the **orientable surfaces**.

Let G be a nonplanar graph. When drawing G on a sphere, some edges of G will cross. The graph G can always be drawn so that only two edges cross at any point of intersection. At each such point of intersection, a handle can be suitably placed on the sphere so that one of these two edges pass over the handle and the intersection of the two edges has been avoided. Consequently, every graph can be embedded on some orientable surface. The smallest nonnegative integer k such that a graph G can be embedded on S_k is called the **genus** of G and is denoted by $\gamma(G)$. Therefore, $\gamma(G) = 0$ if and only if G is planar; while $\gamma(G) = 1$ if and only if G is nonplanar but G can be embedded on the torus. In particular,

$$\gamma(K_5) = 1 \text{ and } \gamma(K_{3,3}) = 1.$$

Figure 5.23(a) shows an embedding of a disconnected graph H on a sphere. In this case, $n = 8$, $m = 9$, and $r = 4$. Thus, $n - m + r = 8 - 9 + 4 = 3$. That $n - m + r \neq 2$ is not particularly surprising as the Euler Identity (Theorem 5.1) requires that H be a *connected* plane graph. Although this is a major reason why we will restrict our attention to connected graphs here, it is not the only reason. There is a desirable property possessed by every embedding of a connected planar graph on a sphere that is possessed by no embedding of a disconnected planar graph on a sphere.

Suppose that G is a graph embedded on a surface S_k, $k \geq 0$. A region of this embedding is a **2-cell** if every closed curve in that region can be continuously deformed in that region to a single point. (Topologically, a region is a 2-cell if it is homeomorphic to a disk.) While the closed curve C in R in the embedding of the graph on a sphere shown in Figure 5.23 can in fact be continuously deformed in R to a single point, the curve C' cannot. Hence, R is not a 2-cell in this embedding.

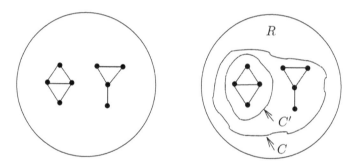

Figure 5.23: An embedding on a sphere that is not a 2-cell embedding

An embedding of a graph G on some surface is a **2-cell embedding** if every region in the embedding is a 2-cell. Consequently, the embedding of the graph shown in Figure 5.23 is not a 2-cell embedding. It turns out, however, that every embedding of a connected graph on a sphere is necessarily a 2-cell embedding. If a connected graph is embedded on a surface S_k where $k > 0$, then the embedding may or may not be a 2-cell embedding, however. For example, the embedding of K_4 in Figure 5.18(b) is not a 2-cell embedding. The curves C and C' shown in Figures 5.24(a) and 5.24(b) cannot be continuously deformed to a single point in the region in which these curves are drawn. On the other hand, the embedding of K_4 shown in Figure 5.18(c) and shown again in Figure 5.24(c) is a 2-cell embedding.

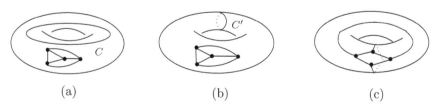

(a)   (b)   (c)

Figure 5.24: Non-2-cell and 2-cell embeddings of K_4 on the torus

The embeddings of K_4, K_5, and $K_{3,3}$ on a torus given in Figures 5.18(c), 5.19(a), and 5.19(b), respectively, are all 2-cell embeddings. Furthermore, in each case, $n - m + r = 0$. As it turns out, if G is a connected graph of order n and size m that is 2-cell embedded on a torus resulting in r regions, then $n - m + r = 0$. This fact together with the Euler Identity (Theorem 5.1) are special cases of a more general result. The mathematician Simon Antoine Jean Lhuilier (1750–1840) spent much of his life working on problems related to the Euler Identity. Lhuilier, like Euler, was from Switzerland and was taught mathematics by one of Euler's former students (Louis Bertrand). Lhuilier saw that the Euler Identity did *not* hold for graphs embedded on spheres containing handles. In fact, he proved a more general form of this identity [139].

Theorem 5.22 (Generalized Euler Identity) *If G is a connected graph of order n and size m that is 2-cell embedded on a surface of genus $k \geq 0$, resulting in r regions, then*

$$n - m + r = 2 - 2k.$$

Proof. We proceed by induction on k. If G is a connected graph of order n and size m that is 2-cell embedded on a surface of genus 0, then G is a plane graph. By the Euler Identity, $n - m + r = 2 = 2 - 2 \cdot 0$. Thus, the basis step of the induction holds.

Assume, for every connected graph G' of order n' and size m' that is 2-cell embedded on a surface S_k $(k \geq 0)$, resulting in r' regions, that

$$n' - m' + r' = 2 - 2k.$$

Let G be a connected graph of order n and size m that is 2-cell embedded on S_{k+1}, resulting in r regions. We may assume, without loss of generality, that no vertex of G lies on any handle of S_{k+1} and that the edges of G are drawn on the handles so that a closed curve can be drawn around each handle that intersects no edge of G more than once.

We now construct a new graph G_1 from G and then construct another graph G_2 from G_1. Let H be one of the $k+1$ handles of S_{k+1}. There are necessarily edges of G on H; for otherwise, the handle belongs to a region R in which case any closed curve around H cannot be continuously deformed in R to a single point, contradicting the assumption that R is a 2-cell. We now draw a closed curve C around H, which intersects some edges of G on H but intersects no edge more than once. Suppose that there are $t \geq 1$ points of intersection of C and the edges on H. These points of intersection will be vertices of G_1, where each of the t edges of G becomes two edges of G_1. Also, the segments of C between vertices become edges of G_1. We add two vertices of degree 2 along C to produce two additional edges of G_1. (This guarantees that the resulting structure will be a graph, not a multigraph.)

The graph just constructed is the desired graph G_1, which has order n_1, size m_1, and r_1 regions. Then

$$n_1 = n + t + 2 \text{ and } m_1 = m + 2t + 2.$$

Since each portion of C that became an edge of G_1 is in a region of G, the addition of such an edge divides that region into two regions, each of which is a 2-cell. Since there are t such edges,

$$r_1 = r + t.$$

We now cut the handle H along C and "patch" the two resulting holes, producing two duplicate copies of the vertices and edges along C (see Figure 5.25). This resulting graph is denoted by G_2, which is now 2-cell embedded on S_k.

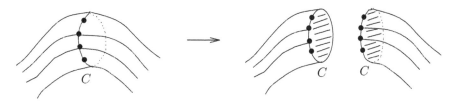

Figure 5.25: Converting a 2-cell embedding of G_1 on S_{k+1}
to a 2-cell embedding of G_1 on S_k

Let G_2 have order n_2, size m_2, and r_2 regions, all of which are 2-cells. Then

$$n_2 = n_1 + t + 2, \ m_2 = m_1 + t + 2, \ \text{and} \ r_2 = r_1 + 2.$$

Furthermore, $n_2 = n + 2t + 4$, $m_2 = m + 3t + 4$, and $r_2 = r + t + 2$. By the induction hypothesis, $n_2 - m_2 + r_2 = 2 - 2k$. Therefore,

$$\begin{aligned} n_2 - m_2 + r_2 &= (n + 2t + 4) - (m + 3t + 4) + (r + t + 2) \\ &= n - m + r + 2 = 2 - 2k. \end{aligned}$$

Therefore, $n - m + r = 2 - 2(k + 1)$.   ∎

The following result [204] was proved by J. W. T. (Ted) Youngs (1910–1970).

Theorem 5.23 *Every embedding of a connected graph G of genus k on S_k, where $k \geq 0$, is a 2-cell embedding.*

With the aid of Theorems 5.22 and 5.23, we have the following.

Corollary 5.24 *If G is a connected graph of order n and size m that is embedded on a surface of genus $\gamma(G)$, resulting in r regions, then*

$$n - m + r = 2 - 2\gamma(G).$$

We now have a corollary of Corollary 5.24.

Theorem 5.25 *If G is a connected graph of order $n \geq 3$ and size m, then*

$$\gamma(G) \geq \frac{m}{6} - \frac{n}{2} + 1.$$

Proof. Suppose that G is embedded on a surface of genus $\gamma(G)$, resulting in r regions. By Corollary 5.24, $n - m + r = 2 - 2\gamma(G)$. Let $R_1, R_2, \ldots, R_r$ be the regions of G and let m_i be the number of edges on the boundary of R_i ($1 \leq i \leq r$). Thus, $m_i \geq 3$. Since every edge is on the boundary of one or two regions, it follows that

$$3r \leq \sum_{i=1}^{r} m_i \leq 2m$$

and so $3r \leq 2m$. Therefore,

$$6 - 6\gamma(G) = 3n - 3m + 3r \leq 3n - 3m + 2m = 3n - m. \tag{5.3}$$

Solving (5.3) for $\gamma(G)$, we have $\gamma(G) \geq \frac{m}{6} - \frac{n}{2} + 1$. ∎

Theorem 5.25 is a generalization of Theorem 5.2, for when G is planar (and so $\gamma(G) = 0$) Theorem 5.25 becomes Theorem 5.2. According to Theorem 5.25,

$$\gamma(K_5) \geq \tfrac{1}{6}, \ \gamma(K_6) \geq \tfrac{1}{2}, \ \text{and} \ \gamma(K_7) \geq 1.$$

This says that all three graphs K_5, K_6, and K_7 are nonplanar. Of course, we already knew by Corollary 5.3 that K_n is nonplanar for every integer $n \geq 5$. We have also seen that $\gamma(K_5) = 1$. Actually, $\gamma(K_7) = 1$ as well. Figure 5.26 shows an embedding of K_7 with vertex set $\{v_1, v_2, \ldots, v_7\}$ on a torus. Because K_6 is nonplanar and K_6 is a subgraph of a graph that can be embedded on a torus, $\gamma(K_6) = 1$.

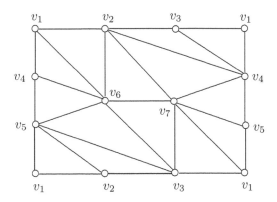

Figure 5.26: An embedding of K_7 on the torus

Applying Theorem 5.25 to a complete graph K_n, $n \geq 3$, we have

$$\gamma(K_n) \geq \frac{\binom{n}{2}}{6} - \frac{n}{2} + 1 = \frac{(n-3)(n-4)}{12}.$$

Since $\gamma(K_n)$ is an integer,

$$\gamma(K_n) \geq \left\lceil \frac{(n-3)(n-4)}{12} \right\rceil.$$

Gerhard Ringel (1919–2008) and J. W. T. Youngs [167] completed a lengthy proof involving many people over a period of many years (see Chapter 8 also) that showed this lower bound for $\gamma(K_n)$ is in fact the value of $\gamma(K_n)$.

Theorem 5.26 *For every integer $n \geq 3$,*

$$\gamma(K_n) = \left\lceil \frac{(n-3)(n-4)}{12} \right\rceil.$$

Ringel [166] also discovered a formula for the genus of every complete bipartite graph.

Theorem 5.27 *For every two integers $r, s \geq 2$,*

$$\gamma(K_{r,s}) = \left\lceil \frac{(r-2)(s-2)}{4} \right\rceil.$$

In particular, Theorem 5.27 implies that a complete bipartite graph G can be embedded on a torus if and only if G is planar or is a subgraph of $K_{4,4}$ or $K_{3,6}$.

There are other kinds of surfaces on which graphs can be embedded. The **Möbius strip** (or **Möbius band**) is a one-sided surface that can be constructed from a rectangular piece of material by giving the rectangle a half-twist (or a rotation through 180°) and then identifying opposite sides of the rectangle (see Figure 5.27). Thus, A represents the same point on the Möbius strip. The Möbius strip is named for the German mathematician August Ferdinand Möbius who, as we noted in Chapter 0, discovered it in 1858 (even though the mathematician Johann Benedict Listing discovered it shortly before Möbius).

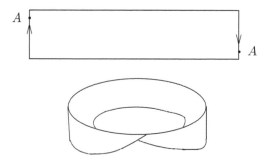

Figure 5.27: The Möbius strip

Certainly every planar graph can be embedded on the Möbius strip. Figure 5.28 shows that $K_{3,3}$ can also be embedded on the Möbius strip.

Of more interest are the nonorientable surfaces (the nonorientable 2-dimensional manifolds), the simplest example of which is the projective plane. The **projective plane** can be represented by identifying opposite sides of a rectangle in the manner shown in Figure 5.29(a). Note that A represents the same point in the projective plane, as does B. Figure 5.29(b) shows the embedding of K_5 on the projective plane.

The projective plane can also be represented by a circle where antipodal pairs of points on the circumference are the same point. Using this representation, we can give the embedding of K_6 on the projective plane shown in Figure 5.30.

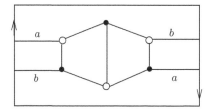

Figure 5.28: An embedding $K_{3,3}$ on the Möbius strip

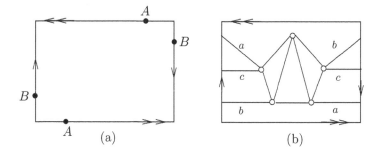

Figure 5.29: An embedding of K_5 on the projective plane

For the embedding of K_5 on the projective plane shown in Figure 5.29(b), $n = 5$, $m = 10$, and $r = 6$; while for the embedding of K_6 shown in Figure 5.30, $n = 6$, $m = 15$, and $r = 10$. In both cases, $n - m + r = 1$. In fact, for any graph of order n and size m that is 2-cell embedded on the projective plane, resulting in r regions,

$$n - m + r = 1.$$

5.6 The Graph Minor Theorem

We have seen by Wagner's theorem (Theorem 5.17) that a graph G is planar if and only if neither K_5 nor $K_{3,3}$ is a minor of G. That is, Wagner's theorem is a **forbidden minor** characterization of planar graphs – in this case two forbidden minors: K_5 and $K_{3,3}$. A natural question to ask is whether a forbidden minor characterization may exist for graphs embedded on other surfaces.

It was shown by Daniel Archdeacon and Philip Huneke [9] that there are exactly 35 forbidden minors for graphs that can be embedded on the projective plane. In recent years, much more general results involving minors have been obtained. The following theorem of Neil Robertson and Paul Seymour [168] has numerous consequences. Its long proof is a consequence of a sequence of several papers that required years to complete.

Theorem 5.28 (**Robertson-Seymour Theorem**) *For every infinite sequence $G_1, G_2, \ldots$ of graphs, there exist graphs G_i and G_j with $i < j$ such that G_i is a minor of G_j.*

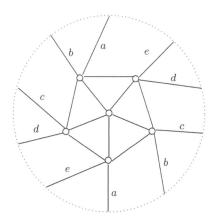

Figure 5.30: An embedding of K_6 on the projective plane

A sequence $G_1, G_2, G_3, \ldots$ of graphs is a **descending chain of proper minors** if G_{i+1} is a proper minor of G_i for every positive integer i. An immediate consequence of Theorem 5.28 is the following.

Corollary 5.29 *There is no infinite descending chain of proper minors.*

Another consequence of Theorem 5.28, however, is one of the major theorems in graph theory. A set S of graphs is said to be **minor-closed** if for every graph G in S, every minor of G also belongs to S.

Theorem 5.30 (**Graph Minor Theorem**) *Let S be a minor-closed set of graphs. Then there exists a finite set M of graphs such that $G \in S$ if and only if no graph in M is a minor of G.*

Proof. Define M to be the set of all graphs F in the complement $\overline{S}$ of S such that every proper minor of F is in S. We claim that this set M has the required properties. First, we show that $G \in S$ if and only if no graph in M is a minor of G.

Suppose, first, that there is a graph $G \in S$ such that some graph F belonging to M is a minor of G. Since $G \in S$ and S is minor-closed, it follows that $F \in S$. This, however, contradicts the assumption that $F \in M$ and so $F \in \overline{S}$.

For the converse, assume to the contrary that there is a graph $G \in \overline{S}$ such that no graph in M is a minor of G. We consider two cases.

Case 1. All of the proper minors of G are in S. Then by the defining property of M, it follows that $G \in M$. Since $G \in M$ and G is a minor of itself, this contradicts our assumption that no graph in M is a minor of G.

Case 2. Some proper minor of G, say G', is not in S. Thus, $G' \in \overline{S}$. Then G' either satisfies the condition of Case 1 or Case 2. If we continue in this manner as long as we remain in Case 2, a chain of proper minors is produced.

If this process terminates, we have the finite sequence

$$G, G' = G^{(1)}, \ldots, G^{(p)},$$

where each graph in the sequence is a proper minor of all those graphs that precede it. Then $G^{(p)} \in M$, which returns us to Case 1. Hence, we have an infinite sequence $G, G' = G^{(1)}, G^{(2)}, \ldots$, where each graph is a proper minor of all those graphs that precede it. This, however, contradicts Corollary 5.29.

It remains only to show that M is finite. Assume, to the contrary, that M is infinite. Let $G_1, G_2, G_3, \ldots$ be any sequence of graphs belonging to M. By the Robertson-Seymour theorem, there are integers i and j with $i < j$ such that G_i is a minor of G_j. However, each graph in M has no proper minor in $\overline{S}$ and consequently no proper minor in M as well. This is a contradiction. ∎

We now return to the question about the existence of a forbidden minor characterization for graphs embeddable on a surface S_k of genus $k \geq 0$. Certainly, if G is a sufficiently small graph (in terms of order and/or size), then G can be embedded on S_k. Hence, if we begin with a graph F that cannot be embedded on S_k and perform successive edge contractions, edge deletions, and vertex deletions, then eventually we arrive at a graph F' that also cannot be embedded on S_k but such that any additional edge contraction, edge deletion, or vertex deletion of F' produces a graph that *can* be embedded on S_k. Such a graph F' is said to be **minimally nonembeddable on** S_k. Consequently, a graph F' is minimally nonembeddable on S_k if F' cannot be embedded on S_k but every proper minor F' can be embedded on S_k. Thus, the set of graphs embeddable on S_k is minor-closed. As a consequence of the Graph Minor Theorem, we have the following.

Theorem 5.31 *For each integer $k \geq 0$, the set of minimally nonembeddable graphs on S_k is finite.*

Although the number of minimally nonembeddable graphs on the torus is finite, it is known that this number exceeds 800.

Exercises for Chapter 5

1. Prove for every planar graph G of order 3 or more that there exists a partition $\{V_1, V_2, V_3\}$ of $V(G)$ such that each subgraph $G_i = G[V_i]$ ($i = 1, 2, 3$) induced by V_i in G is a forest.

2. Give an example of two non-isomorphic maximal planar graphs of the same order.

3. The graph of the tetrahedron (K_4) is a 3-regular maximal planar graph of order 4 and the graph of the icosahedron is a 4-regular maximal planar graph of order 6. The graph $K_5 - e$ is a maximal planar graph of order 5, all of whose vertices have degree 3 or 4. See Figure 5.6. Give an example of a maximal planar graph of order 7 or more all of whose vertices have degree 3 or 4 or show no such graph exists.

4. All four maximal planar graphs of order 3 or more in Figure 5.6 are Hamiltonian. For a maximal planar graph of G of order 3 or more, the graph G^* is obtained from G by placing a vertex v_R in each region R and joining v_R to the three vertices on the boundary of R.

   (a) If G is Hamiltonian, is G^* Hamiltonian?

   (a) If G^* is Hamiltonian, is G Hamiltonian?

5. Let G be a connected cubic plane graph of order n and size m having r regions such that the boundary of each region is either a 5-cycle or a 6-cycle. Determine the number of regions whose boundary is a 5-cycle.

6. Use a discharging method to prove Theorem 5.11: *If G is a maximal planar graph of order 4 or more, then G contains at least one of the following:*

   (1) *a vertex of degree 3, (2) a vertex of degree 4, (3) a vertex of degree 5 that is adjacent to two vertices, each of which has degree 5 or 6.*

7. Determine all connected regular planar graphs G such that the number of regions in a planar embedding of G equals its order.

8. Determine all maximal planar graphs G of order 3 or more such that the number of regions in a planar embedding of G equals its order.

9. Determine whether the graph G shown in Figure 5.31 is nearly maximal planar.

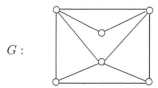

G :

Figure 5.31: The graph G in Exercise 9

10. Show that every graph G of order $n \geq 6$ that contains three spanning trees T_1, T_2, and T_3 such that every edge of G belongs to exactly one of these three trees is nonplanar.

11. If the complement of a nontrivial maximal planar graph G is a spanning tree, then what is the order of G?

12. Consider the plane graph G in Figure 5.32. What is the minimum number of colors needed so that each edge of G is assigned a color and the edges on the boundary of each region of G are colored differently?

13. Use Grinberg's theorem to show that $K_{2,3}$ is not Hamiltonian.

14. Use Grinberg's theorem to show that each of the graphs in Figure 5.33 is not Hamiltonian.

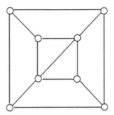

Figure 5.32: The graph G in Exercise 12

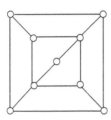 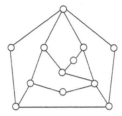

Figure 5.33: Graphs in Exercise 14

15. Let G be the graph shown in Figure 5.34.

   (a) Show that G contains a $K_{3,3}$ as a subgraph.

   (b) Show that G does not contain a subdivision of K_5 as a subgraph.

   (c) Show that K_5 is a minor of G.

G :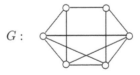

Figure 5.34: The graph G in Exercise 15

16. Let H be the graph shown in Figure 5.35.

   (a) For the order n and size m of H, compare m and $3n - 6$. What does this comparison tell you about the planarity of H?

   (b) Show that H does not contain K_5 as a subgraph.

   (c) Show that either (1) H contains a subdivision of K_5 or (2) K_5 is a minor of H.

17. (a) What is the minimum possible order of a graph G containing only vertices of degree 3 and degree 4 and an equal number of each such that G contains a subdivision of K_5?

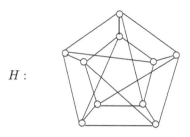

Figure 5.35: The graph H in Exercise 16

(b) Does the graph H of Figure 5.36 contain a subdivision of K_5 or a subdivision of $K_{3,3}$?

(c) Does the graph H of Figure 5.36 contain K_5 or $K_{3,3}$ as a minor?

(d) Is the graph H of Figure 5.36 planar or nonplanar?

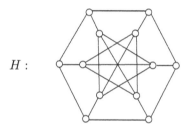

Figure 5.36: The graph H in Exercise 17

18. Prove or disprove: If a graph H is a minor of a planar graph, then H is planar.

19. Determine all connected graphs G of order $n \geq 4$ such that $G \vee K_1$ is outerplanar.

20. Use Theorem 5.18 to prove Theorem 5.21: *The size of every outerplanar graph of order $n \geq 2$ is at most $2n - 3$.*

21. Let $S_{a,b}$ denote the double star in which the degree of the two vertices that are not end-vertices are a and b. Determine all pairs a, b of integers such that $\overline{S}_{a,b}$ is planar.

22. A nonplanar graph G of order 7 has the property that $G - v$ is planar for every vertex v of G.

    (a) Show that G does not contain $K_{3,3}$ as a subgraph.

    (b) Give an example of a graph with this property.

23. A graph G of order 14 and size 48 is 2-cell embedded on the double torus (the surface of genus 2).

    (a) How many regions are there in this embedding?

    (b) Can G be embedded on the torus?

    (c) What is the genus of G?

    (d) Let u and v be two nonadjacent vertices in G. What is the genus of $G + uv$?

24. (a) Show that there is only one regular maximal planar graph G whose order $n \in \{5, 6, \ldots, 11\}$.

    (b) For the graph G in (a), show that $\overline{G}$ has a perfect matching M. Determine the genus of the graph $G + M$.

    (c) Prove that if G is a maximal planar graph G of order $n > 4$ whose complement contains a perfect matching M, then the genus

$$\gamma(G + M) \geq \frac{n}{12}.$$

25. By Theorem 5.26, $\gamma(K_7) = 1$. Let there be an embedding of K_7 on the torus, and let R_1 and R_2 be two neighboring regions. Let G be the graph obtained by adding a new vertex v in R_1 and joining v to the vertices on the boundaries of both R_1 and R_2. What is $\gamma(G)$?

26. The graph H is a certain 6-regular graph of order 12. It is known that $G = H \square K_2$ can be embedded on S_3. What is $\gamma(G)$?

27. A certain graph H of order 12 has 6 vertices of degree 6 and 6 vertices of degree 8. It is known that $G = H \square K_2$ can be embedded on S_5. What is $\gamma(G)$?

28. For a 7-regular graph H of order 12, it is known that $G = H \square K_2$ can be embedded on S_5. What is $\gamma(G)$?

29. It is known that the Petersen graph P is not planar. Thus, P cannot be embedded on the sphere.

    (a) Show that P can be embedded on the torus, however.

    (b) How many regions does P have when it is embedded on the torus?

30. (a) Show that the set $\mathcal{F}$ of forests is a minor-closed family of graphs.

    (b) What are the forbidden minors of $\mathcal{F}$?

31. Prove for each positive integer n that there exists a sequence $G_1, G_2, \ldots, G_n$ of graphs such that if $1 \leq i < j \leq n$, then G_i is not a minor of G_j. How is this related to the Robertson-Seymour theorem?

32. Use the Robertson-Seymour theorem (Theorem 5.28) to show for any infinite sequence $G_1, G_2, G_3, \ldots$ of graphs, that there exist infinitely many pairwise disjoint 2-element sets $\{i, j\}$ of integers with $i < j$ such that G_i is a minor of G_j.

33. Use the Robertson-Seymour theorem (Theorem 5.28) to prove Corollary 5.29: *There is no infinite descending chain of proper minors.*

Chapter 6

Introduction to Vertex Colorings

There is little doubt that the best known and most studied area within graph theory is coloring.

> *Graph coloring is arguably the most popular subject in graph theory.*
>
> *Noga Alon* (1993)

The remainder of this book is devoted to this important subject. Dividing a given set into subsets is a fundamental procedure in mathematics. Often the subsets are required to satisfy some prescribed property – but not always. When the set is associated with a graph in some manner, then we are dealing with graph colorings. With its origins embedded in attempts to solve the famous Four Color Problem (see Chapter 0), graph colorings has become a subject of great interest, largely because of its diverse theoretical results, its unsolved problems, and its numerous applications.

The problems in graph colorings that have received the most attention involve coloring the vertices of a graph. Furthermore, the problems in vertex colorings that have been studied most often are those referred to as proper vertex colorings. We begin with these.

6.1 The Chromatic Number of a Graph

A **proper vertex coloring** of a graph G is an assignment of colors to the vertices of G, one color to each vertex, so that adjacent vertices are colored differently. When it is understood that we are dealing with a proper vertex coloring, we ordinarily refer to this more simply as a **coloring** of G. While the colors used can be elements of any set, actual colors (such as red, blue, green, and yellow) are often chosen only when a small number of colors are being used; otherwise, positive integers (typically $1, 2, \ldots, k$ for some positive integer k) are commonly used for the colors. A reason for using positive integers as colors is that we are often interested in the number of colors being used. Thus, a (proper) coloring can be considered as a function $c : V(G) \to \mathbb{N}$ (where $\mathbb{N}$ is the set of positive integers) such that $c(u) \neq c(v)$ if u and v are adjacent in G. If each color used is one of k given colors, then we refer to the coloring as a k-**coloring**. In a k-coloring, we may then assume that it is the colors $1, 2, \ldots, k$ that are being used. While all k colors are typically used in a k-coloring of a graph, there are occasions when only some of the k colors are used.

Suppose that c is a k-coloring of a graph G, where each color is one of the integers $1, 2, \ldots, k$ as mentioned above. If V_i $(1 \leq i \leq k)$ is the set of vertices in G colored i (where one or more of these sets may be empty), then each nonempty set V_i is called a **color class** and the nonempty elements of $\{V_1, V_2, \ldots, V_k\}$ produce a partition of $V(G)$. Because no two adjacent vertices of G are assigned the same color by c, each nonempty color class V_i $(1 \leq i \leq k)$ is an independent set of vertices of G.

A graph G is k-**colorable** if there exists a k-coloring of G. The minimum positive integer k for which G is k-colorable is the **chromatic number** of G and is denoted by $\chi(G)$. (The symbol χ is the Greek letter *chi*.) The chromatic number of a graph G is therefore the minimum number of independent sets into which $V(G)$ can be partitioned. A graph G with chromatic number k is called a k-**chromatic graph**. Therefore, if $\chi(G) = k$, then there exists a k-coloring of G but not a $(k-1)$-coloring. In fact, a graph G is k-colorable if and only if $\chi(G) \leq k$. Certainly, every graph of order n is n-colorable. Necessarily, if a k-coloring of a k-chromatic graph G is given, then all k colors must be used.

Three different colorings of a graph H are shown in Figure 6.1. The coloring in Figure 6.1(a) is a 5-coloring, the coloring in Figure 6.1(b) is a 4-coloring, and the coloring in Figure 6.1(c) is a 3-coloring. Because the order of G is 9, the graph H is k-colorable for every integer k with $3 \leq k \leq 9$. Since H is 3-colorable, $\chi(H) \leq 3$. There is, however, no 2-coloring of H because H contains triangles and the three vertices of each triangle must be colored differently. Therefore, $\chi(H) \geq 3$ and so $\chi(H) = 3$.

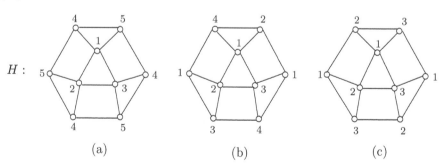

Figure 6.1: Colorings of a graph H

The argument used to verify that the graph H of Figure 6.1 has chromatic number 3 is a common one in graph theory. In general, to show that some graph G has chromatic number k, say, we need to show that there exists a k-coloring of G

(and so $\chi(G) \leq k$) and to show that every coloring of G requires at least k colors (and so $\chi(G) \geq k$).

There is no general formula for the chromatic number of a graph. Consequently, we will often be concerned and must be content with (1) determining the chromatic number of some specific graphs of interest or of graphs belonging to some classes of interest and (2) determining upper and/or lower bounds for the chromatic number of a graph. Certainly, for every graph G of order n,

$$1 \leq \chi(G) \leq n.$$

A rather obvious, but often useful, lower bound for the chromatic number of a graph involves the chromatic numbers of its subgraphs.

Theorem 6.1 *If H is a subgraph of a graph G, then $\chi(H) \leq \chi(G)$.*

Proof. Suppose that $\chi(G) = k$. Then there exists a k-coloring c of G. Since c assigns distinct colors to every two adjacent vertices of G, the coloring c also assigns distinct colors to every two adjacent vertices of H. Therefore, H is k-colorable and so $\chi(H) \leq k = \chi(G)$. ∎

Recall that the **clique number** $\omega(G)$ of a graph G is the order of the largest clique (complete subgraph) of G. The following result is an immediate consequence of Theorem 6.1.

Corollary 6.2 *For every graph G, $\chi(G) \geq \omega(G)$.*

The lower bound for the chromatic number of a graph in Corollary 6.2 is related to a much-studied class of graphs called "perfect graphs", which will be visited in Section 6.3.

Two operations on graphs that are often encountered are the union and join. The chromatic number of a graph that is the union of graphs $G_1, G_2, \ldots, G_k$ can be easily expressed in terms of the chromatic numbers of these k graphs (see Exercise 9).

Proposition 6.3 *For graphs $G_1, G_2, \ldots, G_k$ and $G = G_1 + G_2 + \cdots + G_k$,*

$$\chi(G) = \max\{\chi(G_i) : 1 \leq i \leq k\}.$$

The following is then an immediate consequence of Proposition 6.3.

Corollary 6.4 *If G is a graph with components $G_1, G_2, \ldots, G_k$, then*

$$\chi(G) = \max\{\chi(G_i) : 1 \leq i \leq k\}.$$

There is a result analogous to Corollary 6.4 that expresses the chromatic number of a graph in terms of the chromatic numbers of its blocks (see Exercise 10).

Proposition 6.5 *If G is a connected graph with blocks $B_1, B_2, \ldots, B_k$, then*

$$\chi(G) = \max\{\chi(B_i) : 1 \leq i \leq k\}.$$

Corollary 6.4 and Proposition 6.5 tell us that we can restrict our attention to 2-connected graphs when studying the chromatic number of graphs. In the case of joins, we have the following (see Exercise 11).

Proposition 6.6 *For graphs $G_1, G_2, \ldots, G_k$ and $G = G_1 \vee G_2 \vee \cdots \vee G_k$,*

$$\chi(G) = \sum_{i=1}^{k} \chi(G_i).$$

Every k-partite graph, $k \geq 2$, is k-colorable because the vertices in each partite set can be assigned one of k distinct colors. Thus, if G is a k-partite graph, then $\chi(G) \leq k$. On the other hand, the following statement is a consequence of Proposition 6.6.

> *Every complete k-partite graph has chromatic number k.*

Since the complete graph K_n is trivially a complete n-partite graph, $\chi(K_n) = n$. Furthermore, if G is a graph of order n that is not complete, then assigning the color 1 to two nonadjacent vertices of G and distinct colors to the remaining $n - 2$ vertices of G produces an $(n - 1)$-coloring of G. Therefore:

> *A graph G of order n has chromatic number n if and only if $G = K_n$.*

Since at least two colors are needed to color the vertices of a graph G only when G contains at least one pair of adjacent vertices, it follows that

> *A graph G of order n has chromatic number 1 if and only if $G = \overline{K}_n$.*

Thus, for a graph G to have chromatic number 2, G must have at least one edge. Also, there must be some way to partition $V(G)$ into two independent subsets V_1 (the vertices of G colored 1) and V_2 (the vertices of G colored 2). Since every edge of G must join a vertex of V_1 and a vertex of V_2, the graph G is bipartite. That is:

> *A nonempty graph G has chromatic number 2 if and only if G is bipartite.*

From these observations, we have the following.

Proposition 6.7 *A nontrivial graph G is 2-colorable if and only if G is bipartite.*

By Theorem 1.13, an alternative way to state Proposition 6.7 is the following:

> *If every vertex of a graph G lies on no odd cycle, then $\chi(G) \leq 2$.*

Stated in this manner, Proposition 6.7 can be generalized. The following upper bound is equivalent to one obtained by Stephen C. Locke [143].

Theorem 6.8 *If every vertex of a graph G lies on at most k odd cycles for some nonnegative integer k, then*

$$\chi(G) \leq \left\lceil \frac{1 + \sqrt{8k + 9}}{2} \right\rceil.$$

Proof. If $k = 0$, then G is bipartite. Thus, $\chi(G) \leq 2$ and the theorem follows. Hence, we may assume that $k \geq 1$. Let $t = \lceil (1 + \sqrt{8k+9})/2 \rceil$. Since $k \geq 1$, it follows that $t \geq 3$.

We proceed by induction on the order n of the graph. Since $\chi(G) \leq t$ for all graphs of order t or less, the basis step holds. Assume for an integer n with $n \geq t$ that if H is any graph of order n with the property that every vertex of H lies on at most k odd cycles, then $\chi(H) \leq t$. We show that the statement holds for graphs of order $n+1$. Let G be a graph of order $n+1$ having the property that every vertex of G lies on at most k odd cycles. We show that $\chi(G) \leq t$.

Let v be a vertex of G. Then $G - v$ has order n and every vertex of $G - v$ lies on at most k odd cycles. By the induction hypothesis, $\chi(G - v) \leq t$. Let there be given a t-coloring of $G - v$. Since $t = \lceil (1 + \sqrt{8k+9})/2 \rceil$, it follows that $(1 + \sqrt{8k+9})/2 \leq t$ and so

$$k \leq \tfrac{t^2 - t - 2}{2} = \binom{t}{2} - 1.$$

Since the vertex v lies on at most k odd cycles, there are at most k pairs of distinct colors used to color the neighbors of v on odd cycles containing v. However, since $k \leq \binom{t}{2} - 1$, it follows that there is at least one pair, say {red, blue}, of distinct colors that are not used to color the two neighbors of v in any odd cycle of G containing v. Let G' be the subgraph of $G - v$ induced by those vertices of G colored red or blue. Necessarily, G' is a bipartite graph.

If no red vertex of G' is a neighbor of v in G, then v can be colored red and so $\chi(G) \leq t$. Hence, we may assume that G' contains one or more red vertices that are neighbors of v in G. Let $G_1', G_2', \ldots, G_s'$ ($s \geq 1$) be the components of G' containing a red neighbor of v in G.

We next show that none of these components of G' also contains a blue neighbor of v in G. Assume, to the contrary, that some component G_i' ($1 \leq i \leq s$) contains a red vertex u and a blue vertex w that are both neighbors of v in G. Then G' contains a $u - w$ path P. Since G' is bipartite, P is necessarily of odd length. The cycle C obtained from P by adding the vertex v and the two edges uv and vw is an odd cycle of G containing v where the two neighbors of v on C are colored red and blue, which is impossible. Thus, as claimed, none of the components $G_1', G_2', \ldots, G_s'$ of G' contains a blue neighbor of v in G. Interchanging the colors red and blue in each of these components produces a t-coloring of $G - v$ in which no neighbor of v is colored red. Assigning v the color red yields a t-coloring of G. Hence, $\chi(G) \leq t$. ∎

As a consequence of Proposition 6.7 and Theorem 1.13, it follows that:

A graph G has chromatic number at least 3 if and only if G contains an odd cycle.

Certainly every even cycle is 2-chromatic and the chromatic number of every odd cycle is at least 3. The coloring c defined on the vertices of an odd cycle $C_n = (v_1, v_2, \ldots, v_n, v_1)$ by

$$c(v_i) = \begin{cases} 1 & \text{if } i \text{ is odd and } 1 \leq i < n \\ 2 & \text{if } i \text{ is even} \\ 3 & \text{if } i = n \end{cases}$$

is a 3-coloring. Thus, we have the following.

Proposition 6.9 *For every integer $n \geq 3$,*

$$\chi(C_n) = \begin{cases} 2 & \textit{if } n \textit{ is even} \\ 3 & \textit{if } n \textit{ is odd.} \end{cases}$$

We have already noted that the bound in Theorem 6.8 is sharp when $k = 0$; for if no vertex of a nontrivial graph G lies on an odd cycle, then G is a bipartite graph and $\chi(G) \leq 2$. If G is itself an odd cycle, then $k = 1$ in Theorem 6.8, producing the bound $\chi(G) \leq 3$, which again is sharp. The graph $G = K_4$ has the property that every vertex of G lies on three triangles. Thus, $k = 3$ in Theorem 6.8, yielding $\chi(G) \leq 4$, again a sharp bound.

Many bounds (both upper and lower bounds) have been developed for the chromatic number of a graph. Two of the most elementary bounds for the chromatic number of a graph G involve the independence number $\alpha(G)$, which, recall, is the maximum cardinality of an independent set of vertices of G. The lower bound stated in the following theorem is especially useful.

Theorem 6.10 *If G is a graph of order n, then*

$$\frac{n}{\alpha(G)} \leq \chi(G) \leq n - \alpha(G) + 1.$$

Proof. Suppose that $\chi(G) = k$ and let there be given a k-coloring of G with resulting color classes $V_1, V_2, \ldots, V_k$. Since

$$n = |V(G)| = \sum_{i=1}^{k} |V_i| \leq k\alpha(G),$$

it follows that

$$\frac{n}{\alpha(G)} \leq k = \chi(G).$$

Next, let U be a maximum independent set of vertices of G and assign the color 1 to each vertex of U. Assigning distinct colors different from 1 to each vertex of $V(G) - U$ produces a proper coloring of G. Hence,

$$\chi(G) \leq |V(G) - U| + 1 = n - \alpha(G) + 1. \qquad \blacksquare$$

According to Theorem 6.10, for the complete 3-partite graph $G = K_{1,2,3}$, which has order $n = 6$ and independence number $\alpha(G) = 3$, we have

$$n/\alpha(G) = 2 \leq \chi(G) \leq 4 = n - \alpha(G) + 1.$$

Since G is a complete 3-partite graph, it follows that $\chi(G) = 3$ and so neither bound in Theorem 6.10 is attained in this case. On the other hand, $\omega(G) = 3$ and so $\chi(G) = \omega(G)$.

For the graph G of order 10 shown in Figure 6.2(a), we have $\alpha(G) = 2$ and $\omega(G) = 4$. By Corollary 6.2, $\chi(G) \geq 4$; while, according to Theorem 6.10, $5 \leq \chi(G) \leq 9$. However, the 5-coloring of G in Figure 6.2(b) shows that $\chi(G) \leq 5$ and so $\chi(G) = 5$.

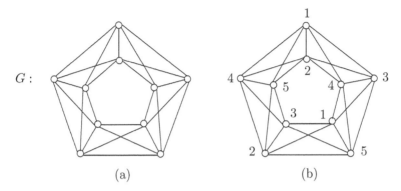

G :

(a) (b)

Figure 6.2: A 5-chromatic graph G with $\alpha(G) = 2$ and $\omega(G) = 4$

Much of Chapter 7 will be devoted to bounds for the chromatic number of a graph.

6.2 Applications of Colorings

There are many problems that can be analyzed and sometimes solved by modeling the situation described in the problem by a graph and defining a vertex coloring of the graph in an appropriate manner. We consider a number of such problems in this section.

From a given group of individuals, suppose that some committees have been formed where an individual may belong to several different committees. A meeting time is to be assigned for each committee. Two committees having a member in common cannot meet at the same time. A graph G can be constructed from this situation in which the vertices are the committees and two vertices are adjacent if the committees have a member in common. Let's look at a specific example of this.

Example 6.11 *At a gathering of eight employees of a company, which we denote by $A = \{a_1, a_2, \ldots, a_8\}$, it is decided that it would be useful to have these individuals meet in committees of three to discuss seven issues of importance to the company. The seven committees selected for this purpose are*

$$A_1 = \{a_1, a_2, a_3\}, \ A_2 = \{a_2, a_3, a_4\}, \ A_3 = \{a_4, a_5, a_6\}, \ A_4 = \{a_5, a_6, a_7\},$$
$$A_5 = \{a_1, a_7, a_8\}, \ A_6 = \{a_1, a_4, a_7\}, \ A_7 = \{a_2, a_6, a_8\}.$$

If each committee is to meet during one of the time periods

$$\text{1-2 pm, 2-3 pm, 3-4 pm, 4-5 pm, 5-6 pm,}$$

then what is the minimum number of time periods needed for all seven committees to meet?

Solution.   No two committees can meet during the same period if some employee belongs to both committees. Define a graph G whose vertex set is

$$V(G) = \{A_1, A_2, \ldots, A_7\},$$

where two vertices A_i and A_j are adjacent if $A_i \cap A_j \neq \emptyset$ (and so A_i and A_j must meet during different time periods). The graph G is shown in Figure 6.3. The answer to the question posed in the example is therefore $\chi(G)$. Since each committee consists of three members and there are only eight employees in all, it follows that the independence number of G is $\alpha(G) = 2$. By Theorem 6.10, $\chi(G) \geq n/\alpha(G) = 7/2$ and so $\chi(G) \geq 4$. Since there is a 4-coloring of G, as shown in Figure 6.3, it follows that $\chi(G) = 4$. Hence, the minimum number of time periods needed for all seven committees to meet is 4. According to the resulting color classes, one possibility for these meetings is

1-2 pm: A_1, A_4;   2-3 pm: A_2, A_5;   3-4 pm: A_3;   4-5 pm: A_6, A_7. ◆

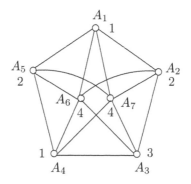

Figure 6.3: The graph G in Example 6.11

Example 6.12 *In a rural community, there are ten children (denoted by c_1, c_2, ..., c_{10}) living in ten different homes who require physical therapy sessions during the week. Ten physical therapists in a neighboring city have volunteered to visit some of these children one day during the week but no child is to be visited twice on the same day. The set of children visited by a physical therapist on any one day is referred to as a tour. It is decided that an optimal number of children to visit on a tour is 4. The following ten tours are agreed upon:*

$$T_1 = \{c_1, c_2, c_3, c_4\}, \quad T_2 = \{c_3, c_5, c_7, c_9\}, \quad T_3 = \{c_1, c_2, c_9, c_{10}\},$$
$$T_4 = \{c_4, c_6, c_7, c_8\}, \quad T_5 = \{c_2, c_5, c_9, c_{10}\}, \quad T_6 = \{c_1, c_4, c_6, c_8\},$$
$$T_7 = \{c_3, c_4, c_8, c_9\}, \quad T_8 = \{c_2, c_5, c_7, c_{10}\}, \quad T_9 = \{c_5, c_6, c_8, c_{10}\},$$
$$T_{10} = \{c_6, c_7, c_8, c_9\}.$$

It would be preferred if all ten tours can take place during Monday through Friday but the physical therapists are willing to work on the weekend if necessary. Is it necessary for someone to work on the weekend?

Solution. A graph G is constructed with vertex set $\{T_1, T_2, \ldots, T_{10}\}$, where T_i is adjacent to T_j $(i \neq j)$ if $T_i \cap T_j \neq \emptyset$. (See Figure 6.4.) The minimum number of days needed for these tours is $\chi(G)$. Since $\{T_2, T_3, T_5, T_7, T_9, T_{10}\}$ induces a maximum clique in G, it follows that $\omega(G) = 6$. By Theorem 6.2, $\chi(G) \geq 6$. There is a 6-coloring of G (see Figure 6.4) and so $\chi(G) = 6$. Thus, visiting all ten children requires six days and it is necessary for some physical therapist to work on the weekend. ♦

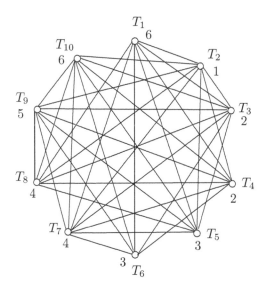

Figure 6.4: The graph G in Example 6.12

Example 6.13 *At a regional airport there is a facility that is used for minor routine maintenance of airplanes. This facility has four locations available for this purpose and so four airplanes can conceivably be serviced at the same time. This facility is open on certain days from 7 am to 7 pm. Performing this maintenance requires $2\frac{1}{2}$ hours per airplane; however, three hours are scheduled for each plane. A certain location may be scheduled for two different planes if the exit time for one plane is the same as the entrance time for the other. On a particular day, twelves airplanes, denoted by $P_1, P_2, \ldots, P_{12}$, are scheduled for maintenance during the indicated time periods:*

P_1 : 11 am - 2 pm;	P_2 : 3 pm - 6 pm;	P_3 : 8 am - 11 am;
P_4 : 1 :30 pm - 4 :30 pm;	P_5 : 1 pm - 4 pm;	P_6 : 2 pm - 5 pm;
P_7 : 9 :30 am - 12 :30 pm;	P_8 : 7 am - 10 am;	P_9 : noon - 3 pm;
P_{10} : 4 pm - 7 pm;	P_{11} : 10 am - 1 pm;	P_{12} : 9 am - noon.

Can a maintenance schedule be constructed for all twelve airplanes?

Solution. A graph G is constructed whose vertex set is the set of airplanes, that is, $V(G) = \{P_1, P_2, \ldots, P_{12}\}$. Two vertices P_i and P_j $(i \neq j)$ are adjacent if their scheduled maintenance periods overlap (see Figure 6.5). Since there are only four locations available for maintenance, the question is whether the graph G is 4-colorable. In fact, $\chi(G) = \omega(G) = 4$, where $\{P_1, P_7, P_{11}, P_{12}\}$ induces a maximum clique in G. Ideally, it would be good if each color class has the same number of vertices (namely three) so that each maintenance crew services the same number of planes during the day. The 4-coloring of G shown in Figure 6.5 has this desired property.   ◆

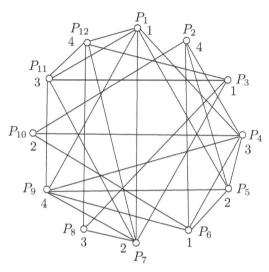

Figure 6.5: The graph G in Example 6.13

Example 6.14 *Two dentists are having new offices designed for themselves. In the common waiting room for their patients, they have decided to have an aquatic area containing fish tanks. Because some fish require a cold water environment while others are more tropical and because some fish are aggressive with other types of fish, not all fish can be placed in a single tank. It is decided to have nine exotic fish, denoted by $F_1, F_2, \ldots, F_9$, where the fish that cannot be placed in the same tank as F_i ($1 \le i \le 9$) are indicated below.*

$F_1 :$ $F_2, F_3, F_4, F_5, F_6, F_8,$	$F_2 :$ $F_1, F_3, F_6, F_7,$	$F_3 :$ $F_1, F_2, F_6, F_7,$
$F_4 :$ $F_1, F_5, F_8, F_9,$	$F_5 :$ $F_1, F_4, F_8, F_9,$	$F_6 :$ $F_1, F_2, F_3, F_7,$
$F_7 :$ $F_2, F_3, F_6, F_9,$	$F_8 :$ $F_1, F_4, F_5, F_9,$	$F_9 :$ $F_4, F_5, F_7, F_8.$

What is the minimum number of tanks required?

Solution. A graph G is constructed with vertex set $V(G) = \{F_1, F_2, \ldots, F_9\}$, where F_i is adjacent to F_j ($i \ne j$) if F_i and F_j cannot be placed in the same tank (see Figure 6.6). Then the minimum number of tanks required to house all fish is $\chi(G)$. In this case, $\omega(G) = 4$, so $\chi(G) \ge 4$. However, $n = 9$ and $\alpha(G) = 2$ and so $\chi(G) \ge 9/2$. Thus, $\chi(G) \ge 5$. A 5-coloring of G is given in Figure 6.6, implying that $\chi(G) \le 5$ and so $\chi(G) = 5$.   ◆

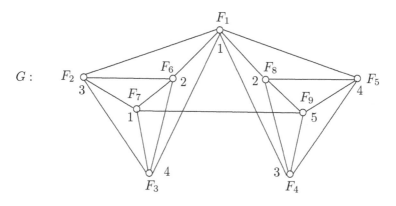

Figure 6.6: The graph of Example 6.14

Example 6.15 *Figure 6.7 shows eight traffic lanes* $L_1, L_1, \ldots, L_8$ *at the intersection of two streets. A traffic light is located at the intersection. During each phase of the traffic light, those cars in lanes for which the light is green may proceed safely through the intersection into certain permitted lanes. What is the minimum number of phases needed for the traffic light so that (eventually) all cars may proceed safely through the intersection?*

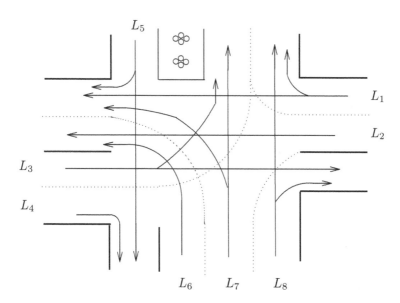

Figure 6.7: Traffic lanes at street intersections in Example 6.15

Solution. A graph G is constructed with vertex set $V(G) = \{L_1, L_2, \ldots, L_8\}$, where L_i is adjacent to L_j ($i \neq j$) if cars in lanes L_i and L_j cannot proceed safely through the intersection at the same time. (See Figure 6.8.) The minimum number of phases needed for the traffic light so that all cars may proceed, in time, through the intersection is $\chi(G)$. Since $\{L_2, L_3, L_5, L_7\}$ induces a maximum clique in G, it follows that $\omega(G) = 4$. By Corollary 6.2, $\chi(G) \geq 4$. Since there is a 4-coloring of G (see Figure 6.8), it follows that $\chi(G) = 4$. For example, since L_6, L_7, and L_8 belong to the same color class, cars in those three lanes may proceed safely through the intersection at the same time.   ♦

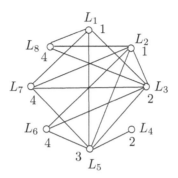

Figure 6.8: The graph of Example 6.15

The solutions of the problems we've just described have something in common. The vertices of the graphs constructed in the solutions are objects where adjacency of two vertices indicates that the objects are incompatible in some way. Each problem involves partitioning the objects into as few subsets as possible so that the elements in each subset are mutually compatible. There is a graph that describes this in general.

Let A be a set and let S be a collection of nonempty subsets of A. The **intersection graph** of S is that graph whose vertices are the elements of S and where two vertices are adjacent if the subsets have a nonempty intersection. The chromatic number of this graph is the minimum number of sets into which the elements of S can be partitioned so that in each set, every two elements of S are disjoint.

For example, suppose that $A = \{1, 2, \ldots, 6\}$ and S is the set of the nine 2-element subsets $\{a, b\}$ of A for which $a + b$ is odd. The corresponding intersection graph is isomorphic to the Cartesian product $C_3 \times C_3$ and is shown in Figure 6.9 embedded on a torus (see Section 5.3). The chromatic number of this graph is 3 (a 3-coloring is also shown in Figure 6.9) and each color class consists of three mutually disjoint 2-element subsets of A.

If the sets defining the intersection graph are closed intervals of real numbers, then the intersection graph is called an **interval graph**. In fact, the graph constructed in Example 6.13 is an interval graph. (Interval graphs will be discussed in more detail in Section 6.3.)

There is a graph that is, in a sense, complementary to an intersection graph. While studying a 1953 article on quadratic forms by Irving Kaplansky (1917–2006), who was a renowned algebraist at the University of Chicago for many years, Martin Kneser became interested in the behavior of partitions of the family of k-element subsets of an n-element set.

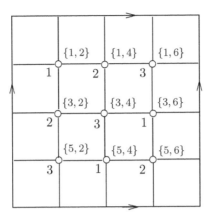

Figure 6.9: $C_3 \times C_3$: An intersection graph embedded on a torus

For positive integers k and n with $n > 2k$, it is possible to partition the k-element subsets of an n-element set, say $S = \{1, 2, \ldots, n\}$, into $n - 2k + 2$ classes such that no pair of disjoint k-element subsets belong to the same class. For example, let S_1 be the class of all k-element subsets of S containing the integer 1 and let S_2 be the class of all k-element subsets of S containing the integer 2 but not containing 1. More generally, for each integer i with $1 \leq i \leq n - 2k + 1$, let S_i be class of all k-element subsets of S containing the integer i but containing no integer j with $1 \leq j < i$. Finally, let S_{n-2k+2} consist of all k-element subsets of the set

$$T = \{n - 2k + 2, n - 2k + 3, \ldots, n\}.$$

For $1 \leq i \leq n - 2k + 1$, every two subsets belonging to S_i contain i and so are not disjoint. Since $|T| = 2k - 1$, no two k-element subsets in S_{n-2k+2} are disjoint. Thus,

$$\{S_1, S_2, \ldots, S_{n-2k+2}\}$$

is a partition of S with the desired properties. Kneser asked whether the k-element subsets of S could be partitioned into $n - 2k + 1$ classes having the same property.

Kneser conjectured that such a partition is not possible and stated this as a problem (Problem 300) in the *Jahresbericht der Deutschen Mathematiker – Vereinigung* in 1955 (see [135]).

Kneser's Conjecture   *Let k and n be positive integers with $n > 2k$. If the k-element subsets of the set $\{1, 2, \ldots, n\}$ are partitioned into $n - 2k + 1$ classes, then at least one of these classes contains two disjoint k-element subsets.*

In 1978 the Hungarian mathematician László Lovász (born in 1948) verified this conjecture using graph theory and at the same time initiated the area of topological combinatorics (see [146]). For positive integers k and n with $n > 2k$, the **Kneser graph** $\mathrm{KG}_{n,k}$ is that graph whose vertices are the k-element subsets of the n-element set $S = \{1, 2, \ldots, n\}$ and where two vertices (k-element subsets) A and B

are adjacent if A and B are disjoint. The graph $\text{KG}_{n,k}$ is therefore a $\binom{n-k}{k}$-regular graph of order $\binom{n}{k}$. In particular, $\text{KG}_{n,1}$ is the complete graph K_n, while $\text{KG}_{5,2}$ is the Petersen graph (see Figure 6.10).

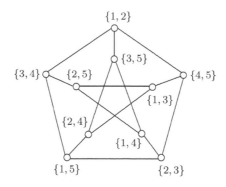

Figure 6.10: The Kneser graph $\text{KG}_{5,2}$ (the Petersen graph)

Kneser's Conjecture can then be stated purely in terms of graph theory, namely:

Kneser's Conjecture   *There exists no $(n - 2k + 1)$-coloring of the Kneser graph* $\text{KG}_{n,k}$.

To see why this is an equivalent formulation of Kneser's Conjecture stated earlier, suppose that there is an $(n - 2k + 1)$-coloring of the Kneser graph $\text{KG}_{n,k}$. Then this implies that there is a partition of the vertex set of $\text{KG}_{n,k}$ into $n - 2k + 1$ independent sets. However, this, in turn, implies that each of the $n - 2k + 1$ color classes (consisting of k-element subsets of $\{1, 2, \ldots, n\}$) contains no pair of disjoint k-element subsets, thereby disproving the conjecture. Indeed, László Lovász [146] proved the following:

Theorem 6.16   *For every two positive integers k and n with $n > 2k$,*

$$\chi(\text{KG}_{n,k}) = n - 2k + 2.$$

There is a subclass of Kneser graphs that is of special interest. For $n \geq 2$, the **odd graph** O_n is that graph whose vertices are the $(n - 1)$-element subsets of $\{1, 2, \ldots, 2n - 1\}$ such that two vertices A and B are adjacent if A and B are disjoint. Consequently, the odd graph O_n is the Kneser graph $\text{KG}_{2n-1,n-1}$. Hence, O_2 is the complete graph K_3 and the graph O_3 is the Petersen graph, while O_4 is a 4-regular graph of order 35. By Theorem 6.16, every odd graph has chromatic number 3.

6.3   Perfect Graphs

In Corollary 6.2 we saw that the clique number $\omega(G)$ of a graph G is a lower bound for $\chi(G)$. While there are many examples of graphs G for which $\chi(G) = \omega(G)$, such as complete graphs and bipartite graphs, there are also many graphs whose chromatic number exceeds its clique number such as the Petersen graph and the odd cycles of length 5 or more. As we are about to see, the chromatic number of a graph can be considerably larger than its clique number.

For a given graph H, a graph G is called H-**free** if no induced subgraph of G is isomorphic to H. In particular, a $K_{1,3}$-free graph is called a **claw-free graph**. We saw in Chapter 1 that a K_3-free graph is commonly called a *triangle-free graph*. Therefore, every bipartite graph is triangle-free, as is the Petersen graph and every cycle of length 4 or more. Consequently, every nonempty triangle-free graph has clique number 2.

The graph G of Figure 6.11 is triangle-free (and so $\omega(G) = 2$) but $\chi(G) = 4$. Hence, $\chi(G)$ exceeds $\omega(G)$ by 2 in this case. This graph is the famous **Grötzsch graph**. It is known to be the unique smallest graph (in terms of order) that is both 4-chromatic and triangle-free. The fact that a graph can be triangle-free and yet have a large chromatic number has been established by a number of mathematicians, including Blanche Descartes [64], John Kelly and Leroy Kelly [133], and Alexander Zykov [206]. The proof of this fact that we present here, however, is due to Jan Mycielski [154].

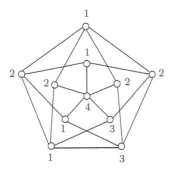

Figure 6.11: The Grötzsch graph: A 4-chromatic triangle-free graph

Theorem 6.17 *For every positive integer k, there exists a triangle-free k-chromatic graph.*

Proof. Since no graph with chromatic number 1 or 2 contains a triangle, the theorem is obviously true for $k = 1$ and $k = 2$. To verify the theorem for $k \geq 3$, we proceed by induction on k. Since $\chi(C_5) = 3$ and C_5 is triangle-free, the statement is true for $k = 3$.

Assume that there exists a triangle-free graph with chromatic number k, where $k \geq 3$. We show that there exists a triangle-free $(k+1)$-chromatic graph. Let H be a triangle-free graph with $\chi(H) = k$, where $V(H) = \{v_1, v_2, \ldots, v_n\}$. We construct a graph G from H by adding $n + 1$ new vertices $u, u_1, u_2, \ldots, u_n$, joining u to each vertex u_i ($1 \leq i \leq n$) and joining u_i to each neighbor of v_i in H. (See Figure 6.12 when $k = 3$ and $H = C_5$, in which case the resulting graph G is the Grötzsch graph of Figure 6.12.)

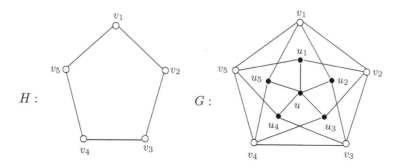

Figure 6.12: The Mycielski construction

We claim that G is a triangle-free $(k+1)$-chromatic graph. First, we show that G is triangle-free. Since $S = \{u_1, u_2, \ldots, u_n\}$ is an independent set of vertices of G and u is adjacent to no vertex of H, it follows that u belongs to no triangle in G. Hence, if there is a triangle T in G, then two of the three vertices of T must belong to H and the third vertex must belong to S, say $V(T) = \{u_i, v_j, v_k\}$. Since u_i is adjacent v_j and v_k, it follows that v_i is adjacent to v_j and v_k. Because v_j and v_k are adjacent in the triangle T, the graph H contains a triangle, which is impossible. Thus, as claimed, G is triangle-free.

Next, we show that $\chi(G) = k + 1$. Since H is a subgraph of G and $\chi(H) = k$, it follows that $\chi(G) \geq k$. Let a k-coloring of H be given and assign to u_i the same color that is assigned to v_i for $1 \leq i \leq n$. Assigning the color $k + 1$ to u produces a $(k+1)$-coloring of G and so $\chi(G) \leq k+1$. Hence, either $\chi(G) = k$ or $\chi(G) = k+1$. Assume, to the contrary, that $\chi(G) = k$. Then there is a k-coloring of G with colors $1, 2, \ldots, k$. We may assume that u is assigned the color k in a k-coloring of G. Necessarily, none of the vertices $u_1, u_2, \ldots, u_n$ is assigned the color k; that is, each vertex of S is assigned one of the colors $1, 2, \ldots, k-1$. Since $\chi(H) = k$, one or more vertices of H are assigned the color k. For each vertex v_i of H colored k, recolor it with the color assigned to u_i. This produces a $(k-1)$-coloring of H, which is impossible. Thus, $\chi(G) = k + 1$.   ∎

If the Mycielski construction (described in the proof of Theorem 6.17) is applied to the Grötzsch graph (Figures 6.11 and 6.12), then a triangle-free 5-chromatic graph of order 23 is produced. Using a computer search, Tommy Jensen and Gordon F. Royle [130] showed that the smallest order of a triangle-free 5-chromatic graph is actually 22. Applying the Mycielski construction to this graph, we can conclude that there is a triangle-free 6-chromatic graph of order 45. Jan Goedgebeur [99] showed, however, that the smallest order of a triangle-free 6-chromatic graph is at least 32 and at most 40.

With the aid of Theorem 6.17, it can be seen that for every two integers ℓ and k with $2 \leq \ell \leq k$, there exists a graph G with $\omega(G) = \ell$ and $\chi(G) = k$ (see Exercise 33). A rather symmetric graph G that is not triangle-free but for which $\chi(G) > \omega(G)$ is shown in Figure 6.13. This graph has order 15 and consists of five mutually vertex-disjoint triangles T_i $(1 \leq i \leq 5)$ where every vertex of T_i is adjacent to every vertex of T_j if either $|i - j| = 1$ or if $\{i, j\} = \{1, 5\}$. (We will visit this graph again in Chapter 7.) This graph G has clique number 6 and independence number 2. By Theorem 6.10, $\chi(G) \geq n/\alpha(G) = 15/2$ and so $\chi(G) \geq 8$. If we color the vertices of T_i $(1 \leq i \leq 5)$ as indicated in Figure 6.13, then it follows that G is 8-colorable and so $\chi(G) \leq 8$. Therefore, $\chi(G) = 8$.

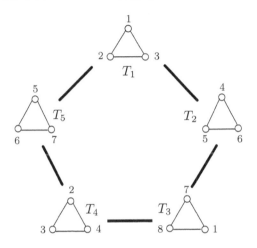

Figure 6.13: A graph G with $\chi(G) = 8$

While much interest has been shown in graphs G for which $\chi(G) > \omega(G)$, even more interest has been shown in graphs G for which not only $\chi(G) = \omega(G)$ but $\chi(H) = \omega(H)$ for *every* induced subgraph H of G. A graph G is called **perfect** if $\chi(H) = \omega(H)$ for every induced subgraph H of G. This definition, introduced in [15], is due to the French graph theorist Claude Berge (1926–2002). As is often the case when a new concept is introduced, it is not initially known whether the concept will lead to interesting results and whether any intriguing characterizations will be forthcoming. For this particular class of graphs, however, all of this occurred.

Certainly, if $G = K_n$, then $\chi(G) = \omega(G) = n$. Furthermore, every induced subgraph H of K_n is also a complete graph and so $\chi(H) = \omega(H)$. Thus, every complete graph is perfect. On the other hand, if $G = \overline{K}_n$ and H is any induced subgraph of G, then $\chi(H) = \omega(H) = 1$. So every empty graph is also perfect. A somewhat more interesting class of perfect graphs are the bipartite graphs.

Theorem 6.18 *Every bipartite graph is perfect.*

Proof. Let G be a bipartite graph and let H be an induced subgraph of G. If H is nonempty, then $\chi(H) = \omega(H) = 2$; while if H is empty, then $\chi(H) = \omega(H) = 1$. In either case, $\chi(H) = \omega(H)$ and so G is perfect. ∎

The next theorem, which is a consequence of a result due to Tibor Gallai [91], describes a related class of perfect graphs.

Theorem 6.19 *Every graph whose complement is bipartite is perfect.*

Proof.   Let G be a graph of order n such that $\overline{G}$ is bipartite. Since the complement of every (nontrivial) induced subgraph of G is also bipartite, to verify that G is perfect, it suffices to show that $\chi(G) = \omega(G)$. Suppose that $\chi(G) = k$ and $\omega(G) = \ell$. Then $k \geq \ell$. Let there be given a k-coloring of G. Then each color class of G consists either of one or two vertices; for if G contains a color class with three or more vertices, then this would imply that $\overline{G}$ has a triangle, which is impossible.

Of the k color classes, suppose that p of these classes consist of a single vertex and that each of the remaining q classes consists of two vertices. Hence, $p + q = k$ and $p + 2q = n$. Let W be the set of vertices of G belonging to a singleton color class. Since every two vertices of W are necessarily adjacent, $G[W] = K_p$ and so $\overline{G}[W] = \overline{K}_p$.

Since no k-coloring of G results in more than q color classes having two vertices, it follows that $\overline{G}$ has a maximum matching M with q edges (see Chapter 4). We claim that for each edge $uv \in M$, either u is adjacent to no vertex of W or v is adjacent to no vertex of W. Suppose that this is not the case. Then we may assume that u is adjacent to some vertex $w_1 \in W$ and v is adjacent to some vertex $w_2 \in W$. Since $\overline{G}$ is triangle-free, $w_1 \neq w_2$. However then, $(M - \{uv\}) \cup \{uw_1, vw_2\}$ is a matching in $\overline{G}$ containing more than $|M|$ edges. This, however, is impossible and so, as claimed, for each edge uv in M either u is adjacent to no vertex of W or v is adjacent to no vertex of W.

Therefore, $\overline{G}$ contains an independent set of at least $p + q = k$ vertices and so $\omega(G) = \ell \geq k$. Hence, $\chi(G) = \omega(G)$.   ∎

From what we've seen, if G is a graph that is either complete or bipartite, then both G and $\overline{G}$ are perfect. Indeed, in 1961 Claude Berge made the following conjecture.

The Perfect Graph Conjecture *A graph is perfect if and only if its complement is perfect.*

In 1972, László Lovász [144] showed that this conjecture is, in fact, true.

Theorem 6.20 (The Perfect Graph Theorem) *A graph is perfect if and only if its complement is perfect.*

We now describe another class of perfect graphs. Recall (from Section 6.2) that a graph G with $V(G) = \{v_1, v_2, \ldots, v_n\}$ is an **interval graph** if there exists a collection S of n closed intervals of real numbers, say

$$S = \{[a_i, b_i] : a_i < b_i, 1 \leq i \leq n\},$$

such that v_i and v_j are adjacent if and only if $[a_i, b_i]$ and $[a_j, b_j]$ have a nonempty intersection. Hence, if G is an interval graph, then every induced subgraph of G is also an interval graph (see Exercise 36).

For example, the graph G in Figure 6.14 is an interval graph as can be seen by considering the five intervals $I_1 = [0, 2]$, $I_2 = [1, 5]$, $I_3 = [3, 6]$, $I_4 = [4, 8]$, $I_5 = [7, 9]$, where v_i and v_j are adjacent in G, $1 \leq i, j \leq 5$, if and only if $I_i \cap I_j \neq \emptyset$. Observe that $\chi(G) = \omega(G) = 3$ for this graph G. This graph is also a perfect graph. Indeed, every interval graph is a perfect graph.

$G:$

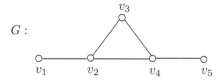

Figure 6.14: An interval graph

Theorem 6.21 *Every interval graph is perfect.*

Proof. Let G be an interval graph with $V(G) = \{v_1, v_2, \ldots, v_n\}$. Since every induced subgraph of an interval graph is also an interval graph, it suffices to show that $\chi(G) = \omega(G)$. Because G is an interval graph, there exist n closed intervals $I_i = [a_i, b_i]$, $1 \le i \le n$, such that v_i is adjacent to v_j ($i \ne j$) if and only if $I_i \cap I_j \ne \emptyset$. We may assume that the intervals (and consequently, the vertices of G) have been labeled so that $a_1 \le a_2 \le \cdots \le a_n$.

We now define a vertex coloring of G. First, assign v_1 the color 1. If v_1 and v_2 are not adjacent (that is, if I_1 and I_2 are disjoint), then assign v_2 the color 1 as well; otherwise, assign v_2 the color 2. Proceeding inductively, suppose that we have assigned colors to $v_1, v_2, \ldots, v_r$ where $2 \le r < n$. We now assign v_{r+1} the smallest color (positive integer) that has not been assigned to any neighbor of v_{r+1} in the set $\{v_1, v_2, \ldots, v_r\}$. Thus, if v_{r+1} is adjacent to no vertex in $\{v_1, v_2, \ldots, v_r\}$, then v_{r+1} is assigned the color 1. This gives a k-coloring of G for some positive integer k and so $\chi(G) \le k$. If $k = 1$, then $G = \overline{K}_n$ and $\chi(G) = \omega(G) = 1$. Hence, we may assume that $k \ge 2$.

Suppose that the vertex v_t has been assigned the color k. Since it was not possible to assign v_t any of the colors $1, 2, \ldots, k-1$, this means that the interval $I_t = [a_t, b_t]$ must have a nonempty intersection with $k-1$ intervals $I_{j_1}, I_{j_2}, \ldots, I_{j_{k-1}}$, where say $1 \le j_1 < j_2 < \cdots < j_{k-1} < t$. Thus, $a_{j_1} \le a_{j_2} \le \ldots \le a_{j_{k-1}} \le a_t$. Since $I_{j_i} \cap I_t \ne \emptyset$ for $1 \le i \le k-1$, it follows that

$$a_t \in I_{j_1} \cap I_{j_2} \cap \cdots \cap I_{j_{k-1}} \cap I_t.$$

Thus, for $U = \{v_{j_1}, v_{j_2}, \ldots, v_{j_{k-1}}, v_t\}$,

$$G[U] = K_k$$

and so $\chi(G) \le k \le \omega(G)$. Since $\chi(G) \ge \omega(G)$, we have $\chi(G) = \omega(G)$, as desired. ∎

In the proof of Theorem 6.21, a vertex coloring c of a graph G with $V(G) = \{v_1, v_2, \ldots, v_n\}$ is defined recursively by $c(v_1) = 1$ and, given that $c(v_i)$ is defined for every integer i with $1 \le i \le r$ for an integer r with $1 \le r < n$, the color $c(v_{r+1})$ is defined as the smallest color not assigned to any neighbor of v_{r+1} among the vertices in $\{v_1, v_2, \ldots, v_r\}$. We will see vertex colorings of graphs defined in this manner again in Chapter 7 along with some useful consequences.

We now consider a more general class of graphs. Recall that a **chord** of a cycle C in a graph is an edge that joins two non-consecutive vertices of C. For example, wz

and xz are chords in the cycle $C = (u, v, w, x, y, z, u)$ in the graph G of Figure 6.15; while in the cycle $C' = (w, x, y, z, w)$ in G, the edge xz is a chord and wz is not. The cycle $C'' = (u, v, w, z, u)$ has no chords. Obviously no triangle contains a chord.

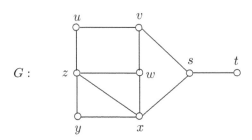

G :

Figure 6.15: Chords in cycles

A graph G is a **chordal graph** if every cycle of length 4 or more in G has a chord. Since the cycle $C'' = (u, v, w, z, u)$ in the graph G of Figure 6.15 contains no chords, the graph G is not a chordal graph.

While every complete graph is a chordal graph, no complete bipartite graph $K_{s,t}$, where $s, t \geq 2$, is chordal, for if u_1 and v_1 belong to one partite set and u_2 and v_2 belong to the other partite set, then the cycle $(u_1, u_2, v_1, v_2, u_1)$ contains no chord. Indeed, no graph having girth 4 or more is chordal. The graphs G_1 and G_2 of Figure 6.16 are chordal graphs. For the subset $S_1 = \{u_1, v_1, x_1\}$ of $V(G_1)$ and the subset $S_2 = \{u_2, w_2, x_2\}$ of $V(G_2)$, let the graph G_3 be obtained by identifying the vertices in the complete subgraph $G_1[S_1]$ with the vertices in the complete subgraph $G_2[S_2]$, where, say, u_1 and u_2 are identified, v_1 and x_2 are identified, and x_1 and w_2 are identified. The graph G_3 shown in Figure 6.16 is also a chordal graph.

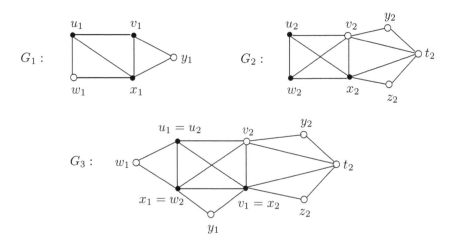

Figure 6.16: Chordal graphs

More generally, suppose that G_1 and G_2 are two graphs containing complete subgraphs H_1 and H_2, respectively, of the same order and G_3 is the graph obtained by identifying the vertices of H_1 with the vertices of H_2 (in a one-to-one manner). If G_3 contains a cycle of length 4 or more having no chord, then C must belong to G_1 or G_2. That is, if G_1 and G_2 are chordal, then G_3 is chordal. Furthermore, if G_3 is chordal, then both G_1 and G_2 are chordal.

Theorem 6.22 *Let G be a graph obtained by identifying two complete subgraphs of the same order in two graphs G_1 and G_2. Then G is chordal if and only if G_1 and G_2 are chordal.*

Proof. We have already noted that if G_1 and G_2 are two chordal graphs containing complete subgraphs H_1 and H_2, respectively, of the same order, then the graph G obtained by identifying the vertices of H_1 with the vertices of H_2 is also chordal. On the other hand, if G_1, say, were not chordal, then it would contain a cycle C of length 4 or more having no chords. However then, C would be a cycle in G having no chords. ∎

We have now observed that every graph obtained by identifying two complete subgraphs of the same order in two chordal graphs is also chordal. These are not only sufficient conditions for a graph to be chordal, they are necessary conditions as well. The following characterization of chordal graphs is due to Andras Hajnal and János Surányi [108] and Gabriel Dirac [69].

Theorem 6.23 *A graph G is chordal if and only if G can be obtained by identifying two complete subgraphs of the same order in two chordal graphs.*

Proof. From our earlier observations, we need only show that every chordal graph can be obtained from two chordal graphs by identifying two complete subgraphs of the same order in these two graphs. If G is complete, say $G = K_n$, then G is chordal and can trivially be obtained by identifying the vertices of $G_1 = K_n$ and the vertices of $G_2 = K_n$ in any one-to-one manner. Hence, we may assume that G is a connected chordal graph that is not complete.

Let S be a minimum vertex-cut of G. Now let V_1 be the vertex set of one component of $G - S$ and let $V_2 = V(G) - (V_1 \cup S)$. Consider the two S-branches

$$G_1 = G[V_1 \cup S] \text{ and } G_2 = G[V_2 \cup S]$$

of G. Consequently, G is obtained by identifying the vertices of S in G_1 and G_2. We now show that $G[S]$ is complete. Since this is certainly true if $|S| = 1$, we may assume that $|S| \geq 2$.

Each vertex v in S is adjacent to at least one vertex in each component of $G - S$, for otherwise $S - \{v\}$ is a vertex-cut of G, which is impossible. Let $u, w \in S$. Hence, there are $u - w$ paths in G_1, where every vertex except u and w belongs to V_1. Among all such paths, let $P = (u, x_1, x_2, \ldots, x_s, w)$ be one of minimum length. Similarly, let $P' = (u, y_1, y_2, \ldots, y_t, w)$ be a $u - w$ path of minimum length where every vertex except u and w belongs to V_2. Hence,

$$C = (u, x_1, x_2, \ldots, x_s, w, y_t, y_{t-1}, \ldots, y_1, u)$$

is a cycle of length 4 or more in G. Since G is chordal, C contains a chord. No vertex x_i $(1 \le i \le s)$ can be adjacent to a vertex y_j $(1 \le j \le t)$ since S is a vertex-cut of G. Furthermore, no non-consecutive vertices of P or of P' can be adjacent due to the manner in which P and P' are defined. Thus, $uw \in E(G)$, implying that $G[S]$ is complete. By Theorem 6.22, G_1 and G_2 are chordal. ∎

With the aid of Theorem 6.23, we now have an even larger class of perfect graphs (see Exercise 40).

Corollary 6.24 *Every chordal graph is perfect.*

Proof. Since every induced subgraph of a chordal graph is also a chordal graph, it suffices to show that if G is a connected chordal graph, then $\chi(G) = \omega(G)$. We proceed by induction on the order n of G. If $n = 1$, then $G = K_1$ and $\chi(G) = \omega(G) = 1$. Assume, therefore, that $\chi(H) = \omega(H)$ for every chordal graph H of order less than n, where $n \ge 2$ and let G be a chordal graph of order $n \ge 2$.

If G is a complete graph, then $\chi(G) = \omega(G) = n$. Hence, we may assume that G is not complete. By Theorem 6.22, G can be obtained from two chordal graphs G_1 and G_2 by identifying two complete subgraphs of the same order in G_1 and G_2. Observe that

$$\chi(G) \le \max\{\chi(G_1), \chi(G_2)\} = k.$$

By the induction hypothesis, $\chi(G_1) = \omega(G_1)$ and $\chi(G_2) = \omega(G_2)$. Thus, $\chi(G) \le \max\{\omega(G_1), \omega(G_2)\} = k$. On the other hand, let S denote the set of vertices in G that belong to G_1 and G_2. Thus, $G[S]$ is complete and no vertex in $V(G_1) - S$ is adjacent to a vertex in $V(G_2) - S$. Hence,

$$\omega(G) = \max\{\omega(G_1), \omega(G_2)\} = k.$$

Thus, $\chi(G) \ge k$. Therefore, $\chi(G) = k = \omega(G)$. ∎

The graph G of Figure 6.17 has clique number 3. Thus, $\chi(G) \ge 3$. In this case, however, $\chi(G) \ne \omega(G)$. Indeed, $\chi(G) = 4$. A 4-coloring of G is shown in Figure 6.17. Thus, G is not perfect. By Corollary 6.24, G is not chordal. In fact, $C = (u, x, v, y, u)$ is a 4-cycle containing no chord. Since G is not perfect, it follows by the Perfect Graph Theorem that $\overline{G}$ is not perfect either. Indeed, $\overline{G} = C_7$ and so $\chi(\overline{G}) = 3$ and $\omega(\overline{G}) = 2$. The graph F of Figure 6.17 is also not chordal; but yet $\chi(H) = \omega(H)$ for every induced subgraph H of F. Hence, the converse of Corollary 6.24 is not true.

We now consider a class of perfect graphs that can be obtained from a given perfect graph. Let G be a graph where $v \in V(G)$. Then the **replication graph** $R_v(G)$ of G (with respect to v) is that graph obtained from G by adding a new vertex v' to G and joining v' to the vertices in the closed neighborhood $N[v]$ of v. In 1972 László Lovász [145] obtained the following result.

Theorem 6.25 (**The Replication Lemma**) *Let G be a graph where $v \in V(G)$. If G is perfect, then $R_v(G)$ is perfect.*

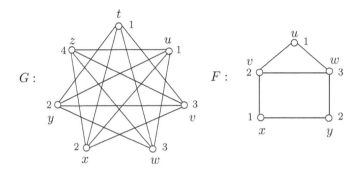

Figure 6.17: Non-chordal graphs

Proof. Let $G' = R_v(G)$. First, we show that $\chi(G') = \omega(G')$. We consider two cases, depending on whether v belongs to a maximum clique of G.

Case 1. v belongs to a maximum clique of G. Then $\omega(G') = \omega(G) + 1$. Since

$$\chi(G') \le \chi(G) + 1 = \omega(G) + 1 = \omega(G'),$$

it follows that $\chi(G') = \omega(G')$.

Case 2. v does not belong to any maximum clique of G. Suppose that $\chi(G) = \omega(G) = k$. Let there be given a k-coloring of G using the colors $1, 2, \ldots, k$. We may assume that v is assigned the color 1. Let V_1 be the color class consisting of the vertices of G that are colored 1. Thus, $v \in V_1$. Since $\omega(G) = k$, every maximum clique of G must contain a vertex of each color. Since v does not belong to a maximum clique, it follows that $|V_1| \ge 2$. Let $U_1 = V_1 - \{v\}$. Because every maximum clique of G contains a vertex of U_1, it follows that $\omega(G - U_1) = \omega(G) - 1 = k - 1$. Since G is perfect, $\chi(G - U_1) = k - 1$. Let a $(k - 1)$-coloring of $G - U_1$ be given, using the colors $1, 2, \ldots, k - 1$. Since V_1 is an independent set of vertices, so is $U_1 \cup \{v'\}$. Assigning the vertices of $U_1 \cup \{v'\}$ the color k produces a k-coloring of G'. Therefore,

$$k = \omega(G) \le \omega(G') \le \chi(G') \le k$$

and so $\chi(G') = \omega(G')$.

It remains to show that $\chi(H) = \omega(H)$ for every induced subgraph H of G'. This is certainly the case if H is a subgraph of G. If H contains v' but not v, then $H \cong G[(V(H) - \{v'\}) \cup \{v\}]$ and so $\chi(H) = \omega(H)$. If H contains both v and v' but $H \ncong G'$, then H is the replication graph of $G[V(H) - \{v'\}]$ and the argument used to show that $\chi(G') = \omega(G')$ can be applied to show that $\chi(H) = \omega(H)$. ∎

In 1961 Claude Berge also conjectured that there are certain conditions that must be satisfied by all perfect graphs and only these graphs. This deeper conjecture became known as:

The Strong Perfect Graph Conjecture A graph G is perfect if and only if neither G nor $\overline{G}$ contains an induced odd cycle of length 5 or more.

After an intensive 28-month assault on this conjecture, its truth was established in 2002 by Maria Chudnovsky, Neil Robertson, Paul Seymour, and Robin Thomas [52].

Theorem 6.26 (The Strong Perfect Graph Theorem) *A graph G is perfect if and only if neither G nor $\overline{G}$ contains an induced odd cycle of length 5 or more.*

Exercises for Chapter 6

1. Show that there is no graph of order 6 and size 13 that has chromatic number 3.

2. The vertices of a graph G are properly colored with three colors in such a way that each vertex is adjacent to vertices colored with only one of the three colors. Show that $\chi(G) \neq 3$. What does this say if $\chi(G) = 3$?

3. Let G be a graph with chromatic number k and let c be a k-coloring of G. For each integer i with $1 \leq i \leq k$, let C_i denote the color class consisting of all vertices colored i. Prove that for each integer i with $1 \leq i \leq k - 1$, there exists a vertex in C_k that is adjacent to a vertex in C_i.

4. It is known that it is possible to color the vertices of a graph G of order 12, size 50, and chromatic number k with k colors so that the number of vertices assigned any of the k colors is the same. Show that $\chi(G) \geq 4$.

5. Let G be a nonempty graph with $\chi(G) = k$. A graph H is obtained from G by subdividing every edge of G. If $\chi(H) = \chi(G)$, then what is k?

6. A balanced coloring of a graph G is an assignment of colors to the vertices of G such that (i) every two adjacent vertices are assigned different colors and (ii) the numbers of vertices assigned any two different colors differ by at most one. The smallest number of colors used in a balanced coloring of G is the balanced chromatic number $\chi_b(G)$ of G.

   (a) Prove that the balanced chromatic number is defined for every graph G.
   (b) Determine $\chi_b(G)$ for the graph G in Figure 6.18.

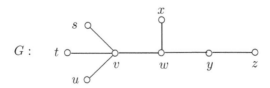

Figure 6.18: The graph in Exercise 6(b)

7. Let G be an r-regular graph, $r \geq 3$, such that $3 \leq \chi(G) \leq r + 1$. Prove or disprove the following.

(a) For every edge e of G, there exists an odd cycle containing e.

(b) For every edge e of G, there exists an odd cycle not containing e.

8. (a) Show for every graph G that $\chi(G) = \max\{\chi(H) : H$ is a subgraph of $G\}$.

   (b) The problem in (a) should suggest a question to you. Ask and answer this question.

9. Prove Proposition 6.3: *For graphs $G_1, G_2, \ldots, G_k$ and $G = G_1 + G_2 + \cdots + G_k$, $\chi(G) = \max\{\chi(G_i) : 1 \le i \le k\}$.*

10. Prove Proposition 6.5: *If G is a nontrivial connected graph with blocks B_1, B_2, ..., B_k, then $\chi(G) = \max\{\chi(B_i) : 1 \le i \le k\}$.*

11. Prove Proposition 6.6: *For graphs $G_1, G_2, \ldots, G_k$ and $G = G_1 \vee G_2 \vee \cdots \vee G_k$, $\chi(G) = \sum_{i=1}^{k} \chi(G_i)$.*

12. (a) Give an example of a graph G, every vertex of which belongs to no more than two odd cycles but most vertices belong to exactly two odd cycles.

    (b) What does Theorem 6.8 say about $\chi(G)$ for any graph G in (a)?

13. (a) Give an example of a graph G, every vertex of which belongs to exactly three odd cycles.

    (b) What does Theorem 6.8 say about the upper bound for $\chi(G)$ for a graph G?

14. To how many odd cycles does each vertex of $G = K_5$ belong? What does Theorem 6.8 say about $\chi(G)$ in this case?

15. For a given positive integer n, determine all graphs of order n for which the two bounds in Theorem 6.10 are equal.

16. Let $k \ge 2$ be an integer. For each integer i with $1 \le i \le 2k + 1$, let G_i be a copy of K_k. The graph G of order $2k^2 + k$ is obtained from the graphs $G_1, G_2, \ldots, G_{2k+1}, G_{2k+2} = G_1$ by joining each vertex in G_i to every vertex in G_{i+1} $(1 \le i \le 2k + 1)$. Compute the bounds $n/\alpha(G)$, $n + 1 - \alpha(G)$, $\omega(G)$ for $\chi(G)$ for an arbitrary $k \ge 2$. Determine $\chi(G)$.

17. For $i = 1, 2, \ldots, 5$, let $G_i = P_4$. A graph G of order n is obtained from the graphs $G_1, G_2, \ldots, G_5, G_6 = G_1$ by joining each vertex of G_i to each vertex of G_{i+1} for $i = 1, 2, \ldots, 5$. Compute the bounds $n/\alpha(G)$, $n + 1 - \alpha(G)$, $\omega(G)$ for $\chi(G)$ for an arbitrary $k \ge 2$. Determine $\chi(G)$.

18. Give an example of a graph G with $\chi(G) = \alpha(G)$ in which no $\chi(G)$-coloring of G results in a color class containing $\alpha(G)$ vertices.

19. Does there exist a k-chromatic graph G for some positive integer k in which no color class of a k-coloring of G contains at least $\alpha(G) - 2$ vertices?

20. Give an example of a graph G of order n for which $n/\alpha(G)$ is an integer and for which $\chi(G)$ is none of the numbers $n/\alpha(G)$, $n - \alpha(G) + 1$, and $\omega(G)$.

21. Prove that if S is a color class resulting from a k-coloring of a k-chromatic graph G, where $k \geq 2$, then there is a component H of $G - S$ such that $\chi(H) = k - 1$.

22. Let $k \geq 2$ be an integer. Prove that if G is a k-colorable graph of order n such that $\delta(G) > \left(\frac{k-2}{k-1}\right) n$, then G is k-chromatic. Is this bound sharp?

23. Prove or disprove: A connected graph G has chromatic number at least 3 if and only if for every vertex v of G, there exist two adjacent vertices u and w in G such that $d(u, v) = d(v, w)$.

24. Recall that an independent set of vertices in a graph G is maximal if it is not properly contained in any other independent set of G.

    (a) Let v be a vertex of a k-chromatic graph G and let $U_1, U_2, \ldots, U_\ell$ be the maximal independent sets of G containing v. Prove that for some k-coloring of G, one of the resulting k color classes is U_i for some i with $1 \leq i \leq k$.

    (b) Let G be a k-chromatic graph, $k \geq 2$, and S an independent set of vertices in G. Prove that $\chi(G[V(G) - S]) = k - 1$ if and only if S is a color class in some k-coloring of G.

25. For a nonempty graph G, let v be a vertex of G and let $U_1, U_2, \ldots, U_\ell$ be the maximal independent sets of G containing v.

    (a) Prove that $\chi(G) = 1 + \min_{1 \leq i \leq \ell} \chi(G[V(G) - U_i])$.

    (b) What is the relationship between $\chi(G)$ and
    $$1 + \max_{1 \leq i \leq \ell} \chi(G[V(G) - U_i])?$$

26. Prove that every graph of order $n = 2k$ having size at least $k^2 + 1$ has chromatic number at least 3.

27. Answer the question asked in Example 6.13 when a certain location cannot be used for two different planes if the exit time for one plane is the same as as the entrance time for the other.

28. Suppose that the two dentists in Example 6.14 had decided to have ten exotic fish, denoted by $F_1, F_2, \ldots, F_{10}$, where the fish that cannot be placed in the same tank as F_i ($1 \leq i \leq 10$) are indicated below.

    F_1: $F_2, F_3, F_4, F_5, F_{10}$ F_2: F_1, F_3, F_6 F_3: F_1, F_2, F_4
    F_4: F_1, F_3, F_5 F_5: F_1, F_4, F_9 F_6: F_2, F_7, F_{10}
    F_7: F_6, F_8, F_{10} F_8: F_7, F_9, F_{10} F_9: F_5, F_8, F_{10}
    F_{10}: F_1, F_6, F_7, F_8, F_9.

What is the minimum number of tanks required?

29. Figure 6.19 shows traffic lanes $L_1, L_1, \ldots, L_7$ at the intersection of two streets. A traffic light is located at the intersection. During a certain phase of the traffic light, those cars in lanes for which the light is green may proceed safely through the intersection in permissible directions. What is the minimum number of phases needed for the traffic light so that (eventually) all cars may proceed through the intersection?

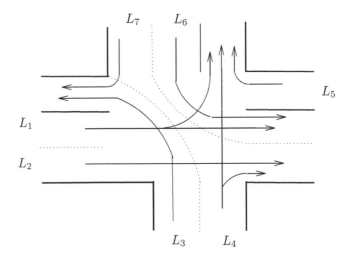

Figure 6.19: Traffic lanes at street intersections in Exercise 29

30. Prove that every graph is an intersection graph.

31. For the graph $G = K_4 - e$, determine the smallest positive integer k for which there is a collection S of subsets of $\{1, 2, \ldots, k\}$ such that the intersection graph of S is isomorphic to G.

32. Determine which connected graphs of order 4 are interval graphs.

33. Show, for every two integers ℓ and k with $2 \le \ell \le k$, that there exists a graph G with $\omega(G) = \ell$ and $\chi(G) = k$

34. Let $n \ge 2$ be an integer. Prove that every $(n-1)$-chromatic graph of order n has clique number $n - 1$.

35. Prove or disprove the following.

   (a) If G is a graph of order $n \ge 3$ with $\chi(G) = n - 2$, then $\omega(G) = n - 2$.
   (b) If G is a graph of sufficiently large order n with $\chi(G) = n - 2$, then $\omega(G) = n - 2$.

36. Prove that every induced subgraph of an interval graph is an interval graph.

37. Let G be a graph with $V(G) = \{v_1, v_2, \ldots, v_n\}$. The **shadow graph** $Shad(G)$ of G is that graph with vertex set $V(G) \cup \{u_1, u_2, \ldots, u_n\}$, where u_i is called the **shadow vertex** of v_i and u_i is adjacent to u_j if v_i is adjacent to v_j and u_i is adjacent to v_j if v_i is adjacent to v_j for $1 \le i, j \le n$. What is the relationship between $\omega(G)$ and $\omega(Shad(G))$ and the relationship between $\chi(G)$ and $\chi(Shad(G))$?

38. Give an example of a regular triangle-free 4-chromatic graph.

39. Show, for every integer $k \ge 3$, that there exists a k-chromatic graph that is not chordal.

40. Prove that every interval graph is a chordal graph.

41. Give an example of a chordal graph that is not an interval graph.

42. Determine whether $\overline{C}_8$ is perfect.

43. Let $\mathcal{G}$ be the set of all nonisomorphic connected graphs of order 25, say $\mathcal{G} = \{G_1, G_2, \ldots, G_k\}$, and let $G = G_1 + G_2 + \cdots + G_k$. Prove or disprove the following.

    (a) For the graph G, $\chi(G) = \omega(G)$.

    (b) The graph G is perfect.

44. Prove that if G is a Hamiltonian-connected graph of order at least 3, then $\chi(G) \ge 3$.

45. Let G be a graph where $v \in V(G)$. We know from Theorem 6.25 that if G is perfect, then the replication graph $R_v(G)$ is perfect. Is the converse true?

46. Prove that there exists no k-chromatic graph G of order $k^2 + 1$ ($k \ge 2$) for which $\chi(\overline{G}) = \chi(G)$.

47. Prove that it is impossible to decompose the complete graph K_{k^3+1} of order $k^3 + 1$ ($k \ge 2$) into three factors F_1, F_2, F_3 such that $\chi(F_i) = k$ for $i = 1, 2, 3$.

48. Let G be a connected graph of order at least 3, where each vertex of G is colored with one of $1, 2, 3$ (not necessarily a proper coloring) and each color is used at least once. Prove that G contains a path on which there are three vertices with distinct colors.

49. Let G be a connected graph of order at least 4, where each vertex of G is colored with one of $1, 2, 3, 4$ (not necessarily a proper coloring) and each color is used at least once. Prove that G contains a tree with at most three end-vertices on which there are four vertices with distinct colors.

Chapter 7

Bounds for the Chromatic Number

In Chapter 6 we were introduced to the chromatic number of a graph, the central concept of this book. While there is no formula for the chromatic number of a graph, we saw that the clique number is a lower bound for the chromatic number of a graph and that there are both a lower bound and an upper bound for the chromatic number of a graph in terms of its order and independence number. There are also upper bounds for the chromatic number of a graph in terms of other parameters of the graph. Several of these will be described in this chapter. Many of these bounds are consequences of an algorithm called the greedy coloring algorithm. We begin this chapter by discussing a class of graphs whose chromatic number decreases when any vertex or edge is removed.

7.1 Color-Critical Graphs

For every k-chromatic graph G with $k \geq 2$ and every vertex v of G, either $\chi(G-v) = k$ or $\chi(G - v) = k - 1$. Furthermore, for every edge e of G, either $\chi(G - e) = k$ or $\chi(G - e) = k - 1$. In fact, if $e = uv$ and $\chi(G - e) = k - 1$, then $\chi(G - u) = k - 1$ and $\chi(G - v) = k - 1$ as well. Graphs that are k-chromatic (but just barely) are often of great interest. A graph G is called **color-critical** if $\chi(H) < \chi(G)$ for every proper subgraph H of G. If G is a color-critical k-chromatic graph, then G is called **critically k-chromatic** or simply **k-critical**. The graph K_2 is the only 2-critical graph. In fact, K_n is n-critical for every integer $n \geq 2$. The odd cycles are the only 3-critical graphs. No characterization of k-critical graphs for any integer $k \geq 4$ has ever been given.

Let G be a k-chromatic graph, where $k \geq 2$, and suppose that H is a k-chromatic subgraph of minimum size in G having no isolated vertices. Then for every proper subgraph F of H, $\chi(F) < \chi(H)$, that is, H is a k-critical subgraph of G. From this observation, it follows that every k-chromatic graph, $k \geq 2$, contains a k-critical subgraph. By Corollary 6.4, every k-critical graph, $k \geq 2$, must be connected; while by Proposition 6.5, every k-critical graph, $k \geq 3$, must be 2-connected and therefore 2-edge-connected. The following theorem tells us even more.

Theorem 7.1 *Every k-critical graph, $k \geq 2$, is $(k-1)$-edge-connected.*

Proof.   The only 2-critical graph is K_2, which is 1-edge-connected; while the only 3-critical graphs are odd cycles, each of which is 2-edge-connected. Since the theorem holds for $k = 2, 3$, we may assume that $k \geq 4$.

Suppose that there is a k-critical graph G, $k \geq 4$, that is not $(k-1)$-edge-connected. This implies that there exists a partition $\{V_1, V_2\}$ of $V(G)$ such that the number of edges joining the vertices of V_1 and the vertices of V_2 is at most $k-2$. Since G is k-critical, the two induced subgraphs

$$G_1 = G[V_1] \text{ and } G_2 = G[V_2]$$

are $(k-1)$-colorable. Let there be given colorings of G_1 and G_2 from the same set of $k-1$ colors and suppose that E' is the set of edges of G that join the vertices in V_1 and the vertices in V_2. It cannot occur that every edge in E' joins vertices of different colors, for otherwise, G itself is $(k-1)$-colorable. Hence, there are edges in E' joining vertices that are assigned the same color. We now show that there exists a permutation of the colors assigned to the vertices of V_1 that results in a proper $(k-1)$-coloring of G_1 in which every edge of E' joins vertices of different colors, which again shows that G is $(k-1)$-colorable, producing a contradiction.

Let $U_1, U_2, \ldots, U_t$ denote the color classes of G_1 for which there is some vertex in U_i $(1 \leq i \leq k-2)$ adjacent to a vertex of V_2. Suppose that there are k_i edges joining the vertices of U_i and the vertices of V_2. Then each $k_i \geq 1$ and

$$\sum_{i=1}^{t} k_i \leq k - 2.$$

If, for every vertex $u_1 \in U_1$, the neighbors of u_1 are assigned a color different from that assigned to u_1, then the color of the vertices in U_1 is not altered. If, on the other hand, some vertex $u_1 \in U_1$ is adjacent to a vertex in V_2 that is colored the same as u_1, then the $k-1$ colors used to color G_1 may be permuted so that no vertex in U_1 is adjacent to a vertex of V_2 having the same color. This is possible since there are at most k_1 colors to avoid when coloring the vertices of U_1 but there are $k-1-k_1 \geq 1$ colors available for this purpose. If, upon giving this new coloring to the vertices of G_1, each vertex $u_2 \in U_2$ is adjacent only to the vertices in V_2 assigned a color different from that of u_2, then no (additional) permutation of the colors of V_1 is performed. Suppose, however, that there is some vertex $u_2 \in U_2$ that is assigned the same color as one of its neighbors in V_2. In this case, we may once again permute the $k-1$ colors used to color the vertices of V_1, where the color assigned to the vertices in U_1 is not changed. This too is possible since there are at most $k_2 + 1$ colors to avoid when coloring the vertices of U_2 but the number of colors available for U_2 is at least

$$(k-1) - (k_2 + 1) \geq (k-1) - (k_2 + k_1) \geq 1.$$

This process is continued until a $(k-1)$-coloring of G is produced, which, as we noted, is impossible. ∎

As a consequence of Theorem 7.1, $\chi(G) \leq 1 + \lambda(G)$ for every color-critical graph G. A related theorem of David W. Matula [148] provides an upper bound for the chromatic number of an arbitrary graph in terms of the edge connectivity of its subgraphs.

Theorem 7.2 *For every graph G,*

$$\chi(G) \leq 1 + \max\{\lambda(H)\},$$

where the maximum is taken over all subgraphs H of G.

Proof. Suppose that F is a color-critical subgraph of G with $\chi(G) = \chi(F)$. By Theorem 7.1,

$$\chi(G) = \chi(F) \leq 1 + \lambda(F) \leq 1 + \max\{\lambda(H)\},$$

where the maximum is taken over all subgraphs H of G. ∎

Since the edge connectivity of a graph never exceeds its minimum degree (by Theorem 2.16), we have the following corollary of Theorem 7.1.

Corollary 7.3 *If G is a color-critical graph, then*

$$\chi(G) \leq 1 + \delta(G).$$

We now consider a consequence of Corollary 7.3. For $n \geq 2$, there is no connected graph of order n having chromatic number 1. Certainly, every tree of order $n \geq 2$ is a connected graph of minimum size having chromatic number 2, while every unicyclic graph of order $n \geq 3$ and containing an odd cycle is a connected graph of minimum size having chromatic number 3. These examples illustrate the following.

Theorem 7.4 *For every two integers n and k with $2 \leq k \leq n$, the minimum size of a connected graph of order n having chromatic number k is $\binom{k}{2} + (n-k)$.*

Proof. Let G be a connected graph of order n having chromatic number k and let H be a k-critical subgraph of G. Suppose that H has order p. Then $k \leq p \leq n$ and G contains $n - p$ vertices not in H. Therefore, there are at least $n - p$ edges of G that are not in H. Since H is k-critical, $\delta(H) \geq k - 1$ by Corollary 7.3. Thus, the size of H is at least $p(k-1)/2$, implying that the size of G is at least

$$\frac{p(k-1)}{2} + (n-p) = p\left(\frac{k-1}{2} - 1\right) + n$$

$$\geq k\left(\frac{k-1}{2} - 1\right) + n = \binom{k}{2} + (n-k).$$

Since the graph G obtained by identifying an end-vertex of P_{n-k+1} with a vertex of K_k is connected, has order n and size $\binom{k}{2} + (n-k)$, and has chromatic number k, this bound is sharp. ∎

We have noted that every k-critical graph, $k \geq 3$, is 2-connected. While every 4-critical graph must be 3-edge-connected (by Theorem 7.1), a 4-critical graph need not be 3-connected, however. The 4-critical graph G of Figure 7.1 has connectivity 2 as $S = \{u, v\}$ is a minimum vertex-cut of G.

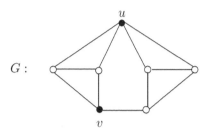

Figure 7.1: A critically 4-chromatic graph

The induced subgraph $G[S]$ for the vertex-cut $S = \{u, v\}$ of the 4-critical graph G of Figure 7.1 is disconnected and is certainly not complete. As it turns out, this is typical. Recall, for a vertex-cut S of a graph G and a component H of $G - S$, that the subgraph $G[V(H) \cup S]$ is referred to as a branch of G at S or an S-branch.

Theorem 7.5 *If S is a vertex-cut of a k-critical graph G, $k \geq 3$, then the subgraph $G[S]$ is not a clique.*

Proof. Since no color-critical graph contains a cut-vertex, $|S| \geq 2$. Let H_1, H_2, $\cdots$, H_s be the components of $G - S$ and let

$$G_i = G[V(H_i) \cup S] \ (1 \leq i \leq s)$$

be the resulting S-branches of G. Since each S-branch is a proper subgraph of G, it follows that $\chi(G_i) \leq k - 1$ for each i $(1 \leq i \leq s)$.

If there exists a $(k - 1)$-coloring of each S-branch G_i in which no two vertices of S are assigned the same color, then by permuting the colors assigned to the vertices of S in each S-branch, if necessary, a $(k-1)$-coloring of G can be produced. Since this is impossible, there is some S-branch in which no $(k-1)$-coloring assigns distinct colors to the vertices of S. This, in turn, implies that $G[S]$ contains two nonadjacent vertices and so $G[S]$ is not a clique. ∎

The following result is a consequence of Theorem 7.5 and its proof.

Corollary 7.6 *Let G be a k-critical graph, $k \geq 3$, where $\kappa(G) = 2$. If $S = \{u, v\}$ is a vertex-cut of G, then $uv \notin E(G)$ and there exists some S-branch G' of G such that $\chi(G' + uv) = k$.*

Proof. If $S = \{u, v\}$, then $uv \notin E(G)$ by Theorem 7.5. As we saw in the proof of Theorem 7.5, there is an S-branch G' of G in which no $(k - 1)$-coloring of G' assigns distinct colors to u and v. Therefore, $G' + uv$ is not $(k - 1)$-colorable and so $\chi(G' + uv) = k$. ∎

7.2 Upper Bounds and Greedy Colorings

Many combinatorial decision problems (those having a "yes" or "no" answer) are difficult to solve but once a solution is revealed, the solution is easy to verify. For example, the problem of determining whether a given graph G is k-colorable for some integer $k \geq 3$ is difficult to solve, that is, it is difficult to determine whether there exists a k-coloring of G. However, it is easy to verify that a given coloring of G is a k-coloring. It is only necessary to show that no more than k distinct colors are used and that adjacent vertices are assigned distinct colors. The collection of all such difficult-to-solve but easy-to-verify problems is denoted by **NP**. The collection of all decision problems that can be solved in polynomial time is denoted by **P**. The problems in the set **NP** have only one property in common with the problems belonging to the set **P**, namely: Given a solution to a problem in either set, the solution can be verified in polynomial time. Thus, $\mathbf{P} \subseteq \mathbf{NP}$. One of the best-known problems in mathematics is whether every problem in the set **NP** also belongs to the set **P**. This problem is called the $\mathbf{P} = \mathbf{NP}$ **problem** and is considered by many as the most important unsolved problem in theoretical computer science. Its importance and fame have only been magnified because of a one million dollar prize offered by the Clay Mathematics Institute for the first correct solution of this problem.

A problem in the set **NP** is called **NP-complete** if a polynomial-time algorithm for a solution would result in polynomial-time solutions for all problems in **NP**. The **NP**-complete problems are among the most difficult in the set **NP** and can be reduced from and to all other **NP**-complete problems in polynomial time. The concept of **NP**-completeness was initiated in 1971 by Stephen Cook [59] who gave an example of the first **NP**-complete problem. The following year Richard M. Karp [132] described some twenty diverse problems, all of them **NP**-complete. It is now known that there are thousands of **NP**-complete problems.

Since determining the chromatic number of a graph is known to be so very difficult, it is not surprising that much of the research emphasis on coloring has centered on finding bounds (both lower bounds and upper bounds) for the chromatic number of a graph. For a graph G of order n with clique number $\omega(G)$ and independence number $\alpha(G)$, we have already seen that $\omega(G)$ and $n/\alpha(G)$ are lower bounds for $\chi(G)$ while $n - \alpha(G) + 1$ is an upper bound for $\chi(G)$. Of course, n is also an upper bound for $\chi(G)$. In particular,

$$\omega(G) \leq \chi(G) \leq n.$$

Bruce Reed [164] showed that $\chi(G)$ can never be closer to n than to $\omega(G)$.

Theorem 7.7 *For every graph G of order n,*

$$\chi(G) \leq \left\lfloor \frac{n + \omega(G)}{2} \right\rfloor.$$

Proof. We proceed by induction on the nonnegative integer $k = |V(G)| - \omega(G)$. If $k = 0$, then G is a graph such that $|V(G)| - \omega(G) = 0$ and so $G = K_n$ for some positive integer n. Thus, $\chi(G) = \omega(G) = n$ and so $\chi(G) = \left\lfloor \frac{n+\omega(G)}{2} \right\rfloor$. This verifies the basis step of the induction.

Assume for a positive integer k that every graph H with $|V(H)| - \omega(H) < k$ satisfies $\chi(H) \leq \left\lfloor \frac{|V(H)|+\omega(H)}{2} \right\rfloor$. Now, let G be a graph of order n such that $n - \omega(G) = k$. We show that $\chi(G) \leq \left\lfloor \frac{n+\omega(G)}{2} \right\rfloor$. If $k = 1$, then $\omega(G) = n - 1$ and so

$$\chi(G) = n - 1 = \left\lfloor \frac{2n-1}{2} \right\rfloor = \left\lfloor \frac{n+\omega(G)}{2} \right\rfloor.$$

Hence, we may assume that $k \geq 2$. Since G is not complete, G contains two nonadjacent vertices u and v. Let $H = G - u - v$. Then H has order $n - 2$ and either $\omega(H) = \omega(G)$ or $\omega(H) = \omega(G) - 1$. In either case, $0 \leq |V(H)| - \omega(H) < k$ and so

$$\chi(H) \leq \left\lfloor \frac{n - 2 + \omega(H)}{2} \right\rfloor$$

by the induction hypothesis. Hence, there exists a $\left\lfloor \frac{n-2+\omega(H)}{2} \right\rfloor$-coloring of H. Assigning u and v the same new color implies that

$$\chi(G) \leq \left\lfloor \frac{n - 2 + \omega(H)}{2} \right\rfloor + 1 = \left\lfloor \frac{n + \omega(H)}{2} \right\rfloor \leq \left\lfloor \frac{n + \omega(G)}{2} \right\rfloor,$$

giving the desired result.   ∎

When assigning colors to the vertices of some graph G, we typically want to use as few colors as possible, namely $\chi(G)$ colors. Since we have mentioned that this is an extraordinarily difficult problem in general, this suggests consideration of the problem of providing a coloring of G that does not use an excessive number of colors (if this is possible). One possible approach is to use a greedy method, which provides a step-by-step strategy for coloring the vertices of a graph such that at each step, an apparent optimal choice for a color of a vertex is made. While this method may not result in coloring the vertices of G using the minimum number of colors, it does provide an upper bound (in fact, several upper bounds) for the chromatic number of G.

Let G be a graph of order n whose n vertices are listed in some specified order. In a greedy coloring of G, the vertices are successively colored with positive integers according to an algorithm that assigns to the vertex under consideration the smallest available color. Hence, if the vertices of G are listed in the order $v_1, v_2, \ldots, v_n$, then the resulting greedy coloring c assigns the color 1 to v_1, that is, $c(v_1) = 1$. If v_2 is not adjacent to v_1, then also define $c(v_2) = 1$; while if v_2 is adjacent to v_1, then define $c(v_2) = 2$. In general, suppose that the first j vertices $v_1, v_2, \ldots, v_j$ in the sequence have been colored, where $1 \leq j < n$, and t is the smallest positive integer not used in coloring any neighbor of v_{j+1} from among $v_1, v_2, \ldots, v_j$. We then define $c(v_{j+1}) = t$. When the algorithm ends, the vertices of G have been assigned colors from the set $\{1, 2, \ldots, k\}$ for some positive integer k. Thus, $\chi(G) \leq k$ and so k is an upper bound for the chromatic number of G. This algorithm is now stated more formally.

The Greedy Coloring Algorithm *Suppose that the vertices of a graph G are listed in the order* $v_1, v_2, \ldots, v_n$.

1. *The vertex* v_1 *is assigned the color 1.*

2. *Once the vertices* $v_1, v_2, \ldots, v_j$ *have been assigned colors, where* $1 \le j < n$, *the vertex* v_{j+1} *is assigned the smallest color that is not assigned to any neighbor of* v_{j+1} *belonging to the set* $\{v_1, v_2, \ldots, v_j\}$

While the greedy coloring algorithm is efficient in the sense that the vertex coloring that it produces, regardless of the order in which its vertices are listed, is done in polynomial time (a polynomial in the order n of the graph), the number of colors in the coloring obtained need not equal or even be close to the chromatic number of the graph. Indeed, there is reason not to be optimistic about finding any efficient algorithm that produces a coloring of each graph where the number of colors is close to the chromatic number of the graph since Michael R. Garey and David S. Johnson [95] have shown that if there should be an efficient algorithm that produces a coloring of every graph G using at most $2\chi(G)$ colors, then there is an efficient algorithm that determines $\chi(G)$ exactly for every graph G.

As an illustration of the greedy coloring algorithm, suppose that we consider the graph C_6 of Figure 7.2. If we list the vertices of C_6 in the order u, w, v, y, z, x, then the greedy coloring algorithm yields the coloring c of G defined by

$$c(u) = 1, \ c(w) = 1, \ c(v) = 2, \ c(y) = 1, \ c(z) = 2, \ c(x) = 2.$$

This gives $\chi(C_6) \le 2$. Of course, $\chi(C_6) = 2$ and so with this ordering of the vertices of C_6, a $\chi(C_6)$-coloring is produced. On the other hand, if the vertices of G are listed in the order u, x, v, w, z, y, then the greedy coloring algorithm yields the coloring c' of G defined by

$$c'(u) = 1, \ c'(x) = 1, \ c'(v) = 2, \ c'(w) = 3, \ c'(z) = 2, \ c'(y) = 3.$$

This gives a 3-coloring of C_6, which, of course, is not the chromatic number of C_6.

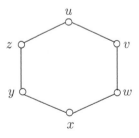

Figure 7.2: The graph C_6

From the second listing of the vertices of C_6, we now see that $\chi(C_6) \le 3$. No greater upper bound for $\chi(C_6)$ is possible using the greedy coloring algorithm.

Theorem 7.8 *For every graph G,*

$$\chi(G) \leq 1 + \Delta(G).$$

Proof. Suppose that the vertices of G are listed in the order $v_1, v_2, \ldots, v_n$ and the greedy coloring algorithm is applied. Then v_1 is assigned the color 1 and for $2 \leq i \leq n$, the vertex v_i is either assigned the color 1 or is assigned the color $k + 1$, where k is the largest integer such that all of the colors $1, 2, \ldots, k$ are used to color the neighbors of v_i in the set $S = \{v_1, v_2, \ldots, v_{i-1}\}$. Since at most $\deg v_i$ neighbors of v_i belong to S, the largest value of k is $\deg v_i$. Hence, the color assigned to v_i is at most $1 + \deg v_i$. Thus,

$$\chi(G) \leq \max_{1 \leq i \leq n} \{1 + \deg v_i\} = 1 + \Delta(G),$$

as desired.   ∎

The following theorem of George Szekeres and Herbert S. Wilf [184] gives a bound for the chromatic number of a graph that is an improvement over that stated in Theorem 7.8 but which is more difficult to compute. This result is also a consequence of Theorem 7.2 and Corollary 7.3.

Theorem 7.9 *For every graph G,*

$$\chi(G) \leq 1 + \max\{\delta(H)\},$$

where the maximum is taken over all subgraphs H of G.

Proof. Let $\chi(G) = k$ and let F be a k-critical subgraph of G. By Corollary 7.3, $\delta(F) \geq k - 1$. Thus,

$$k - 1 \leq \delta(F) \leq \max\{\delta(H)\},$$

where the maximum is taken over all subgraphs H of G. Therefore, $\chi(G) \leq 1 + \max\{\delta(H)\}$.   ∎

Since $\delta(H') \leq \delta(H)$ for each spanning subgraph H' of a graph H, it follows that in Theorem 7.9 we may restrict the subgraphs H of G only to those that are induced.

For an r-regular graph G, both Theorems 7.8 and 7.9 give the same upper bound for $\chi(G)$, namely $1 + r$. On the other hand, if T is a tree of order at least 3, then Theorem 7.8 gives $1 + \Delta(T) \geq 3$ for an upper bound for $\chi(T)$; while Theorem 7.9 gives the improved upper bound $\chi(T) \leq 2$ since every subgraph of a tree contains a vertex of degree 1 or less. Of course, $\chi(T) = 2$ for every nontrivial tree T.

When applying the greedy coloring algorithm to a graph G, there are, in general, less colors to avoid when coloring a vertex if the vertices of higher degree are listed early in the ordering of the vertices of G. The following result is due to Dominic J. A. Welsh and Martin B. Powell [199].

Theorem 7.10 *Let G be a graph of order n whose vertices are listed in the order $v_1, v_2, \ldots, v_n$ so that $\deg v_1 \geq \deg v_2 \geq \cdots \geq \deg v_n$. Then*

$$\chi(G) \leq 1 + \min_{1 \leq i \leq n} \{\max\{i - 1, \deg v_i\}\} = \min_{1 \leq i \leq n} \{\max\{i, 1 + \deg v_i\}\}.$$

Proof. Suppose that $\chi(G) = k$ and let H be a k-critical subgraph of G. Hence, the order of H is at least k and $\delta(H) \geq k - 1$ by Corollary 7.3. Therefore, for $1 \leq i \leq k$,

$$\max\{i, 1 + \deg v_i\} \geq k;$$

while for $k + 1 \leq i \leq n$,

$$\max\{i, 1 + \deg v_i\} \geq k + 1.$$

Consequently, $\chi(G) = k \leq \min_{1 \leq i \leq n} \{\max\{i, 1 + \deg v_i\}\}$. ∎

Let G be a connected graph. The **core** of G is obtained by successively deleting end-vertices until none remain. Thus, if G is a tree, then its core is K_1; while if G is not a tree, then the core of G is the induced subgraph H of maximum order with $\delta(H) \geq 2$. If G is not a tree and H is the core of G, then $\chi(H) = \chi(G)$. The following result by William C. Coffman, S. Louis Hakimi, and Edward Schmeichel [58] gives an upper bound for the chromatic number of a graph in terms of its order and size only.

Theorem 7.11 *Let G be a connected graph of order n and size m that is not a tree. If the core of G is neither a complete graph nor an odd cycle, then*

$$\chi(G) \leq \frac{3 + \sqrt{1 + 8(m - n)}}{2}.$$

Proof. We may assume that $\delta(G) \geq 2$ for if $\delta(G) = 1$, then the core H of G is obtained by reducing m and n by an equal amount and $\chi(H) = \chi(G)$. Hence, we may assume (1) that $\max\{\delta(H)\} \geq 2$ where the maximum is taken over all induced subgraphs H of G and (2) that G is neither a complete graph nor an odd cycle. If G is an even cycle, them $m = n$ and the bound for $\chi(G)$ is 2, which is correct in this case. Thus, we may further assume that $m \geq n + 1$.

Suppose that $d = \max\{\delta(H)\}$ where the maximum is taken over all induced subgraphs H of G. Hence, $d \geq 2$. We consider two cases.

Case 1. $d \geq 4$. In this case, we show that $m \geq n + \binom{d}{2}$. Let k be the smallest integer for which there is an induced subgraph H of G having order $n - k + 1$ and $\delta(H) = d$. Then

$$m \geq |E(H)| \geq \frac{(n - k - 1)d}{2}.$$

We now consider two subcases, depending on whether $k = 1$ or $k \geq 2$.

Subcase 1.1. $k = 1$. Thus, $H = G$ in this case. Since G is not complete, $d \leq n - 2$. Thus,

$$m \geq \frac{nd}{2} = n + \frac{n(d - 2)}{2}$$
$$\geq n + \frac{(d - 2)(d + 2)}{2} \geq n + \binom{d}{2},$$

as desired.

Subcase 1.2. $k \geq 2$. First, we claim that the size of H is at most $m - k$. Let $V(G) - V(H) = S$, where then $|S| = k - 1$. Suppose that the size of the subgraph $G[S]$ is m_1 and the size of the subgraph $[V(G) - V(H), S]$ is m_2. Since $\delta(G) \geq 2$, it follows that

$$2m_1 + m_2 = \sum_{v \in S} \deg_G v \geq 2(k - 1).$$

Since G is connected, $m_2 \geq 1$ and so

$$m_1 + m_2 \geq (k - 1) + \frac{m_2}{2} \geq k.$$

Therefore, as claimed, the size of H is at most $m - k$. That is,

$$m \geq k + |E(H)| \geq k + \frac{(n - k - 1)d}{2}. \tag{7.1}$$

Since the order of H is $n - k - 1$, it follows that $n - k - 1 \geq d + 1$ and so $n - k \geq d$. Because $d \geq 4$, we have $(n - k)(d - 2) \geq d(d - 2)$, which is equivalent to

$$k + \frac{(n - k - 1)d}{2} \geq n + \binom{d}{2}. \tag{7.2}$$

By (7.1) and (7.2), $m \geq n + \binom{d}{2}$ in this subcase as well.

Since $d \geq 4$ and $m \geq n + \binom{d}{2}$, it follows that $d^2 - d - 2(m - n) \leq 0$ and so $d \leq \frac{1 + \sqrt{1 + 8(m - n)}}{2}$. Thus,

$$\chi(G) \leq 1 + d \leq \frac{3 + \sqrt{1 + 8(m - n)}}{2}.$$

Case 2. $2 \leq d \leq 3$. Since $\chi(G) \leq 1 + d \leq 4$, the bound in the theorem is correct if $m - n \geq 3$. Hence, we may assume that $1 \leq m - n \leq 2$. If $n \geq 5$, then $d = 2$ and $\chi(G) \leq 3$, giving the correct bound. On the other hand, if $n \leq 4$, then $\chi(G) \leq 3$ since $G \neq K_4$, again giving the correct bound. ∎

Dennis Paul Geller [96] gave a *lower* bound for the chromatic number of a graph in terms of its order and size only.

Theorem 7.12 *If G is a graph of order n and size m, then*

$$\chi(G) \geq \frac{n^2}{n^2 - 2m}.$$

Proof. Suppose that $\chi(G) = k$ and let c be a k-coloring of G resulting in color classes $V_1, V_2, \ldots, V_k$ with $|V_i| = n_i$ for $1 \leq i \leq k$. Then the largest possible size of G occurs when G is a complete k-partite graph with partite sets $V_1, V_2, \ldots, V_k$ and the cardinalities of these partite sets are as equal as possible (or each $|V_i|$ is as close to $\frac{n}{k}$ as possible for $1 \leq i \leq k$). This implies that

$$m \leq \binom{k}{2} \frac{n^2}{k^2}$$

and so
$$2m \leq \frac{(k-1)n^2}{k}.$$

Thus,
$$\frac{n^2}{n^2 - 2m} \leq \frac{n^2}{n^2 - \frac{(k-1)n^2}{k}} = k = \chi(G),$$

giving the desired result. ∎

We have seen that $\chi(K_n) = n$ and $\chi(\overline{K}_n) = 1$. So, $\chi(K_n) + \chi(\overline{K}_n) = n + 1$. By Theorem 6.19, if G is a graph such that $\overline{G}$ is a nonempty bipartite graph, then $\chi(\overline{G}) = 2$, while $\chi(G) = \omega(G) \leq n - 1$. Thus, $\chi(G) + \chi(\overline{G}) \leq n + 1$. Edward A. Nordhaus and Jerry W. Gaddum [155] showed that this inequality holds for all graphs G of order n when they established two pairs of inequalities involving the sum and product of the chromatic numbers of a graph and its complement. Such inequalities established for any parameter have become known as **Nordhaus-Gaddum inequalities**. The following proof is due to Hudson V. Kronk (see [49]).

Theorem 7.13 (**Nordhaus-Gaddum Theorem**) *If G is a graph of order n, then*

(i) $2\sqrt{n} \leq \chi(G) + \chi(\overline{G}) \leq n + 1$

(ii) $n \leq \chi(G) \cdot \chi(\overline{G}) \leq \left(\frac{n+1}{2}\right)^2$.

Proof. Suppose that $\chi(G) = k$ and $\chi(\overline{G}) = \ell$. Let a k-coloring c of G and an ℓ-coloring $\overline{c}$ of $\overline{G}$ be given. Using these colorings, we obtain a coloring of K_n. With each vertex v of G (and of $\overline{G}$), we associate the ordered pair $(c(v), \overline{c}(v))$. Since every two vertices of K_n are either adjacent in G or in $\overline{G}$, they are assigned different colors in that subgraph of K_n. Thus, this is a coloring of K_n using at most $k\ell$ colors. Therefore,
$$n = \chi(K_n) \leq k\ell = \chi(G) \cdot \chi(\overline{G}).$$

This establishes the lower bound in (ii). Since the geometric mean of two positive real numbers never exceeds their arithmetic mean, it follows that
$$\sqrt{n} \leq \sqrt{\chi(G) \cdot \chi(\overline{G})} \leq \frac{\chi(G) + \chi(\overline{G})}{2}. \tag{7.3}$$

Consequently,
$$2\sqrt{n} \leq \chi(G) + \chi(\overline{G}),$$

which verifies the lower bound in (i).

To verify the upper bound in (i), let $p = \max\{\delta(H)\}$, where the maximum is taken over all subgraphs H of G. Hence, the minimum degree of every subgraph of G is at most p. By Theorem 7.9, $\chi(G) \leq 1 + p$.

We claim that the minimum degree of every subgraph of $\overline{G}$ is at most $n - p - 1$. Assume, to the contrary, that there is a subgraph H of G such that $\delta(\overline{H}) \geq n - p$ for the subgraph $\overline{H}$ in $\overline{G}$. Thus, every vertex of H has degree $p - 1$ or less in G. Let F

be a subgraph of G such that $\delta(F) = p$. So every vertex of F has degree p or more. This implies that no vertex of F belongs to H. Since the order of F is at least $p+1$, the order of H is at most $n - (p + 1) = n - p - 1$. This, however, contradicts the fact that $\delta(\overline{H}) \geq n - p$. Thus, as claimed, the minimum degree of every subgraph of $\overline{G}$ is at most $n - p - 1$. By Theorem 7.9, $\chi(\overline{G}) \leq 1 + (n - p - 1) = n - p$ and so

$$\chi(G) + \chi(\overline{G}) \leq (1 + p) + (n - p) = n + 1.$$

This verifies the upper bound in (i). By (7.3),

$$\chi(G) \cdot \chi(\overline{G}) \leq \left(\frac{n+1}{2}\right)^2,$$

verifying the final inequality.   ∎

Bonnie M. Stewart [182] and Hans-Joachim Finck [84] showed that no improvement in Theorem 7.13 is possible.

Theorem 7.14   *Let n be a positive integer. For every two positive integers a and b such that*

$$2\sqrt{n} \leq a + b \leq n + 1 \text{ and } n \leq ab \leq \left(\tfrac{n+1}{2}\right)^2,$$

there is a graph G of order n such that $\chi(G) = a$ and $\chi(\overline{G}) = b$.

Proof.   Let $n_1, n_2, \ldots, n_a$ be positive integers such that $\sum_{i=1}^{a} n_i = n$ and $n_1 \leq n_2 \leq \ldots \leq n_a = b$. Since $a + b - 1 \leq n \leq ab$, such integers n_i $(1 \leq i \leq a)$ exist. The graph $G = K_{n_1, n_2, \ldots, n_a}$ has order n and $\overline{G} = K_{n_1} + K_{n_2} + \cdots + K_{n_a}$. Hence, $\chi(G) = a$ and $\chi(\overline{G}) = b$.   ∎

We saw in Theorem 7.11 that for a connected graph G of order n and size m that is not a tree,

$$\chi(G) \leq \frac{3 + \sqrt{1 + 8(m - n)}}{2},$$

provided that the core of G is neither a complete graph nor an odd cycle. Complete graphs and odd cycles are exceptional graphs in another upper bound for the chromatic number. While $1 + \Delta(G)$ is an upper bound for the chromatic number of a connected graph G, Rowland Leonard Brooks [29] showed that the instances where $\chi(G) = 1 + \Delta(G)$ are rare.

Theorem 7.15   (Brooks's Theorem)   *For every connected graph G that is not an odd cycle or a complete graph,*

$$\chi(G) \leq \Delta(G).$$

Proof.   Let $\chi(G) = k \geq 2$ and let H be a k-critical subgraph of G. Thus, H is 2-connected and $\Delta(H) \leq \Delta(G)$. If $H = K_k$ or H is an odd cycle, then $G \neq H$ since G is neither an odd cycle nor a complete graph. Since G is connected, $\Delta(G) \geq k$ if $H = K_k$; while $\Delta(G) \geq 3$ if H is an odd cycle. If $H = K_k$, then $k = \chi(H) = \chi(G) \leq$

$\Delta(G)$; while if H is an odd cycle, then $3 = \chi(H) = \chi(G) \leq \Delta(G)$. Therefore, in both cases, $\chi(G) \leq \Delta(G)$, as desired. Hence, we may assume that H is a k-critical subgraph that is neither an odd cycle nor a complete graph. This implies that $k \geq 4$.

Suppose that H has order n. Since $\chi(G) = k \geq 4$ and H is not complete, $n > k$ and so $n \geq 5$. Since H is 2-connected, either H is 3-connected or H has connectivity 2. We consider these two cases.

Case 1. H is 3-connected. Since H is not complete, there are two vertices u and w of H such that $d_H(u, w) = 2$. Let (u, v, w) be a $u - w$ geodesic in H. Since H is 3-connected, $H - u - w$ is connected. Let $v = u_1, u_2, \ldots, u_{n-2}$ be the vertices of $H - u - w$, so listed that each vertex u_i $(2 \leq i \leq n - 2)$ is adjacent to some vertex preceding it. Let $u_{n-1} = u$ and $u_n = w$. Consequently, for each set

$$U_j = \{u_1, u_2, \ldots, u_j\}, \ 1 \leq j \leq n,$$

the induced subgraph $H[U_j]$ is connected.

We now apply a greedy coloring to H with respect to the reverse ordering

$$w = u_n, u = u_{n-1}, u_{n-2}, \ldots, u_2, u_1 = v \tag{7.4}$$

of the vertices of H. Since w and u are not adjacent, each is assigned the color 1. Furthermore, each vertex u_i $(2 \leq i \leq n - 2)$ is assigned the smallest color in the set $\{1, 2, \ldots, \Delta(H)\}$ that was not used to color a neighbor of u_i that preceded it in the sequence (7.4). Since each vertex u_i has at least one neighbor following it in the sequence (7.4), u_i has at most $\Delta(H) - 1$ neighbors preceding it in the sequence and so a color is available for u_i. Moreover, the vertex $u_1 = v$ is adjacent to two vertices colored 1 (namely $w = u_n$ and $u = u_{n-1}$) and so at most $\Delta(H) - 1$ colors are assigned to the neighbors of v, leaving a color for v. Hence,

$$\chi(G) = \chi(H) \leq \Delta(H) \leq \Delta(G). \tag{7.5}$$

Case 2. $\kappa(H) = 2$. We claim that H contains a vertex x such that $2 < \deg_H x < n - 1$. Suppose that this is not the case. Then every vertex of H has degree 2 or $n - 1$. Because $\chi(H) \geq 4$, it follows that H cannot contain only vertices of degree 2; and because H is not complete, H cannot contain only vertices of degree $n - 1$. If H contains vertices of both degrees 2 and $n - 1$ and no others, then either

$$H = K_1 + \left(\frac{n-1}{2}\right) K_2 \quad \text{or} \quad H = K_{1,1,n-2}.$$

In both cases, $\chi(H) = 3$ and H is not critical, which is impossible. Thus, as claimed, H contains a vertex x such that $2 < \deg_H x < n - 1$.

Since $\kappa(H) = 2$, either $\kappa(H - x) = 2$ or $\kappa(H - x) = 1$. If $\kappa(H - x) = 2$, then x belongs to no minimum vertex-cut of H, which implies that H contains a vertex y such that $d_H(x, y) = 2$. Proceeding as in Case 1 with $u = x$ and $w = y$, we see that there is a coloring of H with at most $\Delta(H)$ colors and so once again we have (7.5), that is, $\chi(G) \leq \Delta(G)$.

Finally, we may assume that $\kappa(H - x) = 1$. Thus, $H - x$ contains end-blocks B_1 and B_2, containing cut-vertices x_1 and x_2, respectively, of $H - x$. Since H is 2-connected, there exist vertices $y_1 \in V(B_1) - \{x_1\}$ and $y_2 \in V(B_2) - \{x_2\}$ such that x is adjacent to both y_1 and y_2. Proceeding as in Case 1 with $u = y_1$ and $w = y_2$, we obtain a coloring of H with at most $\Delta(H)$ colors, once again giving us (7.5) and so $\chi(G) \leq \Delta(G)$.   ∎

We have now seen several bounds for the chromatic number of a graph G. The clique number $\omega(G)$ is the best known and simplest lower bound for $\chi(G)$, while $1 + \Delta(G)$ is the best known and simplest upper bound for $\chi(G)$. If $\Delta(G) \geq 3$ and G is not complete, then $\Delta(G)$ is an improved upper bound for $\chi(G)$ by Theorem 7.15. Bruce Reed [164] conjectured that there is an upper bound for $\chi(G)$ involving the average of the two numbers $\omega(G)$ and $1 + \Delta(G)$.

Reed's Conjecture   *For every graph G,*

$$\chi(G) \leq \left\lceil \frac{\omega(G) + 1 + \Delta(G)}{2} \right\rceil.$$

As we saw in Theorem 6.10, for a graph G of order n, the number $n - \alpha(G) + 1$ is also an upper bound for $\chi(G)$. By Theorem 7.7, $(n + \omega(G))/2$ is also an upper bound for $\chi(G)$. Robert C. Brigham and Ronald D. Dutton [28] proved that this upper bound for $\chi(G)$ can be reduced by $(\alpha(G) - 1)/2$.

Theorem 7.16   *For every graph G of order n,*

$$\chi(G) \leq \frac{\omega(G) + n + 1 - \alpha(G)}{2}.$$

Proof.   We proceed by induction on n. When $n = 1$, $G = K_1$ and $\chi(G) = \omega(G) = \alpha(G) = 1$ and so

$$\chi(G) = \frac{\omega(G) + n + 1 - \alpha(G)}{2}.$$

Thus, the basis step holds for the induction.

Assume that the inequality holds for all graphs of order less than n where $n \geq 2$ and let G be a graph of order n. If $G = \overline{K}_n$, then $\chi(G) = \omega(G) = 1$ and $\alpha(G) = n$; so,

$$\chi(G) = \frac{\omega(G) + n + 1 - \alpha(G)}{2}.$$

Hence, we may assume that $G \neq \overline{K}_n$. Thus, $1 \leq \alpha(G) \leq n - 1$. Let V_0 be a maximum independent set of vertices in G. Therefore, $|V_0| = \alpha(G)$. Let $G_1 = G - V_0$, where $\alpha(G_1) = \alpha_1$ and $\omega(G_1) = \omega_1$. Furthermore, let $V(G_1) = V_1$, where, then, $|V_1| = n - \alpha(G)$. We consider two cases.

Case 1. G_1 is a complete graph. Thus, $V(G)$ can be partitioned into V_0 and V_1, where $G[V_0] = \overline{K}_{\alpha(G)}$ and $G_1 = G[V_1] = K_{n-\alpha(G)}$. Therefore, either $\chi(G) = \omega(G) = n - \alpha(G)$ or $\chi(G) = \omega(G) = n - \alpha(G) + 1$. We now consider these two subcases.

Subcase 1.1. $\chi(G) = \omega(G) = n - \alpha(G)$. So,

$$\begin{aligned}\chi(G) &= n - \alpha(G) = \frac{(n - \alpha(G)) + (n - \alpha(G))}{2} \\ &= \frac{\omega(G) + n - \alpha(G)}{2} < \frac{\omega(G) + n + 1 - \alpha(G)}{2}.\end{aligned}$$

Subcase 1.2. $\chi(G) = \omega(G) = n - \alpha(G) + 1$. Here

$$\begin{aligned}\chi(G) &= n - \alpha(G) + 1 = \frac{(n - \alpha(G) + 1) + (n - \alpha(G) + 1)}{2} \\ &= \frac{\omega(G) + (n - \alpha(G) + 1)}{2}.\end{aligned}$$

Case 2. G_1 is not a complete graph. In this case, $\alpha_1 \geq 2$. Since $\chi(G) \leq \chi(G_1) + 1$, it follows by the induction hypothesis that

$$\begin{aligned}\chi(G) &\leq \chi(G_1) + 1 \leq \frac{\omega_1 + (n - \alpha(G)) + 1 - \alpha_1}{2} + 1 \\ &\leq \frac{\omega(G) + (n - \alpha(G)) + 1 - \alpha_1}{2} + 1 \leq \frac{\omega(G) + (n - \alpha(G) + 1)}{2},\end{aligned}$$

completing the proof. ∎

Applying Theorem 7.16 to both a graph G of order n and its complement $\overline{G}$ (where then $\omega(\overline{G}) = \alpha(G)$ and $\alpha(\overline{G}) = \omega(G)$), we have

$$\chi(G) \leq \frac{\omega(G) + n + 1 - \alpha(G)}{2}$$

and

$$\chi(\overline{G}) \leq \frac{\omega(\overline{G}) + n + 1 - \alpha(\overline{G})}{2}.$$

Adding these inequalities gives us an alternative proof of the major inequality stated in the Nordhaus-Gaddum Theorem (Theorem 7.10): For every graph G of order n,

$$\chi(G) + \chi(\overline{G}) \leq n + 1.$$

7.3 Upper Bounds and Oriented Graphs

Bounds for the chromatic number of a graph G can also be given in terms of the length $\ell(D)$ of a longest (directed) path in an orientation D of G. Suppose that G is a k-chromatic graph, where a k-coloring of G is given using the colors $1, 2, \ldots, k$. An orientation D of G can be constructed by directing each edge uv of G from u to v if the color assigned to u is smaller than the color assigned to v. Thus, the length of every directed path in D is at most $k - 1$. In particular, $\ell(D) \leq \chi(G) - 1$. So there exists an orientation D of G such that $\chi(G) \geq 1 + \ell(D)$. On the other hand, Tibor Gallai [93], Bernard Roy [174], and L. M. Vitaver [192] independently discovered the following result.

Theorem 7.17 (The Gallai-Roy-Vitaver Theorem) *For every orientation D of a graph G,*

$$\chi(G) \leq 1 + \ell(D).$$

Proof. Let D be an orientation of G and let D' be a spanning acyclic subdigraph of D of maximum size. A coloring c is defined on G by assigning to each vertex v of G the color 1 plus the length of a longest path in D' whose terminal vertex is v. Then, as we proceed along the vertices in any directed path in D', the colors are strictly increasing.

Let (u, v) be an arc of D. If (u, v) belongs to D', then $c(u) < c(v)$. On the other hand, if (u, v) is not in D', then adding (u, v) to D' creates a directed cycle, which implies that $c(v) < c(u)$. Consequently, $c(u) \neq c(v)$ for every two adjacent vertices u and v of G and so

$$c : V(G) \rightarrow \{1, 2, \ldots, 1 + \ell(D)\}$$

is a proper coloring of G. Therefore, $\chi(G) \leq 1 + \ell(D)$. $\blacksquare$

The following result is therefore a consequence of Theorem 7.17 and the observation that precedes it.

Corollary 7.18 *Let G be a graph and let $\ell = \min\{\ell(D)\}$, where the minimum is taken over all orientations D of G. Then*

$$\chi(G) = 1 + \ell.$$

There is also a corollary to Corollary 7.18.

Corollary 7.19 *Every orientation of a graph G contains a directed path with at least $\chi(G)$ vertices.*

Not only does every orientation of a k-chromatic graph G contain a directed path with at least k vertices but for each k-coloring of G, every orientation of G has a directed path containing a vertex of each color. The following even stronger result is due to Hao Li [140], but the proof we present is due to Gerard J. Chang, Li-Da Tong, Jing-Ho Yan, and Hong-Gwa Yeh [36].

Theorem 7.20 *Let G be a connected k-chromatic graph. For every k-coloring of G and for every vertex v of G, there exists a path P in G with initial vertex v such that for each of the k colors, there is a vertex on P assigned that color.*

Proof. We proceed by induction on the chromatic number of a graph. If $\chi(G) = 1$, then $G = K_1$ and the result follows trivially. Therefore, the basis step of the induction is true.

For an integer $k \geq 2$, assume for every connected graph H with $\chi(H) = k - 1$ that for every $(k - 1)$-coloring of H and for every vertex x of H, there exists a path in H with initial vertex x such that for each of the $k - 1$ colors, there is a vertex on the path assigned that color. Let G be a connected k-chromatic graph and let a k-coloring c of G be given with colors from the set $S = \{1, 2, \ldots, k\}$. Furthermore, let v be a vertex of G and suppose that $c(v) = j$, where $j \in S$. Let V_j be the color class consisting of those vertices of G assigned the color j. Then there exists a component H of $G - V_j$ such that $\chi(H) = k - 1$.

Let u be a vertex of H whose distance from v is minimum and let P' be a $v - u$ geodesic in G. By the induction hypothesis, there is a path P'' in H with initial vertex u such that for each color $i \in S - \{j\}$, there is a vertex x on P'' in H such that $c(x) = i$. The path P consisting of P' followed by P'' has the desired property. ∎

Not only is there an upper bound for the chromatic number of a graph G in terms of the length of a longest path in an orientation D of G, an upper bound can be given with the aid of orientations and the cycles of G.

Let D be an acyclic orientation of a graph G that is not a forest. For each cycle C of G, there are then $a(C)$ edges of C oriented in one direction and $b(C)$ edges oriented in the opposite direction for some positive integers $a(C)$ and $b(C)$ with $a(C) \geq b(C)$. We will refer to each of these $a(C)$ edges of C as a **forward edge** and each of the $b(C)$ edges as a **backward edge**. Let $r(D)$ denote the maximum of $a(C)/b(C)$ over all cycles C of G, that is,

$$r(D) = \max \left\{ \frac{a(C)}{b(C)} \right\},$$

where the maximum is taken over all cycles C of G. Then

$$\frac{b(C)}{a(C)} \leq \frac{a(C)}{b(C)} \leq r(D)$$

for each cycle C of G. Also, if W is a $u - v$ walk in G, where

$$W = (u = v_0, v_1, \ldots, v_k = v),$$

then $v_i v_{i+1}$ is a **forward edge** of W if (v_i, v_{i+1}) is an arc of D; while $v_i v_{i+1}$ is a **backward edge** of W if (v_{i+1}, v_i) is an arc of D. If there are $k \geq 2$ occurrences of an edge xy of G on W, then xy is a forward edge and/or backward edge of W a total of k times.

The length of a (directed) path P in D is denoted by $\ell(P)$. The following theorem is due to George James Minty, Jr. [151].

Theorem 7.21 *Let G be a graph that is not a forest. Then $\chi(G) \leq k$ for some integer $k \geq 2$ if and only if there exists some acyclic orientation D of G such that $r(D) \leq k - 1$.*

Proof. First, suppose that $\chi(G) \leq k$ and let there be given a k-coloring c of G using the colors in the set $\{1, 2, \ldots, k\}$. Let D be the orientation of G obtained by directing each edge uv of G from u to v if $c(u) < c(v)$. Then D is an acyclic orientation of G and the length of each directed path in D is at most $k - 1$. Let C be a cycle of G such that $r(D) = a(C)/b(C)$, where C is the underlying graph of C' in D. Since C' is acyclic, C' can be decomposed into an even number $2t$ of directed paths $P_1, P_1', P_2, P_2', \ldots, P_t, P_t'$ where as we proceed around C' in some direction, the paths $P_1, P_2, \ldots, P_t$ are oriented in some direction and the paths $P'_1, P'_2, \ldots, P'_t$ are oriented in the opposite direction. Suppose that

$$a(C) = \sum_{i=1}^{t} \ell(P_i) \text{ and } b(C) = \sum_{i=1}^{t} \ell(P'_i).$$

Since $\ell(P_i) \leq k - 1$ and $\ell(P'_i) \geq 1$ for each i $(1 \leq i \leq t)$, it follows that

$$r(D) = \frac{a(C)}{b(C)} \leq \frac{t(k-1)}{t} = k - 1,$$

as desired.

We now verify the converse. Let D be an acyclic orientation of a graph G that is not a forest such that $r(D) \leq k - 1$ for some integer $k \geq 2$. We show that there exists a k-coloring of G. Select a vertex u of D. For a vertex v of G distinct from u, let W be a $u - v$ walk in G. Suppose that W has $a = a(W)$ forward edges and $b = b(W)$ backward edges and define

$$s(W) = a(W) - b(W)(k - 1).$$

Next, define a function f on $V(G)$ by

$$f(v) = \max\{s(W)\}$$

over all $u - v$ walks W in G. We claim that $f(v) = s(P)$ for some $u - v$ path P in G, for suppose that this is not case. Let W be a $u - v$ walk of minimum length such that $f(v) = s(W)$. By assumption, W is not a path. Observe that no edge of G can occur consecutively in W. Thus, W contains a cycle C in G. Suppose that C has a' forward edges and b' backward edges. By assumption, $a'/b' \leq k - 1$ and so $a' - b'(k - 1) \leq 0$. Hence, by deleting the edges of C from W, we obtain a $u - v$ walk W' of smaller length such that

$$s(W') = s(W) - (a' - b'(k - 1)) \geq s(W).$$

Thus, $s(W') = s(W)$, contradicting the defining property of W. Since there are only finitely many $u - v$ paths in G, the function f is well-defined.

Let v_1 and v_2 be two adjacent vertices of G. We may assume that $v_1 v_2$ is directed from v_1 to v_2. We claim that

$$0 < |f(v_1) - f(v_2)| < k.$$

First, we show that $f(v_1) \neq f(v_2)$. Assume, to the contrary, that $f(v_1) = f(v_2)$. Let P_i be a $u - v_i$ path of minimum length such that $f(v_i) = s(P_i) = a_i - b_i(k - 1)$, where a_i edges of P_i are forward and b_i edges of P_i are backward for $i = 1, 2$. Thus, $a_1 - b_1(k - 1) = a_2 - b_2(k - 1)$ or, equivalently,

$$a_1 - a_2 = (b_1 - b_2)(k - 1). \tag{7.6}$$

If P_1 does not contain v_2, then the path P consisting of P_1 followed by v_2 has $s(P) > s(P_1)$, which implies that $s(P_2) > s(P_1)$, which is a contradiction. Thus, P_1

must contain v_2. The $v_2 - v_1$ subpath P_1' of P_1 contains $a_1 - a_2$ forward edges and $b_1 - b_2$ backward edges. The cycle C' consisting of P_1' followed by v_2 has $a_1 - a_2 + 1$ forward edges and $b_1 - b_2$ backward edges. By assumption,

$$\frac{a_1 - a_2 + 1}{b_1 - b_2} \leq k - 1.$$

It then follows by (7.6) that $a_1 - a_2 + 1 \leq (b_1 - b_2)(k - 1) = a_1 - a_2$, which is impossible. Therefore, $f(v_1) \neq f(v_2)$ for every two adjacent vertices v_1 and v_2 of G, as claimed.

Next we show that $|f(v_1) - f(v_2)| < k$ for every two adjacent vertices v_1 and v_2 of G. Assume, to the contrary, that $|f(v_1) - f(v_2)| \geq k$ for some pair v_1, v_2 of adjacent vertices of G. We now consider four cases.

Case 1. P_1 *does not contain* v_2 *and* P_2 *does not contain* v_1. Let P_2' be the $u - v_2$ path obtained by following P_1 by v_2. Then $s(P_2') = s(P_1) + 1$. Hence, $f(v_2) = s(P_2) \geq s(P_2') > f(v_1)$ and so by our assumption that $|f(v_1) - f(v_2)| \geq k$, we have $f(v_2) \geq f(v_1) + k$. Let P_1' be the $u - v_1$ path obtained by following P_2 by v_1. Then $s(P_1') = s(P_2) - (k - 1)$. Thus,

$$f(v_1) = s(P_1) > s(P_1') = f(v_2) - (k - 1).$$

Hence,

$$f(v_2) + 1 < f(v_1) + k \leq f(v_2),$$

which is impossible.

Case 2. P_2 *contains* v_1 *but* P_1 *does not contain* v_2. As in Case 1, $f(v_2) \geq f(v_1) + k$. If P' is the $u - v_1$ subpath of P_2, then $s(P') = s(P_1)$. Let P be the $v_1 - v_2$ subpath of P_2. Since $f(v_2) \geq f(v_1) + k$ and $k \geq 2$, it follows that P is not the path (v_1, v_2). Hence, there is a cycle C consisting of P followed by v_1. Then

$$\frac{a_2 - a_1}{b_2 - b_1 + 1} \leq k - 1$$

and so $a_2 - a_1 - (b_2 - b_1)(k - 1) \leq k - 1$. Thus,

$$0 < f(v_2) - f(v_1) \leq (k - 1),$$

a contradiction.

Case 3. P_1 *contains* v_2 *but* P_2 *does not contain* v_1. Let P_1' be the $u - v_1$ path obtained by following P_2 by v_1. Thus, $s(P_1') = s(P_2) - (k - 1)$ and so

$$f(v_1) = s(P_1) \geq s(P_1') = s(P_2) - (k - 1) = f(v_2) - (k - 1).$$

Thus, $f(v_2) \leq f(v_1) + (k - 1)$. Because $|f(v_1) - f(v_2)| \geq k$, it follows that $f(v_1) > f(v_2)$ and so $f(v_1) \geq f(v_2) + k$. Let C be the cycle obtained by following the $v_2 - v_1$ subpath of P_1 by v_2. Then

$$\frac{a_1 - a_2 + 1}{b_1 - b_2} \leq k - 1$$

and so $a_1 - a_2 + 1 \le (b_1 - b_2)(k - 1)$. Hence,

$$f(v_2) - f(v_1) = [a_2 - b_2(k - 1)] - [a_1 - b_1(k - 1)] \ge 1,$$

which contradicts $f(v_1) > f(v_2)$.

　　Case 4. P_1 *contains* v_2 *and* P_2 *contains* v_1. Let P_2' be the $u - v_1$ subpath of P_2 and let P_1' be the $u - v_2$ subpath of P_1. Then $s(P_2') \le f(v_1)$ and $s(P_1') \le f(v_2)$. Suppose that the $v_1 - v_2$ subpath of P_2 has a forward edges and b backward edges, while the $v_2 - v_1$ subpath of P_1 has a' forward edges and b' backward edges. Then

$$
\begin{aligned}
f(v_2) &= s(P_2) = s(P_2') + a - b(k - 1) \qquad\qquad (7.7)\\
f(v_1) &= s(P_1) = s(P_1') + a' - b'(k - 1).
\end{aligned}
$$

Let P be the $u - v_2$ path obtained by following P_2' by v_2.

　　If the $v_1 - v_2$ subpath of P_2 is (v_1, v_2), then

$$f(v_2) = s(P_2) = s(P_2') + 1 \le f(v_1) + 1.$$

Since $|f(v_1) - f(v_2)| \ge k$, it follows that $f(v_2) < f(v_1)$. If the $v_1 - v_2$ subpath of P_2 has at least two edges, then let C be the cycle obtained by following the $v_1 - v_2$ subpath of P_2 by v_1. Then

$$\frac{a}{b + 1} \le k - 1.$$

Thus,

$$a - b(k - 1) \le k - 1. \qquad\qquad (7.8)$$

Since $f(v_2) = s(P_2') + a - b(k - 1)$ and $s(P_2') \le f(v_1)$, it follows by (7.7) and (7.8) that

$$f(v_2) \le f(v_1) + a - b(k - 1) \le f(v_1) + (k - 1).$$

Thus, $f(v_2) - f(v_1) \le k - 1$. By assumption, $|f(v_1) - f(v_2)| \ge k$ and so $f(v_2) < f(v_1)$. Therefore, in each case,

$$f(v_2) < f(v_1). \qquad\qquad (7.9)$$

　　If the $v_2 - v_1$ subpath of P_1 is (v_2, v_1), then

$$f(v_1) = s(P_1) = s(P_1') - (k - 1) \le f(v_2) - (k - 1).$$

Thus, $f(v_2) - f(v_1) \ge k - 1$ and so $f(v_2) > f(v_1)$, which contradicts (7.9). If the $v_2 - v_1$ subpath of P_1 has at least two edges, then let C' be the cycle obtained by following the $v_2 - v_1$ subpath of P_1 by v_2. Then

$$\frac{a' + 1}{b'} \le k - 1$$

and so $a' - b'(k - 1) \le -1$. Since $f(v_1) = s(P_1') + a' - b'(k - 1)$ and $f(v_2) \ge s(P_1')$, it follows that

$$
\begin{aligned}
f(v_1) &= s(P_1') + a' - b'(k - 1) \le f(v_2) + a' - b'(k - 1)\\
&\le f(v_2) - 1 < f(v_2).
\end{aligned}
$$

Therefore, $f(v_2) > f(v_1)$, which again contradicts (7.9).

Consequently, as claimed, $0 < |f(v_1) - f(v_2)| < k$ for every two adjacent vertices v_1 and v_2 of G. We now define a coloring c of G by $c(u) = 1$ and for each vertex v distinct from u we define $c(v)$ as the color in $\{1, 2, \ldots, k\}$ such that $c(v) \equiv (1 + f(v)) \pmod k$. Since c is a proper k-coloring of G, it follows that $\chi(G) \leq k$. ∎

Letting $k = 1 + \lceil r(D) \rceil$ in Theorem 7.21, we have the following result.

Corollary 7.22 *For every graph G that is not a forest, there exists an acyclic orientation D of G such that*

$$\chi(G) \leq 1 + \lceil r(D) \rceil.$$

We saw in the proof of Theorem 7.21 that if D is an acyclic orientation of a graph G that is not a forest, the length of whose longest directed path is $\ell(D)$, then every cycle C of G gives rise to a subdigraph C' that can be decomposed into $2t$ directed paths $P_1, P_1', P_2, P_2', \ldots, P_t, P_t'$ where, as we proceed about C' in some direction, the paths $P_1, P_2, \ldots, P_t$ are oriented in some direction and the paths $P_1', P_2', \ldots, P_t'$ are oriented in the opposite direction. Since

$$a(C) = \sum_{i=1}^{t} \ell(P_i) \leq t\ell(D) \text{ and } b(C) = \sum_{i=1}^{t} \ell(P_i') \geq t,$$

it follows that

$$r(D) = \frac{a(C)}{b(C)} \leq \frac{t\ell(D)}{t} = \ell(D).$$

Therefore, the Gallai-Roy-Vitaver Theorem (Theorem 7.17) is a consequence of Corollary 7.22.

7.4 The Chromatic Number of Cartesian Products

Since the Cartesian product $G \square H$ of two graphs G and H contains subgraphs that are isomorphic to both G and H, it is a consequence of Theorem 6.1 that

$$\chi(G \square H) \geq \max\{\chi(G), \chi(H)\}.$$

In this case, more can be said.

Theorem 7.23 *For every two graphs G and H,*

$$\chi(G \square H) = \max\{\chi(G), \chi(H)\}.$$

Proof. As observed above, $\chi(G \square H) \geq \max\{\chi(G), \chi(H)\}$. Let

$$k = \max\{\chi(G), \chi(H)\}.$$

Then there exist both a k-coloring c' of G and a k-coloring c'' of H. We define a k-coloring c of $G \,\square\, H$ by assigning the vertex (u, v) of $G \,\square\, H$ the color $c(u, v)$, where $0 \leq c(u, v) \leq k - 1$ and $c(u, v) \equiv (c'(u) + c''(v)) \pmod{k}$. To show that c is a proper coloring, let (x, y) be a vertex adjacent to (u, v) in $G \,\square\, H$. Then either $u = x$ and $vy \in E(H)$ or $v = y$ and $ux \in E(G)$, say the former. Then (u, y) is adjacent to (u, v). Hence, $0 \leq c(u, y) \leq k - 1$ and $c(u, y) \equiv (c'(u) + c''(y)) \pmod{k}$. Since $vy \in E(H)$, it follows that $c''(v) \not\equiv c''(y) \pmod{k}$ and $c(u, v) \equiv c'(u) + c''(v) \not\equiv c'(u) + c''(y) \equiv c(u, y) \pmod{k}$.   ∎

As a consequence of Theorem 7.23, we have the following.

Corollary 7.24   *For every nonempty graph G,*

$$\chi(G \,\square\, K_2) = \chi(G).$$

The Cartesian product $G \,\square\, K_2$ of a graph G and K_2 is a special case of a more general class of graphs. Let G be a graph with $V(G) = \{v_1, v_2, \ldots, v_n\}$ and let α be a permutation of the set $S = \{1, 2, \ldots, n\}$. By the **permutation graph** $P_\alpha(G)$ we mean the graph of order $2n$ obtained from two copies of G, where the second copy of G is denoted by G' and the vertex v_i in G is denoted by u_i in G' and v_i is joined to the vertex $u_{\alpha(i)}$ in G'. The edges $v_i u_{\alpha(i)}$ are called the **permutation edges** of $P_\alpha(G)$. Therefore, if α is the identity map on S, then $P_\alpha(G) = G \,\square\, K_2$. Since G is a subgraph of $P_\alpha(G)$ for every permutation α of S, it follows by Theorem 6.1 that $\chi(P_\alpha(G)) \geq \chi(G)$.

For example, consider the graph $G = C_5$ of Figure 7.3 and the permutation $\alpha = (1)(2354)$ of the set $\{1, 2, 3, 4, 5\}$. Then the graph $P_\alpha(C_5)$ is also shown in Figure 7.3. Redrawing $P_\alpha(C_5)$, we see that this is, in fact, the Petersen graph. Thus, $\chi(C_5) = \chi(P_\alpha(C_5)) = 3$. All four permutation graphs of C_5 appeared on the cover of the book *Graph Theory* by Frank Harary [115].

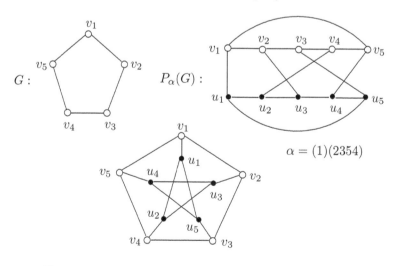

Figure 7.3: The Petersen graph as a permutation graph

The examples we've seen thus far might suggest that if G is a nonempty graph with $V(G) = \{v_1, v_2, \ldots, v_n\}$ and α is a permutation on the set $S = \{1, 2, \ldots, n\}$, then $\chi(G) = \chi(P_\alpha(G))$. This, however, is not the case.

Let G be the graph of order 5 shown in Figure 7.4, where $V(G) = \{v_1, v_2, \ldots, v_5\}$ and let $\alpha = (1324)(5)$. The permutation graph $P_\alpha(G)$ is also shown in Figure 7.4.

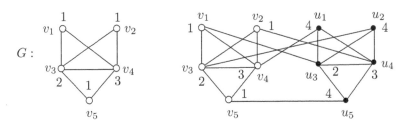

Figure 7.4: A permutation graph $P_\alpha(G)$ with $\chi(P_\alpha(G)) > \chi(G)$

Certainly, $\chi(G) = 3$ for the graph G of Figure 7.4. Therefore, $\chi(P_\alpha(G)) \geq 3$ for the permutation graph $P_\alpha(G)$ of Figure 7.4. We claim that $\chi(P_\alpha(G)) > \chi(G)$. Suppose that $\chi(P_\alpha(G)) = 3$. Then there exists a 3-coloring c of $P_\alpha(G)$. We may assume that $c(v_1) = c(v_2) = c(v_5) = 1$, $c(v_3) = 2$, and $c(v_4) = 3$. Since none of u_3, u_4, and u_5 can be colored 1, two of these vertices must be colored either 2 or 3. Since they are mutually adjacent, this is a contradiction. The 4-coloring of $P_\alpha(G)$ in Figure 7.4 shows that $\chi(P_\alpha(G)) = 4$. No permutation graph of the graph G of Figure 7.4 can have chromatic number greater than 4, however, according to the following theorem of Gary Chartrand and Joseph B. Frechen [40]:

Theorem 7.25 *For every graph G and every permutation graph $P_\alpha(G)$ of G,*

$$\chi(G) \leq \chi(P_\alpha(G)) \leq \left\lceil \frac{4\chi(G)}{3} \right\rceil.$$

Proof. Let G be a graph of order n and let $P_\alpha(G)$ be a permutation graph of G. Since G is a subgraph of $P_\alpha(G)$, it follows that $\chi(G) \leq \chi(P_\alpha(G))$. It remains therefore to establish the upper bound for $\chi(P_\alpha(G))$.

Suppose that $\chi(G) = k$. If $k = 1$, then $\chi(P_\alpha(G)) = 2 = \left\lceil \frac{4k}{3} \right\rceil$. Thus, we may assume that $k \geq 2$. Suppose that ϵ is the identity permutation on the set $S = \{1, 2, \ldots, n\}$. Then $P_\alpha(G) = P_\epsilon(G) = G \,\square\, K_2$. By Corollary 7.24, $\chi(P_\epsilon(G)) = \chi(G) = k$.

We now show that for every permutation graph $P_\alpha(G)$ of G, there exists a $\left\lceil \frac{4k}{3} \right\rceil$-coloring of $P_\alpha(G)$. We begin with a k-coloring of G with the color classes $V_1, V_2, \ldots, V_k$, where $c(v) = i$ for each $v \in V_i$ for $1 \leq i \leq k$ and the same k-coloring of G' with color classes $V_1', V_2' \ldots, V_k'$. We now consider two cases according to whether $\left\lceil \frac{4k}{3} \right\rceil$ is even or odd.

Case 1. $\left\lceil \frac{4k}{3} \right\rceil$ *is even, say* $\left\lceil \frac{4k}{3} \right\rceil = 2\ell$ *for some positive integer ℓ.* We assign to each vertex of the set V_i, $1 \leq i \leq \ell$, the color i; while we assign to each vertex of the set V_i', $1 \leq i \leq \ell$, the color $i + \ell$. For $j = 1, 2, \ldots, k - \ell$, we assign the color $\ell + j$ to the vertices of $V_{\ell+j}$ that are not adjacent to any vertices of V'_j and assign the color $2\ell + 1 - j$ to the vertices of $V_{\ell+j}$ otherwise. In a similar manner, we assign the color $j = 1, 2, \ldots, k - \ell$ to the vertices of $V'_{\ell+j}$ not adjacent to any vertices of V_j and assign the color $\ell + 1 - j$ to the vertices of $V'_{\ell+j}$ otherwise. Since $\lceil \frac{4k}{3} \rceil = 2\ell$, it follows that $\frac{4k}{3} \leq 2\ell$ and so $k \leq 3\ell/2$. Hence, there are sufficiently many colors for this coloring. Because this coloring of $P_\alpha(G)$ is a proper coloring, $\chi(P_\alpha(G)) \leq 2\ell = \lceil \frac{4n}{3} \rceil$.

Case 2. $\lceil \frac{4k}{3} \rceil$ *is odd, say* $\lceil \frac{4k}{3} \rceil = 2\ell + 1$ *for some positive integer* ℓ. We assign to each vertex of the set V_i, $1 \leq i \leq \ell + 1$, the color i; while we assign the color $i + \ell + 1$ to the vertices of the set V'_i, $1 \leq i \leq \ell$. For a vertex of $V_{\ell+j}$ for $2 \leq j \leq k - \ell$, we assign the color $\ell + j$ if it is not adjacent to any vertices of V'_{j-1} and assign the color $2\ell + 3 - j$ otherwise. For a vertex $V'_{\ell+j}$ for $1 \leq j \leq k - \ell$, we assign the color j if it is adjacent to no vertex of V_j and assign the color $\ell + 2 - j$ otherwise. The fact that $2\ell + 1 \geq 4k/3$ gives $k \leq (6\ell + 3)/4$ and assures us that there are enough colors to accomplish the coloring. Since this coloring of $P_\alpha(G)$ is a proper coloring, $\chi(P_\alpha(G)) \leq 2\ell + 1 = \lceil \frac{4n}{3} \rceil$.   ∎

We now show that the upper bound for $\chi(P_\alpha(G))$ is sharp. For an integer $k \geq 2$, let $S = \{1, 2, \ldots, k\}$ and let $G = K_{n_1, n_2, \ldots, n_k}$ be a complete k-partite graph in which $n_i = k$ for $1 \leq i \leq k$. Denote the partite sets of G by $V_1, V_2, \ldots, V_k$, where $V_i = \{v_{i_1}, v_{i_2}, \ldots, v_{i_k}\}$ for $1 \leq i \leq k$. Then $\chi(G) = k$. Let G' be a second copy of G with corresponding partite sets $V'_1, V'_2, \ldots, V'_k$, where $V'_i = \{v'_{i_1}, v'_{i_2}, \ldots, v'_{i_k}\}$ for $1 \leq i \leq k$. Let α be the permutation defined on the set $S \times S = \{(i, j) : 1 \leq i, j \leq k\}$ by

$$\alpha(i, j) = (j, i),$$

that is, the vertex v_{ij} in G is joined to the vertex v'_{ji} in G'. (See Figure 7.5 for the case when $k = 3$ in which only the permutation edges are shown.)

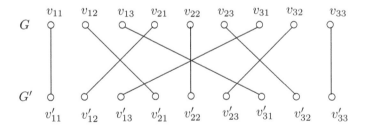

Figure 7.5: The permutation edges in $P_\alpha(K_{3,3,3})$

Because of the construction of $P_\alpha(G)$, it follows that (1) if any color is assigned to a vertex of V_i, then this color cannot be used for any vertex in V_j $(i \neq j)$ and (2) if a color a is assigned to all of the vertices in some set V_i and a color b is assigned to all of the vertices in some set V'_j, then $a \neq b$. We claim that $\chi(P_\alpha(G)) = \lceil \frac{4k}{3} \rceil$. Assume, to the contrary, that $\chi(P_\alpha(G)) = \ell < \lceil \frac{4k}{3} \rceil$ and let there be given an ℓ-coloring of $P_\alpha(G)$. Thus, $\ell < 4k/3$. Suppose that r of the sets V_i have the same color assigned to each vertex of the set and that s of the sets V'_j have the same color assigned to each vertex of the set. As we noted in (2), each of the r colors is distinct from each of the s colors. Thus, $r + s \leq \chi(P_\alpha(G))$ and at least two colors are used for each of the remaining $k - r$ sets of V'_j. Thus, at least $r + 2(k - r)$ colors are used to color the vertices of G and at least $s + 2(k - s)$ colors are used to color the vertices of G'. Hence,

$$r + 2(k - r) \leq \ell \text{ and } s + 2(k - s) \leq \ell$$

and so

$$r \geq 2k - \ell \text{ and } s \geq 2k - \ell.$$

Therefore,

$$\chi(P_\alpha(G)) \geq r + s > 4k - 2(4k/3) = 4k/3 > \ell = \chi(P_\alpha(G)),$$

which is a contradiction.

Let $G = K_{9,9,\ldots,9}$ be the complete 9-partite graph with partite sets $V_i = \{v_{i1}, v_{i2}, \ldots, v_{i9}\}$, $1 \leq i \leq 9$. For the set $S = \{1, 2, \ldots, 9\}$ and the permutation α on $S \times S$ defined by $\alpha((i, j)) = (j, i)$, it follows that

$$\chi(P_\alpha(G)) = \left\lceil \frac{4\chi(G)}{3} \right\rceil = 12.$$

The 12-coloring of $P_\alpha(G)$ described in the proof of Theorem 7.25 is shown in Figure 7.6.

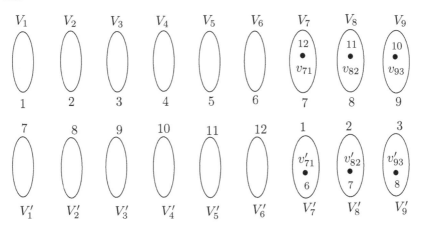

Figure 7.6: A 12-coloring of $P_\alpha(K_{9,9,\ldots,9})$

Exercises for Chapter 7

1. Let G be a k-chromatic graph where $k \geq 2$.

   (a) Prove for every vertex v of G that either $\chi(G-v) = k$ or $\chi(G-v) = k-1$.

   (b) Prove for every edge e of G that either $\chi(G-e) = k$ or $\chi(G-e) = k-1$.

2. Let G be a k-chromatic graph such that $\chi(G-e) = k-1$ for some edge $e = uv$ of G. Prove that $\chi(G-u) = \chi(G-v) = k-1$.

3. Show that the odd cycles are the only 3-critical graphs.

4. Let G be a graph of order $n \geq 3$ such that $G \neq K_n$. Suppose that $\chi(G-v) < \chi(G)$ for every vertex v of G. Either give an example of a graph G such that $G \neq K_n$ and $\chi(G-u-w) < \chi(G-u)$ for every two vertices u and w of G or show that no such graph G exists.

5. Determine all k-critical graphs with $k \geq 3$ such that $G - v$ is $(k-1)$-critical for every vertex v of G.

6. Prove or disprove: For every integer $k \geq 3$, there exists a triangle-free, k-critical graph.

7. Prove or disprove: If G and H are color-critical graphs, then $G \vee H$ is color-critical.

8. It has been mentioned that every k-critical graph, $k \geq 3$, is 2-connected. Show that there exists a k-critical graph having connectivity 2 for every integer $k \geq 3$.

9. Prove or disprove the following.

   (a) There exists no graph G with $\chi(G) = 3$ without isolated vertices such that $\chi(G-v) = 2$ for exactly 75% of the vertices v of G.

   (b) There exists no graph G with $\chi(G) = 3$ without isolated vertices and containing a 3-critical component such that $\chi(G-v) = 2$ for exactly 75% of the vertices v of G.

10. Let G be a k-critical graph of order n. Prove that if G is perfect, then $k = n$.

11. By Theorem 6.10, it follows that for every graph G of order n,

$$\frac{n}{\alpha(G)} \leq \chi(G) \leq n - \alpha(G) + 1.$$

Prove or disprove: The chromatic number of a graph G can never be closer to $n - \alpha(G) + 1$ than to $\frac{n}{\alpha(G)}$.

12. Show, for every connected graph G of order n and diameter d, that

$$\chi(G) \leq n - d + 1.$$

13. (a) Show, for every k-chromatic graph G, that there exists an ordering of the vertices of G such that the greedy coloring algorithm gives a k-coloring of G.

   (b) Show, for every positive integer p, that there exists a graph G and an ordering of the vertices of G such that the greedy coloring algorithm gives a k-coloring of G, where $k = p + \chi(G)$.

14. For each integer $k \geq 3$, give an example of a regular k-chromatic graph G such that $G \neq K_k$.

15. (a) What upper bound for $\chi(G)$ is given in Theorem 7.9 for the graph G in Figure 7.7?

   (b) What is $\chi(G)$ for this graph G?

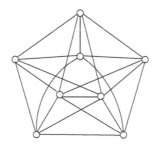

Figure 7.7: The graph G in Exercise 15

16. For the double star T containing two vertices of degree 4, what upper bound for $\chi(T)$ is given by Theorems 7.8, 7.9, and 7.10?

17. We have seen that $\chi(T) = 2$ for every nontrivial tree T. Prove, for every integer $k \geq 2$, that there exists a tree T_k with $\Delta(T_k) = k$ and an ordering s of the vertices of T_k that produces a greedy coloring of T_k using $k + 1$ colors.

18. Let T be the tree of Figure 7.8.

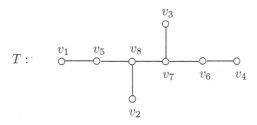

Figure 7.8: The tree T in Exercise 18

   (a) What is the greedy coloring c produced by the ordering $s : v_1, v_2, \ldots, v_8$ of the vertices of T?

(b) Does there exist a different ordering of the vertices of T giving a greedy coloring that uses fewer colors?

(c) Does there exist a different ordering of the vertices of T giving a greedy coloring that uses more colors?

19. Since the upper bound $(3+\sqrt{1+8(m-n)})/2$ for $\chi(G)$ given in Theorem 7.11 for a connected graph G of order n and size m that is not a tree is not always an integer, it follows that

$$\chi(G) \le \left\lfloor \frac{3+\sqrt{1+8(m-n)}}{2} \right\rfloor . \qquad (7.10)$$

(a) How does the upper bound in (7.10) compare with $\chi(G)$ for the wheel $G = W_5 = C_5 \vee K_1$ of order 6?

(b) Find a 5-chromatic graph G for which the upper bound in (7.10) gives the exact value of $\chi(G)$.

(c) The question asked in (b) should suggest another question to you. Ask and answer this question.

20. What does the bound in Theorem 7.12 say for a complete k-partite graph $K_{n_1,n_2,\ldots,n_k}$ of order n where $n_1 = n_2 = \cdots = n_k$?

21. By Theorem 7.13, $\chi(G) \cdot \chi(\overline{G}) \ge n$ for every graph G of order n.

(a) Show that if G is a perfect graph of order n, then $\omega(G) \cdot \omega(\overline{G}) \ge n$.

(b) Show that there are graphs G of order n for which $\omega(G) \cdot \omega(\overline{G}) < n$.

22. Recall that the Nordhaus-Gaddum Theorem (Theorem 7.13) says, in part, that if G is a graph of order n, then $\chi(G)\chi(\overline{G}) \ge n$.

(a) Prove that the complete graph K_n is factored into three nonempty graphs G_1, G_2 and G_3, then $\chi(G_1)\chi(G_2)\chi(G_3) \ge n$.

(b) Prove that there are infinitely many integers n for which K_n can be factored into three factors G_1, G_2 and G_3 such that

$$\chi(G_1)\chi(G_2)\chi(G_3) = n.$$

23. What proposed upper bound does Reed's conjecture give for $\chi(G)$ when
(a) $G = K_n$,
(b) $G = C_n$, where $n \ge 5$ is odd, and
(c) G is the Petersen graph?

24. (a) Use the fact that the chromatic number of K_4 is 4 to show that every orientation of K_4 has a directed path of length 3.

(b) Determine the minimum length of a longest path among all orientations of Grötzsch graph shown in Figure 7.9.

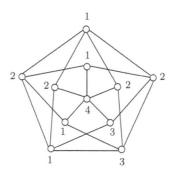

Figure 7.9: The Grötzsch graph in Exercise 24

25. Use Corollary 7.18 to determine the chromatic number of every nonempty bipartite graph.

26. Let $G = W_5 = C_5 \vee K_1$ be the wheel of order 6. Prove for every acyclic orientation D of G that there always exists some cycle C' of G where in D more than twice as many edges of C' are oriented in one direction than in the opposite direction.

27. What is the smallest upper bound that Corollary 7.22 gives for $\chi(G)$ when $G = K_n$ for $n \geq 3$?

28. Let G be the graph in Figure 7.10. Among all orientations of G, what is the minimum length of a longest directed path?

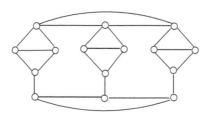

Figure 7.10: The graph in Exercise 28

29. For a graph H and the 3-cube Q_3, let $G = H \square Q_3$. If $\chi(G) = 5$, what is $\chi(H)$?

30. (a) Find the smallest positive integer n for which there exists a permutation graph $P_\alpha(C_n)$ such that $\chi(P_\alpha(C_n)) \neq \chi(C_n)$.

    (b) Show for every permutation graph $P_\alpha(C_5)$ that $\chi(P_\alpha(C_5)) = \chi(C_5)$.

    (c) Parts (a) and (b) should suggest a question to you. Ask and answer such a question.

31. Consider the complete 3-partite graph $G = K_{2,2,2}$ where $V(G)=\{v_1, v_2, \ldots, v_6\}$ such that its three partite sets are $V_1 = \{v_1, v_2\}$, $V_2 = \{v_3, v_4\}$, and $V_3 = \{v_5, v_6\}$.

    (a) For the permutation $\alpha_1 = (1)(2\ 4\ 6\ 5\ 3)$, determine $\chi(P_{\alpha_1}(G))$ and provide a $\chi(P_{\alpha_1}(G))$-coloring of the graph $P_{\alpha_1}(G)$.

    (b) Repeat (a) for the permutation $\alpha_2 = (1)(2\ 3\ 5)(4\ 6)$.

Chapter 8

Coloring Graphs on Surfaces

Many historical events related to the famous Four Color Problem are described in Chapter 0. It is this problem that would lead to graph coloring parameters and problems and play a major role in the development of graph theory. In this chapter, the Four Color Problem is revisited and some of these events are reviewed, together with some conjectures, concepts, and results, which initially had hoped to provide added insight into and possibly even a solution to this problem.

8.1 The Four Color Problem

The origin of graph colorings has been traced back to 1852 when the Four Color Problem was first posed. While the history of this problem is discussed in some detail in Chapter 0, we give a brief review here of a few of the key events that took place during this period.

In 1852 Francis Guthrie (1831–1899), a former graduate of University College London, observed that the counties of England could be colored with four colors so that neighboring counties were colored differently. This led him to ask whether the counties of every map (real or imagined) can be colored with four or fewer colors so that every two neighboring counties are colored differently. Francis mentioned this problem to his younger brother Frederick, who at the time was taking a class from the well-known mathematician Augustus De Morgan. With the approval of his brother, Frederick mentioned this problem to De Morgan, who considered the problem to be new but was unable to solve it. Despite De Morgan's great interest in the problem, few other mathematicians who were aware of the problem seemed to share this interest. A quarter century passed with little activity on the problem.

At a meeting of the London Mathematical Society in 1878, the great mathematician Arthur Cayley inquired about the status of this Four Color Problem. This revived interest in the problem and would lead to an 1879 article written by the British lawyer Alfred Bray Kempe containing a proposed proof that every map can be colored with four or fewer colors so that neighboring counties are colored differently. For the next ten years, the Four Color Problem was considered to be solved. However, an 1890 article by the British mathematician Percy John Heawood presented a map and a partial coloring of the counties of the map, which Heawood showed was a counterexample to the technique used by Kempe. Although this counterexample did not imply that there were maps requiring five or more colors, it did show that Kempe's method was unsuccessful. Nevertheless, Heawood was able to use Kempe's technique to prove that every map could be colored with five or fewer colors. (Heawood accomplished even more in his paper, which will be discussed in Section 8.4.)

The Four Color Problem can be stated strictly in terms of plane (or planar) graphs, rather than in terms of maps. Let G be a plane graph. Then G is k-**region colorable** if each region of G can be assigned one of k given colors so that neighboring (adjacent) regions are colored differently. Since it was believed by many that the question posed in the Four Color Problem had an affirmative answer, this led to the following.

The Four Color Conjecture Every plane graph is 4-region colorable.

There is another, even more popular statement of the Four Color Conjecture, which involves the coloring of vertices. Let G be a plane graph. The **planar dual** (or, more simply, the **dual**) G^* of G can be constructed by first placing a vertex in each region of G. This set of vertices is $V(G^*)$. Two distinct vertices of G^* are then joined by an edge for each edge on the boundaries of the regions corresponding to these vertices of G^*. Furthermore, a loop is added at a vertex of G^* for each bridge of G on the boundary of the corresponding region. Each edge of G^* is drawn so that it crosses its associated edge of G but crosses no other edge of G or of G^*. Thus, the dual G^* is planar. Since G^* may contain parallel edges and possibly loops (perhaps even parallel loops), G^* is a multigraph and may not be a graph. The dual G^* has the properties that its order is the same as the number of regions of G and the number of regions of G^* is the order of G. Both G and G^* have the same size. If each set of parallel edges in G^* is replaced by a single edge and all loops are deleted, a graph G' results, called the **dual graph** of G. A plane graph G, its planar dual G^*, and its dual graph G' are shown in Figure 8.1.

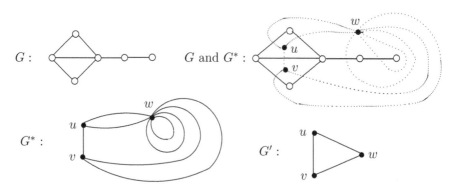

Figure 8.1: A graph, its dual, and its dual graph

A few observations regarding duals and dual graphs of plane graphs are useful. First, if G is a *connected* plane graph, then $(G^*)^* = G$, a property expected of dual operations. Second, every connected plane graph is the dual graph of some connected plane graph. And, finally, a plane graph G is k-region colorable for some positive integer k if and only if its dual graph G' is k-colorable. Hence, the Four Color Conjecture can now be rephrased.

The Four Color Conjecture Every planar graph is 4-colorable.

Thus, what Heawood proved with the aid of Kempe's proof technique is the following.

Theorem 8.1 (**The Five Color Theorem**) *Every planar graph is 5-colorable.*

Proof. We proceed by induction on the order n of the graph. Clearly, the result is true if $1 \le n \le 5$. Assume that every planar graph of order $n-1$ is 5-colorable, where $n \ge 6$, and let G be planar graph of order n. We show that G is 5-colorable.

By Corollary 5.4, G contains a vertex v with $\deg v \le 5$. Since $G - v$ is a planar graph of order $n-1$, it follows by the induction hypothesis that $G-v$ is 5-colorable. Let there be given a 5-coloring of $G - v$, where the colors used are denoted by 1, 2, 3, 4, 5. If one of these colors is not used to color the neighbors of v, then this color can be assigned to v, producing a 5-coloring of G. Hence, we may assume that $\deg v = 5$ and that all five colors are used to color the neighbors of v.

Let there be a planar embedding of G and suppose that v_1, v_2, v_3, v_4, v_5 are the neighbors of v arranged cyclically about v. We may assume that v_i has been assigned the color i for $1 \le i \le 5$. Let H be the subgraph of $G - v$ induced by the set of vertices colored 1 or 3. Thus, $v_1, v_3 \in V(H)$. If v_1 and v_3 should belong to different components of H, then by interchanging the colors of the vertices belonging to the component H_1 of H containing v_1, a 5-coloring of G can be produced by assigning the color 1 to v.

Suppose then that v_1 and v_3 belong to the same component of H. This implies that $G - v$ contains a $v_1 - v_3$ path P, every vertex of which is colored 1 or 3. The path P and the path (v_1, v, v_3) in G produce a cycle, enclosing either v_2 or both v_4 and v_5. In particular, this implies that there is no $v_2 - v_4$ path in $G - v$, every vertex of which is colored 2 or 4.

Let F be the subgraph of $G - v$ induced by the set of vertices colored 2 or 4, and let F_2 be the component of F containing v_2. Necessarily, $v_4 \notin V(F_2)$. By interchanging the colors of the vertices of F_2, a 5-coloring of G can be produced by assigning the color 2 to v.   ∎

In the 19th century, neither Kempe nor Heawood had access to upper bounds for the chromatic number of a planar graph that would be established decades later. Of course, the upper bound $1 + \Delta(G)$ for the chromatic number of a planar graph G is of no value since every star $K_{1,n-1}$, for example, is planar and the bound $1 + \Delta(G)$ only tells us that $\chi(K_{1,n-1}) \le 1 + \Delta(K_{1,n-1}) = n$. By Theorem 7.8, however, the chromatic number of a planar graph G is bounded above by $1 + \max\{\delta(H)\}$ over all subgraphs H of G. Since $\delta(H) \le 5$ for every subgraph H of G (see Corollary 5.4), it follows that $\max\{\delta(H)\} \leq 5$ and so $\chi(G) \leq 6$. Hence, "The Six Color Theorem" follows immediately from this bound. After the publication of Heawood's paper in 1890, it was known that the chromatic number of every planar graph is at most 5 but it was not known whether even a single planar graph had chromatic number 5. Despite major attempts to settle this question by many people (much of which and many of whom are described in Chapter 0), it would take another 86 years to resolve the issue, when in 1976 Kenneth Appel and Wolfgang Haken announced that they had been successful in providing a computer-aided proof of what had once been one of the most famous unsolved problems in mathematics.

Theorem 8.2 (The Four Color Theorem) *Every planar graph is 4-colorable.*

8.2   The Conjectures of Hajós and Hadwiger

We know that $\chi(G) \geq \omega(G)$ for every graph G and that this inequality can be strict. Indeed, Theorem 6.17 states that for every integer $k \geq 3$, there is a graph G such that $\chi(G) = k$ and $\omega(G) = 2$. Even though K_k need not be present in a k-chromatic graph G, it has been thought (and conjectured) over the years that K_k may be indirectly present in G. Of course, K_k is present in a k-chromatic graph for $k = 1, 2$. This is not true for $k = 3$, however. Indeed, every odd cycle of order at least 5 is 3-chromatic but none of these graphs contains K_3 as a subgraph. All of these do contain a subdivision of K_3, however. Since every 3-chromatic graph contains an odd cycle, it follows that if G is a graph with $\chi(G) \geq 3$, then G must contain a subdivision of K_3. In 1952 Gabriel A. Dirac [67] showed that the corresponding result is true as well for graphs having chromatic number at least 4.

Theorem 8.3 *If G is a graph with $\chi(G) \geq 4$, then G contains a subdivision of K_4.*

Proof.   We proceed by induction on the order $n \geq 4$ of G. The basis step of the induction follows since K_4 is the only graph of order 4 having chromatic number at least 4. For an integer $n \geq 5$, assume that every graph of order n' with $4 \leq n' < n$ having chromatic number at least 4 contains a subdivision of K_4. Let G be a graph of order n such that $\chi(G) \geq 4$. We show that G contains a subdivision of K_4.

Let H be a critically 4-chromatic subgraph of G. If the order of H is less than n, then it follows by the induction hypothesis that H (and G as well) contains a subdivision of K_4. Hence, we may assume that H has order n. Therefore, H is 2-connected.

Suppose first that $\kappa(H) = 2$ and $S = \{u, v\}$ is a vertex-cut of H. By Corollary 7.5, $uv \notin E(H)$ and H contains an S-branch H' such that $\chi(H' + uv) = 4$. Because the order of $H' + uv$ is less than n, it follows by the induction hypothesis that $H' + uv$ contains a subdivision F of K_4. If F does not contain the edge uv, then H' and therefore G contains F. Hence, we may assume that F contains the edge uv. In this case, let H'' be an S-branch of H distinct from H'. Because S is a minimum vertex-cut, both u and v are adjacent to vertices in each component of $H - S$. Hence, H'' contains a $u - v$ path P. Replacing the edge uv in F by P produces a subdivision of K_4 in H.

We may now assume that H is 3-connected. Let $w \in V(H)$. Then $H - w$ is 2-connected and so contains a cycle C. Let w_1, w_2, and w_3 be three vertices belonging to C. With the aid of Corollary 2.21, H contains $w - w_i$ paths P_i $(1 \le i \le 3)$ such that every two of these paths have only w in common and w_i is the only vertex of P_i on C. Then C and the paths P_i $(1 \le i \le 3)$ produce a subdivision of K_4 in H.

Since H contains a subdivision of K_4, the graph G does as well. ∎

Consequently, for $2 \le k \le 4$, every k-chromatic graph contains a subdivision of K_k. In 1961 György Hajós (1912–1972) [109] conjectured that this is true for every integer $k \ge 2$.

Hajós's Conjecture If G is a k-chromatic graph, where $k \ge 2$, then G contains a subdivision of K_k.

In 1979 Paul Catlin (1948–1995) [35] constructed a family of graphs showing that Hajós's Conjecture is false for every integer $k \ge 7$. For example, recall the graph G of Figure 6.13 of order 15 (shown again in Figure 8.2) consisting of five mutually vertex-disjoint triangles T_i $(1 \le i \le 5)$, where every vertex of T_i is adjacent to every vertex of T_j if either $|i - j| = 1$ or if $\{i, j\} = \{1, 5\}$. It was shown that $\omega(G) = 6$ and $\chi(G) = 8$. Since $\omega(G) = 6$, the graph G does not contain K_8 as a subgraph (or even K_7). We claim, in fact, that G does not contain a subdivision of K_8 either, for suppose that H is such a subgraph of G. Then H contains eight vertices of degree 7 and all other vertices of H have degree 2.

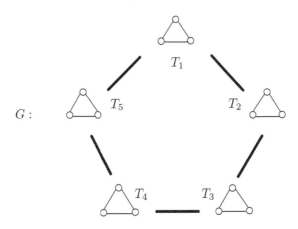

Figure 8.2: A counterexample to Hajós's Conjecture for $k = 8$

First, we show that no triangle T_i $(1 \le i \le 5)$ can contain exactly one vertex of degree 7 in H. Suppose that the triangle T_1, say, contains exactly one such vertex v. Then v is the initial vertex of seven paths to the remaining seven vertices of degree 7 in H, where every two of these paths have only v in common. Since v is adjacent only to six vertices outside of T_1, this is impossible.

Next, we show that no triangle can contain exactly two vertices of degree 7 in H. Suppose that the triangle T_1, say, contains exactly two such vertices, namely u and v. Then u is the initial vertex of six paths to six vertices outside of T_1. Necessarily, these six paths must contain all six vertices in T_2 and T_5. However, this is true of the vertex v as well. This implies that the six vertices of T_2 and T_5 are the remaining vertices of degree 7 in H. Therefore, there are two vertices u_5 and v_5 in T_5 and two vertices u_2 and v_2 in T_2 such that the interior vertices of some $u_5 - u_2$ path, $u_5 - v_2$ path, $v_5 - u_2$ path, and $v_5 - v_2$ path contain only vertices of T_3 and T_4. Since these four paths must be internally disjoint and since T_3 (and T_4) contains only three vertices, this is impossible.

Therefore, no triangle T_i ($1 \leq i \leq 5$) can contain exactly one or exactly two vertices of degree 7 in H. This, however, says that no triangle T_i ($1 \leq i \leq 5$) can contain exactly three vertices of degree 7 in H either. Thus, as claimed, G does not contain a subdivision of K_8. Hence, the graph G is a counterexample to Hajós's Conjecture for $k = 8$.

Let k be an integer such that $k \geq 9$ and consider the graph $F = G + K_{k-8}$. Then $\chi(F) = k$. We claim that F does not contain a subdivision of K_k, for suppose that H is such a subgraph in F. Delete all vertices of K_{k-8} that belong to H, arriving at a subgraph H' of G. This says that H' contains a subdivision of K_8, which is impossible. Hence, Hajós's Conjecture is false for all integers $k \geq 8$. As we noted, Catlin showed that this conjecture is false for every integer $k \geq 7$ (see Exercise 2).

Recall that a graph G is perfect if $\chi(H) = \omega(H)$ for every induced subgraph H of G. Furthermore, for a graph G and a vertex v of G, the replication graph $R_v(G)$ of G is that graph obtained from G by adding a new vertex v' to G and joining v' to every vertex in the closed neighborhood $N[v]$ of v. We saw in Theorem 6.25 that if G is perfect, then $R_v(G)$ is perfect for every $v \in V(G)$. Carsten Thomassen [187] showed that there is a connection between perfect graphs and Hajós's Conjecture.

A graph G is perfect if and only if every replication graph of G satisfies Hajós's Conjecture.

Recall also that a graph H is a minor of a graph G if H can be obtained from G by a sequence of contractions, edge deletions, and vertex deletions (in any order). Furthermore, from Theorem 5.15, if a graph G contains a subdivision of a graph H, then H is a minor of a graph G. In particular, if a k-chromatic graph G contains a subdivision of K_k, then K_k is a minor of G. Of course, we have seen that for $k \geq 7$, a k-chromatic graph need not contain a subdivision of K_k. This does not imply, however, that a k-chromatic graph need not contain K_k as a minor. Indeed, years before Hajós's Conjecture, on December 15, 1942 Hugo Hadwiger (1908–1981) made the following conjecture during a lecture he gave at the University of Zürich in Switzerland.

Hadwiger's Conjecture Every k-chromatic graph contains K_k as a minor.

A published version of Hadwiger's lecture appeared in a 1943 article [106]. This paper not only contained the conjecture but three theorems of interest, namely:

(1) Hadwiger's Conjecture is true for $1 \leq k \leq 4$.

(2) If G is a graph with $\delta(G) \geq k - 1$ where $1 \leq k \leq 4$, then G contains K_k as a minor.

(3) If G is a connected graph of order n and size m that has K_k as a minor, then $m \geq n + \binom{k}{2} - k$ (see Exercise 4).

In 1937 Klaus Wagner [197] proved that every planar graph is 4-colorable if and only if every 5-chromatic graph contains K_5 as a minor. That is, Wagner had shown the equivalence of the Four Color Conjecture and Hadwiger's Conjecture for $k = 5$ six years before Hadwiger stated his conjecture. In his 1943 paper [106], Hadwiger mentioned that his conjecture for $k = 5$ implies the Four Color Conjecture and referenced Wagner's paper but he did not refer to the equivalence. Hadwiger's Conjecture can therefore be considered as a generalization of the Four Color Conjecture.

Using the Four Color Theorem, Neil Robertson, Paul Seymour, and Robin Thomas [169] verified Hadwiger's Conjecture for $k = 6$. While Hadwiger's Conjecture is true for $k \leq 6$, it is open for every integer $k > 6$.

The **Hadwiger number** $had(G)$ of a graph G has been defined as the greatest positive integer k for which K_k is a minor of G. In this context, Hadwiger's Conjecture can be stated as:

For every graph G, $\chi(G) \leq had(G)$.

A proof of the general case of Hadwiger's Conjecture would give, as a corollary, a new proof of the Four Color Theorem.

8.3 Chromatic Polynomials

During the period that the Four Color Problem was unsolved, which spanned more than a century, many approaches were introduced with the hopes that they would lead to a solution of this famous problem. In 1912 George David Birkhoff [22] defined a function $P(M, \lambda)$ that gives the number of proper λ-colorings of a map M for a positive integer λ. As we will see, $P(M, \lambda)$ is a polynomial in λ for every map M and is called the chromatic polynomial of M. Consequently, if it could be verified that $P(M, 4) > 0$ for every map M, then this would have established the truth of the Four Color Conjecture.

In 1932 Hassler Whitney [202] expanded the study of chromatic polynomials from maps to graphs. While Whitney obtained a number of results on chromatic polynomials of graphs and others obtained results on the roots of chromatic polynomials of planar graphs, this never led to a proof of the Four Color Conjecture.

Renewed interest in chromatic polynomials of graphs occurred in 1968 when Ronald C. Read [163] wrote a survey paper on chromatic polynomials.

For a graph G with $V(G) = \{v_1, v_2, \ldots, v_n\}$, say, and a positive integer λ, the number of different proper λ-colorings of G is denoted by $P(G, \lambda)$ and is called the **chromatic polynomial** of G. Two λ-colorings c and c' of G from the same set $\{1, 2, \ldots, \lambda\}$ of λ colors are considered different if $c(v) \neq c'(v)$ for some vertex v of G. Obviously, if $\lambda < \chi(G)$, then $P(G, \lambda) = 0$. By convention, $P(G, 0) = 0$. Indeed, we have the following.

Proposition 8.4 *Let G be a graph. Then $\chi(G) = k$ if and only if k is the smallest positive integer for which $P(G, k) > 0$.*

As an example, we determine the number of ways that the vertices of the graph G of Figure 8.3 can be colored from the set $\{1, 2, 3, 4, 5\}$. The vertex v can be assigned any of these 5 colors, while w can be assigned any color other than the color assigned to v. That is, w can be assigned any of the 4 remaining colors. Both u and t can be assigned any of the 3 colors not used for v and w. Therefore, the number $P(G, 5)$ of 5-colorings of G is $5 \cdot 4 \cdot 3^2 = 180$. More generally, $P(G, \lambda) = \lambda(\lambda - 1)(\lambda - 2)^2$ for every integer λ.

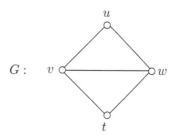

$$G: \quad$$

Figure 8.3: A graph G with $P(G, \lambda) = \lambda(\lambda - 1)(\lambda - 2)^2$

There are some classes of graphs G for which $P(G, \lambda)$ can be easily computed.

Theorem 8.5 *For every positive integer λ,*

(a) $P(K_n, \lambda) = \lambda(\lambda - 1)(\lambda - 2) \cdots (\lambda - n + 1) = \lambda^{(n)}$,

(b) $P(\overline{K}_n, \lambda) = \lambda^n$.

In particular, if $\lambda \geq n$ in Theorem 8.5(a), then

$$P(K_n, \lambda) = \lambda^{(n)} = \frac{\lambda!}{(\lambda - n)!}.$$

We now determine the chromatic polynomial of C_4 in Figure 8.4. There are λ choices for the color of v_1. The vertices v_2 and v_4 must be assigned colors different from the that assigned to v_1. The vertices v_2 and v_4 may be assigned the same color or may be assigned different colors. If v_2 and v_4 are assigned the same color, then there are $\lambda - 1$ choices for that color. The vertex v_3 can then be assigned any color except the color assigned to v_2 and v_4. Hence, the number of distinct λ-colorings of C_4 in which v_2 and v_4 are colored the same is $\lambda(\lambda - 1)^2$.

If, on the other hand, v_2 and v_4 are colored differently, then there are $\lambda - 1$ choices for v_2 and $\lambda - 2$ choices for v_4. Since v_3 can be assigned any color except the two colors assigned to v_2 and v_4, the number of λ-colorings of C_4 in which v_2 and v_4 are colored differently is $\lambda(\lambda - 1)(\lambda - 2)^2$. Hence, the number of distinct

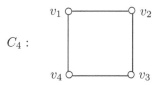

$C_4:$

Figure 8.4: The chromatic polynomial of C_4

λ-colorings of C_4 is

$$
\begin{aligned}
P(C_4, \lambda) &= \lambda(\lambda - 1)^2 + \lambda(\lambda - 1)(\lambda - 2)^2 \\
&= \lambda(\lambda - 1)(\lambda^2 - 3\lambda + 3) \\
&= \lambda^4 - 4\lambda^3 + 6\lambda^2 - 3\lambda \\
&= (\lambda - 1)^4 + (\lambda - 1).
\end{aligned}
$$

The preceding example illustrates an important observation. Suppose that u and v are nonadjacent vertices in a graph G. The number of λ-colorings of G equals the number of λ-colorings of G in which u and v are colored differently plus the number of λ-colorings of G in which u and v are colored the same. Since the number of λ-colorings of G in which u and v are colored differently is the number of λ-colorings of $G + uv$ while the number of λ-colorings of G in which u and v are colored the same is the number of λ-colorings of the graph H obtained by identifying u and v (an elementary homomorphism), it follows that

$$P(G, \lambda) = P(G + uv, \lambda) + P(H, \lambda).$$

This observation is summarized below:

Theorem 8.6 *Let G be a graph containing nonadjacent vertices u and v and let H be the graph obtained from G by identifying u and v. Then*

$$P(G, \lambda) = P(G + uv, \lambda) + P(H, \lambda).$$

Note that if G is a graph of order $n \geq 2$ and size $m \geq 1$, then $G + uv$ has order n and size $m + 1$ while H has order $n - 1$ and size at most m.

Of course, the equation stated in Theorem 8.6 can also be expressed as

$$P(G + uv, \lambda) = P(G, \lambda) - P(H, \lambda).$$

In this context, Theorem 8.6 can be rephrased in terms of an edge deletion and an elementary contraction.

Corollary 8.7 *Let G be a graph containing adjacent vertices u and v and let F be the graph obtained from G by identifying u and v. Then*

$$P(G, \lambda) = P(G - uv, \lambda) - P(F, \lambda).$$

By systematically applying Theorem 8.6 to pairs of nonadjacent vertices in a graph G, we eventually arrive at a collection of complete graphs. We now illustrate this. Suppose that we wish to compute the chromatic polynomial of the graph G of Figure 8.5. For the nonadjacent vertices u and v of G and the graph H obtained by identifying u and v, it follows by Theorem 8.6 that the chromatic polynomial of G is the sum of the chromatic polynomials of $G + uv$ and H.

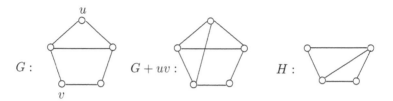

Figure 8.5: $P(G, \lambda) = P(G + uv, \lambda) + P(H, \lambda)$

At this point it is useful to adopt a convention introduced by Alexander Zykov [206] and utilized later by Ronald Read [163]. Rather than repeatedly writing the equation that appears in the statement of Theorem 8.6, we represent the chromatic polynomial of a graph by a drawing of the graph and indicate on the drawing which pair u, v of nonadjacent vertices will be separately joined by an edge and identified. So, for the graph G of Figure 8.5, we have

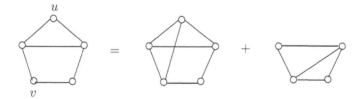

Continuing in this manner, as shown in Figure 8.6, we obtain

$$P(G, \lambda) = \lambda^5 - 6\lambda^4 + 14\lambda^3 - 15\lambda^2 + 6\lambda.$$

Using this approach, we see that the chromatic polynomial of every graph is the sum of chromatic polynomials of complete graphs. A consequence of this observation is the following.

Theorem 8.8   *The chromatic polynomial $P(G, \lambda)$ of a graph G is a polynomial in λ.*

There are some interesting properties possessed by the chromatic polynomial of every graph. In fact, if G is a graph of order n and size m, then the chromatic polynomial $P(G, \lambda)$ of G can be expressed as

$$P(G, \lambda) = c_0 \lambda^n + c_1 \lambda^{n-1} + c_2 \lambda^{n-2} + \cdots + c_{n-1} \lambda + c_n,$$

Figure 8.6: $P(G, \lambda) = P(G + uv, \lambda) + P(H, \lambda)$

where $c_0 = 1$ (and so $P(G, \lambda)$ is a polynomial of degree n with leading coefficient 1), $c_1 = -m$, $c_i \geq 0$ if i is even with $0 \leq i \leq n$, and $c_i \leq 0$ if i is odd with $1 \leq i \leq n$. Since $P(G, 0) = 0$, it follows that $c_n = 0$.

The following theorem is due to Hassler Whitney (see [202]).

Theorem 8.9 *Let G be a graph of order n and size m. Then $P(G, \lambda)$ is a polynomial of degree n with leading coefficient 1 such that the coefficient of λ^{n-1} is $-m$, and whose coefficients alternate in sign.*

Proof. We proceed by induction on m. If $m = 0$, then $G = \overline{K}_n$ and $P(G, \lambda) = \lambda^n$, as we have seen. Then $P(\overline{K}_n, \lambda) = \lambda^n$ has the desired properties.

Assume that the result holds for all graphs whose size is less than m, where $m \geq 1$. Let G be a graph of size m and let $e = uv$ be an edge of G. By Corollary 8.7,

$$P(G, \lambda) = P(G - e, \lambda) - P(F, \lambda),$$

where F is the graph obtained from G by identifying u and v. Since $G - e$ has order n and size $m - 1$, it follows by the induction hypothesis that

$$P(G - e, \lambda) = a_0 \lambda^n + a_1 \lambda^{n-1} + a_2 \lambda^{n-2} + \cdots + a_{n-1}\lambda + a_n,$$

where $a_0 = 1$, $a_1 = -(m-1)$, $a_i \geq 0$ if i is even with $0 \leq i \leq n$, and $a_i \leq 0$ if i is odd with $1 \leq i \leq n$. Furthermore, since F has order $n-1$ and size m', where $m' \leq m-1$, it follows that

$$P(F, \lambda) = b_0 \lambda^{n-1} + b_1 \lambda^{n-2} + b_2 \lambda^{n-3} + \cdots + b_{n-2} \lambda + b_{n-1},$$

where $b_0 = 1$, $b_1 = -m'$, $b_i \geq 0$ if i is even with $0 \leq i \leq n-1$, and $b_i \leq 0$ if i is odd with $1 \leq i \leq n-1$. By Corollary 8.7,

$$
\begin{aligned}
P(G, \lambda) &= P(G - e, \lambda) - P(F, \lambda) \\
&= (a_0 \lambda^n + a_1 \lambda^{n-1} + a_2 \lambda^{n-2} + \cdots + a_{n-1} \lambda + a_n) - \\
&\quad (b_0 \lambda^{n-1} + b_1 \lambda^{n-2} + b_2 \lambda^{n-3} + \cdots + b_{n-2} \lambda + b_{n-1}) \\
&= a_0 \lambda^n + (a_1 - b_0) \lambda^{n-1} + (a_2 - b_1) \lambda^{n-2} + \cdots \\
&\quad + (a_{n-1} - b_{n-2}) \lambda + (a_n - b_{n-1}).
\end{aligned}
$$

Since $a_0 = 1$, $a_1 - b_0 = -(m-1) - 1 = -m$, $a_i - b_{i-1} \geq 0$ if i is even with $2 \leq i \leq n$, and $a_i - b_{i-1} \leq 0$ if i is odd with $1 \leq i \leq n$, $P(G, \lambda)$ has the desired properties and the theorem follows by mathematical induction.   ∎

Suppose that a graph G contains an end-vertex v whose only neighbor is u. Then, of course, $P(G - v, \lambda)$ is the number of λ-colorings of $G - v$. The vertex v can then be assigned any of the λ colors except the color assigned to u. This observation gives the following.

Theorem 8.10 *If G is a graph containing an end-vertex v, then*

$$P(G, \lambda) = (\lambda - 1) P(G - v, \lambda).$$

One consequence of this result is the following.

Corollary 8.11 *If T is a tree of order $n \geq 1$, then*

$$P(T, \lambda) = \lambda(\lambda - 1)^{n-1}.$$

Proof. We proceed by induction on n. For $n = 1$, $T = K_1$ and certainly $P(K_1, \lambda) = \lambda$. Thus, the basis step of the induction is true. Suppose that $P(T', \lambda) = \lambda(\lambda - 1)^{n-2}$ for every tree T' of order $n - 1 \geq 1$ and let T be a tree of order n. Let v be an end-vertex of T. Thus, $T - v$ is a tree of order $n - 1$. By Theorem 8.10 and the induction hypothesis,

$$P(T, \lambda) = (\lambda - 1) P(T - v, \lambda) = (\lambda - 1) \left[\lambda(\lambda - 1)^{n-2} \right] = \lambda(\lambda - 1)^{n-1},$$

as desired.   ∎

Two graphs are **chromatically equivalent** if they have the same chromatic polynomial. By Theorems 8.6 and 8.10, two chromatically equivalent graphs must have the same order, the same size, and the same chromatic number. By Corollary 8.11, every two trees of the same order are chromatically equivalent. It is not known under what conditions two graphs are chromatically equivalent in general. A graph G is **chromatically unique** if $P(H, \lambda) = P(G, \lambda)$ implies that $H \cong G$. Here too, it is not known the conditions under which a graph is chromatically unique.

8.4   The Heawood Map-Coloring Problem

As mentioned in Section 8.1 and reported in more detail in Chapter 0, it was Percy John Heawood's 1890 article [121] in which he described an error that occurred in Alfred Bray Kempe's attempted proof of the Four Color Theorem. As a consequence of Heawood's discovery of an irreparable error by Kempe, it was no longer known in 1890 whether the chromatic number of every graph that can be embedded on the sphere was at most 4. Indeed, it was not known how large the chromatic number of a planar graph could be. However, as we noted, Heawood was able to use Kempe's proof technique to give a proof of the Five Color Theorem (presented in Section 8.1). Heawood was not content with this, however. He introduced much more in his paper and it is these added accomplishments for which he will forever be remembered. Heawood became interested in the largest chromatic number of a graph that could be embedded on certain surfaces, namely the orientable surface S_k, $k \geq 0$, (obtained by attaching k handles to a sphere). Thus, S_0 is the sphere and S_1 is the torus. The **chromatic number of a surface** S_k is defined by

$$\chi(S_k) = \max\{\chi(G)\},$$

where the maximum is taken over all graphs G that can be embedded on S_k. That $\chi(S_0) = 4$ is the Four Color Theorem. Heawood was successful in determining the chromatic number of the torus.

Theorem 8.12 $\chi(S_1) = 7$.

Proof.   In Figure 5.28, we saw that the complete graph K_7 can be embedded on the torus. Since $\chi(K_7) = 7$, it follows that $\chi(S_1) \geq 7$.

Now let G be a graph that can be embedded on the torus. Among the subgraphs of G, let H be one having the largest minimum degree. We show that $\delta(H) \leq 6$. Suppose that H has order n and size m. If $n \leq 7$, then certainly $\delta(H) \leq 6$. Hence, we may assume that $n > 7$.

Since G is embeddable on the torus, so is H. Thus, $\gamma(H) \leq 1$. It therefore follows by Theorem 5.25 that

$$1 \geq \gamma(H) \geq \frac{m}{6} - \frac{n}{2} + 1$$

and so $m \leq 3n$. Hence,

$$n\delta(H) \leq \sum_{v \in V(H)} \deg_H v = 2m \leq 6n$$

and so $\delta(H) \leq 6$. Therefore, in any case, $\delta(H) \leq 6$. By Theorem 7.9,

$$\chi(G) \leq 1 + \delta(H) \leq 7.$$

Hence, $\chi(S_1) = 7$.   ∎

In his important paper, Heawood obtained an upper bound for the chromatic number of S_k for every positive integer k.

Theorem 8.13   *For every nonnegative integer k,*

$$\chi(S_k) \leq \left\lfloor \frac{7 + \sqrt{1 + 48k}}{2} \right\rfloor.$$

Proof.   Let G be a graph that is embeddable on the surface S_k and let

$$h = \frac{7 + \sqrt{1 + 48k}}{2}.$$

Hence, $1 + 48k = (2h - 7)^2$. Solving for $h - 1$, we have

$$h - 1 = 6 + \frac{12(k - 1)}{h}. \tag{8.1}$$

Among the subgraphs of G, let H be one having the largest minimum degree. We show that $\delta(H) \leq h - 1$. Suppose that H has order n and size m. If $n \leq h$, then $\delta(H) \leq h - 1$. Hence, we may assume that $n > h$. Since G is embeddable on S_k, so is H. Therefore, $\gamma(H) \leq k$. By Theorem 5.25,

$$k \geq \gamma(H) \geq \frac{m}{6} - \frac{n}{2} + 1.$$

Thus, $m \leq 3n + 6(k - 1)$. We therefore have

$$n\delta(H) \leq \sum_{v \in V(H)} \deg_H v = 2m \leq 6n + 12(k - 1)$$

and so, by (8.1),

$$\delta(H) \leq 6 + \frac{12(k - 1)}{n} \leq 6 + \frac{12(k - 1)}{h} = h - 1.$$

Hence, $\delta(H) \leq h - 1$ in any case. By Theorem 7.9,

$$\chi(G) \leq 1 + \delta(H) \leq h = \frac{7 + \sqrt{1 + 48k}}{2},$$

giving the desired result.   ■

In fact, Heawood was under the impression that he had shown that

$$\chi(S_k) = \left\lfloor \frac{7 + \sqrt{1 + 48k}}{2} \right\rfloor \tag{8.2}$$

for every positive integer k, but, in fact, all he had established was the bound given in Theorem 8.13. It was not unusual during the period surrounding Heawood's paper for mathematicians to write and present arguments in a more casual style, which made it easier for errors and omissions to occur. Indeed, the year following the publication of Heawood's paper, Lothar Heffter [123] wrote a paper in which he drew attention to the incomplete nature of Heawood's argument. Heffter was able to show that equality held in (8.2) not only for $k = 1$ but for $1 \leq k \leq 6$ and some other values of k as well. To verify equality in (8.2) for every positive integer k, it would be necessary to show, for every positive integer k, that there is a graph G_k that is embeddable on S_k for which

$$\chi(G_k) = \left\lfloor \frac{7 + \sqrt{1 + 48k}}{2} \right\rfloor.$$

The question whether equality held in (8.2) for every positive integer k would become a famous problem.

The Heawood Map-Coloring Problem For every positive integer k, is it true that

$$\chi(S_k) = \left\lfloor \frac{7 + \sqrt{1 + 48k}}{2} \right\rfloor?$$

There was a great deal of confusion surrounding this famous problem and the origin of this confusion is also unknown. For example, in their famous book *What is Mathematics?*, Richard Courant and Herbert E. Robbins [60] reported that

$$\chi(S_k) = \left\lfloor \frac{7 + \sqrt{1 + 48k}}{2} \right\rfloor$$

for every positive integer k. Whether the belief that this is true led Courant and Robbins to include this premature statement in their book or whether writing this statement in their book led to mathematical folklore is not known. Indeed, this was not even known to Courant and Robbins. There were reports that the Heawood Map-Coloring Problem had been solved as early as the early 1930s in Göttingen in Germany.

Solving the Heawood Map-Coloring Problem would require the work of many mathematicians and another 78 years. However, primarily through the efforts of Gerhard Ringel and J. W. T. (Ted) Youngs [167], this problem was finally settled.

Theorem 8.14 (The Heawood Map-Coloring Theorem) *For every positive integer* k,

$$\chi(S_k) = \left\lfloor \frac{7 + \sqrt{1 + 48k}}{2} \right\rfloor.$$

Proof. By Theorem 8.13,

$$\chi(S_k) \leq \left\lfloor \frac{7 + \sqrt{1 + 48k}}{2} \right\rfloor.$$

It therefore remains to verify the reverse inequality. Let

$$n = \left\lfloor \frac{7 + \sqrt{1 + 48k}}{2} \right\rfloor.$$

Thus, $n \leq (7 + \sqrt{1 + 48k})/2$. Solving this inequality in terms of k, we have

$$k \geq (n - 3)(n - 4)/12.$$

Since k is an integer, it follows by Theorem 5.26 that

$$k \geq \left\lceil \frac{(n - 3)(n - 4)}{12} \right\rceil = \gamma(K_n)$$

and so $\gamma(K_n) \leq k$. Since K_n is embeddable on $S_{\gamma(K_n)}$ and $\gamma(K_n) \leq k$, it follows that K_n is embeddable on S_k. In addition, since $\chi(K_n) = n$, we have $\chi(S_{\gamma(K_n)}) \geq n$. That is,

$$\chi(S_k) \geq n = \left\lceil \frac{7 + \sqrt{1 + 48k}}{2} \right\rceil,$$

giving the desired result.   ∎

Exercises for Chapter 8

1. It was once thought that the regions of every map can be colored with four or fewer colors because no map contains five mutually adjacent regions. Show that there exist maps that do not contain four mutually adjacent regions. Does this mean that every such map can be colored with three or fewer colors?

2. Show that there exists a 7-chromatic graph that does not contain a subdivision of K_7.

3. Determine the chromatic number of a graph G with the properties (1) G is not bipartite, (2) $\delta(G) \geq 4$ and (3) G does not contain K_4 as a minor.

4. Prove that if G is a connected graph of order n and size m that has K_k as a minor, then $m \geq n + \binom{k}{2} - k$.

5. Prove that the chromatic polynomial of every graph can be expressed as the sum and difference of the chromatic polynomials of empty graphs.

6. (a) Determine $P(C_6, \lambda)$ by repeated application of Theorem 8.6.

   (b) Use the polynomial obtained in (a) to determine $P(C_6, 2)$. Explain why this answer is not surprising.

7. (a) Determine $P(K_{2,2,2}, \lambda)$ by repeated application of Theorem 8.6.

   (b) Use the polynomial obtained in (a) to determine $P(K_{2,2,2}, 3)$. Explain why this answer is not surprising.

8. We have seen that $P(K_{2,2}, \lambda) = \lambda^{(4)} + 2\lambda^{(3)} + \lambda^{(2)}$ and $P(K_{2,2,2}, \lambda) = \lambda^{(6)} + 3\lambda^{(5)} + 3\lambda^{(4)} + \lambda^{(3)}$. That is, the coefficients of $\lambda^{(4)}, \lambda^{(3)}, \lambda^{(2)}$ are those of x^4, x^3, x^2 in $(x^2 + x)^2$ and the coefficients of $\lambda^{(6)}, \lambda^{(5)}, \lambda^{(4)}, \lambda^{(3)}$ are those of x^6, x^5, x^4, x^3 in $(x^2 + x)^3$. Based on these observations, what conjecture would you make of $P(K_{2,2,2,2}, \lambda)$?

9. Prove that $P(C_n, \lambda) = (\lambda - 1)^n + (-1)^n(\lambda - 1)$ for each integer $n \geq 3$.

10. Let u and v be two vertices of a complete graph K_n, $n \geq 3$, and let G be the graph obtained from deleting uv from K_n. Use Theorem 8.6 to determine $P(G, \lambda)$.

11. Let K_n be a complete graph of order $n \geq 4$. Let G be the graph obtained by removing two adjacent edges from K_n and let H be the graph obtained by removing two nonadjacent edges from K_n. Determine $P(G, \lambda)$ and $P(H, \lambda)$.

12. Let G be a maximal planar graph of order $n \geq 3$ (embedded in the plane) with chromatic polynomial $P(G, \lambda)$. A plane graph H is obtained from G by placing a new vertex v in each region of G and joining v to the vertices on the boundary of this region. Express $P(H, \lambda)$ in terms of $P(G, \lambda)$.

13. We know that every two trees of the same order are chromatically equivalent.

    (a) Which unicyclic graphs of the same order are chromatically equivalent?

    (b) How many distinct chromatic polynomials are there for unicyclic graphs of order $n \geq 3$?

14. Prove that if G is a graph with components $G_1, G_2, \ldots, G_k$, then

$$P(G, \lambda) = \prod_{i=1}^{k} P(G_i, \lambda).$$

15. (a) Prove that if G is a nontrivial connected graph, then $P(G, \lambda) = \lambda g(\lambda)$, where $g(0) \neq 0$.

    (b) Prove that a graph G has exactly k components if and only if

$$P(G, \lambda) = \lambda^k f(\lambda),$$

    where $f(\lambda)$ is a polynomial with $f(0) \neq 0$.

16. Show that if F is a forest of order n with k components, then

$$P(F, \lambda) = \lambda^k (\lambda - 1)^{n-k}.$$

17. Prove that if G is a connected graph with blocks $B_1, B_2, \ldots, B_r$, then

$$P(G, \lambda) = \frac{\prod_{i=1}^{r} P(B_i, \lambda)}{\lambda^{r-1}}.$$

18. It has been stated that if G and H are two chromatically equivalent graphs, then G and H have the same order, the same size, and the same chromatic number. Show that the converse of this statement is false.

19. Prove for each integer $r \geq 2$ that $K_{r,r}$ is chromatically unique.

20. Let G be a graph. Prove that if $P(G, \lambda) = \lambda(\lambda - 1)^{n-1}$, then G is a tree of order n.

21. Prove that if G is a connected graph of order n, then $P(G, \lambda) \le \lambda(\lambda - 1)^{n-1}$.

22. Prove or disprove: The polynomial $\lambda^4 - 3\lambda^3 + 3\lambda^2$ is the chromatic polynomial of some graph.

23. Prove or disprove: The graph G in Figure 8.7 is chromatically unique.

$G:$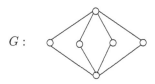

Figure 8.7: The graph G in Exercise 23

24. Prove or disprove: If G is a graph such that $\chi(G) \le \chi(S_k)$ for some positive integer k, then G can be embedded on S_k.

25. Let M be a perfect matching in the graph $G = K_{16}$. Can the graph $G - M$ be embedded on the torus?

26. The complete graphs K_5, K_6, and K_7 have the property that each can be embedded on S_1 but not on S_0. Also, $\chi(S_1) = 7$. The complete graph K_8 can be embedded on S_2 but not on S_1. Also, $\chi(S_2) = 8$. The complete graph K_9 can be embedded on S_3 but not on S_2, and $\chi(S_3) = 9$. Is it the case, for every positive integer k, that $\chi(S_k)$ is the largest order of a complete graph that can be embedded on S_k but not on S_{k-1}?

Chapter 9

Restricted Vertex Colorings

When attempting to properly color the vertices of a graph G (often with a restricted number of colors), there may be instances when (1) there is only one choice for the color of each vertex of G except for the names of the colors, (2) every vertex of G has some preassigned restriction on the choice of a color that can be used for the vertex, or (3) some vertices of G have been given preassigned colors and the remaining vertices must be colored according to these restrictions. Colorings with such restrictions are explored in this chapter.

9.1 Uniquely Colorable Graphs

Suppose that G is a k-chromatic graph. Then every k-coloring of G produces a partition of $V(G)$ into k independent subsets (color classes). If every two k-colorings of G result in the same partition of $V(G)$ into color classes, then G is called **uniquely k-colorable** or simply **uniquely colorable**. Trivially, the complete graph K_n is uniquely colorable. In fact, every complete k-partite graph, $k \geq 2$, is uniquely colorable.

Certainly every 1-chromatic graph is uniquely colorable. Moreover, let there be given a 2-coloring of a nontrivial connected bipartite graph G with the colors 1 and 2. Then if some vertex v of G is assigned the color 1, say, then only the vertices of G whose distance from v is even can be colored 1 as well. Therefore, every nontrivial connected bipartite graph is uniquely 2-colorable. Such a graph is shown in Figure 9.1(a). The necessity of the condition that a uniquely colorable bipartite graph is connected is shown in Figures 9.1(b) and 9.1(c), where two 2-colorings of a disconnected bipartite graph H result in different partitions of $V(H)$.

Each of the graphs G_1 and G_2 of Figure 9.2 is 3-chromatic. Since the vertex u in G_1 is adjacent to the remaining four vertices of G_1, it follows that whatever color is assigned to u cannot be assigned to any other vertex of G_1. Since $G_1 - u$ is a path (and therefore a nontrivial connected bipartite graph), there is a unique 2-coloring of $G_1 - u$ except for the names of the colors. Thus, G_1 is uniquely colorable. In fact, every 3-coloring of G_1 results in the partition $\{\{u\}, \{v, x\}, \{w, y\}\}$ of $V(G_1)$ into

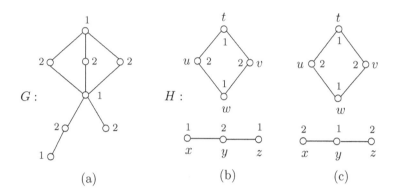

Figure 9.1: A graph that is uniquely 2-colorable and another that is not

color classes. The unique 3-coloring of G_1 (except for the names of the colors) is shown in Figure 9.2(a). On the other hand, the graph G_2 is not uniquely colorable since there are five 3-colorings of G_2 that result in five different partitions of $V(G_2)$ into color classes. Two of these are shown in Figures 9.2(b) and 9.2(c), resulting in the partitions $\{\{u\}, \{v, x\}, \{w, y\}\}$ and $\{\{v\}, \{w, y\}, \{u, x\}\}$, respectively.

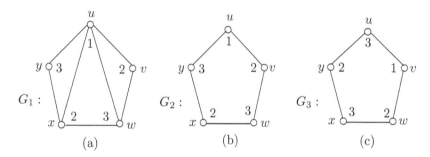

Figure 9.2: A graph that is uniquely 3-colorable and another that is not

Recall (see Section 8.3) that the chromatic polynomial $P(G, \lambda)$ of a graph G is the number of distinct λ-colorings of G. Thus, a k-chromatic graph G is uniquely k-colorable if and only if $P(G, k) = k!$.

We have noted that every uniquely colorable bipartite graph must be connected. In fact, Dorwin Cartwright and Frank Harary [33] showed that in every k-coloring of a uniquely k-colorable graph G, where $k \geq 2$, the subgraph of G induced by any two color classes must also be a connected bipartite graph.

Theorem 9.1 *In every k-coloring of a uniquely k-colorable graph G, where $k \geq 2$, the subgraph of G induced by the union of every two color classes of G is connected.*

Proof. Assume, to the contrary, that there exist two color classes V_1 and V_2 in some k-coloring of G such that $H = G[V_1 \cup V_2]$ is disconnected. We may assume that the vertices in V_1 are colored 1 and those in V_2 are colored 2. Let H_1 and H_2 be two components of H. If H_1, say, is trivial, then the color assigned to the vertex of H_1 can be changed to the other color, resulting in a contradiction. If H_1 is nontrivial, then the colors 1 and 2 of the vertices in H_1 can be interchanged to produce a new partition of $V(G)$ into color classes, producing a contradiction. ∎

As a consequence of Theorem 9.1, every uniquely k-colorable graph, $k \geq 2$, is connected. In fact, Gary Chartrand and Dennis Paul Geller [41] showed that even more can be said.

Theorem 9.2 *Every uniquely k-colorable graph is $(k-1)$-connected.*

Proof. The result is trivial for $k = 1$ and, by Theorem 9.1, the result follows for $k = 2$ as well. Hence, we may assume that $k \geq 3$. Let G be a uniquely k-colorable graph, where $k \geq 3$. If $G = K_k$, then G is $(k-1)$-connected; so we may assume that G is not complete. Assume, to the contrary, that G is not $(k-1)$-connected. Hence, there exists a vertex-cut W of G with $|W| = k - 2$.

Let there be given a k-coloring of G. Consequently, there are at least two colors, say 1 and 2, not used to color any vertices of W. Let V_1 be the color class consisting of the vertices colored 1 and V_2 the set of the vertices colored 2. By Theorem 9.1, $H = G[V_1 \cup V_2]$ is connected. Hence, H is a subgraph of some component G_1 of $G - W$. Let G_2 be another component of $G - W$. Assigning some vertex of G_2 the color 1 produces a new k-coloring of G that results in a new partition of $V(G)$ into color classes, contradicting our assumption that G is uniquely k-colorable. ∎

We then have an immediate corollary of Theorem 9.2.

Corollary 9.3 *If G is a uniquely k-colorable graph, then $\delta(G) \geq k - 1$.*

Much of the interest in uniquely colorable graphs has been directed towards planar graphs. Since every complete graph is uniquely colorable, each complete graph K_n, $1 \leq n \leq 4$, is a uniquely colorable planar graph. Indeed, each complete graph K_n, $1 \leq n \leq 4$, is a uniquely colorable maximal planar graph. Since the complete 3-partite graph $K_{2,2,2}$ (the graph of the octahedron) is also uniquely colorable, $K_{2,2,2}$ is a uniquely 3-colorable maximal planar graph (see Figures 9.3(a)). The graph G in Figure 9.3(b) is also a uniquely 3-colorable maximal planar graph. The fact that the 3-colorable maximal planar graphs shown in Figures 9.3 are also uniquely colorable is not surprising, as Chartrand and Geller [41] observed.

Theorem 9.4 *If G is a 3-colorable maximal planar graph, then G is uniquely 3-colorable.*

Proof. The result is obvious if $G = K_3$, so we may assume that the order of G is at least 4. Let there be a planar embedding of G. Every edge lies on the boundaries of two distinct triangular regions of G. Let T denote a triangle that is the boundary of some region in the embedding and assign the colors 1, 2, 3 to the vertices of T. Let v be a vertex of G not on T. Then there exists a sequence $T = T_0, T_1, \ldots, T_k$ of $k + 1 \geq 2$ triangles in G, each the boundary of a region of G, such that T_i and T_{i+1} share a common edge for $0 \leq i \leq k - 1$ and such that v belongs to T_k but

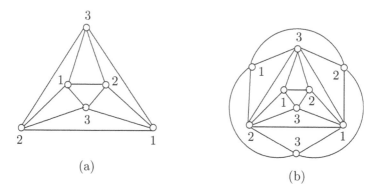

(a)

(b)

Figure 9.3: Uniquely 3-colorable maximal planar graphs

not to T_i $(0 \leq i \leq k - 1)$. Once the vertices of the triangles $T = T_0, T_1, \ldots, T_j$ $(0 \leq j \leq k - 1)$ have been assigned colors, the vertex in T_{j+1} that is not in T_j is then uniquely determined and so G is uniquely 3-colorable. ∎

The two 3-colorable maximal planar graphs in Figures 9.3 have another property in common. There are both Eulerian. That this is a characteristic of all maximal planar 3-chromatic graphs was first observed by Percy John Heawood [122] in 1898.

Theorem 9.5 *A maximal planar graph G of order 3 or more has chromatic number 3 if and only if G is Eulerian.*

Proof. Let there be given a planar embedding of G. Suppose first that G is not Eulerian. Then G contains a vertex v of odd degree $k \geq 3$. Let $N(v) = \{v_1, v_2, \ldots, v_k\}$, where the vertices $v_1, v_2, \ldots, v_k$ occur in clockwise order about v. Since the boundary of every region of G is a triangle, it follows that $v_i v_{i+1} \in E(G)$ for $i = 1, 2, \ldots, k - 1$ and $v_k v_1 \in E(G)$. Thus, $C = (v_1, v_2, \ldots, v_k, v_1)$ is an odd cycle in G. Because v is adjacent to every vertex of C, it follows that $\chi(G) = 4$, which is a contradiction.

We verify the converse by induction on the order of a maximal planar Eulerian graph. If the order of a maximal planar Eulerian graph G is 3, then $G = K_3$ and $\chi(G) = 3$. Assume that every maximal planar Eulerian graph of order k has chromatic number 3 for an integer $k \geq 3$ and let G be a maximal planar Eulerian graph of order $k + 1$. Let there be given a planar embedding of G and let uw be an edge of G. Then uw is on the boundary of two (triangular) regions of G. Let x be the third vertex on the boundary of one of these regions and y the third vertex on the boundary of the other region. Suppose that

$$N(x) = \{u = x_1, x_2, \ldots, x_k = w\} \text{ and } N(y) = \{u = y_1, y_2, \ldots, y_\ell = w\},$$

where k and ℓ are even, such that $C = (x_1, x_2, \ldots, x_k, x_1)$ and $C' = (y_1, y_2, \ldots, y_\ell, y_1)$ are even cycles. Let G' be the graph obtained from G by (1) deleting x, y, and uw from G and (2) adding a new vertex z and joining z to every vertex of C and C'. Then G' is a maximal planar Eulerian graph of order k. By the induction hypothesis, $\chi(G') = 3$. According to Theorem 9.4, G' is uniquely colorable. Since z is adjacent to every vertex of C and C', we may assume that z is colored 1 and that the vertices of C and C' alternate in the colors 2 and 3. From the 3-coloring of G', a 3-coloring of G can be given where every vertex of $V(G) - \{x, y\}$ is assigned the same color as in G' and x and y are colored 1. Thus, $\chi(G) = 3$.   ∎

Since the boundary of every region in a planar embedding of a maximal planar graph G of order 3 or more is a triangle, it follows that $\chi(G) = 3$ or $\chi(G) = 4$. While K_4 is both maximal planar and uniquely 4-colorable, the graph G of Figure 9.4 is maximal planar and 4-chromatic but is not uniquely 4-colorable. For example, interchanging the colors of the vertices u and v in G produces a 4-coloring of G that results in a new partition of $V(G)$ into color classes.

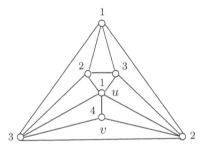

Figure 9.4: A 4-chromatic maximal planar graph
that is not uniquely 4-colorable

On the other hand, Chartrand and Geller [41] showed that every uniquely 4-colorable planar graph must be maximal planar.

Theorem 9.6 *Every uniquely 4-colorable planar graph is maximal planar.*

Proof. Let G be a uniquely 4-colorable planar graph of order $n \geq 4$ and let there be given a 4-coloring of G. Denote the (unique) color classes resulting from this 4-coloring by V_1, V_2, V_3, and V_4, where $|V_i| = n_i$ for $1 \leq i \leq 4$ and so $n = n_1 + n_2 + n_3 + n_4$. By Theorem 9.1, each of the induced subgraphs $G[V_i \cup V_j]$ is connected, where $1 \leq i < j \leq 4$. Thus, the size of this subgraph is at least $n_i + n_j - 1$. Summing these over all six pairs i, j with $1 \leq i < j \leq 4$, we obtain

$$3(n_1 + n_2 + n_3 + n_4) - 6 = 3n - 6.$$

Hence, the size of G is at least $3n - 6$. However, since the size of every planar graph of order $n \geq 3$ is at most $3n - 6$, the size of G is exactly $3n - 6$, implying that G is maximal planar.   ∎

For a planar graph G with chromatic number k, consider the following two statements:

$$\text{If } G \text{ is maximal planar, then } G \text{ is uniquely colorable.} \qquad (9.1)$$
$$\text{If } G \text{ is uniquely colorable, then } G \text{ is maximal planar.} \qquad (9.2)$$

By Theorem 9.5, (9.1) is true if $k = 3$, while by Theorem 9.6, (9.2) is true if $k = 4$. However, if the values of k are interchanged, then neither (9.1) nor (9.2) is true. For example, the 4-chromatic graph of Figures 9.4 is a maximal planar graph that is not uniquely colorable. Also the 3-chromatic graph of Figures 9.2(a) is uniquely colorable but is not maximal planar.

Recall that a graph G is outerplanar if there exists a planar embedding of G so that every vertex of G lies on the boundary of the exterior region. By Theorem 5.19, every nontrivial outerplanar graph contains at least two vertices of degree at most 2. Since every subgraph of an outerplanar graph is outerplanar, it follows by Theorem 7.8 that for every outerplanar graph G,

$$\chi(G) \leq 1 + \max\{\delta(H)\} \leq 3, \tag{9.3}$$

where the maximum is taken over all subgraphs H of G. Since there are many 3-chromatic outerplanar graphs, the bound in (9.3) is sharp.

Recall also that every maximal outerplanar graph of order $n \geq 3$ is 2-connected and that the size of every maximal outerplanar graph of order $n \geq 2$ is $2n - 3$. We now show for an outerplanar graph G with $\chi(G) = 3$ that both (9.1) and (9.2) are true with "maximal planar" replaced by "maximal outerplanar".

Theorem 9.7 *An outerplanar graph G of order $n \geq 3$ is uniquely 3-colorable if and only if G is maximal outerplanar.*

Proof. Let G be a uniquely 3-colorable outerplanar graph of order $n \geq 3$ and let there be given a 3-coloring of G. Furthermore, let the (unique) color classes resulting from this 3-coloring be denoted by V_1, V_2, and V_3, where $|V_i| = n_i$ for $1 \leq i \leq 3$ and so $n = n_1 + n_2 + n_3$. By Theorem 9.1, each of the induced subgraphs

$$G[V_1 \cup V_2], \ G[V_1 \cup V_3], \ \text{and} \ G[V_2 \cup V_3]$$

is connected. Thus, the sizes of these three subgraphs are at least $n_1 + n_2 - 1$, $n_1 + n_3 - 1$, and $n_2 + n_3 - 1$, respectively. Consequently, the size of G is at least

$$(n_1 + n_2 - 1) + (n_1 + n_3 - 1) + (n_2 + n_3 - 1) = 2(n_1 + n_2 + n_3) - 3 = 2n - 3.$$

Since the size of an outerplanar graph cannot exceed $2n - 3$, it follows that the size of G is $2n - 3$ and so G is maximal outerplanar.

For the converse, let G be a maximal outerplanar graph of order $n \geq 3$ and let there be given an outerplanar embedding of G such that the boundary of every region of G is a triangle except possibly the exterior region. As we noted, $\chi(G) = 3$. Assign the colors 1, 2, 3 to the vertices of some triangle T of G. For every vertex v of G not on T, there exists a sequence

$$T = T_0, T_1, \ldots, T_k \ (k \geq 1)$$

of triangles such that T_i and T_{i+1} share a common edge for each i with $0 \leq i \leq k-1$ and v belongs to T_k but to no triangle T_i with $0 \leq i \leq k - 1$. The only uncolored vertex of T_1 has its color uniquely determined by the other two vertices of T_1.

Proceeding successively through the vertices of $T_1, T_2, \ldots, T_k$ not already colored, we have that the color of the vertex v is uniquely determined and so G is uniquely 3-colorable. ∎

Exercise 22 of Chapter 6 states that if G is a k-colorable graph, $k \geq 2$, of order n such that $\delta(G) > \left(\frac{k-2}{k-1}\right) n$, then G is k-chromatic. Béla Bollobás [23] showed that with only a slightly stronger minimum degree condition, the graph G must be uniquely k-colorable.

Theorem 9.8 *If G is a k-colorable graph, $k \geq 2$, of order n such that*

$$\delta(G) > \left(\frac{3k-5}{3k-2}\right) n, \tag{9.4}$$

then G is uniquely k-colorable.

Proof. We proceed by induction on k. First, we consider the case $k = 2$. Let G be a 2-colorable graph of order n such that $\delta(G) > n/4$. Hence, G is bipartite. We claim that G is connected. Suppose that G is disconnected. Then G contains a component H of order $p \leq n/2$. Since $\delta(G) > n/4$, it follows that $\delta(H) > p/2$. Since H has a partite set of order at most $p/2$, this contradicts the assumption that $\delta(H) > p/2$. Thus, G is connected and so G is uniquely 2-colorable.

Assume, for an integer $k \geq 3$, that if G' is a $(k-1)$-colorable graph of order n' such that

$$\delta(G') > \left(\frac{3(k-1)-5}{3(k-1)-2}\right) n',$$

then G' is uniquely $(k-1)$-colorable. Let G be a k-colorable graph of order n such that

$$\delta(G) > \left(\frac{3k-5}{3k-2}\right) n.$$

We show that G is uniquely k-colorable.

Let v be any vertex of G and let $G_v = G[N(v)]$ be a subgraph of order p. Since v is adjacent to every vertex of G_v, no vertex of G_v is assigned the color of v in any k-coloring of G and so G_v is $(k-1)$-colorable. Since $\deg_G v = p \geq \delta(G)$, it follows that

$$p > \left(\frac{3k-5}{3k-2}\right) n.$$

Hence,

$$\frac{p}{3k-5} > \frac{n}{3k-2}$$

and so

$$p - \left(\frac{3}{3k-2}\right) n > p - \left(\frac{3}{3k-5}\right) p = \left(\frac{3k-8}{3k-5}\right) p.$$

For a vertex u in G_v,

$$\deg_{G_v} u > \left(\frac{3k-5}{3k-2}\right) n - (n-p) = p - \left(\frac{3}{3k-2}\right) n$$

$$> \left(\frac{3k-8}{3k-5}\right) p = \left(\frac{3(k-1)-5}{3(k-1)-2}\right) p.$$

Since G_v is a $(k-1)$-colorable graph, it follows by the induction hypothesis that G_v is uniquely $(k-1)$-colorable.

Now let there be given a k-coloring c of G and let x be an arbitrary vertex of G. As we have seen, G_x is uniquely $(k-1)$-colorable and so $\chi(G_x) = k-1$. Since x is adjacent to every vertex in G_x, the vertex x must be assigned a color different from all those colors assigned to vertices in G_x. Thus, there is only one available color for x and so $c(x)$ is uniquely determined, implying that G is uniquely k-colorable. ∎

From Exercise 22 of Chapter 6 and Theorem 9.8, it follows that if G is a k-colorable graph, $k \geq 2$, of order n such that $\delta(G) > \left(1 - \frac{1}{k-1}\right)n$, then $\chi(G) = k$; while if $\delta(G) > \left(1 - \frac{1}{k-\frac{2}{3}}\right)n$, then G is uniquely k-colorable.

The bound in (9.4) of Theorem 9.8 is sharp. Let $r \geq 1$. For $k = 2$, let $G = 2K_{r,r}$ and for $k \geq 3$, let $G = F + 2K_{r,r}$, where F is the complete $(k-2)$-partite graph each of whose partite sets consists of $3r$ vertices. Then G is a regular graph of order $n = r(3k-2)$ such that $\deg v = r(3k-5)$ for all $v \in V(G)$. Then $\chi(G) = k$ and

$$\delta(G) = r(3k-5) = \left(\frac{3k-5}{3k-2}\right)n.$$

However, G is not uniquely k-colorable since the two colors assigned to the vertices of $2K_{r,r}$ can be interchanged in one of the two copies of $K_{r,r}$.

9.2   List Colorings

In recent decades there has been increased interest in colorings of graphs in which the color of each vertex is to be chosen from a specified list of allowable colors. Let G be a graph for which there is an associated set $L(v)$ of permissible colors for each vertex v of G. The set $L(v)$ is commonly called a **color list** for v. A **list coloring** of G is then a proper coloring c of G such that $c(v) \in L(v)$ for each vertex v of G. A list coloring is also referred to as a **choice function**. If

$$\mathfrak{L} = \{L(v) :\ v \in V(G)\}$$

is a set of color lists for the vertices of G and there exists a list coloring for this set $\mathfrak{L}$ of color lists, then G is said to be $\mathfrak{L}$-**choosable** or $\mathfrak{L}$-**list-colorable**. A graph G is k-**choosable** or k-**list-colorable** if G is $\mathfrak{L}$-choosable for every collection $\mathfrak{L}$ of lists $L(v)$ for the vertices v of G such that $|L(v)| \geq k$ for each vertex v. The **list chromatic number** $\chi_\ell(G)$ of G is the minimum positive integer k such that G is k-choosable. Then $\chi_\ell(G) \geq \chi(G)$. The concept of list colorings was introduced by Vadim Vizing [195] in 1976 and, independently, by Paul Erdős, Arthur L. Rubin, and Herbert Taylor [75] in 1979.

Suppose that G is a graph with $\Delta(G) = \Delta$. By Theorem 7.8 if we let

$$L(v) = \{1, 2, \ldots, \Delta, 1 + \Delta\}$$

for each vertex v of G, then for these color lists there is always a list coloring of G. Indeed, if $V(G) = \{v_1, v_2, \ldots, v_n\}$ and $\mathfrak{L} = \{L(v_i) :\ 1 \leq i \leq n\}$ is a collection of color lists for G where each set $L(v_i)$ consists of any $1+\Delta$ colors, then a greedy coloring of G produces a proper coloring and so G is $\mathcal{L}$-choosable. Therefore, $\chi_\ell(G) \leq 1 + \Delta(G)$. Summarizing these observations, we have the following.

Theorem 9.9 *For every graph G,*

$$\chi(G) \leq \chi_\ell(G) \leq 1 + \Delta(G).$$

We now consider some examples. First, $\chi(C_4) = 2$ and so $\chi_\ell(C_4) \geq 2$. Consider the cycle C_4 of Figure 9.5 and suppose that we are given any four color lists $L(v_i)$, $1 \leq i \leq 4$, with $|L(v_i)| = 2$. Let $L(v_1) = \{a, b\}$. We consider three cases.

Case 1. $a \in L(v_2) \cap L(v_4)$. In this case, assign v_1 the color b and v_2 and v_4 the color a. Then there is at least one color in $L(v_3)$ that is not a. Assigning v_3 that color gives C_4 a list coloring for this collection of lists.

Case 2. The color a belongs to exactly one of $L(v_2)$ and $L(v_4)$, say $a \in L(v_2) - L(v_4)$. If there is some color $x \in L(v_2) \cap L(v_4)$, then assign v_2 and v_4 the color x and v_1 the color a. There is at least one color in $L(v_3)$ different from x. Assign v_3 that color. Hence, there is a list coloring of C_4 for this collection of lists. Next, suppose that there is no color belonging to both $L(v_2)$ and $L(v_4)$. If $a \in L(v_3)$, then assign a to both v_1 and v_3. There is a color available for both v_2 and v_4. If $a \notin L(v_3)$, then assign v_1 the color a, assign v_2 the color y in $L(v_2)$ different from a, assign v_3 any color z in $L(v_3)$ different from y, and assign v_4 any color in $L(v_4)$ different from z. This is a list coloring for C_4.

Case 3. $a \notin L(v_2) \cup L(v_4)$. Then assign v_1 the color a and v_3 any color from $L(v_3)$. Hence, there is an available color from $L(v_2)$ and $L(v_4)$ to assign to v_2 and v_4, respectively. Therefore, there is a list coloring of C_4 for this collection of lists.

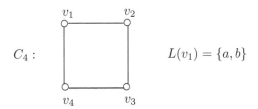

Figure 9.5: The graph C_4 is 2-list colorable

Actually, $\chi_\ell(C_n) = 2$ for every even integer $n \geq 4$. Before showing this, however, it is useful to show that $\chi_\ell(T) = 2$ for every nontrivial tree T.

Theorem 9.10 *Every tree is 2-choosable. Furthermore, for every tree T, for a vertex u of T, and for a collection $\mathcal{L} = \{L(v) : v \in V(T)\}$ of color lists of size 2, where $a \in L(u)$, there exists an $\mathcal{L}$-list-coloring of T in which u is assigned color a.*

Proof. We proceed by induction on the order of the tree. The result is obvious for a tree of order 1 or 2. Assume that the statement is true for all trees of order

k, where $k \geq 2$. Let T be a tree of order $k + 1$ and let $\mathcal{L} = \{L(v) : v \in V(T)\}$ be a collection of color lists of size 2. Let $u \in V(T)$ and suppose that $a \in L(u)$. Let x be an end-vertex of T such that $x \neq u$ and let $\mathcal{L}' = \{L(v) : v \in V(T - x)\}$. Let y be the neighbor of x in T. By the induction hypothesis, there exists an $\mathcal{L}'$-list-coloring c' of $T - x$ in which u is colored a. Now let $b \in L(x)$ such that $b \neq c'(y)$. Then the coloring c defined by

$$c(v) = \begin{cases} b & \text{if } v = x \\ c'(v) & \text{if } v \neq x \end{cases}$$

is an $\mathcal{L}$-list-coloring of T in which u is colored a. ∎

Theorem 9.11 *Every even cycle is 2-choosable.*

Proof. We already know that C_4 is 2-choosable. Let C_n be an n-cycle, where $n \geq 6$ is even. Suppose that $C_n = (v_1, v_2, \ldots, v_n, v_1)$. Let there be given a collection $\mathcal{L} = \{L(v_i) : 1 \leq i \leq n\}$ of color lists of size 2 for the vertices of C_n. We show that C_n is $\mathcal{L}$-list-colorable. We consider two cases.

Case 1. All of the color lists are the same, say $L(v_i) = \{1, 2\}$ for $1 \leq i \leq n$. If we assign the color 1 to v_i for odd i and the color 2 to v_i for even i, then C_n is $\mathcal{L}$-list-colorable.

Case 2. The color lists in $\mathcal{L}$ are not all the same. Then there are adjacent vertices v_i and v_{i+1} in G such that $L(v_i) \neq L(v_{i+1})$. Thus, there exists a color $a \in L(v_{i+1}) - L(v_i)$. The graph $C_n - v_i$ is a path of order $n - 1$. Let $\mathcal{L}' = \{L(v) : v \in V(C_n - v_i)\}$. By Theorem 9.10, there exists an $\mathcal{L}'$-list-coloring c' of $C_n - v_i$ in which $c'(v_{i+1}) = a$. Let $b \in L(v_i)$ such that $b \neq c'(v_{i-1})$. Then the coloring c defined by

$$c(v) = \begin{cases} b & \text{if } v = v_i \\ c'(v) & \text{if } v \neq v_i \end{cases}$$

is an $\mathcal{L}$-list-coloring of G. ∎

Since the chromatic number of every odd cycle is 3, the list chromatic number of every odd cycle is at least 3. Indeed, every odd cycle is 3-choosable (see Exercise 16).

We have seen that all trees and even cycles are 2-choosable. Of course, these are both classes of bipartite graphs. Not every bipartite graph is 2-choosable, however. To illustrate this, we consider $\chi_\ell(K_{3,3})$, where $K_{3,3}$ is shown in Figure 9.6(a). First, we show that $\chi_\ell(K_{3,3}) \leq 3$. Let there be given lists $L(v_i)$, $1 \leq i \leq 6$, where $|L(v_i)| = 3$. We consider two cases.

Case 1. Some color occurs in two or more of the lists $L(v_1)$, $L(v_2)$, $L(v_3)$ or in two or more of the lists $L(v_4)$, $L(v_5)$, $L(v_6)$, say color a occurs in $L(v_1)$ and $L(v_2)$. Then assign v_1 and v_2 the color a and assign v_3 any color in $L(v_3)$. Then there is an available color in $L(v_i)$ for v_i ($i = 4, 5, 6$).

Case 2. The sets $L(v_1)$, $L(v_2)$, $L(v_3)$ are pairwise disjoint as are the sets $L(v_4)$, $L(v_5)$, $L(v_6)$. Let $a_1 \in L(v_1)$ and $a_2 \in L(v_2)$. If none of the sets $L(v_4)$, $L(v_5)$, $L(v_6)$ contain both a_1 and a_2, then let a_3 be any color in $L(v_3)$. Then there is an available color for each of v_4, v_5, v_6 to construct a proper coloring of $K_{3,3}$. If exactly one of the sets $L(v_4)$, $L(v_5)$, $L(v_6)$ contains both a_1 and a_2, then select a color $a_3 \in L(v_3)$ so that none of $L(v_4)$, $L(v_5)$, $L(v_6)$ contains all of a_1, a_2, a_3. By assigning v_3 the color a_3, we see that there is an available color for each of v_4, v_5, and v_6.

Hence, as claimed, $\chi_\ell(K_{3,3}) \leq 3$. We show in fact that $\chi_\ell(K_{3,3}) = 3$. Consider the sets $L(v_i)$, $1 \leq i \leq 6$, shown in Figures 9.6(b). Assume, without loss of generality, that v_1 is colored 1. Then v_4 must be colored 2 and v_5 must be colored 3. Whichever color is chosen for v_3 is the same color as that of either v_4 or v_5. This produces a contradiction. Hence, $K_{3,3}$ is not 2-choosable and so $\chi_\ell(K_{3,3}) = 3$.

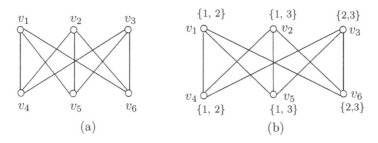

(a)   (b)

Figure 9.6: The graph $K_{3,3}$ is 3-choosable

The graph $G = K_{3,3}$ shows that it is possible for $\chi_\ell(G) > \chi(G)$. In fact, $\chi_\ell(G)$ can be considerably larger than $\chi(G)$.

Theorem 9.12 *If r and k are positive integers such that $r \geq \binom{2k-1}{k}$, then*

$$\chi_\ell(K_{r,r}) \geq k + 1.$$

Proof. Assume, to the contrary, that $\chi_\ell(K_{r,r}) \leq k$. Then $K_{r,r}$ is k-choosable. Let U and W be the partite sets of $K_{r,r}$, where

$$U = \{u_1, u_2, \ldots, u_r\} \text{ and } W = \{w_1, w_2, \ldots, w_r\}.$$

Let $S = \{1, 2, \ldots, 2k - 1\}$. There are $\binom{2k-1}{k}$ distinct k-element subsets of S. Assign these color lists to $\binom{2k-1}{k}$ vertices of U and to $\binom{2k-1}{k}$ vertices of W. Any remaining vertices of U and W are assigned any of the k-element subsets of S. For $i = 1, 2, \ldots, r$, choose a color $a_i \in L(u_i)$ and let $T = \{a_i : 1 \leq i \leq r\}$. We consider two cases.

Case 1. $|T| \leq k - 1$. Then there exists a k-element subset S' of S that is disjoint from T. However, $L(u_j) = S'$ for some j with $1 \leq j \leq r$. This is a contradiction.

Case 2. $|T| \geq k$. Hence, there exists a k-element subset T' of T. Thus, $L(w_j) = T'$ for some j with $1 \leq j \leq r$. Whichever color from $L(w_j)$ is assigned for w_j, this color has been assigned to some vertex u_i. Thus, u_i and w_j have been assigned the same color and $u_i w_j$ is an edge of $K_{r,r}$. This is a contradiction. ∎

Graphs that are 2-choosable have been characterized. A Θ-**graph** consists of two vertices u and v connected by three internally disjoint $u - v$ paths. The graph

$\Theta_{i,j,k}$ is the Θ-graph whose three internally disjoint $u-v$ paths have lengths i, j, and k. The **core** of a graph is obtained by successively removing end-vertices until none remain. The following is due to Erdös, Rubin, and Taylor [75].

Theorem 9.13 *A connected graph G is 2-choosable if and only if its core is K_1, an even cycle, or $\Theta_{2,2,2k}$ for some $k \geq 1$.*

According to the Four Color Theorem, the chromatic number of a planar graph is at most 4. In 1976 Vizing [195] and in 1979 Erdös, Rubin, and Taylor [75] conjectured that the maximum list chromatic number of a planar graph is 5. In 1993, Margit Voigt [196] gave an example of a planar graph of order 238 that is not 4-choosable. In 1994 Carsten Thomassen [186] completed the verification of this conjecture. To show that every planar graph is 5-choosable, it suffices to verify this result for maximal planar graphs. In fact, it suffices to verify this result for a slightly more general class of graphs.

Recall that a planar graph G is nearly maximal planar if there exists a planar embedding of G such that the boundary of every region is a cycle, at most one of which is not a triangle. If G is a nearly maximal planar graph, then we may assume that there is a planar embedding of G such that the boundary of every interior region is a triangle, while the boundary of the exterior region is a cycle of length 3 or more. Therefore, every maximal planar graph is nearly maximal planar (see Figure 9.7(a)). Also, every wheel is nearly maximal planar (see Figure 9.7(b)). The graphs in Figures 9.7(c) and 9.7(d) (where the graph in Figure 9.7(d) is redrawn in Figure 9.7(e)) are nearly maximal planar.

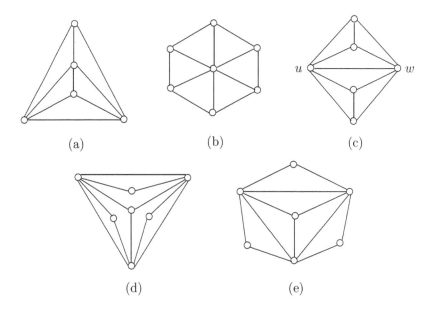

Figure 9.7: Nearly maximal planar graphs

Theorem 9.14 *Every planar graph is 5-choosable.*

Proof. It suffices to verify the theorem for nearly maximal planar graphs. In fact, we verify the following somewhat stronger statement by induction on the order of nearly maximal planar graphs:

> Let G be a nearly maximal planar graph of order $n \geq 3$ such that the boundary of its exterior region is a cycle C (of length 3 or more) and such that $\mathfrak{L} = \{L(v) : v \in V(G)\}$ is a collection of prescribed color lists for G with $|L(v)| \geq 3$ for each $v \in V(C)$ and $|L(v)| \geq 5$ for each $v \in V(G) - V(C)$. If x and y are any two consecutive vertices on C with $a \in L(x)$ and $b \in L(y)$ where $a \neq b$, then there exists an $\mathfrak{L}$-list-coloring of G with x and y colored a and b, respectively.

The statement is certainly true for $n = 3$. Assume for an integer $n \geq 4$ that the statement is true for all nearly maximal planar graphs of order less than n satisfying the conditions in the statement and let G be a nearly maximal planar graph of order n the boundary of whose exterior region is the cycle C and such that

$$\mathfrak{L} = \{L(v) : v \in V(G)\}$$

is a collection of color lists for G for which $|L(v)| \geq 3$ for each $v \in V(C)$ and $|L(v)| \geq 5$ for each $v \in V(G) - V(C)$. Let x and y be any two consecutive vertices on C and suppose that $a \in L(x)$ and $b \in L(y)$ where $a \neq b$. We show that there exists an $\mathfrak{L}$-list-coloring c of G in which x and y are colored a and b, respectively. We consider two cases, according to whether C has a chord.

Case 1. The cycle C has a chord uw. The cycle C contains two $u - w$ paths P' and P'', exactly one of which, say P', contains both x and y. Let C' be the cycle determined by P' and uw and let G' be the nearly maximal planar subgraph of G induced by those vertices lying on or interior to C'. Let

$$\mathfrak{L}' = \{L(v) : v \in V(G')\}.$$

By the induction hypothesis, there is an $\mathfrak{L}'$-list coloring c' of G' in which x and y are colored a and b, respectively. Suppose that $c'(u) = a'$ and $c'(w) = b'$.

Let C'' be the cycle determined by P'' and uw and let G'' be the nearly maximal planar subgraph of G induced by those vertices lying on or interior to C''. Furthermore, let

$$\mathfrak{L}'' = \{L(v) : v \in V(G'')\}.$$

Again, by the induction hypothesis, there is an $\mathfrak{L}''$-list coloring c'' of G'' such that $c''(u) = c'(u) = a'$ and $c''(w) = c'(w) = b'$. Now the coloring c of G defined by

$$c(v) = \begin{cases} c'(v) & \text{if } v \in V(G') \\ c''(v) & \text{if } v \in V(G'') \end{cases}$$

is an $\mathfrak{L}$-list-coloring of G.

Case 2. The cycle C has no chord. Let v_0 be the vertex on C that is adjacent to x such that $v_0 \neq y$ and let

$$N(v_0) = \{x, v_1, v_2, \ldots, v_k, z\},$$

where z is on C. Since G is nearly maximal planar, we may assume that $xv_1, v_k z \in E(G)$ and $v_i v_{i+1} \in E(G)$ for $i = 1, 2, \ldots, k - 1$ (see Figure 9.8).

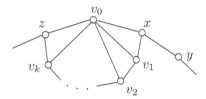

Figure 9.8: A step in the proof of Theorem 9.14

Let P be the $x - z$ path on C that does not contain v_0 and let

$$P^* = (x, v_1, v_2, \ldots, v_k, z).$$

Furthermore, let C^* be the cycle determined by P and P^*. Then $G - v_0$ is a nearly maximal planar graph of order $n - 1$ in which C^* is the boundary of the exterior region. Since $|L(v_0)| \geq 3$, there are (at least) two colors a^* and b^* in $L(v_0)$ different from a. We now define a collection $\mathfrak{L}^*$ of color lists $L^*(v)$ for the vertices v of $G - v_0$ by

$$L^*(v) = L(v) \text{ if } v \neq v_i \ (1 \leq i \leq k)$$

and

$$L^*(v_i) = L(v_i) - \{a^*, b^*\} \ (1 \leq i \leq k)$$

and let

$$\mathfrak{L}^* = \{L^*(v) : \ v \in V(G - v_0)\}.$$

Hence, $|L^*(v)| \geq 3$ for $v \in V(C^*)$ and $|L^*(v)| \geq 5$ for $v \in V(G^*) - V(C^*)$. By the induction hypothesis, there is an $\mathfrak{L}^*$-list coloring of $G - v_0$ with x and y colored a and b, respectively. Since at least one of the colors a^* and b^* has not been assigned to z, one of these colors is available for v_0, producing an $\mathfrak{L}$-list coloring of G. Thus, G is $\mathfrak{L}$-choosable and so is 5-choosable. ∎

As we mentioned, in 1993 Margit Voigt gave an example of a planar graph of order 238 that is not 4-choosable. In 1996 Maryam Mirzakhani (1977–2017) [152] gave an example of a planar graph of order 63 that is not 4-choosable. Mirzakhani was the recipient of the 2014 Fields Medal, the most prestigious award in mathematics. We now describe the Mirzakhani graph and verify that it is, in fact, not 4-choosable.

First, let H be the planar graph of order 17 shown in Figure 9.9(a). For each vertex u of H, a list $L(u)$ of three or four colors is given in Figure 9.9(b), where $L(u) \subseteq \{1, 2, 3, 4\}$. In fact, if $\deg_H u = 4$, then $L(u) = \{1, 2, 3, 4\}$; while if $\deg_H v \neq 4$, then $|L(u)| = 3$. Let $\mathcal{L} = \{L(u) : u \in V(H)\}$. We claim that H is not $\mathcal{L}$-choosable.

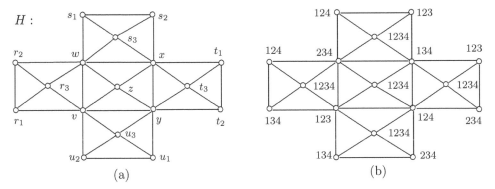

Figure 9.9: A planar graph of order 17

Lemma 9.15 *The planar graph H of Figure 9.9(a) with the set $\mathcal{L}$ of color lists in Figure 9.9(b) is not $\mathcal{L}$-choosable.*

Proof. Assume, to the contrary, that H is $\mathcal{L}$-choosable. Then there exists a 4-coloring c of H such that $c(u) \in L(u)$ for each vertex u in H. Since each vertex of degree 4 in H is adjacent to vertices assigned either two or three distinct colors, it follows that each vertex of degree 4 in H is adjacent to two (nonadjacent) vertices assigned the same color. We claim that $c(x) = 1$ or $c(w) = 2$. Suppose that $c(x) \neq 1$ and $c(w) \neq 2$. Then there are two possibilities. Suppose first that $c(x) = 3$ and $c(w) = 4$. Then either $c(s_1) = 3$ or $c(s_2) = 4$. This is impossible, however, since $3 \notin L(s_1)$ and $4 \notin L(s_2)$. Next, suppose that $c(x) = 4$ and $c(w) = 3$. Then either $c(v) = 4$ or $c(y) = 3$. This is impossible as well since $4 \notin L(v)$ and $3 \notin L(y)$. Hence, as claimed, $c(x) = 1$ or $c(w) = 2$. We consider these two cases.

Case 1. $c(x) = 1$. Since none of the vertices t_1, t_2, and y can be assigned the color 1, it follows that $c(t_1) = c(y) = 2$. Since none of the vertices u_1, u_2, and v can be assigned the color 2, it follows that $c(u_1) = c(v) = 3$. Therefore, none of the vertices r_1, r_2, and w can be assigned the color 3. Thus, $c(r_1) = c(w) = 4$. Hence, $c(x) = 1$, $c(y) = 2$, $c(v) = 3$, and $c(w) = 4$, which is impossible.

Case 2. $c(w) = 2$. Proceeding as in Case 1, we first see that $c(r_2) = c(v) = 1$. From this, it follows that $c(u_2) = c(y) = 4$. Next, we find that $c(t_2) = c(x) = 3$. Hence, $c(v) = 1$, $c(w) = 2$, $c(x) = 3$, and $c(y) = 4$, again an impossibility.

Therefore, as claimed, the graph H is not $\mathcal{L}$-choosable for the set $\mathcal{L}$ of lists described in Figure 9.9(b). ∎

Let H_1, H_2, H_3, and H_4 be four copies of H. For $i = 1, 2, 3, 4$, the color i in the color list of every vertex of H_i is replaced by 5, and the color i is then added to the color list of each vertex not having degree 4. The graphs H_i $(i = 1, 2, 3, 4)$ and the color lists of their vertices are shown in Figure 9.10.

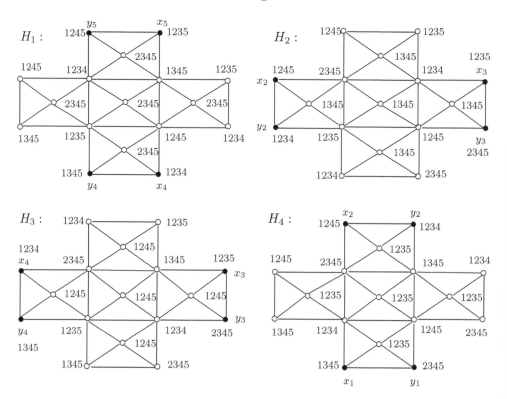

Figure 9.10: The graphs H_i $(i = 1, 2, 3, 4)$

The **Mirzakhani graph** G (a planar graph of order 63) is constructed from the graphs H_i $(i = 1, 2, 3, 4)$ of Figure 9.10 by identifying the two vertices labeled x_i and the two vertices labeled y_i for $i = 2, 3, 4$ and adding a new vertex p with $L'(p) = \{1, 2, 3, 4\}$ and joining p to each vertex of each copy H_i of H whose degree is not 4. The Mirzakhani graph is shown in Figure 9.11 along with the resulting color lists for each vertex. Let $\mathfrak{L}' = \{L'(u) : u \in V(G)\}$. We show that G is not $\mathfrak{L}'$-choosable.

Theorem 9.16 *The Mirzakhani graph (Figure 9.11) is not 4-choosable.*

Proof. Let $L'(u)$ be the color list for each vertex u in G shown in Figure 9.11 and let $\mathfrak{L}' = \{L'(u) : u \in V(G)\}$. We claim that G is not $\mathfrak{L}'$-choosable. Suppose, to the contrary, that G is $\mathfrak{L}'$-choosable. Then there is a coloring c' such that $c'(u) \in L'(u)$ for each $u \in V(G)$. Since the graph H of Figure 9.9(a) is not $\mathfrak{L}$-choosable for the set $\mathfrak{L}$ of lists in Figure 9.9(b), the only way for G to be $\mathfrak{L}'$-choosable is that $c'(v_i) = i$

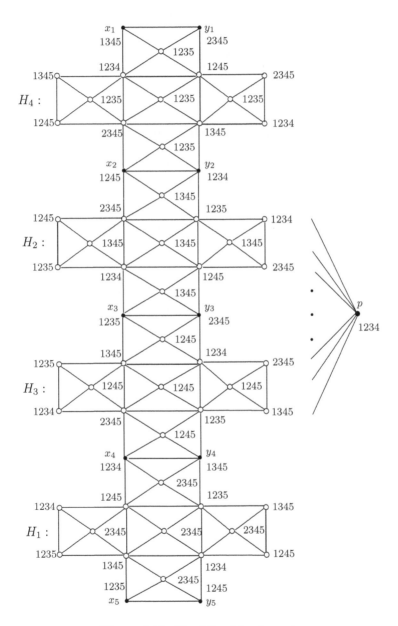

Figure 9.11: The Mirzakhani graph:
A non-4-choosable planar graph of order 63

for some $v_i \in V(H_i)$ for $i = 1, 2, 3, 4$. However then, regardless of the value of $c'(p)$, the vertex p is adjacent to a vertex in G having the same color as p, producing a contradiction, Thus, as claimed, G is not $\mathcal{L}'$-choosable. Because $|L'(u)| = 4$ for each $u \in V(G)$, it follows that G is not 4-choosable.   ∎

Since the Mirzakhani graph has chromatic number 3, it follows that a 3-colorable planar graph need not be 4-choosable. Noga Alon and Michael Tarsi [4] did show that every bipartite graph is 3-choosable, however.

9.3   Precoloring Extensions of Graphs

A special case of a list coloring of a graph G is one where there is a proper subset W of $V(G)$ for which $|L(w)| = 1$ for each $w \in W$, $L(w_1) \cap L(w_2) = \emptyset$ whenever $w_1 w_2 \in E(G)$, $\cup_{w \in W} L(w) \subseteq L(v)$ for each vertex $v \in V(G) - W$, and $|L(v)| \geq k$ for some integer $k \geq \chi(G)$. Under these conditions, of course, every vertex of W must be assigned a specific color and the question is whether, beginning with this proper coloring of a subgraph of G, it is possible to extend this coloring, under certain conditions, to produce a proper coloring of G itself. That is, when attempting to provide a proper coloring of a given graph G, it might be reasonable to begin with a coloring of some of the vertices of G so that the colors assigned to two adjacent vertices are different. The question then is how to complete a coloring of G from this initial coloring.

By a **precoloring** of a graph G, we mean a coloring $p : W \to \mathbb{N}$ of a nonempty subset W of $V(G)$ such that $p(u) \neq p(v)$ if $u, v \in W$ and $uv \in E(G)$. If k distinct colors are used in a precoloring p, then p is called a k-**precoloring**. What we are interested in is whether the coloring p of W can be extended to a coloring $c : V(G) \to \mathbb{N}$ such that $c(w) = p(w)$ for each $w \in W$. With no further restriction, every coloring p of W can be extended to a coloring c of G by letting

$$A \subseteq \mathbb{N} - \{p(v) : v \in W\}$$

such that $|A| = |V(G)| - |W|$, letting

$$c : V(G) - W \to A$$

be a bijective function, and defining

$$c(w) = p(w) \text{ for all } w \in W.$$

Thus, the coloring p of W can be extended to the coloring c of G. What we are primarily interested in, however, is whether a k-precoloring of G, where $k \geq \chi(G)$, can be extended to a k-coloring of G or perhaps to an ℓ-coloring where $\ell \geq k$ and ℓ exceeds k by as little as possible.

First, note that if the two end-vertices of the path P_{2k} for any positive integer k are assigned the same color in a 2-precoloring of P_{2k}, then this cannot be extended to a 2-coloring of P_{2k}. On the other hand, if the goal is to extend this 2-precoloring of P_{2k} to a 3-precoloring of P_{2k}, then it can be done.

Also, if, in the planar graph G in Figure 9.12, $p(x) = p(u_1) = 1$ in a 4-precoloring p of $W = \{x, u_1\}$, then p *cannot* be extended to a 4-coloring of G. On the other hand, if p is a 5-precoloring of W, then p *can* be extended to a 5-coloring of G.

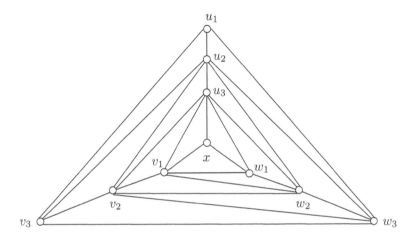

Figure 9.12: The graph G

For a set W of vertices in a graph G, the number $d(W)$ is the minimum distance between any two vertices of W, that is,

$$d(W) = \min\{d(w, w') : w, w' \in W, w \neq w'\}.$$

Carsten Thomassen [186] asked the following question:

If G is a planar graph and W is a set of vertices of G such that $d(W) \geq 100$, does a 5-coloring of W always extend to a 5-coloring of G?

If "5-coloring" is replaced by "4-coloring" and "$d(W) \geq 100$" is replaced by "$d(W) \geq 3$", then the answer to this question is *no*. For example, if we were to assign the color 1 to the vertices in the set $W = \{w_1, w_2, w_3, w_4\}$ in the planar graph G of Figure 9.13, then this 4-coloring of W cannot be extended to a 4-coloring of G.

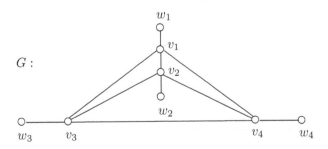

Figure 9.13: A planar graph G

Michael Albertson [1] showed that if W is a set of vertices in a planar graph G such that $d(W) \geq 4$, then every 5-coloring of W can be extended to a 5-coloring of G. Indeed, Albertson proved the following more general result.

Theorem 9.17 *Let G be a k-colorable graph and let W be a set of vertices of G such that $d(W) \geq 4$. Then every $(k+1)$-coloring of W can be extended to a $(k+1)$-coloring of G.*

Proof. Let $p : W \to \{1, 2, \ldots, k+1\}$ be a $(k+1)$-precoloring and let $c' : V(G) \to \{1, 2, \ldots, k\}$ be a k-coloring of G. We use the k-coloring c' of G to define a $(k+1)$-coloring c of G such that $c(w) = p(w)$ for every $w \in W$. First, if $w \in W$, then define $c(w) = p(w)$. Next, if $u \in V(G) - W$ such that u is a neighbor of (necessarily exactly one vertex) $w \in W$ and $c'(w) = p(w)$, then define $c(u) = k + 1$. In all other cases, we define $c(u) = c'(u)$ for $u \in V(G) - W$. Hence, only vertices v in G for which $c(v) = k + 1$ are either those vertices $v \in W$ with $c'(v) = k + 1$ or are neighbors of a vertex $w \in W$ such that $c(w) = c'(v) \neq k + 1$. Hence, if $x, y \in V(G)$ such that $c(x) = c(y) = k + 1$, then $d(x, y) \geq 2$ since $d(W) \geq 4$. Therefore, c is a proper $(k+1)$-coloring of G that is an extension of the $(k+1)$-coloring p of W. ∎

The **corona** $cor(G)$ of a graph G is that graph obtained from G by adding a new vertex w' to G for each vertex w of G and joining w' to w. If G has order n and size m, then $cor(G)$ has order $2n$ and size $m + n$. For example, the graph G of Figure 9.13 is the corona of K_4.

Let M be the Mirzakhani graph (shown in Figure 9.11) and let W' be the set of end-vertices of the corona of M. Furthermore, let

$$\mathcal{L}' = \{L'(w) : w \in W\}$$

be the set of color lists of the vertices of W shown in Figure 9.11. Since $|L'(w)| = 4$ and $L'(w) \subseteq \{1, 2, 3, 4, 5\}$ for each $w \in V(M)$, there is exactly one color in $\{1, 2, 3, 4, 5\}$ that does not belong to $L'(w)$. Define $p(w')$ to be this color. In order to extend this 5-coloring p of W' to a 5-coloring c of G, each vertex w of M must be assigned a color $c(w)$ such that $c(w) \in L'(w)$. However, this is only possible if M is $\mathcal{L}'$-choosable, which, as we saw in the proof of Theorem 9.16, is not the case. Since $d(W') = 3$, this shows that the condition "$d(W') \geq 4$" in the statement of Theorem 9.17 is necessary.

According to Theorem 9.17 then, if W is any set of vertices in a planar graph G with $d(W) \geq 4$, then any 5-coloring of W can be extended to a 5-coloring of G. As we saw with the corona $cor(M)$ of the Mirzakhani graph M, there is a 5-coloring of the set W' of end-vertices of $cor(G)$ that cannot be extended to a 5-coloring of $cor(G)$. In this case, $d(W') = 3$. No matter how large a positive integer k may be, there exists a planar graph G and a set W of the vertices of G for which $d(W) \geq k$ such that some 4-coloring of W cannot be extended to a 4-coloring of G. In order to see this, we first consider the following theorem of John Perry Ballantine [10].

Theorem 9.18 *Let G be a maximal planar graph of order 5 or more that contains exactly two vertices u and v of odd degree. Every 4-coloring of G must assign the same color to u and v.*

Proof. Let there be given a planar embedding of G and a 4-coloring of G using colors from the set $S = \{1, 2, 3, 4\}$. An edge whose incident vertices are colored i and j is referred to as an ij-edge, while a region (necessarily a triangular region) whose incident vertices are colored i, j, and k is called an ijk-region. For colors i, j, and k, let r_i denote the number of regions incident with a vertex colored i, let r_{ij} denote the number of regions incident with an ij-edge, and let r_{ijk} denote the number of ijk-regions. Thus,

$$r_1 = r_{123} + r_{124} + r_{134},$$

for example. A similar equation holds for r_2, r_3, and r_4.

Let i and j be two fixed colors of S, where k and ℓ are the remaining two colors. Every ij-edge is incident with (1) two ijk-regions, (2) two $ij\ell$-regions, or (3) one ijk-region and one $ij\ell$-region. This implies that r_{ij} is even. Since

$$r_{ij} = r_{ijk} + r_{ij\ell},$$

it follows that r_{ijk} and $r_{ij\ell}$ are of the same parity. Since the number of regions in G incident with a vertex x is $\deg x$, it follows that for each color i, the number r_i is the sum of the degrees of the vertices of G colored i.

Assume, to the contrary, that u and v are assigned distinct colors, say u is colored 1 and v is colored 2. Therefore, r_1 and r_2 are odd and r_3 and r_4 are even. Because $r_1 = r_{123} + r_{124} + r_{134}$ and every two of r_{123}, r_{124}, and r_{134} are of the same parity, these three numbers are all odd. Therefore, r_{123}, r_{134}, and r_{234} are odd as well. However, $r_3 = r_{123} + r_{134} + r_{234}$ is even, which is impossible. ∎

As a consequence of Theorem 9.18, we then have the following.

Corollary 9.19 *Let G be a maximal planar graph of order 5 or more containing exactly two odd vertices u and v and let W be any subset of $V(G)$ containing u and v. Then no 4-coloring of W that assigns distinct colors to u and v can be extended to a 4-coloring of G.*

Since there are planar graphs containing exactly two odd vertices that are arbitrarily far apart (see Exercise 29), there are planar graphs G containing a set W of vertices such that $d(W)$ is arbitrarily large and for which some 4-coloring of W cannot be extended to a 4-coloring of G.

As we saw, there are planar graphs G and sets W of vertices of G with $d(W) = 3$ for which some 5-coloring of W cannot be extended to a 5-coloring of G. Such is not the case for 6-colorings, however, as Michael Albertson [1] showed.

Theorem 9.20 *Let G be a planar graph containing a set W of vertices such that $d(W) \geq 3$. Every 6-coloring of W can be extended to a 6-coloring of G.*

Proof. Let p be a 6-coloring of W and let $G' = G - W$. For each $x \in V(G')$, let $L(x)$ be a subset of $\{1, 2, \ldots, 6\}$ such that $|L(x)| = 5$ and $p(w) \notin L(x)$ if some vertex $w \in W$ is adjacent to x. Surely at most one vertex of W is adjacent to x. Since G' is planar, it follows by Theorem 9.14 that G' is 5-choosable. Hence, there exists a

5-coloring c of G' such that $c(x) \in L(x)$ for each $x \in V(G')$. Defining $c(w) = p(w)$ for each $w \in W$ produces a 6-coloring c of G that is an extension of p.   ∎

More generally, for a coloring of a set W of vertices of a graph G, Michael Albertson and Emily Moore [2] obtained the following result with $d(W) \geq 3$.

Theorem 9.21 *Let G be a k-chromatic graph and $W \subseteq V(G)$ such that $d(W) \geq 3$. Then every $(k+1)$-coloring of W can be extended to an ℓ-coloring of G such that*

$$\ell \leq \left\lceil \frac{3k+1}{2} \right\rceil .$$

Proof.   Suppose that a $(k+1)$-coloring of W is given using the colors $1, 2, \ldots, k+1$. Since $\chi(G) = k$, the set $V(G) - W$ can be partitioned into k independent sets $V_1, V_2, \ldots, V_k$. Because the distance between every two vertices of W is at least 3, every vertex in each set V_i $(1 \leq i \leq k)$ is adjacent to at most one vertex of W. Hence, for every integer i with $1 \leq i \leq \lfloor (k+1)/2 \rfloor$, each vertex of V_i can be colored with one of the two colors $2i - 1$ and $2i$ in such a way that no vertex of V_i is assigned the same color as a neighbor of this vertex in W. For every integer i with $1 \leq i \leq k - \lfloor (k+1)/2 \rfloor$, each vertex in the set $V_{\lfloor (k+1)/2 \rfloor + i}$ can be assigned the color $k + 1 + i$. Thus, the total number of colors used is

$$(k+1) + k - \left\lfloor \frac{k+1}{2} \right\rfloor = \left\lceil \frac{k+1}{2} \right\rceil + \left\lfloor \frac{k+1}{2} \right\rfloor + k - \left\lfloor \frac{k+1}{2} \right\rfloor = \left\lceil \frac{3k+1}{2} \right\rceil ,$$

giving the desired result.   ∎

To see that the bound for ℓ given in Theorem 9.21 is sharp, let G be the corona of the complete k-partite graph $K_{k+1,k+1,\ldots,k+1}$ having partite sets $V_1, V_2, \cdots, V_k$ and let W be the set of end-vertices of G. Then $\chi(G) = k$ and $d(W) = 3$. Now consider the coloring of W in which the $k+1$ vertices of W adjacent to the vertices in V_i $(1 \leq i \leq k)$ are colored with distinct colors from the set $\{1, 2, \ldots, k+1\}$. The vertices in $\lfloor (k+1)/2 \rfloor$ of the sets $V_1, V_2, \cdots, V_k$ can be colored with two colors each from the set $\{1, 2, \ldots, k+1\}$, while the vertices in each of the remaining sets must be colored with a color from the set

$$\{k+2, k+3, \ldots, k+1 + (k - \lfloor (k+1)/2 \rfloor)\}$$

(see Figure 9.14 for $k = 4$). Hence, a total of

$$(k+1) + k - \left\lfloor \frac{k+1}{2} \right\rfloor = \left\lceil \frac{3k+1}{2} \right\rceil$$

colors is needed, establishing the sharpness of the bound for ℓ given in Theorem 9.21.

There is no theorem analogous to Theorem 9.21 that provides a reasonable bound for ℓ when $d(W) = 2$. For example, suppose that the partite sets of the complete k-partite graph $K_{k+1,k+1,\ldots,k+1}$ are $V_1, V_2, \cdots, V_k$. For each set V_i $(1 \leq i \leq k)$, we add an independent set of $k+1$ new vertices and join each of these vertices to the

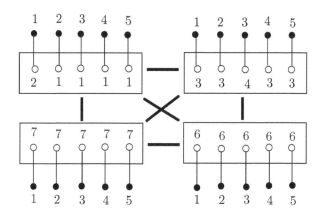

Figure 9.14: Illustrating the sharpness of the bound
for ℓ given in Theorem 9.21

vertices of V_i, resulting in a new graph G. Then $\chi(G) = k$. Let W be the set of
all newly added vertices; so $d(W) = 2$. We now assign the colors $1, 2, \ldots, k+1$ to
the vertices of W joined to the vertices of V_i for each integer i $(1 \le i \le k)$. Then
k new colors are needed to color the remaining vertices of G, which of course was
required without the $(k+1)$-coloring of W.

Exercises for Chapter 9

1. Let G be a noncomplete graph of order n and let k be an integer with $\chi(G) <$
   $k < n$. Show that there exist two k-colorings of G that result in distinct
   partitions of $V(G)$ into k color classes.

2. We know by the Four Color Theorem that no planar graph is 5-chromatic.
   Prove that even if the Four Color Theorem were false, there would exist no
   uniquely 5-colorable planar graph.

3. State and prove a characterization of uniquely 2-colorable graphs.

4. Determine $\chi(G)$ for the planar graph G in Figure 9.15. Is G uniquely colorable?

G :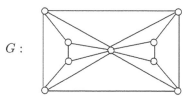

Figure 9.15: The graph G in Exercise 4

5. Let G be a planar graph with $\chi(G) = 3$. By Theorem 9.4, if G is maximal planar, then G is uniquely colorable. Is the converse true?

6. Show, for each positive integer n and partition $\{n_1, n_2, \ldots, n_k\}$ of n into k positive integers (so $n = n_1 + n_2 + \cdots + n_k$), there exists a uniquely k-colorable graph G with color classes $V_1, V_2, \ldots, V_k$ such that $|V_i| = n_i$ for $1 \leq i \leq k$.

7. Let G and H be two graphs. Prove that the join $G \vee H$ of G and H is uniquely colorable if and only if G and H are uniquely colorable.

8. Characterize those graphs G for which $G \square K_2$ is uniquely colorable.

9. Prove or disprove: If G is a nontrivial uniquely k-colorable graph of order n such that one of the k resulting color classes consists of a single vertex v, then $\deg v = n - 1$.

10. By Theorem 9.1, it follows that in every 3-coloring of a uniquely 3-colorable graph G, the subgraph of G induced by the union of every two color classes of G is connected. If there is a 3-coloring of a 3-chromatic graph G such that the subgraph of G induced by the union of every two color classes of G is connected, does this imply that G is uniquely 3-colorable?

11. Prove that if G is a uniquely k-colorable graph of order n and size m, then

$$m \geq \frac{(k-1)(2n-k)}{2}.$$

12. Give an example of a 3-chromatic graph G that is not uniquely colorable for which every 3-coloring of G results in one of three distinct partitions of $V(G)$ into color classes of two vertices each.

13. In any k-coloring of a uniquely k-colorable graph G, $k \geq 2$, there is a unique partition of $V(G)$ into k color classes. Does there exist a connected graph G with $\chi(G) = k \geq 3$ such that there are exactly two distinct partitions of $V(G)$ into k color classes?

14. In showing that the bound given in (9.4) of Theorem 9.8 is sharp, an example of a k-colorable graph G of order n was given such that $\delta(G) = \left(\frac{3k-5}{3k-2}\right) n$ but G is not uniquely k-colorable. In fact, only $k - 2$ color classes of G are uniquely determined in any k-coloring of G. Give an example of a k-colorable graph H of order n for which $\delta(H) = \left(\frac{2k-5}{2k-3}\right) n$ and such that in any k-coloring of H, only $k - 3$ color classes are uniquely determined.

15. Use the greedy coloring algorithm to show that $\chi_\ell(G) \leq 1 + \Delta(G)$ for every graph G (Theorem 9.9).

16. By Theorem 9.9, $\chi_\ell(G) \leq 1 + \Delta(G)$ for every graph G. Therefore, if G is an odd cycle, then G is 3-choosable. Use Theorem 9.10, rather than Theorem 9.9, to prove that every odd cycle is 3-choosable.

17. Prove that $\chi_\ell(K_{2,3}) = 2$.

18. Prove that $\chi_\ell(K_{2,4}) = 3$.

19. Prove that $\chi_\ell(K_{3,27}) > 3$.

20. Prove that the list-chromatic number of $P_n \,\square\, K_2$ is 3 for every integer $n \geq 4$.

21. We have seen that $\chi_\ell(K_{3,3}) = 3$. By Theorem 9.12, $\chi_\ell(K_{10,10}) > 3$. Give a proof of this special case of Theorem 9.12.

22. Show that the Mirzakhani graph (Figure 9.11) has chromatic number 3.

23. It is known that the minimum degree of every induced subgraph of an outer-planar graph is at most 2. Use this fact to prove that every outerplanar graph is 3-choosable.

24. Use the fact that every planar graph contains a vertex of degree 5 or less to prove that every planar graph is 6-choosable.

25. Suppose for $G = K_{3,3}$ that a set $\mathfrak{L} = \{L(v) : v \in V(G)\}$ of color lists is given for the vertices v in G, where $|L(u)| = 3$ for one vertex u in G and $|L(w)| = 2$ for all other vertices w in G. Does there exist a list coloring of G for these lists?

26. For the bipartite 3-cube Q_3 of Figure 9.16, let the lists $L(v_i)$, $1 \leq i \leq 8$, be given as indicated and let $\mathcal{L} = \{L(v_i) : 1 \leq i \leq 8\}$.

$$L(v_1) = \{1,2\}, \; L(v_2) = \{1,3\}, \; L(v_3) = \{2,3\}, \; L(v_4) = \{2,3\},$$
$$L(v_5) = \{1,2\}, \; L(v_6) = \{1,3\}, \; L(v_7) = \{2,3\}, \; L(v_8) = \{2,3\}.$$

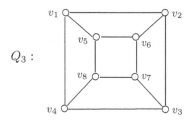

Figure 9.16: The graph in Exercise 26

(a) Is Q_3 $\mathcal{L}$-list colorable?

(b) Is Q_3 2-list colorable?

(c) If $\mathcal{L}' = \{L'(v_i) : 1 \leq i \leq 8\}$, where $L'(v_i) = L'(v_j)$ for $1 \leq i, j \leq 8$ and $|L'(v_i)| = 2$ for $1 \leq i \leq 8$, is Q_3 $\mathcal{L}'$-list colorable?

27. Prove or disprove: Every 2-coloring of a set W of vertices of a tree T with $d(W) \geq 3$ can be extended to a 3-coloring of T.

28. Prove or disprove: Every k-coloring, $k \geq 3$, of a set W of vertices of a tree T can be extended to a $(k + 1)$-coloring of T.

29. Prove, for each positive integer k, that there exists a maximal planar graph G, a set W of vertices of G with $d(W) \geq k$, and a 4-coloring of W that cannot be extended to a 4-coloring of G.

30. For a graph G and the set $S = \{1, 2, \ldots, k\}$, where $k \in \mathbb{N}$, suppose that $L(v) \subseteq S$ for each $v \in V(G)$, where $|L(v)| = 1$ for all v belonging to some subset W of $V(G)$, $|\bigcup_{v \in W} L(v)| = \ell$, and $L(v) = S$ for all $v \in V(G) - W$. If G is $\mathcal{L}$-list colorable where $\mathcal{L} = \{L(v) : v \in V(G)\}$, then what does this say about precoloring extensions of the graph G?

Chapter 10

Edge Colorings

The only graph colorings we have considered thus far have been vertex colorings and region colorings in the case of plane graphs. There is a third coloring, however, which we will discuss in this chapter: edge colorings. As with vertex colorings where the primary emphasis has been on proper vertex colorings, the customary requirement for edge colorings is that adjacent edges be colored differently, resulting in proper edge colorings. This too will be our focus in the current chapter. We will see that the subject of edge colorings is closely related to matchings and factorizations and that this area has applications to problems of scheduling.

10.1   The Chromatic Index and Vizing's Theorem

An **edge coloring** of a graph G is an assignment of colors to the edges of G, one color to each edge. If adjacent edges are assigned distinct colors, then the edge coloring is a **proper edge coloring**. Since proper edge colorings are the most common edge colorings, when we refer to an edge coloring of a graph, we will mean a proper edge coloring unless stated otherwise. In this chapter, we will be concerned only with proper edge colorings.

Since a proper edge coloring of a nonempty graph G is a proper vertex coloring of its line graph $L(G)$, edge colorings of graphs is the same subject as vertex colorings of line graphs. Because investigating vertex colorings of line graphs provides no apparent advantage to investigating edge colorings of graphs, we will study this subject strictly in terms of edge colorings.

A proper edge coloring that uses colors from a set of k colors is a k-**edge coloring**. Thus, a k-edge coloring of a graph G can be described as a function $c : E(G) \to [k] = \{1, 2, \ldots, k\}$ such that $c(e) \neq e(f)$ for every two adjacent edges e and f in G. A graph G is k-**edge colorable** if there exists a k-edge coloring of G. In Figure 10.1(a), a 5-edge coloring of a graph H is shown; while in Figures 10.1(b) and 10.1(c), a 4-edge coloring and a 3-edge coloring of H are shown.

As with vertex colorings, we are often interested in edge colorings of (nonempty) graphs using a minimum number of colors. The **chromatic index** (or **edge chro-**

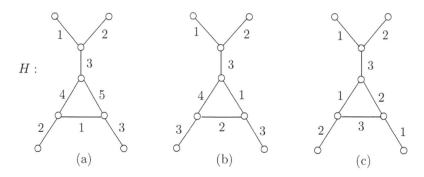

Figure 10.1: Edge colorings of a graph

matic number) $\chi'(G)$ of a graph G is the minimum positive integer k for which G is k-edge colorable. Furthermore, $\chi'(G) = \chi(L(G))$ for every nonempty graph G.

If a graph G is k-edge colorable for some positive integer k, then $\chi'(G) \leq k$. In particular, since the graph H of Figure 10.1 is 3-edge colorable, $\chi'(H) \leq 3$. On the other hand, since H contains three mutually adjacent edges (indeed, several such sets of three edges), at least three distinct colors are required in any edge coloring of H and so $\chi'(H) \geq 3$. Therefore, $\chi'(H) = 3$.

Let there be given a k-edge coloring of a nonempty graph G using the colors $1, 2, \ldots, k$ and let E_i $(1 \leq i \leq k)$ be the set of edges of G assigned the color i. Then the nonempty sets among $E_1, E_2, \ldots, E_k$ of $E(G)$ are the **edge color classes** of G for the given k-edge coloring. Thus, the nonempty sets in $\{E_1, E_2, \ldots, E_k\}$ produce a partition of $E(G)$ into edge color classes. Since no two adjacent edges of G are assigned the same color in a (proper) edge coloring of G, every nonempty edge color class consists of an independent set of edges of G. Indeed, the chromatic index of G is the minimum number of independent sets of edges into which $E(G)$ can be partitioned. Also, if $\chi'(G) = k$ for some graph G, then every k-edge coloring of G results in k nonempty edge color classes.

Recall (from Chapter 4) that the **edge independence number** $\alpha'(G)$ of a nonempty graph G is the maximum number of edges in an independent set of edges of G. Furthermore, if the order of G is n, then $\alpha'(G) \leq n/2$. The following gives a simple yet useful lower bound for the chromatic index of a graph and is an analogue to the lower bound for the chromatic number of a graph presented in Theorem 6.10.

Theorem 10.1 *If G is a graph of size $m \geq 1$, then*

$$\chi'(G) \geq \frac{m}{\alpha'(G)}.$$

Proof. Suppose that $\chi'(G) = k$ and that $E_1, E_2, \ldots, E_k$ are the edge color classes in a k-edge coloring of G. Thus, $|E_i| \leq \alpha'(G)$ for each i $(1 \leq i \leq k)$. Hence,

$$m = |E(G)| = \sum_{i=1}^{k} |E_i| \leq k\alpha'(G)$$

and so $\chi'(G) = k \geq \frac{m}{\alpha'(G)}$. ■

Since every edge coloring of a graph G must assign distinct colors to adjacent edges, for each vertex v of G it follows that $\deg v$ colors must be used to color the edges incident with every vertex v in G. Therefore,

$$\chi'(G) \geq \Delta(G) \tag{10.1}$$

for every nonempty graph G.

In the graph G of order $n = 7$ and size $m = 10$ of Figure 10.2, $\Delta(G) = 3$. Hence, by (10.1), $\chi'(G) \geq 3$. On the other hand, $X = \{uz, vx, wy\}$ is an independent set of three edges of G and so $\alpha'(G) \geq 3$. Because, $\alpha'(G) \leq n/2 = 7/2$, it follows that $\alpha'(G) = 3$. By Theorem 10.1, $\chi'(G) \geq m/\alpha'(G) = 10/3$ and so $\chi'(G) \geq 4$. The 4-edge coloring of G in Figure 10.2 shows that $\chi'(G) \leq 4$ and so $\chi'(G) = 4$.

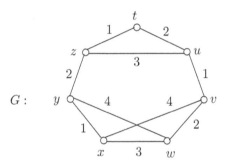

Figure 10.2: A graph with chromatic index 4

While $\Delta(G)$ is a rather obvious lower bound for the chromatic index of a nonempty graph G, the Russian graph theorist Vadim G. Vizing [194] established a remarkable upper bound for the chromatic index of a graph. Vizing's theorem, published in 1964, must be considered the major theorem in the area of edge colorings. Vizing's theorem was rediscovered in 1966 by Ram Prakash Gupta [104].

Theorem 10.2 (**Vizing's Theorem**) *For every nonempty graph G,*

$$\chi'(G) \leq 1 + \Delta(G).$$

Proof. Suppose that the theorem is false. Then among all those graphs H for which $\chi'(H) \geq 2 + \Delta(H)$, let G be one of minimum size. Let $\Delta = \Delta(G)$. Thus, G is not $(1 + \Delta)$-edge colorable. On the other hand, if $e = uv$ is an edge of G, then $G - e$ is $(1 + \Delta(G - e))$-edge colorable. Since $\Delta(G - e) \leq \Delta(G)$, the graph $G - e$ is $(1 + \Delta)$-edge colorable.

Let there be given a $(1 + \Delta)$-edge coloring of $G - e$. Hence, with the exception of e, every edge of G is assigned one of $1 + \Delta$ colors such that adjacent edges are colored differently. For each edge $e' = uv'$ of G incident with u (including the edge e), we define the **dual color** of e' as any of the $1 + \Delta$ colors that is not used to color the edges incident with v'. (See Figure 10.3.) Since $\deg v' \leq \Delta$, there is always

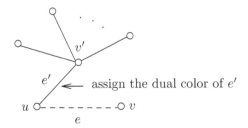

Figure 10.3: A step in the proof of Theorem 10.2

at least one available color for the dual color of the edge uv'. It may occur that distinct edges have the same dual color.

Denote the edge e by $e_0 = uv_0$ as well (where then $v_0 = v$) and suppose that e_0 has dual color α_1. (Thus, α_1 is not the color of any edge incident with v.) Necessarily, some edge $e_1 = uv_1$ incident with u is colored α_1, for otherwise the color α_1 could be assigned to e, producing a $(1 + \Delta)$-edge coloring of G.

Let α_2 be the dual color of e_1. (Thus, no edge incident with v_1 is colored α_2.) If there should be some edge incident with u that is colored α_2, then denote this edge by $e_2 = uv_2$ and let its dual color be denoted by α_3. (See Figure 10.4.) Proceeding in this manner, we then construct a sequence $e_0, e_1, \ldots, e_k$ $(k \geq 1)$ containing a maximum number of distinct edges, where $e_i = uv_i$ for $0 \leq i \leq k$. Consequently, the final edge e_k of this sequence is colored α_k and has dual color α_{k+1}. Therefore, each edge e_i $(0 \leq i \leq k)$ is colored α_i and has dual color α_{i+1}.

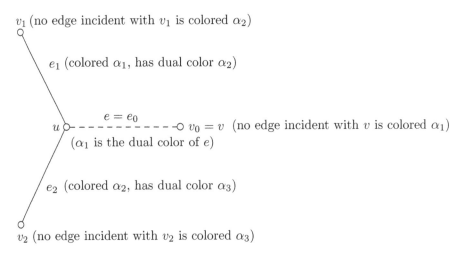

Figure 10.4: A step in the proof of Theorem 10.2

We claim that there is some edge incident with u that is colored α_{k+1}. Suppose that this is not the case. Then each of the edges $e_0, e_1, \ldots, e_k$ can be assigned its dual color, producing a $(1 + \Delta)$-edge coloring of G. This, however, is impossible.

Thus, as claimed, there is an edge e_{k+1} incident with u that is colored α_{k+1}.

Since the sequence $e_0, e_1, \ldots, e_k$ contains the maximum number of distinct edges, it follows that $e_{k+1} = e_j$ for some j with $1 \leq j \leq k$ and so $\alpha_{k+1} = \alpha_j$. Since the color assigned to e_k is not the same as its dual color, it follows that $\alpha_{k+1} \neq \alpha_k$. Therefore, $1 \leq j < k$. Let $j = t + 1$ for $0 \leq t < k - 1$. Hence, $\alpha_{k+1} = \alpha_{t+1}$ and so e_k and e_t have the same dual color.

There must be a color β used to color an edge incident with v in $G - e$ that is not used to color any edge incident with u. If this were not the case, then there would be $\deg_G u - 1 \leq \Delta - 1$ colors used to color the edges incident with u or v, leaving two or more colors available for e. Assigning e one of these colors produces a $(1 + \Delta)$-edge coloring of G, resulting in a contradiction.

The color β must also be assigned to some edge incident with v_i for each i with $1 \leq i \leq k$. If this were not the case, then there would exist a vertex v_r with $1 \leq r \leq k$ such that no edge incident with v_r is colored β. However, we could then change the color of e_r to β and color each edge e_i $(0 \leq i < r)$ with its dual color to obtain a $(1 + \Delta)$-edge coloring of G, which is impossible.

Let P be a path of maximum length with initial vertex v_k whose edges are alternately colored β and α_{k+1}, and let Q be a path of maximum length with initial vertex v_t whose edges are alternately colored β and $\alpha_{t+1} = \alpha_{k+1}$. Suppose that P is a $v_k - x$ path and Q is a $v_t - y$ path. We now consider four cases depending on whether the vertices x and y belong to the set $\{v_0, v_1, \ldots, v_{k-1}, u\}$.

Case 1. $x = v_r$ *for some integer r with* $0 \leq r \leq k - 1$. Since α_{k+1} is the dual color of e_k, no edge incident with v_k is colored α_{k+1} and so the initial edge of P must be colored β. We have seen that for every integer i with $0 \leq i \leq k$, there is an edge incident with v_i that is colored β. Because of the defining property of P, the color of the terminal edge of P cannot be α_{k+1}. This implies that no edge incident with v_r is colored α_{k+1} and so both the initial and terminal edges of P are colored β. Unless $v_r = v_t$, the vertex v_t is not on P as no edge incident with v_t is colored α_{k+1}.

We now interchange the colors β and α_{k+1} of the edges of P. If $r = 0$, then e can be colored β; otherwise, $r > 0$ and no edge incident with v_r is colored β and the dual color of e_i with $0 \leq i < r$ is not changed. Then the edge e_r can be colored β and each edge e_i with $0 \leq i < r$ can be colored with its dual color. This, however, results in a $(1 + \Delta)$-edge coloring of G, which is impossible.

Case 2. $y = v_r$ *for some integer r with* $0 \leq r \leq k$ *where* $r \neq t$. As in Case 1, the initial and terminal edges of Q must also be colored β and no edge incident with v_r is colored α_{k+1}. Furthermore, Q does not contain the vertex v_k unless $v_r = v_k$. We now interchange the colors β and α_{k+1} of the edges of Q. If $r < t$, then we proceed as in Case 1. On the other hand, if $r > t$, we change the color of e to β if $t = 0$; while if $t > 0$, we change the color of e_t to β and color each edge e_i $(0 \leq i < t)$ with its dual color. This implies that G is $(1 + \Delta)$-edge colorable, producing a contradiction.

Case 3. *Either* (1) $x \neq v_r$ *for* $0 \leq r \leq k - 1$ *and* $x \neq u$ *or* (2) $y \neq v_r$ *for* $r \neq t$ *and* $y \neq u$. Since (1) and (2) are similar, we consider (1) only. Upon interchanging the colors β and α_{k+1} of the edges of P, the edge incident with v_k is colored β. Furthermore, the dual color of each edge e_i $(0 \leq i < k)$ has not been altered. Thus, e_k is colored β and each edge e_i $(0 \leq i < k)$ is colored with its dual color, producing a contradiction.

Case 4. $x = y = u$. Necessarily, the initial edges of P and Q are colored β and the terminal edges of P and Q are colored α_{k+1}. Since no edge incident with u is colored β, the paths P and Q cannot be edge-disjoint, for this would imply that u is incident with two distinct edges having the same color (namely α_{k+1}), which is impossible. Thus, P and Q have the same terminal edge and so there is a first edge f that P and Q have in common. Since f is adjacent to another edge of P and another edge of Q, there are three mutually adjacent edges of G belonging to P or Q and so there are adjacent edges of $G - e$ that are colored the same. Since this is impossible, this case cannot occur.   ∎

While multiple edges have no effect on the chromatic number of a graph, quite obviously they can greatly influence the chromatic index of a graph. For example, Theorem 10.1 holds for multigraphs as well, that is, if G is a multigraph of size $m \geq 1$, then

$$\chi'(G) \geq \frac{m}{\alpha'(G)}. \tag{10.2}$$

For a multigraph G, we write $\mu(G)$ for the **maximum multiplicity** of G (also called the **strength** of G), which is the maximum number of edges joining the same pair of vertices of G. Vizing [194] and, independently, Gupta [104] found an upper bound for $\chi'(G)$ in terms of $\Delta(G)$ and $\mu(G)$.

Theorem 10.3   *For every nonempty multigraph G,*

$$\chi'(G) \leq \Delta(G) + \mu(G).$$

For a graph G, Theorem 10.3 reduces to $\chi'(G) \leq \Delta(G) + 1$, which is Theorem 10.2. Claude Elwood Shannon (1916–2001) found an upper bound for the chromatic index of a multigraph G in terms of $\Delta(G)$ alone [180].

Theorem 10.4 (Shannon's Theorem) *If G is a multigraph, then*

$$\chi'(G) \leq \frac{3\Delta(G)}{2}.$$

Proof.   Suppose that the theorem is false. Among all multigraphs H with $\chi'(H) > 3\Delta(H)/2$, let G be one of minimum size. Let $\Delta(G) = \Delta$ and $\mu(G) = \mu$. Suppose that $\chi'(G) = k$. Hence, $\chi(G - f) = k - 1$ for every edge f of G. By Theorem 10.3, $k \leq \Delta + \mu$ and, by assumption, $k > 3\Delta/2$.

Let u and v be vertices of G such that there are μ edges joining them. Let e be one of the edges joining u and v. Thus, $\chi'(G - e) = k - 1$. Hence, there exists a $(k - 1)$-edge coloring of $G - e$. The number of colors not used in coloring the edges incident with u is at least $(k - 1) - (\Delta - 1) = k - \Delta$. Similarly, the number of colors not used in coloring the edges incident with v is at least $k - \Delta$ as well.

Each of these $k - \Delta$ or more colors not used to color an edge incident with u must be used to color an edge incident with v, for otherwise there is a color available for e, contradicting our assumption that $\chi'(G) = k$. Similarly, each of the $k - \Delta$ or more colors not used to color an edge incident with v must be used to color an edge incident with u. Hence, the total number of colors used to color the edges incident with u or v is at least

$$2(k - \Delta) + \mu - 1 \le k - 1.$$

Since $3\Delta/2 < k \le \Delta + \mu$, it follows that $\mu > \Delta/2$ and so

$$2(k - \Delta) + (\Delta/2) - 1 < 2(k - \Delta) + \mu - 1 \le k - 1.$$

Therefore,

$$2k - (3\Delta/2) - 1 < k - 1,$$

implying that $k < 3\Delta/2$, which is a contradiction.   ∎

There are occasions when the upper bound for the chromatic index of a multigraph given by Shannon's theorem is an improvement over that provided by Theorem 10.3. For the multigraph G of Figure 10.5(a), $\Delta(G) = 7$ and $\mu(G) = 4$. Thus, $7 \le \chi(G) \le 11$ by Theorem 10.3. By Shannon's theorem, $\chi'(G) \le (3 \cdot 7)/2$ and so $\chi'(G) \le 10$. Since the size of G is 16 and at most two edges of G can be assigned the same color, it follows by Theorem 10.1 for multigraphs (10.2) that $\chi'(G) \ge 16/2 = 8$. In fact, $\chi'(G) = 8$ as the 8-edge coloring of G in Figure 10.5(b) shows.

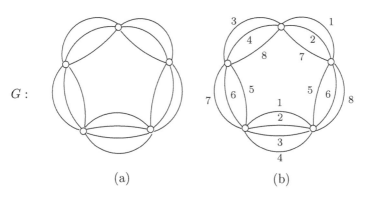

G:

(a)   (b)

Figure 10.5: A multigraph G with $\Delta(G) = 7$, $\mu(G) = 4$, and $\chi'(G) = 8$

10.2   Class One and Class Two Graphs

By Vizing's theorem, it follows that for every nonempty graph G, either $\chi'(G) = \Delta(G)$ or $\chi'(G) = 1 + \Delta(G)$. A graph G belongs to or is of **Class one** if $\chi'(G) = \Delta(G)$ and is of **Class two** if $\chi'(G) = 1 + \Delta(G)$. Consequently, a major question in the area of edge colorings is that of determining to which of these two classes a given graph belongs.

While we will see many graphs of Class one and many graphs of Class two, it turns out that it is much more likely that a graph is of Class one. Paul Erdős and Robin Wilson [77] proved the following, where the set of graphs of order n is denoted by $\mathcal{G}_n$ and the set of Class one graphs having order n is denoted by $\mathcal{G}_{n,1}$.

Theorem 10.5 *Almost every graph is of Class one, that is,*

$$\lim_{n \to \infty} \frac{|\mathcal{G}_{n,1}|}{|\mathcal{G}_n|} = 1.$$

We now look at a few well-known graphs and classes of graphs to determine whether they are of Class one or of Class two. We begin with the cycles. Since the cycle C_n ($n \geq 3$) is 2-regular, $\chi'(C_n) = 2$ or $\chi'(C_n) = 3$. If n is even, then the edges may be alternately colored 1 and 2, producing a 2-edge coloring of C_n. If n is odd, then $\alpha'(C_n) = (n-1)/2$. Since the size of C_n is n, it follows by Theorem 10.1 that $\chi'(C_n) \geq n/\alpha'(C_n) = 2n/(n-1) > 2$ and so $\chi'(C_n) = 3$. Therefore,

$$\chi'(C_n) \quad = \quad \begin{cases} 2 & \text{if } n \text{ is even} \\ 3 & \text{if } n \text{ is odd}. \end{cases}$$

Since $\Delta(C_n) = 2$, it follows that C_n is of Class one if n is even and of Class two if n is odd.

We now turn to complete graphs. Since K_n is $(n-1)$-regular, either $\chi'(K_n) = n - 1$ or $\chi'(K_n) = n$. If n is even, then it follows by Theorem 4.15 that K_n is 1-factorable, that is, K_n can be factored into $n - 1$ 1-factors $F_1, F_2, \ldots, F_{n-1}$. By assigning each edge of F_i ($1 \leq i \leq n - 1$) the color i, an $(n-1)$-edge coloring of K_n is produced. If n is odd, then $\alpha'(K_n) = (n-1)/2$. Since the size m of K_n is $n(n-1)/2$, it follows by Theorem 10.1 that $\chi'(K_n) \geq m/\alpha'(K_n) = n$. Thus, $\chi'(K_n) = n$. In summary,

$$\chi'(K_n) \quad = \quad \begin{cases} n - 1 & \text{if } n \text{ is even} \\ n & \text{if } n \text{ is odd}. \end{cases}$$

Consequently, the chromatic index of every nonempty complete graph is an odd integer. Since $\Delta(K_n) = n - 1$, it follows, as with the cycles C_n, that K_n is of Class one if n is even and of Class two if n is odd.

Of course, both the cycles and complete graphs are regular graphs. For an r-regular graph G, either $\chi'(G) = r$ or $\chi'(G) = r + 1$. If $\chi'(G) = r$, then there is an r-edge coloring of G, resulting in r color classes $E_1, E_2, \ldots, E_r$. Since every vertex v of G has degree r, the vertex v is incident with exactly one edge in each set E_i ($1 \leq i \leq r$). Therefore, each color class E_i is a perfect matching and G is 1-factorable. Conversely, if G is 1-factorable, then $\chi'(G) = r$.

Theorem 10.6 *A regular graph G is of Class one if and only if G is 1-factorable.*

We saw in Section 4.3 that the Petersen graph P is not 1-factorable and so it is of Class two, that is, $\chi'(P) = 4$. The formulas mentioned above for the chromatic index of cycles and complete graphs are immediate consequences of Theorem 10.6, as is the following.

Corollary 10.7 *Every regular graph of odd order is of Class two.*

We have already seen that the even cycles are of Class one. The four graphs shown in Figure 10.6 are of Class one as well. The even cycles and the four graphs of Figure 10.6 are all bipartite. These graphs serve as illustrations of a theorem due to Denés König [136].

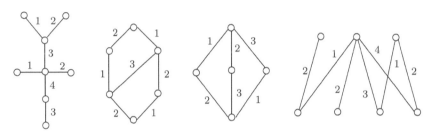

Figure 10.6: Some Class one graphs

Theorem 10.8 (König's Theorem) *If G is a nonempty bipartite graph, then*

$$\chi'(G) = \Delta(G).$$

Proof. Suppose that the theorem is false. Then among the counterexamples, let G be one of minimum size. Thus, G is a bipartite graph such that $\chi'(G) = \Delta(G)+1$. Let $e \in E(G)$, where $e = uv$. Necessarily, u and v belong to different partite sets of G. Then $\chi'(G - e) = \Delta(G - e)$. Now $\Delta(G - e) = \Delta(G)$, for otherwise G is $\Delta(G)$-edge colorable.

Let there be given a $\Delta(G)$-edge coloring of $G - e$. Each of the $\Delta(G)$ colors must be assigned to an edge incident either with u or with v in $G - e$, for otherwise this color could be assigned to e producing a $\Delta(G)$-edge coloring of G. Because $\deg_{G-e} u < \Delta(G)$ and $\deg_{G-e} v < \Delta(G)$, there is a color α of the $\Delta(G)$ colors not used in coloring the edges of $G - e$ incident with u and a color β of the $\Delta(G)$ colors not used in coloring the edges of $G - e$ incident with v. Then $\alpha \neq \beta$ and, furthermore, some edge incident with v is colored α and some edge incident with u is colored β.

Let P be a path of maximum length having initial vertex v whose edges are alternately colored α and β. The path P cannot contain u, for otherwise P has odd length, implying that the initial and terminal edges of P are both colored α. This is impossible, however, since u is incident with no edge colored α. Interchanging the colors α and β of the edges of P produces a new $\Delta(G)$-edge coloring of $G - e$ in which neither u nor v is incident with an edge colored α. Assigning e the color α produces a $\Delta(G)$-edge coloring of G, which is a contradiction.   ∎

We have seen that if G is a graph of size m, then any partition of $E(G)$ into independent sets must contain at least $\frac{m}{\alpha'(G)}$ sets. If v is a vertex with deg $v = \Delta(G)$, then each of the $\Delta(G)$ edges incident with v must belong to distinct independent sets. Thus, $\frac{m}{\alpha'(G)} \geq \Delta(G)$ and so $m \geq \Delta(G) \cdot \alpha'(G)$. If $m > \Delta(G) \cdot \alpha'(G)$, then we can say more.

Theorem 10.9 *If G is a graph of size m such that*

$$m > \alpha'(G)\Delta(G),$$

then G is of Class two.

Proof.   By Theorem 10.1, $\chi'(G) \geq \frac{m}{\alpha'(G)}$. Thus,

$$\chi'(G) \geq \frac{m}{\alpha'(G)} > \frac{\Delta(G) \cdot \alpha'(G)}{\alpha'(G)} = \Delta(G),$$

which implies that $\chi(G) = 1 + \Delta(G)$ and so G is of Class two.   ∎

If G is a graph of order n, then $\alpha'(G) \leq \lfloor \frac{n}{2} \rfloor$. Therefore, the largest possible value of $\Delta(G) \cdot \alpha'(G)$ is $\Delta(G) \cdot \lfloor \frac{n}{2} \rfloor$. A graph G of order n and size m is called **overfull** if $m > \Delta(G) \cdot \lfloor \frac{n}{2} \rfloor$. If n is even, then $\lfloor n/2 \rfloor = n/2$ and

$$2m = \sum_{v \in V(G)} \deg v \leq n\Delta(G).$$

Therefore, $m \leq \Delta(G) \cdot (n/2) = \Delta(G) \cdot \lfloor \frac{n}{2} \rfloor$ and so G is not overfull. Thus, no graph of even order is overfull.

Since $\alpha'(G) \leq \lfloor \frac{n}{2} \rfloor$ for every graph G of order n, Theorem 10.9 has an immediate corollary (see Exercise 7).

Corollary 10.10 *Every overfull graph is of Class two.*

We now look at two problems whose solutions involve edge colorings.

Example 10.11 *A community, well known for having several professional tennis players who train there, holds a charity tennis tournament each year, which alternates between men and women tennis players. During the coming year, women tennis players will be featured and the professional players Alice, Barbara, and Carrie will be in charge. Two tennis players from each of two local tennis clubs have been invited to participate as well. Debbie and Elizabeth will participate from Woodland Hills Tennis Club and Frances and Gina will participate from Mountain Meadows Tennis Club. No two professionals will play each other in the tournament and no two players from the same tennis club will play each other; otherwise, every two of the seven players will play each other. If no player is to play two matches on the same day, what is the minimum number of days needed to schedule this tournament?*

Solution. We construct a graph H with $V(H) = \{$A, B, $\cdots$, G$\}$ whose vertices correspond to the seven tennis players. Two vertices x and y are adjacent in H if x and y are to play a tennis match against each other. The graph H is shown in Figure 10.7. The answer to the question posed is the chromatic index of H. The order of H is $n = 7$ and the degrees of its vertices are $5, 5, 5, 5, 4, 4, 4$. Thus, $\Delta(H) = 5$ and the size of H is $m = 16$. Since

$$16 = m > \Delta(H) \cdot \left\lfloor \frac{n}{2} \right\rfloor = 15,$$

the graph H is overfull. By Corollary 10.10, H is of Class two and so $\chi'(H) = 1 + \Delta(H) = 6$. A 6-edge coloring of H is also shown in Figure 10.7. This provides us with a schedule for the tennis tournament taking place over a minimum of six days. ◆

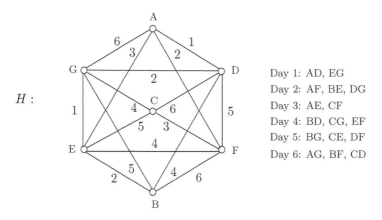

Figure 10.7: The graph H in Example 10.11 and a 6-edge coloring of H

Example 10.12 *One year it is decided to have a charity tennis tournament consisting entirely of double matches. Five tennis players (denoted by A, B, C, D, E) have agreed to participate. Each pair $\{W, X\}$ of tennis players will play a match against every other pair $\{Y, Z\}$ of tennis players, where then $\{W, X\} \cap \{Y, Z\} = \emptyset$, but no 2-person team is to play two matches on the same day. What is the minimum number of days needed to schedule such a tournament? Give an example of such a tournament using a minimum number of days.*

Solution. We construct a graph G whose vertex set is the set of 2-element subsets of $\{A, B, C, D, E\}$. Thus, the order of G is $\binom{5}{2} = 10$. Two vertices $\{W, X\}$ and $\{Y, Z\}$ are adjacent if these sets are disjoint. The graph G is shown in Figure 10.8. Thus, G is the Petersen graph, or equivalently the Kneser graph $KG_{5,2}$ (see Section 6.2). To answer the question, we determine the chromatic index of G. Since the Petersen graph is known to be of Class two, it follows that $\chi'(G) = 1 + \Delta(G) = 4$. A 4-edge coloring of G is given in Figure 10.8 together with a possible schedule of tennis matches over a period of four days. ◆

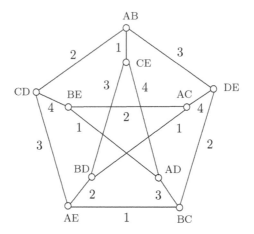

Day 1: AB-CE, AC-BD, AE-BC, AD-BE
Day 2: AB-CD, AC-BE, AE-BD, BC-DE
Day 3: AB-DE, AD-BC, AE-CD, BD-CE
Day 4: AC-DE, AD-CE, BE-CD

Figure 10.8: The Petersen graph G in Example 10.12 and a 4-edge coloring of G

As we have seen, the size of a graph G of Class one and having order n cannot exceed $\Delta(G) \cdot \left\lfloor \frac{n}{2} \right\rfloor$. The size of any overfull graph of order n exceeds this number and is therefore of Class two. There are related subgraphs such that if a graph G should contain one of these, then G must also be of Class two.

A subgraph H of odd order n' and size m' of a graph G is an **overfull subgraph** of G if

$$m' > \Delta(G) \cdot \left\lfloor \frac{n'}{2} \right\rfloor = \Delta(G) \cdot \frac{n' - 1}{2}.$$

Actually, if H is an overfull subgraph of G, then $\Delta(H) = \Delta(G)$ (see Exercise 8). This says that H is itself of Class two. Not only is an overfull subgraph of a graph G of Class two, G itself is of Class two.

Theorem 10.13 *Every graph having an overfull subgraph is of Class two.*

Proof. Let H be an overfull subgraph of a graph G. As we observed, $\Delta(H) = \Delta(G)$ and H is of Class two; so $\chi'(H) = 1 + \Delta(H)$. Thus,

$$\chi'(G) \geq \chi'(H) = 1 + \Delta(H) = 1 + \Delta(G)$$

and so $\chi'(G) = 1 + \Delta(G)$.   ∎

The following result provides a useful property of overfull subgraphs of a graph.

Theorem 10.14 *Let G be an r-regular graph of even order $n = 2k$, where $\{V_1, V_2\}$ is a partition of $V(G)$ such that $|V_1| = n_1$ and $|V_2| = n_2$ are odd. Suppose that $G_1 = G[V_1]$ is an overfull subgraph of G. Then $G_2 = G[V_2]$ is also an overfull*

subgraph of G. Furthermore, if k is odd, then $r < k$; while if k is even, then
$r < k - 1$.

Proof. Let the size of G_i be m_i for $i = 1, 2$. Since G_1 is overfull, $m_1 > r\left(\frac{n_1-1}{2}\right)$.
We show that $m_2 > r\left(\frac{n_2-1}{2}\right)$. Now,

$$m_2 = \frac{rn}{2} - (rn_1 - m_1) = \frac{rn}{2} - rn_1 + m_1.$$

Since $2m_1 > rn_1 - r$, it follows that

$$\begin{aligned} 2m_2 &= rn - 2rn_1 + 2m_1 > rn - 2rn_1 + rn_1 - r \\ &= r(n - n_1 - 1) = r(n_2 - 1). \end{aligned}$$

Hence, $m_2 > r\left(\frac{n_2-1}{2}\right)$ and G_2 is overfull. Therefore, both G_1 and G_2 are overfull.
Now, either $n_1 \le k$ or $n_2 \le k$, say the former. If k is even, then $n_1 \le k - 1$.
Since G_1 is overfull,

$$m_1 > r\left(\frac{n_1-1}{2}\right) \text{ and } 2m_1 > r(n_1 - 1).$$

Hence,

$$2\binom{n_1}{2} \ge 2m_1 > r(n_1 - 1) \text{ and so } n_1(n_1 - 1) > r(n_1 - 1).$$

Therefore, $r < n_1$. Thus, $r < k$ if k is odd and $r < k - 1$ if k is even. ∎

In the definition of an overfull subgraph H of order n' and size m' in a graph
G, we have $m' > \Delta(G) \cdot \frac{n'-1}{2}$. As we saw in Theorem 10.13, this implies that G
is of Class two. On the other hand, if G should contain a subgraph H of order
n' and size m' such that $m' > \Delta(H) \cdot \frac{n'-1}{2}$, where $\Delta(H) < \Delta(G)$, then H is an
overfull graph and so H is of Class two. This need not imply that G is of Class two,
however. For example, $C_5 \,\Box\, K_2$ is of Class one but C_5 is of Class two.

While every graph containing an overfull subgraph must be of Class two, a graph
can be of Class two without containing any overfull subgraph. The Petersen graph
P (which is 3-regular of order 10) contains no overfull subgraph; yet we saw that P
is of Class two.

The following conjecture is due to Amanda G. Chetwynd and Anthony J. W.
Hilton [51].

The Overfull Conjecture Let G be a graph of order n such that $\Delta(G) > n/3$.
Then G is of Class two if and only if G contains an overfull subgraph.

Let v be a vertex of the Petersen graph P. Then $P - v$ has order $n = 9$ and
$\Delta(P - v) = \frac{n}{3} = 3$. Even though $P - v$ is of Class two, it has no overfull subgraph.
Hence, if the Overfull Conjecture is true, the resulting theorem cannot be improved
in general.

We encountered the following conjecture in Chapter 4, which we saw is true for
sufficiently large even integers.

The 1-Factorization Conjecture If G is an r-regular graph of even order n such
that $r \ge n/2$, then G is 1-factorable.

The Overfull Conjecture implies the 1-Factorization Conjecture.

Theorem 10.15 *If the Overfull Conjecture is true, then so too is the 1-Factorization Conjecture.*

Proof. Assume, to the contrary, that the Overfull Conjecture is true but the 1-Factorization Conjecture is false. Then there exists an r-regular graph G of even order n such that $r \geq n/2$ such that G is not 1-factorable. Thus, G is of Class two. By the Overfull Conjecture, G contains an overfull subgraph G_1.

Let G_2 be the subgraph of G induced by $V(G) - V(G_1)$. By Theorem 10.14, G_2 is also overfull. At least one of G_1 and G_2 has order at most $n/2$. Suppose that G_1 has order at most $n/2$. Again, by Theorem 10.14, if $n/2$ is odd, then $r < n/2$; while if $n/2$ is even, then $r < (n/2) - 1$. Since $r \geq n/2$, a contradiction is produced. ∎

10.3   Tait Colorings

The Scottish physicist Peter Guthrie Tait (1831–1901) was one of many individuals who played a role in the story of the Four Color Problem (see Chapter 0). In addition to his interest in physics, Tait was also interested in mathematics and golf. His son Frederick (better known as Freddie Tait) shared his father's interest in golf and became the finest amateur golfer of his time.

Peter Tait became acquainted with the Four Color Problem through Arthur Cayley and became interested in Alfred Bray Kempe's solution of the problem. In fact, Tait felt that Kempe's solution was too lengthy and came up with several solutions of his own. Unfortunately, as in the case of Kempe's solution, none of Tait's solutions proved to be correct. Nevertheless, he presented his solutions to the Royal Society of Edinburgh on 15 March 1880 and published his work in the Proceedings of the Society. Later that year, Tait came up with another idea, which he believed would lead to a solution of the Four Color Problem. Even though his idea was not useful in producing a solution, it did lead to a new type of graph coloring, namely the subject of this chapter: edge colorings.

It was known that the Four Color Conjecture could be verified if it could be shown that every cubic bridgeless plane graph was 4-region colorable. Tait [185] showed that this problem could be looked at from another perspective.

Theorem 10.16 (**Tait's Theorem**) *A bridgeless cubic plane graph G is 4-region colorable if and only if G is 3-edge colorable.*

Proof. Suppose first that G is 4-region colorable. Let a 4-region coloring of G be given, where the colors are the four elements of $\mathbb{Z}_2 \times \mathbb{Z}_2$. Thus, each color can be expressed as (a, b) or ab, where $a, b \in \{0, 1\}$. The four colors used are then $c_0 = 00$, $c_1 = 01$, $c_2 = 10$, and $c_3 = 11$. Addition of colors is defined as coordinate-wise addition in $\mathbb{Z}_2$. For example, $c_1 + c_3 = 01 + 11 = 10 = c_2$. Since every element of $\mathbb{Z}_2 \times \mathbb{Z}_2$ is self-inverse, the sum of two distinct elements of $\mathbb{Z}_2 \times \mathbb{Z}_2$ is never $c_0 = 00$.

We now define an edge coloring of G. Since G is bridgeless, each edge e lies on the boundary of two distinct regions. Define the color of e as the sum of the colors of the two regions having e on their boundary. Thus, no edge of G is assigned the color c_0 and a 3-edge coloring of G is produced. It remains to show that this

3-edge coloring is proper. Let e_1 and e_2 be two adjacent edges of G and let v be the vertex incident with e_1 and e_2. Then v is incident with a third edge e_3 as well. For $1 \leq i < j \leq 3$, suppose that e_i and e_j are on the boundary of the region R_{ij}. Since the colors of R_{13} and R_{23} are different, the sum of the colors of R_{13} and R_{23} and the sum of the colors of R_{12} and R_{23} are different and so this 3-edge coloring is proper.

We now turn to the converse. Suppose that G is 3-edge colorable. Let a 3-edge coloring of G be given using the colors $c_1 = 01$, $c_2 = 10$, and $c_3 = 11$, as described above. This produces a partition of $E(G)$ into three perfect matchings E_1, E_2, and E_3, where E_i $(1 \leq i \leq 3)$ is the set of edges colored c_i.

Let G_1 be the spanning subgraph of G with edge set $E(G_1) = E_1 \cup E_2$ and let G_2 be the spanning subgraph of G with edge set $E(G_2) = E_2 \cup E_3$. Thus, both G_1 and G_2 are 2-regular spanning subgraphs of G and so each of G_1 and G_2 is the disjoint union of even cycles. For $i = 1, 2$, every region of G_i is the union of regions of G. For each cycle C in the graph G_i $(i = 1, 2)$, every region of G either lies interior or exterior to C. Furthermore, since G is bridgeless, each edge of G_i $(i = 1, 2)$ belongs to a cycle C' of G_i and is on the boundary of two distinct regions of G_i, one of which lies interior to C' and the other exterior to C'.

We now define a 4-region coloring of G. We assign to a region R of G the color $a_1 a_2$ with $a_i \in \{0, 1\}$ for $i = 1, 2$, where $a_i = 0$ if R lies interior to an even number of cycles in G_i and $a_i = 1$ otherwise. Figure 10.9(a) shows a 3-edge coloring of a cubic graph G, Figure 10.9(b) shows the cycles of G_1, Figure 10.9(c) shows the cycles of G_2, and Figure 10.9(d) shows the resulting 4-region coloring of G as defined above.

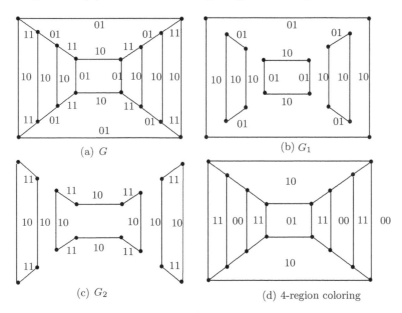

(a) G 　　　　　　　　　　　　　(b) G_1

(c) G_2 　　　　　　　　　　　　(d) 4-region coloring

Figure 10.9: A step in the proof of Theorem 10.16

It remains to show that this 4-region coloring of G is proper, that is, every two adjacent regions of G are assigned different colors. Let R_1 and R_2 be two adjacent regions of G. Thus, there is an edge e that lies on the boundary of both R_1 and R_2. If e is colored c_1 or c_2, then e lies on a cycle in G_1; while if e is colored c_2 or c_3, then e lies on a cycle in G_2. Thus, e lies on a cycle in G_1, a cycle in G_2 or both. Let C be a cycle in G_1, say, containing the edge e. Exactly one of R_1 and R_2 lies interior to C, while for every other cycle C' of G_1, either R_1 and R_2 are both interior to C' or both exterior to C'. Hence, the first coordinate of the colors of R_1 and R_2 are different. Therefore, the colors of every two adjacent regions of G differ in the first coordinate or the second coordinate or both. Hence, this 4-region coloring of G is proper.   ∎

In 1884 Tait wrote that every cubic graph is 1-factorable but this result was *not true without limitation*. Julius Petersen interpreted Tait's statement to mean that every cubic *bridgeless* graph is 1-factorable. However, in 1898 Petersen showed that even with this added hypothesis, such a graph need not be 1-factorable. Petersen did this by giving an example of a cubic bridgeless graph that is not 1-factorable: the Petersen graph. As was mentioned in Chapter 4, however, Petersen did prove that every cubic bridgeless graph does contain a 1-factor (see Theorem 4.13).

Eventually 3-edge colorings of cubic graphs became known as **Tait colorings**. Certainly, a cubic graph G has a Tait coloring if and only if G is 1-factorable. Every Hamiltonian cubic graph necessarily has a Tait coloring, for the edges of a Hamiltonian cycle can be alternately colored 1 and 2, with the remaining edges (constituting a perfect matching) colored 3. Tait believed that every 3-connected cubic planar graph is Hamiltonian. If Tait was correct, this would mean then that every 3-connected cubic planar graph is 3-edge colorable. However, as we are about to see, this implies that every 2-connected cubic planar graph is 3-edge colorable. But the 2-connected cubic graphs are precisely the connected bridgeless cubic graphs and so by Tait's theorem, the Four Color Conjecture would be true.

Theorem 10.17 *If every 3-connected cubic planar graph is 3-edge colorable, then every 2-connected cubic planar graph is 3-edge colorable.*

Proof. Suppose that the statement is false. Then all 3-connected cubic planar graphs are 3-edge colorable, but there exist cubic planar graphs having connectivity 2 that are not 3-edge colorable. Among the cubic planar graphs having connectivity 2 that are not 3-edge colorable, let G be one of minimum order n. Certainly n is even and since there is no such graph of order 4, it follows that $n \geq 6$. As we saw in Theorem 2.17, $\kappa(G) = \lambda(G) = 2$. This implies that every minimum edge-cut of G consists of two nonadjacent edges of G.

Let $\{u_1v_1, x_1y_1\}$ be a minimum edge-cut of G. Thus, the vertices u_1, v_1, x_1, and y_1 are distinct and G has the appearance shown in Figure 10.10, where F_1 and H_1 are the two components of $G - u_1v_1 - x_1y_1$.

Suppose that $u_1x_1, v_1y_1 \notin E(G)$. Then $F_1 + u_1x_1$ and $H_1 + v_1y_1$ are 2-connected cubic planar graphs of order less than n and so are 3-colorable. Let 3-edge colorings of $F_1 + u_1x_1$ and $H_1 + v_1y_1$ be given using the colors 1, 2, and 3. Now permute the colors of the edges in both $F_1 + u_1x_1$ and $H_1 + v_1y_1$, if necessary, so that both u_1x_1 and v_1y_1 are assigned the color 1. Deleting the edges u_1x_1 and v_1y_1, adding

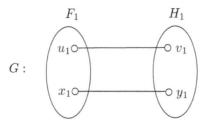

Figure 10.10: A step in the proof of Theorem 10.17

the edges u_1v_1 and x_1y_1, and assigning both u_1v_1 and x_1y_1 the color 1 results in a 3-edge coloring of G, which is impossible.

Thus, we may assume that at least one of u_1x_1 and v_1y_1 is an edge of G, say $u_1x_1 \in E(G)$. Then u_1 is adjacent to a vertex u_2 in F_1 that is different from x_1, and x_1 is adjacent to a vertex x_2 in F_1 that is different from u_1. Since $\kappa(G) = 2$, it follows that $u_2 \neq x_2$. We then have the situation shown in Figure 10.11, where $\{u_2u_1, x_2x_1\}$ is a minimum edge-cut of G, and F_2 and H_1 are the two components of $G - u_1 - x_1$.

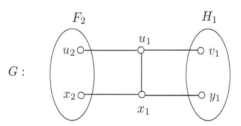

Figure 10.11: A step in the proof of Theorem 10.17

Suppose that $u_2x_2, v_1y_1 \notin E(G)$. Then $F_2 + u_2x_2$ and $H_1 + v_1y_1$ are 2-connected cubic planar graphs of order less than n and so are 3-edge colorable. Let 3-edge colorings of $F_2 + u_2x_2$ and $H_1 + v_1y_1$ be given using the colors 1, 2, and 3. Now permute the colors of the edges in both $F_2 + u_2v_2$ and $H_1 + v_1y_1$, if necessary, so that u_2x_2 is colored 2 and v_1y_1 is colored 1. Deleting the edges u_2x_2 and v_1y_1, assigning u_1u_2 and x_1x_2 the color 2, assigning u_1v_1 and x_1y_1 the color 1, and assigning u_1x_1 the color 3 produces a 3-edge coloring of G, which is a contradiction.

Thus, we may assume that at least one of u_2x_2 and v_1y_1 is an edge of G. Continuing in this manner, we have a sequence $\{u_1, x_1\}, \{u_2, x_2\}, \ldots, \{u_k, x_k\}$, $k \geq 1$, of pairs of vertices of F_1 such that $u_kx_k \notin E(G)$ and $u_ix_i \in E(G)$ for $1 \leq i < k$ and a sequence

$$\{v_1, y_1\}, \{v_2, y_2\}, \ldots, \{v_\ell, y_\ell\} \ (\ell \geq 1)$$

of pairs of vertices of H_1 such that $v_\ell y_\ell \notin E(G)$ and $v_iy_i \in E(G)$ for $1 \leq i < \ell$, as shown in Figure 10.12, where F_k and H_ℓ are the two components of

$$G - (\{u_1, u_2, \ldots, u_{k-1}\} \cup \{x_1, x_2, \ldots, x_{k-1}\} \cup \{v_1, v_2, \ldots, v_{\ell-1}\} \cup \{y_1, y_2, \ldots, y_{\ell-1}\}).$$

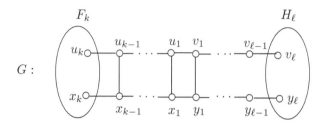

$$G:$$

Figure 10.12: A step in the proof of Theorem 10.17

Since $F_k + u_k x_k$ and $H_\ell + v_\ell y_\ell$ are 2-connected cubic planar graphs of order less than n, each is 3-edge colorable. Let 3-edge colorings of $F_k + u_k x_k$ and $H_\ell + v_\ell y_\ell$ be given using the colors 1, 2, and 3. Permute the colors 1, 2, and 3 of the edges in both $F_k + u_k x_k$ and $H_\ell + v_\ell y_\ell$, if necessary, so that both $u_k x_k$ and $v_\ell y_\ell$ are colored 1 if k and ℓ are of the same parity and $u_k x_k$ is colored 2 and $v_\ell y_\ell$ is colored 1 if k and ℓ are of the opposite parity. Deleting the edges $u_k x_k$ and $v_\ell y_\ell$, alternating the colors 1 and 2 along the paths

$$(v_\ell, v_{\ell-1}, \ldots, v_1, u_1, u_2, \ldots, u_k) \text{ and } (y_\ell, y_{\ell-1}, \ldots, y_1, x_1, x_2, \ldots, x_k),$$

and assigning the color 3 to the edges

$$u_1 x_1, \ u_2 x_2, \ \ldots, \ u_k x_k, \ v_1 y_1, \ v_2 y_2, \ \ldots, \ v_\ell y_\ell$$

produces a 3-edge coloring of G, again a contradiction.   ∎

As a consequence of Theorems 10.16 and 10.17, every planar graph is 4-colorable – *provided* Tait was correct that every 3-connected cubic planar graph is Hamiltonian. In 1946, however, William Tutte found a 3-connected cubic graph that was not Hamiltonian. This graph (the **Tutte graph**), shown in Figure 10.13, was encountered earlier (in Section 3.3), where it was shown that it is not Hamiltonian.

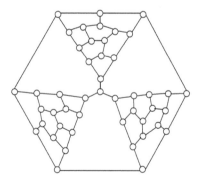

Figure 10.13: The Tutte graph

Despite the fact that Tait was wrong and that the 3-edge colorability of bridgeless cubic planar graphs was never used to prove the Four Color Theorem, the eventual verification of the Four Color Theorem did instead show that every bridgeless cubic planar graph is 3-edge colorable.

Corollary 10.18 *Every bridgeless cubic planar graph is of Class one.*

An immediate consequence of Corollary 10.18 is that there are planar graphs G of Class one with $\Delta(G) = 3$. In fact, there are planar graphs G of Class two with $\Delta(G) = 3$. Indeed, one can say more. For every integer k with $2 \leq k \leq 5$, there is a planar graph of Class one and a planar graph of Class two, both having maximum degree k (see Exercise 15). This may be as far as the story goes, however, for in 1965 Vadim Vizing [194] proved the following.

Theorem 10.19 *If G is a planar graph with $\Delta(G) \geq 8$, then G is of Class one.*

In 2001 Daniel Sanders and Yue Zhao [177] resolved one of the two missing cases.

Theorem 10.20 *If G is a planar graph with $\Delta(G) = 7$, then G is of Class one.*

Thus, only one case remains. Is it true that every planar graph with maximum degree 6 is of Class one? Vizing has conjectured that such is the case.

Vizing's Planar Graph Conjecture Every planar graph with maximum degree 6 is of Class one.

While every bridgeless cubic planar graph is of Class one, there are many cubic graphs that are of Class two. First, every cubic graph containing a bridge is of Class two (see Exercise 20). Furthermore, every bridgeless cubic graph of Class two is necessarily non-Hamiltonian and nonplanar. An example of a bridgeless cubic graph of Class two is the Petersen graph, shown in Figure 10.14(a).

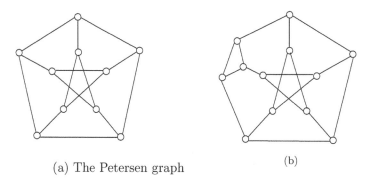

(a) The Petersen graph   (b)

Figure 10.14: Bridgeless cubic graphs of Class two

Recall that the girth of a graph G that is not a forest is the length of a smallest cycle in G and that the girth of the Petersen graph is 5. The **cyclic edge-connectivity** of a graph is the smallest number of edges whose removal results in a disconnected graph, each component of which contains a cycle. The cyclic edge-connectivity of the Petersen graph is 5. The graph shown in Figure 10.14(b) and

constructed from the Petersen graph is another bridgeless cubic graph of Class two
(see Exercise 22).

There is a class of bridgeless cubic graphs of Class two that are of special interest.
A **snark** is a cubic graph of Class two that has girth at least 5 and cyclic edge-
connectivity at least 4. The Petersen graph is therefore a snark. The girth and
cyclic edge-connectivity requirements are present in the definition to rule out trivial
examples. The term "snark" was coined for these graphs in 1976 by Martin Gardner,
a longtime popular writer for the magazine *Scientific American*. (We encountered
him in Chapter 0.) Gardner borrowed this word from Lewis Carroll (the pen-
name of the mathematician Charles Lutwidge Dodgson), well known for his book
Alice's Adventures in Wonderland. One century earlier, in 1876, Carroll wrote a
nonsensical poem titled *The Hunting of the Snark* in which a group of adventurers
are in pursuit of a legendary and elusive beast: the snark. Gardner chose to call
these graphs "snarks" because just as Carroll's snarks were difficult to find, so too
were these graphs difficult to find (at least for the first several years).

The oldest snark is the Petersen graph, discovered in 1891. In 1946 Danilo
Blanuša discovered two more snarks, both of order 18 (see Figure 10.15(a)). The
Descartes snark (discovered by William Tutte in 1948) has order 210. George
Szekeres discovered the **Szekeres snark** of order 50 in 1973. Until 1973 these
were the only known snarks. In 1975, however, Rufus Isaacs described two infinite
families of snarks, one of which essentially contained all previously known snarks
while the second family was completely new. This second family contained the
so-called **flower snarks**, one example of which is shown in Figure 10.15(b). In
addition, Isaacs found a snark that belonged to neither family. This **double-star
snark** is shown in Figure 10.15(c).

All of the snarks shown in Figure 10.15 appear to have a certain resemblance
to the Petersen graph. In fact, William Tutte conjectured that every snark has the
Petersen graph as a minor. Neil Robertson, Daniel Sanders, Paul Seymour, and
Robin Thomas announced in 2001 that they had verified this conjecture.

Theorem 10.21 *Every snark has the Petersen graph as a minor.*

In 1969 Mark E. Watkins [198] introduced a class of graphs generalizing the
Petersen graph. Each of these is a special permutation graph of a cycle C_n. The
generalized Petersen graph $P(n, k)$ with $1 \leq k < n/2$ has vertex set

$$\{u_i :\ 0 \leq i \leq n - 1\} \cup \{v_i :\ 0 \leq i \leq n - 1\}$$

and edge set

$$\{u_i u_{i+1} :\ 0 \leq i \leq n - 1\} \cup \{u_i v_i :\ 0 \leq i \leq n - 1\} \cup \{v_i v_{i+k} :\ 0 \leq i \leq n - 1\},$$

where $u_n = u_0$ and $v_n = v_0$. The graphs $P(n, 1)$ are therefore prisms and $P(5, 2)$ is
the Petersen graph. The graphs $P(7, k)$ for $k = 1, 2, 3$ are shown in Figure 10.16.

Frank Castagna and Geert Prins [34] proved the following in 1972.

Theorem 10.22 *The only generalized Petersen graph that is not Tait colorable is
the Petersen graph.*

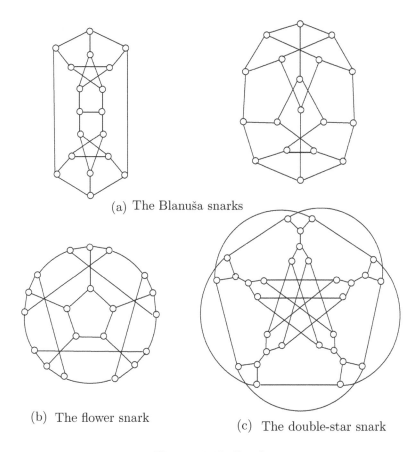

(a) The Blanuša snarks

(b) The flower snark

(c) The double-star snark

Figure 10.15: Snarks

10.4 Nowhere-Zero Flows

As we saw in the preceding section, Tait colorings are edge colorings that are intimately tied to colorings of the regions of bridgeless cubic plane graphs. There are labelings of the arcs of orientations of bridgeless cubic plane graphs that also have a connection to colorings of the regions of such graphs.

For an oriented graph D, a **flow** on D is a function $\phi : E(D) \to \mathbb{Z}$ such that for each vertex v of D,

$$\sigma^+(v; \phi) = \sum_{(v,w) \in E(D)} \phi(v,w) = \sum_{(w,v) \in E(D)} \phi(w,v) = \sigma^-(v; \phi). \qquad (10.3)$$

That is, for each vertex v of D, the sum of the flow values of the arcs directed away from v equals the sum of the flow values of the arcs directed towards v. The property (10.3) of ϕ is called the **conservation property**. Since $\sigma^+(v; \phi) - \sigma^-(v; \phi) = 0$ for every vertex v of D, the sum of the flow values of the arcs incident with v is even. This, in turn, implies that the number of arcs incident with v having an odd flow

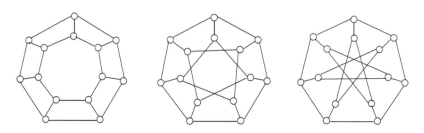

Figure 10.16: The generalized Petersen graphs $P(7, k)$ for $k = 1, 2, 3$

value is even.

For an integer $k \geq 2$, if a flow ϕ on an oriented graph D has the property that $|\phi(e)| < k$ for every arc e of D, then ϕ is called a k-**flow** on D. Furthermore, if $0 < |\phi(e)| < k$ for every arc e of D (that is, $\phi(e)$ is never 0), then ϕ is called a **nowhere-zero k-flow** on D. Hence, a nowhere-zero k-flow on D has the property that

$$\phi(e) \in \{\pm 1, \pm 2, \ldots, \pm(k-1)\}$$

for every arc e of D. It is the nowhere-zero k-flows for particular values of k that will be of special interest to us.

Those graphs for which some orientation has a nowhere-zero 2-flow can be described quite easily.

Theorem 10.23   *A nontrivial connected graph G has an orientation with a nowhere-zero 2-flow if and only if G is Eulerian.*

Proof.   Let G be an Eulerian graph and let C be an Eulerian circuit of G. Direct the edges of C in the direction of C, producing an Eulerian digraph D, where then od $v = $ id v for every vertex v of D. Then the function ϕ defined by $\phi(e) = 1$ for each arc e of D satisfies the conservation property and so ϕ is a nowhere-zero 2-flow on D.

Conversely, suppose that G is a nontrivial connected graph that is not Eulerian. Then G contains a vertex u of odd degree. Let D be any orientation of G. Then any function ϕ defined on $E(D)$ for which $\phi(e) \in \{-1, 1\}$ has an odd number of arcs incident with v having an odd flow value. Thus, ϕ is not a nowhere-zero 2-flow.   ∎

The graph $G = C_4$ of Figure 10.17 is obviously Eulerian and therefore has an orientation D_1 with a nowhere-zero 2-flow. However, D_1 is not the only orientation of G with this property. The orientations D_2 and D_3 of G also have a nowhere-zero 2-flow.

The graph and digraphs shown in Figure 10.17 may suggest that the existence of a nowhere-zero k-flow for some integer $k \geq 2$ on all orientations of a graph G depends only on the existence of a nowhere-zero k-flow on a single orientation of G. This is, in fact, what happens.

Theorem 10.24   *Let G be a graph. If some orientation of G has a nowhere-zero k-flow, where $k \geq 2$, then every orientation of G has a nowhere-zero k-flow.*

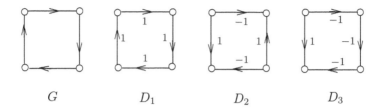

$$G \qquad\qquad D_1 \qquad\qquad D_2 \qquad\qquad D_3$$

Figure 10.17: Oriented graphs with a nowhere-zero 2-flow

Proof. Let D be an orientation of G having a nowhere-zero k-flow ϕ. Thus, $\sigma^+(v; \phi) = \sigma^-(v; \phi)$ for each vertex v of D. Let D' be the orientation of G obtained by reversing the direction of some arc $f = (x, y)$ of D, resulting in the arc $f' = (y, x)$ of D'. We now define the function $\phi' : E(D') \to \{\pm 1, \pm 2, \dots, \pm(k-1)\}$ by

$$\phi'(e) \;=\; \begin{cases} \phi(e) & \text{if } e \neq f' \\ -\phi(f) & \text{if } e = f'. \end{cases}$$

For $v \in V(D) - \{x, y\}$, $\sigma^+(v; \phi') = \sigma^-(v; \phi')$. Also,

$$\sigma^+(x; \phi') \;=\; \sigma^+(x; \phi) - \phi(f) = \sigma^-(x; \phi) + \phi'(f') = \sigma^-(x; \phi')$$
$$\sigma^+(y; \phi') \;=\; \sigma^+(y; \phi) + \phi'(f') = \sigma^-(y; \phi) + (-\phi(f)) = \sigma^-(y; \phi').$$

Thus, ϕ' is a nowhere-zero k-flow of D'.

Now, if D'' is any orientation of G, then D'' can be obtained from D by a sequence of arc reversals in D. Since a nowhere-zero k-flow can be defined on each orientation, as described above, at each step of the sequence, a nowhere-zero k-flow can be defined on D''. ∎

According to Theorem 10.24, the property of an orientation (indeed of *all* orientations) of a graph G having a nowhere-zero k-flow is a characteristic of G rather than a characteristic of its orientations. We therefore say that a **graph G has a nowhere-zero k-flow** if every orientation of G has a nowhere-zero k-flow. As a consequence of the proof of Theorem 10.24, we also have the following.

Theorem 10.25 *If G is a graph having a nowhere-zero k-flow for some $k \geq 2$, then there is an orientation D of G and a nowhere-zero k-flow on D all of whose flow values are positive.*

By Theorem 10.23, the graph G of Figure 10.18 does not have a nowhere-zero 2-flow. Trivially, it has an orientation D possessing a 2-flow. This orientation does have a nowhere-zero 3-flow, however. By Theorem 10.24, G itself has a nowhere-zero 3-flow.

If D is an orientation of a graph G with a flow ϕ and a is an integer, then $a\phi$ is also a flow on D. In fact, if ϕ is a k-flow and $a \geq 1$, then $a\phi$ is an ak-flow. Indeed, if ϕ is a nowhere-zero k-flow, then $a\phi$ is a nowhere-zero ak-flow. More generally, we have the following (see Exercise 25).

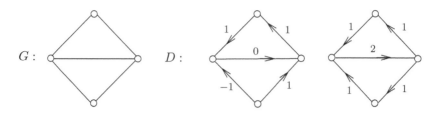

Figure 10.18: A graph with a nowhere-zero 3-flow

Theorem 10.26 *If ϕ_1 and ϕ_2 are flows on an orientation D of a graph, then every linear combination of ϕ_1 and ϕ_2 is also a flow on D.*

Let G be a graph with a nowhere-zero k-flow for some integer $k \geq 2$ and let ϕ be a nowhere-zero k-flow defined on some orientation D of G. Then $\sigma^+(v; \phi) = \sigma^-(v; \phi)$ for every vertex v of D. Let $S = \{v_1, v_2, \ldots, v_t\}$ be a proper subset of $V(G)$ and $T = V(G) - S$. Define

$$\sigma^+(S; \phi) = \sum_{(u,v)\in[S,T]} \phi(u,v)$$

and

$$\sigma^-(S; \phi) = \sum_{(v,u)\in[T,S]} \phi(v,u).$$

Since ϕ is a flow on D,

$$\sum_{i=1}^{t} \sigma^+(v_i; \phi) = \sum_{i=1}^{t} \sigma^-(v_i; \phi). \tag{10.4}$$

For every arc (v_a, v_b) with $v_a, v_b \in S$, the flow value $\phi(v_a, v_b)$ occurs in both the left and the right sums in (10.4). Cancelling all such terms in (10.4), we are left with $\sigma^+(S; \phi)$ on the left side of (10.4) and $\sigma^-(S; \phi)$ on the right side of (10.4). Thus, $\sigma^+(S; \phi) = \sigma^-(S; \phi)$. In summary, we have the following.

Theorem 10.27 *Let ϕ be a nowhere-zero k-flow defined on some orientation D of a graph G and let S be a nonempty proper subset of $V(G)$. Then the sum of the flow values of the arcs directed from S to $V(G) - S$ equals the sum of the flow values of the arcs directed from $V(G) - S$ to S.*

A fundamental question concerns determining those nontrivial connected graphs having a nowhere-zero k-flow for some integer $k \geq 2$. Suppose that ϕ is a nowhere-zero k-flow defined on some orientation D of a nontrivial connected graph G. If G should contain a bridge $e = uv$, where S and $V(G) - S$ are the vertex sets of the components of $G - e$, then the conclusion of Theorem 10.27 cannot occur. As a consequence of this observation, we have the following.

Corollary 10.28 *No graph with a bridge has a nowhere-zero k-flow for any integer $k \geq 2$.*

The nowhere-zero k-flows of bridgeless planar graphs will be of special interest to us because of their connection with region colorings. Let D be an orientation of a bridgeless plane graph G. Suppose that c is a k-region coloring of G for some integer $k \geq 2$. Thus, for each region R of G, we may assume that the color $c(R)$ is one of the colors $1, 2, \ldots, k$. For each edge uv of G belonging to the boundaries of two regions R_1 and R_2 of G and the arc $e = (u, v)$ of D, define

$$\phi(e) = c(R_1) - c(R_2),$$

where R_1 is the region that lies to the right of e and R_2 is the region that lies to the left of e (see Figure 10.19).

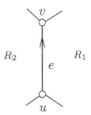

Figure 10.19: Defining $\phi(e)$ for $e = (u, v)$

A 4-region coloring of the plane graph of Figure 10.20(a) is given in that figure. A resulting nowhere-zero 4-flow of an orientation D of G is shown in Figure 10.20(b).

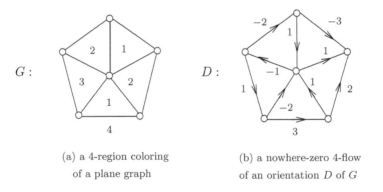

(a) a 4-region coloring   (b) a nowhere-zero 4-flow
of a plane graph   of an orientation D of G

Figure 10.20: Constructing a nowhere-zero flow from
a region coloring of a plane graph

For bridgeless plane graphs, k-region colorability and the existence of nowhere-zero k-flows for an integer $k \geq 2$ are equivalent.

Theorem 10.29 *For an integer $k \geq 2$, a bridgeless plane graph G is k-region colorable if and only if G has a nowhere-zero k-flow.*

Proof. First, let there be given a k-region coloring c of G and let D be an orientation of G. For each arc $e = (u, v)$ of D, let R_1 be the region of G that lies to the right of e and R_2 the region of G that lies to the left of e. Define an integer-valued function ϕ of $E(D)$ by $\phi(e) = c(R_1) - c(R_2)$. We show that ϕ is a nowhere-zero k-flow. Since uv is not a bridge of G, $c(R_1) \neq c(R_2)$ and since $1 \leq c(R) \leq k$ for each region R of G, it follows that $\phi(e) \in \{\pm 1, \pm 2, \ldots, \pm(k-1)\}$.

It remains only to show that $\sigma^+(v; \phi) = \sigma^-(v; \phi)$ for each vertex v of D. Suppose that $\deg_G v = t$ and that $v_1, v_2, \ldots, v_t$ are the neighbors of v as we proceed about v in some direction. For $i = 1, 2, \ldots, t$, let R_i denote the region having vv_i and vv_{i+1} on its boundary. Thus, in D each edge vv_i $(1 \leq i \leq t)$ is either the arc (v, v_i) or the arc (v_i, v). Let $c(R_i) = c_i$ for $i = 1, 2, \ldots, t$. Let $j \in \{1, 2, \ldots, t\}$ and consider $c(R_j) = c_j$. If $(v, v_j), (v, v_{j+1}) \in E(D)$, then c_j and $-c_j$ occur in $\sigma^+(v; \phi)$ and c_j does not occur in $\sigma^-(v; \phi)$. The situation is reversed if $(v_j, v), (v_{j+1}, v) \in E(D)$. If $(v, v_{j+1}), (v_j, v) \in E(D)$, then the term c_j occurs in both $\sigma^+(v; \phi)$ and $\sigma^-(v; \phi)$; while if $(v_{j+1}, v), (v, v_j) \in E(D)$, then the term $-c_j$ occurs in both $\sigma^+(v; \phi)$ and $\sigma^-(v; \phi)$. Thus, $\sigma^+(v; \phi) = \sigma^-(v; \phi)$ and ϕ is a nowhere-zero k-flow.

Next, suppose that G is a bridgeless plane graph having a nowhere-zero k-flow. This implies that for a given orientation D of G, there exists a nowhere-zero k-flow ϕ of D. By definition then, $\sigma^+(v; \phi) = \sigma^-(v; \phi)$ for every vertex v of D.

We now consider directed closed curves in the plane that do not pass through any vertex of D. Such closed curves may enclose none, one, or several vertices of G. For a directed closed curve C, we define the number $\sigma(C; \phi)$ to be the sum of terms $\phi(e)$ or $-\phi(e)$ for each occurrence of an arc e crossed by C. In particular, as we proceed along C in the direction of C and cross an arc e, we contribute $\phi(e)$ to $\sigma(C; \phi)$ if e is directed to the right of C and contribute $-\phi(e)$ to $\sigma(C; \phi)$ if e is directed to the left of C.

If C is a directed simple closed curve in the plane that encloses no vertex of D, then for each occurrence of an arc e crossed by C that is directed to the right of C, there is an occurrence of e crossed by C that is directed to the left of C. Hence, in this case, $\sigma(C; \phi) = 0$. If C encloses a single vertex v, then because $\sigma^+(v; \phi) = \sigma^-(v; \phi)$, it follows that $\sigma(C; \phi) = 0$ here as well. (See Figure 10.21 for example.)

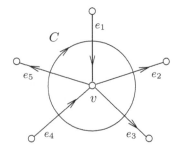

$$\sigma(C; \phi) = (\phi(e_1) + \phi(e_4)) - (\phi(e_2) + \phi(e_3) + \phi(e_5)) = 0$$

Figure 10.21: Computing $\sigma(C; \phi)$

Suppose now that C is a directed simple closed curve in the plane that encloses two or more vertices. Let $S = \{v_1, v_2, \cdots, v_s\}$, $s \geq 2$, be the set of vertices of D lying interior to C. Because, by Theorem 10.27, the sum of the flow values of the arcs directed from S to $V(G) - S$ equals the sum of the flow values of the arcs directed from $V(G) - S$ to S, it follows that $\sigma(C; \phi) = 0$ as well.

If C is a directed closed curve that is not a simple closed curve, then C is a union of directed simple closed curves $C_1, C_2, \ldots, C_r$ and so

$$\sigma(C; \phi) = \sum_{i=1}^{s} \sigma(C_i; \phi) = 0.$$

Consequently, $\sigma(C; \phi) = 0$ for *every* directed closed curve C in the plane.

We now show that there is a proper coloring of the regions of G using the colors $1, 2, \ldots, k$. Assign the color k to the exterior region of G. Let R be some interior region in G. Choose a point A in the exterior region and a point B in R and let P be an open curve directed from A to B so that P passes through no vertices of D. The number $\sigma(P; \phi)$ is defined as the sum (addition performed modulo k) of the numbers $\phi(e)$ or $-\phi(e)$, for each occurrence of an arc e crossed by P, where either $\phi(e)$ or $-\phi(e)$ is contributed to the sum $\sigma(P; \phi)$ according to whether e is directed to the right or to the left of P, respectively. The least positive integer in the equivalence class containing $\sigma(P; \phi)$ in the ring of integers $\mathbb{Z}_k$ is the color $c(R)$ assigned to R. Thus, $c(R) \in \{1, 2, \ldots, k\}$.

We now show that the color $c(R)$ assigned to R is well-defined. Suppose that the color assigned to R by the curve P above is $c(R) = a$. Let Q be another directed open curve from A to B and let $\sigma(Q; \phi) = b$. We claim that $a = b$. If $\tilde{Q}$ is the directed open curve from B to A obtained by reversing the direction of Q, then $\sigma(\tilde{Q}; \phi) = -b$. Now let C be the directed closed curve obtained by following P by $\tilde{Q}$. Then, as we saw, $\sigma(C; \phi) = 0$. But $\sigma(C; \phi) = \sigma(P; \phi) + \sigma(\tilde{Q}; \phi) = a - b$. So $a - b = 0$ and $a = b$. Thus, the color $c(R)$ of R defined in this manner is, in fact, well-defined.

It remains to show that the region coloring c of G is proper. Let R' and R'' be two adjacent regions of D, where e' is an arc on the boundaries of both R' and R''. Let B' be a point in R' and let B'' be a point in R''. Furthermore, let P' be a directed open curve from A to B' that does not cross e' and suppose that P'' is a directed open curve from A to B'' that extends P' to B'' so that e' is the only additional arc crossed by P''. Therefore,

$$\sigma(P''; \phi) = \sigma(P'; \phi) + \phi(e')$$

or

$$\sigma(P''; \phi) = \sigma(P'; \phi) - \phi(e').$$

Since $\phi(e') \not\equiv 0 \pmod{k}$, it follows that the colors assigned to R' and R'' are distinct. Hence, the region coloring c of G is proper. ∎

Letting $k = 4$ in Theorem 10.29, we have the following corollary of the Four Color Theorem.

Corollary 10.30 *Every bridgeless planar graph has a nowhere-zero 4-flow.*

While every bridgeless planar graph has a nowhere-zero 4-flow, it is not the case that every bridgeless nonplanar graph has a nowhere-zero 4-flow.

Theorem 10.31 *The Petersen graph does not have a nowhere-zero 4-flow.*

Proof. Suppose that the Petersen graph P has a nowhere-zero 4-flow. Then there exists an orientation D of P and a nowhere-zero 4-flow ϕ on D such that $\phi(e) \in \{1, 2, 3\}$ for every arc e of D (by Theorem 10.25). Since $\sigma^+(v; \phi) = \sigma^-(v; \phi)$ for every vertex v of D, the only possible flow values of the three arcs incident with v are $1, 1, 2$ and $1, 2, 3$. In particular, this implies that every vertex of D is incident with exactly one arc having flow value 2. Thus, the arcs of D with flow value 2 correspond to a 1-factor F of P. The remaining arcs of D then correspond to a 2-factor H of P. Because the Petersen graph is not Hamiltonian and has girth 5, the 2-factor H must consist of two disjoint 5-cycles. Every vertex v of H is incident with one or two arcs having flow value 1. Furthermore, if a vertex v of H is incident with two arcs having flow value 1, then these two arcs are either both directed towards v or both directed away from v. A vertex v of H is said to be of *type I* if there is an arc having flow value 1 directed towards v; while v is of *type II* if there is an arc having flow value 1 directed away from v. Consequently, each vertex v of H is either of type I or of type II, but not both. Moreover, the vertices in each of the 5-cycles of H must alternate between type I and type II, which is impossible for an odd cycle.   ∎

The Petersen graph does have a nowhere-zero 5-flow, however (see Exercise 26). The Petersen graph plays an important role among cubic graphs as Neil Robertson, Daniel Sanders, Paul Seymour, and Robin Thomas verified in a series of papers [170, 171, 172, 175, 176].

Theorem 10.32 *Every bridgeless cubic graph not having the Petersen graph as a minor has a nowhere-zero 4-flow.*

There is a question of determining the smallest positive integer k such that every bridgeless graph has a nowhere-zero k-flow. In this connection, William Tutte [188] conjectured the following.

Conjecture 10.33 *Every bridgeless graph has a nowhere-zero 5-flow.*

While this conjecture is still open, Paul Seymour [179] did establish the following.

Theorem 10.34 *Every bridgeless graph has a nowhere-zero 6-flow.*

Tait colorings deal exclusively with cubic graphs, of course, and it is this class of graphs that has received added attention when studying nowhere-zero flows. By Theorem 10.23, no cubic graph has a nowhere-zero 2-flow. In the case of nowhere-zero 3-flows, it is quite easy to determine which cubic graphs have these.

Theorem 10.35 *A cubic graph G has a nowhere-zero 3-flow if and only if G is bipartite.*

Proof. Let G be a cubic bipartite graph with partite sets U and W. Since G is regular, G contains a 1-factor F. Orient the edges of F from U to W and assign each arc the value 2. Orient all other edges of G from W to U and assign each of these arcs the value 1. This is a nowhere-zero 3-flow.

For the converse, let G be a cubic graph having a nowhere-zero 3-flow. By Theorem 10.25, there is an orientation D on which is defined a nowhere-zero 3-flow having only the values 1 and 2. In particular, the values must be 1, 1, 2 for the three arcs incident with each vertex of D. The arcs having the value 2 form a subdigraph H of D and the underlying graph of H is a 1-factor of G. Let U be the set of vertices of G where each vertex of U has outdegree 1 in H and let W be the remaining vertices of G. Every arc of D not in H must then have a value of 1 and be directed from a vertex of W to a vertex of U. Thus, G is a bipartite graph with partite sets U and W. ∎

Those cubic graphs having a nowhere-zero 4-flow depend only on their chromatic indexes.

Theorem 10.36 *A cubic graph G has a nowhere-zero 4-flow if and only if G is of Class one.*

Proof. Let G be a cubic graph that has a nowhere-zero 4-flow. By Theorem 10.25, there exists an orientation D of G and a nowhere-zero 4-flow ϕ on D such that $\phi(e) \in \{1, 2, 3\}$ for each arc e of D. Then each vertex is incident with arcs having either the values 1, 1, 2 or 1, 2, 3. In particular, each vertex of D is incident with exactly one arc having the value 2. The arcs having the value 2 produce a 1-factor in G and so the remaining arcs of D produce a 2-factor of G. If the 2-factor is a Hamiltonian cycle of G, then G is of Class one. Hence, we may assume that this 2-factor consists of two or more mutually disjoint cycles. Let C be one of these cycles and let H be the orientation of C in D.

Suppose that C is an r-cycle and let $S = V(C)$. Since each arc of D joining a vertex of S and a vertex of $V(D) - S$ has flow value 2 and since $\sigma^+(S, \phi) = \sigma^-(S; \phi)$ by Theorem 10.27, it follows that there is an even number t of arcs joining the vertices of S and the vertices of $V(D) - S$. Since this set of arcs is independent, it follows that $G[S]$ contains t vertices of degree 2 and $r - t$ vertices of degree 3. Because $G[S]$ has an even number of odd vertices, $r - t$ is even. However, since t is also even, r is even as well and C is an even cycle. Thus, the 2-factor of G is the union of even cycles and so G is of Class one.

We now verify the converse. Let G be a cubic graph of Class one. Therefore, G is 3-edge colorable and 1-factorable into factors F_1, F_2, F_3, where F_i is the 1-factor $(1 \leq i \leq 3)$ whose edges are colored i. Every two of these three 1-factors produce a 2-factor consisting of a union of disjoint cycles. Since the edges of each cycle alternate in colors, the cycles are even and each 2-factor is bipartite. Let G_1 be the 2-factor obtained from F_1 and F_3 and G_2 the 2-factor obtained from F_2 and F_3. Now let D be an orientation of G and let D_i be the resulting orientation of G_i $(i = 1, 2)$. Since each component of each graph G_i is Eulerian, it follows by Theorem 10.23 that there is a nowhere-zero 2-flow ϕ_i on D_i. For $e \in E(D) - E(D_i)$, define $\phi_i(e) = 0$ for $i = 1, 2$. Then ϕ_i is a 2-flow on D. By Theorem 10.26, the function ϕ defined by $\phi = \phi_1 + 2\phi_2$ is also a flow on D. Because $\phi(e) \in \{\pm 1, \pm 2, \pm 3\}$, it follows that ϕ is a nowhere-zero 4-flow on D.   ∎

Since the Petersen graph is a cubic graph that is of Class two, it follows by Theorem 10.36 that it has no nowhere-zero 4-flow (which we also saw in Theorem 10.31). We noted that the Petersen graph does have a nowhere-zero 5-flow, however, which has been conjectured to be true for every bridgeless graph.

There is another concept and conjecture related to bridgeless graphs which ultimately returns us to snarks.

If G is a connected bridgeless plane graph, then the boundary of every region is a cycle and every edge of G lies on the boundaries of two regions. Thus, if S is the set of cycles of G that are the boundaries of the regions of G, then every edge of G belongs to exactly two elements of S. For which other graphs G is there a collection S of cycles of G such that every edge of G belongs to exactly two cycles of S?

A **cycle double cover** of a graph G is a set (actually a multiset) S of not necessarily distinct cycles of G such that every edge of G belongs to exactly two cycles of S. Certainly no cycle of G can appear more than twice in S. Also, if G contains a bridge e, then e belongs to no cycle and G can contains no cycle double cover. If G is Eulerian, then G contains an Eulerian circuit and therefore a set S' of cycles such that every edge of G belongs to exactly one cycle of S'. If S is the set of cycles of G that contains each cycle of S' twice, then S is a cycle double cover of G.

That every bridgeless graph has a cycle double cover was conjectured by Paul Seymour [178] in 1979. George Szekeres [183] conjectured this for cubic graphs even earlier – in 1973.

The Cycle Double Cover Conjecture   Every nontrivial connected bridgeless graph has a cycle double cover.

As we have seen, this conjecture is true for all nontrivial connected bridgeless planar graphs and for all Eulerian graphs. Initially, it may seem apparent that the Cycle Double Cover Conjecture is true, for if we were to replace each edge of a bridgeless graph G by two parallel edges then the resulting multigraph H is Eulerian. This implies that there is a set S of cycles of H such that each edge of H belongs to exactly one cycle in S. This, in turn, implies that each edge of G belongs to exactly two cycles in S, completing the proof. This argument is faulty, however, for one or more cycles of H in S may be 2-cycles, which do not correspond to cycles of G. Nevertheless, no counterexample to the Cycle Double Cover Conjecture is known. If the Cycle Double Cover Conjecture is false, then there exists a minimum counterexample, namely, a connected bridgeless graph of minimum size having no cycle double cover. Francois Jaeger (1947–1997) proved that a minimum counterexample to the Cycle Double Cover Conjecture must be a snark [129]. However, all known snarks possess a cycle double cover.

10.5 List Edge Colorings

In Section 9.2 we encountered the topic of list colorings. In a list coloring of a graph G, there is a list (or set) of available colors for each vertex of G, with the goal being to select a color from each list so that a proper vertex coloring of G results. One of the major problems concerns the determination of the smallest positive integer k such that if every list contains k or more colors, then a proper vertex coloring of G can be constructed. This smallest positive integer k is called the *list chromatic number* $\chi_\ell(G)$ of G. In this section we consider the edge analogue of this concept.

Let G be a nonempty graph and for each edge e of G, let $L(e)$ be a list (or set) of colors. Furthermore, let $\mathfrak{L} = \{L(e) : e \in E(G)\}$. The graph G is $\mathfrak{L}$-**edge choosable** (or $\mathfrak{L}$-**list edge colorable**) if there exists a proper edge coloring c of G such that $c(e) \in L(e)$ for every edge e of G. For a positive integer k, a nonempty graph G is k-**edge choosable** (or k-**list edge colorable**) if for every set $\mathfrak{L} = \{L(e) : e \in E(G)\}$ where each $|L(e)| \geq k$, the graph G is $\mathfrak{L}$-edge choosable. The **list chromatic index** $\chi'_\ell(G)$ is the minimum positive integer k for which G is k-edge choosable. Necessarily then, $\chi'(G) \leq \chi'_\ell(G)$ for every nonempty graph G. Applying a greedy edge coloring to a nonempty graph G (see Exercise 30) gives

$$\chi'_\ell(G) \leq 2\Delta(G) - 1.$$

Since the graph K_3 is of Class two, $\chi'(K_3) = 1 + \Delta(K_3) = 3$. Thus, $\chi'_\ell(K_3) \geq 3$. However, if e_1, e_2, and e_3 are the three edges of K_3 and $\mathfrak{L}(e_1)$, $\mathfrak{L}(e_2)$, and $\mathfrak{L}(e_3)$ are three sets of three or more colors each, then three distinct colors $c(e_1), c(e_2)$ and $c(e_3)$ can be chosen such that $c(e_i) \in \mathfrak{L}(e_i)$ for $1 \leq i \leq 3$ and so K_3 is 3-edge choosable. Therefore, $\chi'_\ell(K_3) = 3$. Although it would be natural now to expect to see an example of a graph G with $\chi'(G) < \chi'_\ell(G)$, we know of no graph with this property.

While the following conjecture was made independently by Vadim Vizing, Ram Prakash Gupta, and Michael Albertson and Karen Collins (see [107]), it first appeared in print in a 1985 paper by Béla Bollobás and Andrew J. Harris [24].

The List Coloring Conjecture For every nonempty graph G,

$$\chi'(G) = \chi'_\ell(G).$$

The identical conjecture has been made for multigraphs as well. Since the list chromatic index of a graph equals the list chromatic number of its line graph, the List Coloring Conjecture can also be stated as $\chi_\ell(G) = \chi(G)$ for every line graph G.

In 1979 Jeffrey Howard Dinitz had already conjectured that $\chi'(G) = \chi'_\ell(G)$ when G is a regular complete bipartite graph. By Theorem 10.8, it is known that $\chi'(K_{r,r}) = r$.

The Dinitz Conjecture For every positive integer r, $\chi'_\ell(K_{r,r}) = r$.

Fred Galvin [94] not only verified the Dinitz Conjecture, he showed that the List Coloring Conjecture is true for all bipartite graphs (indeed, for all bipartite

multigraphs). The proof of this result that we present is based on a proof of Tomaž
Slivnik [181], which in turn is based on Galvin's proof. We begin with a theorem of
Slivnik. First, we introduce some notation. Let G be a nonempty bipartite graph
with partite sets U and W. For each edge e of G, let u_e denote the vertex of U
incident with e and let w_e denote the vertex of W incident with e. For adjacent
edges e and f, it therefore follows that $u_e = u_f$ or $w_e = w_f$, but not both.

Lemma 10.37 *Let G be a nonempty bipartite graph and let $c : E(G) \to \mathbb{N}$ be an
edge coloring of G. For each edge e of G, let $\sigma_G(e)$ be the sum*

$$\sigma_G(e) \;=\; 1 + |\{f \in E(G) : u_e = u_f \text{ and } c(f) > c(e)\}|$$
$$+ \; |\{f \in E(G) : w_e = w_f \text{ and } c(f) < c(e)\}|$$

and let $L(e)$ be a set of $\sigma_G(e)$ colors. If

$$\mathfrak{L} = \{L(e) : e \in E(G)\},$$

then G is $\mathfrak{L}$-edge choosable.

Proof. We proceed by induction on the size m of G. Since the result holds if
$m = 1$, the basis step for the induction is true. Assume that the statement of the
theorem is true for all nonempty bipartite graphs of size less than m, where $m \geq 2$,
and let G be a nonempty bipartite graph of size m on which is defined an edge
coloring c and where the numbers $\sigma_G(e)$, the sets $L(e)$, and the set $\mathfrak{L}$ are defined in
the statement of the theorem.

Let A be a set of edges of G. A matching $M \subseteq A$ is said to be **optimal** (in A)
if the following is satisfied:

For every edge $e \in A - M$, there is an edge $f \in M$ such that either

(i) $u_e = u_f$ and $c(f) > c(e)$ or   (ii) $w_e = w_f$ and $c(f) < c(e)$.

We now show (by induction on the size of A) that for every $A \subseteq E(G)$, there
is an optimal matching $M \subseteq A$. First, observe that if A itself is a matching, then
$M = A$ is vacuously optimal. If $|A| = 1$, then $M = A$ is optimal and the basis step
of this induction is satisfied. Assume for an integer k with $1 < k \leq m$ that for each
set A' of edges of G with $|A'| = k - 1$, there is an optimal matching in A'. Let A
be a set of edges of G with $|A| = k$. We show that A contains an optimal matching
M.

An edge e belonging to A is U-**maximum** if there is no edge f in A for which
$u_e = u_f$ and $c(f) > c(e)$, while e is W-**maximum** if there is no edge f in A for
which $w_e = w_f$ and $c(f) > c(e)$. An edge $e \in A$ that is both U-maximum and
W-maximum is called c-**maximum**. Consequently, an edge $e \in A$ is c-maximum if
$c(f) < c(e)$ for every edge f adjacent to e. We consider two cases.

Case 1. Every U-maximum edge in A is W-maximum. Let

$$M = \{e \in A : e \text{ is } c\text{-maximum}\}.$$

We claim that M is optimal. Since no two edges of M can be adjacent, M is a matching. Let $e \in A - M$. Since e is not c-maximum, e is not U-maximum. So there exists an edge $f \in A$ for which $c(f)$ is maximum and $u_e = u_f$. This implies that f is U-maximum and is consequently c-maximum as well. Therefore, $f \in M$ and $c(f) > c(e)$. Hence, M is optimal.

Case 2. There exists an edge g in A that is U-maximum but not W-maximum. Because g is not W-maximum, there exists an edge $h \in A$ such that $w_h = w_g$ and $c(h) > c(g)$. We consider the set $A - \{h\}$ which consists of $k - 1$ edges. By the induction hypothesis, there is an optimal matching M in $A - \{h\}$. Hence, for every edge e in the set $(A - \{h\}) - M = A - (M \cup \{h\})$, there is an edge $f \in M$ for which either (i) $u_e = u_f$ and $c(f) > c(e)$ or (ii) $w_e = w_f$ and $c(f) < c(e)$.

We show that M is optimal in the set A. First, we establish the existence of an edge f in A such that either (i) $u_f = u_h$ and $c(f) > c(h)$ or (ii) $w_f = w_h$ and $c(f) < c(h)$. We consider two subcases.

Subcase 2.1. $g \notin M$. Then $g \in A - (M \cup \{h\})$. By the induction hypothesis, there is an edge $f \in M$ for which either (i) $u_g = u_f$ and $c(f) > c(g)$ or (ii) $w_g = w_f$ and $c(f) < c(g)$. Since g is U-maximum, (i) cannot occur and so (ii) must hold. Thus, $c(f) < c(g) < c(h)$ and M is optimal.

Subcase 2.2. $g \in M$. Then g is the desired edge f and, again, M is optimal.

Therefore, M is optimal in either subcase and consequently, for every set A of edges of G, there exists an optimal matching $M \subseteq A$.

Now select a color $a \in \cup_{e \in E(G)} L(e)$ and let

$$A = \{e \in E(G) : a \in L(e)\}.$$

As verified above, there is an optimal matching M in A. Let $G' = G - M$. For each edge e in G', let $L'(e) = L(e) - \{a\}$. If $e \in E(G) - A$, then $a \notin L(e)$ and

$$|L'(e)| = |L(e)| = \sigma_G(e) \geq \sigma_{G'}(e).$$

On the other hand, if $e \in A - M$, then $a \in L(e)$. Since M is optimal in A, there is an edge $f \in M$ such that either (i) $u_e = u_f$ and $c(f) > c(e)$ or (ii) $w_e = w_f$ and $c(f) < c(e)$. Thus,

$$|L'(e)| = |L(e)| - 1 = \sigma_G(e) - 1 \geq \sigma_{G'}(e).$$

Let $\mathcal{L}' = \{L'(e) : e \in E(G')\}$.

Since the size of G' is less than that of G, it follows by the induction hypothesis that G' is $\mathcal{L}'$-edge choosable. Thus, there exists a proper edge coloring $c' : E(G') \to \mathbb{N}$ of G' such that $c'(e) \in L'(e)$ for every edge e of G'. Define $c : E(G) \to \mathbb{N}$ by

$$c(e) = \begin{cases} c'(e) & \text{if } e \in E(G') \\ a & \text{if } e \in M. \end{cases}$$

Then $c(e) \in L(e)$ for every edge e of G and $c(e) \neq c(f)$ for every two adjacent edges e and f of G. Hence, c is a proper edge coloring and G is $\mathcal{L}$-edge choosable. ∎

From Lemma 10.37, we can now present a proof of Galvin's theorem.

Theorem 10.38 (**Galvin's Theorem**)   *If G is a bipartite graph, then*

$$\chi'_\ell(G) = \chi'(G).$$

Proof.   Since G is bipartite, it follows by Theorem 10.8 that $\chi'(G) = \Delta(G) = \Delta$. Thus, there exists a proper edge coloring $c : E(G) \to \{1, 2, \ldots, \Delta\}$. For each edge e of G, let $L(e)$ be a list of colors such that

$$
\begin{aligned}
|L(e)| \quad &= \quad \sigma_G(e) = 1 + |\{f \in E(G) : u_e = u_f \text{ and } c(f) > c(e)\}| \\
&\quad + |\{f \in E(G) : w_e = w_f \text{ and } c(f) < c(e)\}| \\
&\leq \quad 1 + (\chi'(G) - c(e)) + (c(e) - 1) = \chi'(G).
\end{aligned}
$$

Let $\mathfrak{L} = \{L(e) : e \in E(G)\}$. By Lemma 10.37, G is $\mathfrak{L}$-edge choosable. Therefore, G is $\chi'(G)$-edge choosable and so $\chi'_\ell(G) \leq \chi'(G)$. Since $\chi'(G) \leq \chi'_\ell(G)$, we have the desired result.   ∎

Since every bipartite graph is of Class one, it follows that the list chromatic index of every bipartite graph G equals $\Delta(G)$.

10.6   Total Colorings of Graphs

We now consider colorings that assign colors to both the vertices and the edges of a graph. A **total coloring** of a graph G is an assignment of colors to the vertices and edges of G such that distinct colors are assigned to (1) every two adjacent vertices, (2) every two adjacent edges, and (3) every incident vertex and edge. A k-**total coloring** of a graph G is a total coloring of G from a set of k colors. A graph G is k-**total colorable** if there is a k-total coloring of G. The **total chromatic number** $\chi''(G)$ of a graph G is the minimum positive integer k for which G is k-total colorable.

If c is a total coloring of a graph G and v is a vertex of G with $\deg v = \Delta(G)$, then c must assign distinct colors to the $\Delta(G)$ edges incident with v as well as to v itself. This implies that

$$\chi''(G) \geq 1 + \Delta(G) \text{ for every graph } G.$$

However, in the 1960s Mehdi Behzad [11] and Vadim Vizing [193] independently conjectured, similar to the upper bound for the chromatic index established by Vizing, that the total chromatic number cannot exceed this lower bound by more than 1. This conjecture has become known as the Total Coloring Conjecture.

The Total Coloring Conjecture   For every graph G,

$$\chi''(G) \leq 2 + \Delta(G).$$

Even though it is not known if $2 + \Delta(G)$ is an upper bound for the total chromatic number of every graph, the number $2 + \chi'_\ell(G)$ *is* an upper bound.

Theorem 10.39 *For every graph G,*

$$\chi''(G) \leq 2 + \chi'_\ell(G).$$

Proof. Suppose that $\chi'_\ell(G) = k$. By Theorem 10.2,

$$\begin{aligned} \chi(G) &\leq 1 + \Delta(G) \leq 1 + \chi'(G) \leq 1 + \chi'_\ell(G) \\ &< 2 + \chi'_\ell(G) = 2 + k. \end{aligned}$$

Thus, G is $(k+2)$-colorable. Let a $(k+2)$-coloring c of G be given. For each edge $e = uv$ of G, let $L(e)$ be a list of $k+2$ colors and let

$$L'(e) = L(e) - \{c(u), c(v)\}.$$

Since $|L'(e)| \geq k$ for each edge e of G and $\chi'_\ell(G) = k$, it follows that there is a proper edge coloring c' of G such that $c'(e) \in L'(e)$ and so $c'(e) \notin \{c(u), c(v)\}$. Hence, the total coloring c'' of G defined by

$$c''(x) = \begin{cases} c(x) & \text{if } x \in V(G) \\ c'(x) & \text{if } x \in E(G) \end{cases}$$

is a $(k+2)$-total coloring of G and so

$$\chi''(G) \leq 2 + k = 2 + \chi'_\ell(G),$$

as desired. ∎

The List Coloring Conjecture (see Section 10.5) states that $\chi'(G) = \chi'_\ell(G)$ for every nonempty graph G. If this conjecture is true, then $\chi'_\ell(G) \leq 1 + \Delta(G)$ by Vizing's theorem (Theorem 10.2) and so $\chi''(G) \leq 3 + \Delta(G)$ by Theorem 10.39. In 1998 Michael Molloy and Bruce Reed [153] established the existence of a constant c such that $c + \Delta(G)$ is an upper bound for $\chi''(G)$ for every graph G. In particular, they proved the following.

Theorem 10.40 *For every graph G,*

$$\chi''(G) \leq 10^{26} + \Delta(G).$$

Just as the chromatic index of a nonempty graph G equals the chromatic number of its line graph $L(G)$, the total chromatic number of G also equals the chromatic number of a related graph.

The **total graph** $T(G)$ of a graph G is that graph for which $V(T(G)) = V(G) \cup E(G)$ and such that two distinct vertices x and y of $T(G)$ are adjacent if x and y are adjacent vertices of G, adjacent edges of G, or an incident vertex and edge. It therefore follows that

$$\chi''(G) = \chi(T(G)) \text{ for every graph } G.$$

A graph G and its total graph are shown in Figure 10.22, together with a total coloring of G and a vertex coloring of $T(G)$. In this case, $\chi''(G) = \chi(T(G)) = 4$.

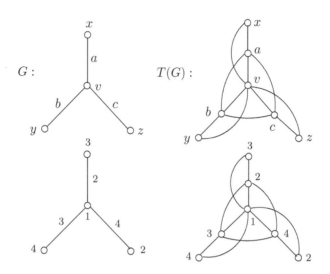

Figure 10.22: Total graphs and total colorings

Exercises for Chapter 10

1. (a) Determine the upper bounds for $\chi'(G)$ given by Theorems 10.3 and 10.4 for the multigraph G shown in Figure 10.23.

   (b) Determine $\chi'(G)$ for the multigraph G shown in Figure 10.23.

Figure 10.23: The multigraph G in Exercise 1

2. Let G be a graph with $\Delta(G) = \Delta$. Suppose that v_1 and v_2 are two nonadjacent vertices of G such that $\deg v_i < \Delta/2$ for $i = 1, 2$. How does $\chi'(G + v_1v_2)$ compare with $\chi'(G)$?

3. Prove for every graph G (whether G is of Class one or of Class two) that $G \square K_2$ is of Class one.

4. What can be said about the Class of a graph obtained by adding a pendant edge to a vertex of maximum degree in a graph?

5. Use the fact that every r-regular bipartite graph is 1-factorable to give an alternative proof of Theorem 10.8: *If G is a nonempty bipartite graph, then $\chi'(G) = \Delta(G)$.*

6. Show that Corollary 10.7 is also a corollary of Theorem 10.9.

7. Prove Corollary 10.10: *Every overfull graph is of Class two.*

8. Show that if H is an overfull subgraph of a graph G, then $\Delta(H) = \Delta(G)$.

9. A 5-regular graph G is connected and has cut-vertices but no bridges. Furthermore, every end-block of G has odd order. Prove that G is of Class two.

10. Let G be a graph of odd order $2k + 1 \geq 5$ such that $2k$ of the vertices of G have the same degree r and the remaining vertex of G has degree s for positive integers r, s with $r > s$. Determine whether G is of Class one or Class two.

11. Show that the condition $\Delta(G_1) = \Delta(G_2) = r$ is needed in Theorem 10.14.

12. A nonempty graph G is of Type one if $\chi'(L(G)) = \omega(L(G))$ and of Type two if $\chi'(L(G)) = 1 + \omega(L(G))$. Prove or disprove: Every nonempty graph is either of Type one or of Type two.

13. The **total deficiency** of a graph G of order n and size m is the number $n \cdot \Delta(G) - 2m$. Prove that if G is a graph of odd order whose total deficiency is less than $\Delta(G)$, then G is of Class two.

14. Prove that every cubic graph having connectivity 1 is of Class two.

15. Show that for every integer k with $2 \leq k \leq 5$, there is a planar graph of Class one and a planar graph of Class two, both having maximum degree k.

16. For a positive integer k, let H be a $2k$-regular graph of order $4k + 1$. Let G be obtained from H by removing a set of $k - 1$ independent edges from H. Prove that G is of Class two.

17. In Figure 10.7, a solution of Example 10.11 is given by providing a 6-edge coloring of the graph H shown in that figure. Give a characteristic of the resulting tennis schedule which might not be considered ideal and correct this deficiency by giving a different 6-edge coloring of H.

18. In Figure 10.8, a solution of Example 10.12 is given by providing a 4-edge coloring of the graph G shown in that figure. Give a characteristic of the resulting tennis schedule which might not be considered ideal and correct this deficiency by giving a different 4-edge coloring of G.

19. Give an example of a cubic planar graph with a bridge that has no Tait coloring.

20. (a) Give an example of a planar cubic graph containing a bridge.

    (b) Give an example of a nonplanar cubic graph containing a bridge.

    (c) Prove that every cubic graph containing a bridge is of Class two.

21. For the bridgeless cubic graph G shown in Figure 10.24, determine:

(a) whether G is planar.

(b) whether G is Hamiltonian.

(c) the girth of G.

(d) the cyclic edge-connectivity of G.

(e) to which class G belongs.

(f) whether G is a snark.

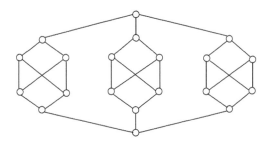

Figure 10.24: The graph in Exercise 21

22. (a) Determine the chromatic index of the graph G shown in Figure 10.25.

(b) Is this graph a snark?

$G:$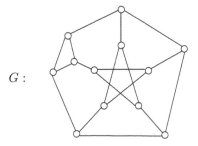

Figure 10.25: The graph in Exercise 22

23. The following is from the Lewis Carroll poem *The Hunting of the Snark*:

> *Taking Three as the subject to reason about,*
> *A convenient number to state,*
> *We add Seven, and Ten, and then multiply out*
> *By One Thousand diminished by Eight.*
> *The result we proceed to divide, as you see,*
> *By Nine Hundred and Ninety Two:*
> *Then subtract Seventeen, and the answer must be*
> *Exactly and perfectly true.*

If in this poem, taking Thirty-Three as "the subject to reason about" (rather than Three), what is the "answer"?

24. Let G be a graph having a nowhere-zero k-flow for some $k \geq 2$. Prove that for each partition $\{E_1, E_2\}$ of $E(G)$, there exists a nowhere-zero k-flow ϕ such that $\phi(e) > 0$ if and only if $e \in E_1$.

25. Prove Theorem 10.26: *If ϕ_1 and ϕ_2 are flows on an orientation D of a graph G, then every linear combination of ϕ_1 and ϕ_2 is also a flow on D.*

26. Show that the Petersen graph has a nowhere-zero 5-flow.

27. Prove that every bridgeless graph containing an Eulerian trail has a nowhere-zero 3-flow.

28. Can a graph that is not a cycle have a cycle double cover consisting only of three Hamiltonian cycles?

29. Show that K_6 has a cycle double cover consisting only of Hamiltonian cycles.

30. Let G be a nonempty graph. Show that $\chi'_\ell(G) \leq 2\Delta(G) - 1$ by applying a greedy edge coloring to G.

31. For the two graphs $G_1 = K_4$ and $G_2 = K_5$, determine $\chi''(G_i)$ for $i = 1, 2$ and express $\chi''(G_i)$ in terms of $\Delta(G_i)$.

32. (a) Show for every graph G that $\chi''(G) \leq \chi(G) + \chi'(G)$.

    (b) Give an example of a connected graph G such that $\chi''(G) = \chi(G) + \chi'(G)$.

33. For each integer $k \geq 2$, prove that the minimum positive integer n such that every k-edge (non-proper) coloring of K_n results in a monochromatic subgraph G of K_n with $\chi(G) \geq 3$ is $n = 2^k + 1$.

Chapter 11

Ramsey Theory

There are instances when we are interested in edge colorings of graphs that do not require adjacent edges to be assigned distinct colors. Of course, in these cases such colorings are not proper edge colorings. In many of these occasions, we are concerned with coloring the edges of complete graphs or certain complete multi-partite graphs. Often we have a fixed number of colors and seek the smallest order of a complete or complete multi-partite graph where each edge coloring of this graph with the prescribed number of colors results in a desired subgraph, all of whose edges are colored the same (a **monochromatic subgraph**). There are many variations of this. The best-known problems of these types deal with the topic of Ramsey numbers of graphs.

11.1 Ramsey Numbers

Frank Plumpton Ramsey (1903–1930) was a British philosopher, economist, and mathematician. Ramsey's first major work was his 1925 paper "The Foundations of Mathematics", in which he intended to improve upon *Principia Mathematica* by Bertrand Russell and Alfred North Whitehead. His second paper "On a Problem of Formal Logic" [161] contained a result, a restricted version of which is the following.

Theorem 11.1 (**Ramsey's Theorem**) *For any $k + 1 \geq 3$ positive integers t, n_1, n_2, ..., n_k, there exists a positive integer N such that if each of the t-element subsets of the set $\{1, 2, ..., N\}$ is colored with one of the k colors $1, 2, ..., k$, then for some integer i with $1 \leq i \leq k$, there is a subset S of $\{1, 2, ..., N\}$ containing n_i elements such that every t-element subset of S is colored i.*

In order to see a connection that Ramsey's theorem has with graph theory, suppose that $\{1, 2, ..., N\}$ is the vertex set of the complete graph K_N. In the case where $t = 2$, each 2-element subset of the set $\{1, 2, ..., N\}$ is assigned one of the colors $1, 2, ..., k$, that is, there is a k-edge coloring of K_N. It is this case of Ramsey's theorem in which we have a special interest and which is stated next.

Theorem 11.2 (**Ramsey's Theorem**) *For any $k \geq 2$ positive integers n_1, $n_2, \ldots, n_k$, there exists a positive integer N such that for every k-edge coloring of K_N, there is a complete subgraph K_{n_i} of K_N for some integer i $(1 \leq i \leq k)$ such that every edge of K_{n_i} is colored i.*

In fact, our primary interest in Ramsey's theorem is the case where $k = 2$. In a **red-blue edge coloring** (or simply a **red-blue coloring**) of a graph G, every edge of G is colored red or blue. Adjacent edges may be colored the same; in fact, this is probably necessary. Indeed, in a red-blue coloring of G, it is possible that all edges are colored red or all edges are colored blue. For two graphs F and H, the **Ramsey number** $R(F, H)$ is the minimum order n of a complete graph such that for every red-blue coloring of K_n, there is either a subgraph isomorphic to F all of whose edges are colored red (a **red** F) or a subgraph isomorphic to H all of whose edges are colored blue (a **blue** H). Certainly

$$R(F, H) = R(H, F)$$

for every two graphs F and H. The Ramsey number $R(F, F)$ is thus the minimum order n of a complete graph such that if every edge of K_n is colored with one of two given colors, then there is a subgraph isomorphic to F all of whose edges are colored the same (a **monochromatic** F). The Ramsey number $R(F, F)$ is sometimes called the **Ramsey number of the graph** F. Also, sometimes $R(F, F)$ is denoted by $R(F)$. We begin with perhaps the best-known Ramsey number, namely the Ramsey number of K_3.

Theorem 11.3 $R(K_3, K_3) = 6$.

Proof.   Let there be given a red-blue coloring of K_6. Consider some vertex v_1 of K_6. Since v_1 is incident with five edges, it follows by the Pigeonhole Principle that at least three of these five edges are colored the same, say red. Suppose that v_1v_2, v_1v_3, and v_1v_4 are red edges. If any of the edges v_2v_3, v_2v_4, and v_3v_4 is colored red, then we have a red K_3; otherwise, all three of these edges are colored blue, producing a blue K_3. Hence, $R(K_3, K_3) \leq 6$. On the other hand, let $V(K_5) = \{v_1, v_2, \ldots v_5\}$ and define a red-blue coloring of K_5 by coloring each edge of the 5-cycle $(v_1, v_2, \ldots, v_5, v_1)$ red and the remaining edges blue (see Figure 11.1, where bold edges are colored red and dashed edges are colored blue).

   Since this red-blue coloring produces neither a red K_3 nor a blue K_3, it follows that $R(K_3, K_3) \geq 6$ and so $R(K_3, K_3) = 6$.   ∎

Theorem 11.3 provides the answer to a popular recreational question.

> *In any gathering of people, every two of whom are either acquaintances or strangers, what is the smallest positive integer n such that in any gathering of n people, there are either three mutual acquaintances or three mutual strangers?*

This situation can be modeled by a graph of order n, in fact by K_n, where the vertices are the people, together with a red-blue coloring of K_n, where a red edge,

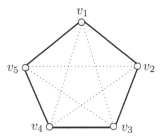

Figure 11.1: A red-blue coloring of K_5

say, indicates that the two people are acquaintances and a blue edge indicates that the two people are strangers. By Theorem 11.3 the answer to the question asked above is 6.

As an example of a Ramsey number $R(F, H)$, where neither F nor H is complete, we determine $R(F, H)$ for the graphs F and H shown in Figure 11.2 .

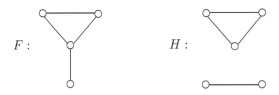

Figure 11.2: Determining $R(F, H)$

Example 11.4 *For the graphs F and H shown in Figure 11.2,*

$$R(F, H) = 7.$$

Proof. Since the red-blue coloring of K_6 in which the red graph is $2K_3$ and the blue graph is $K_{3,3}$ has neither a red F nor a blue H, it follows that $R(F, H) \geq 7$. Now let there be given a red-blue coloring of K_7. Since $R(K_3, K_3) = 6$ by Theorem 11.3, K_7 contains a monochromatic K_3. Let U be the vertex set of a monochromatic K_3 and let W be the set consisting of the remaining four vertices of K_7. We consider two cases.

Case 1. The monochromatic K_3 with vertex set U is blue. If any edge joining two vertices of W is blue, then there is a blue H; otherwise, there is a red F.

Case 2. The monochromatic K_3 with vertex set U is red. If any edge joining a vertex of U and a vertex of W is red, then there is a red F. Otherwise, every edge joining a vertex of U and a vertex of W is blue. If any edge joining two vertices of W is blue, then there is a blue H; otherwise, there is a red F. ∎

The Ramsey number $R(F, H)$ of two graphs F and H can be defined without regard to edge colorings. The Ramsey number $R(F, H)$ can be defined as the smallest positive integer n such that for every graph G of order n, either G contains a subgraph isomorphic to F or its complement $\overline{G}$ contains a subgraph isomorphic to H. Assigning the color red to each edge of G and the color blue to each edge of $\overline{G}$ returns us to our initial definition of $R(F, H)$.

Historically, it is the Ramsey numbers $R(K_s, K_t)$ that were the first to be studied. The numbers $R(K_s, K_t)$ are commonly expressed as $R(s, t)$ as well and are referred to as the **classical Ramsey numbers**. By Ramsey's theorem, $R(s, t)$ exists for every two positive integers s and t. We begin with some observations. First, as observed above, $R(s, t) = R(t, s)$ for every two positive integers s and t. Also,

$$R(1, t) = 1 \text{ and } R(2, t) = t$$

for every positive integer t; and by Theorem 11.3, $R(3, 3) = 6$.

Indeed, the Ramsey number $R(F, H)$ exists for every two graphs F and H. In fact, if F has order s and H has order t, then

$$R(F, H) \leq R(s, t).$$

The existence of the Ramsey numbers $R(s, t)$ was also established in 1935 by Paul Erdös and George Szekeres [76], where an upper bound for $R(s, t)$ was obtained as well. Recall, for positive integers k and n with $k \leq n$, the combinatorial identity:

$$\binom{n}{k-1} + \binom{n}{k} = \binom{n+1}{k}. \tag{11.1}$$

Theorem 11.5 *For every two positive integers s and t, the Ramsey number $R(s, t)$ exists; in fact,*

$$R(s, t) \leq \binom{s+t-2}{s-1}.$$

Proof. We proceed by induction on $n = s + t$. We have already observed that $R(1, t) = 1$ and $R(2, t) = t$ for every positive integer t. Hence, $R(s, t) \leq \binom{s+t-2}{s-1}$ when $n = s + t \leq 5$. Thus, we may assume that $s \geq 3$ and $t \geq 3$ and so $n \geq 6$. Suppose that $R(s', t')$ exists for all positive integers s' and t' such that $s' + t' < k$ where $k \geq 6$ and that

$$R(s', t') \leq \binom{s'+t'-2}{s'-1}.$$

We show for positive integers s and t with $s, t \geq 3$ and $k = s + t$ that

$$R(s, t) \leq \binom{s+t-2}{s-1}.$$

By the induction hypothesis, the Ramsey numbers $R(s-1, t)$ and $R(s, t-1)$ exist and

$$R(s-1, t) \leq \binom{s+t-3}{s-2} \text{ and } R(s, t-1) \leq \binom{s+t-3}{s-1}.$$

Since

$$\binom{s+t-3}{s-2} + \binom{s+t-3}{s-1} = \binom{s+t-2}{s-1}$$

by (11.1), it follows that

$$R(s-1,t) + R(s,t-1) \le \binom{s+t-2}{s-1}.$$

Let there be given a red-blue coloring of K_n, where $n = R(s-1,t) + R(s,t-1)$. We show that K_n contains either a red K_s or a blue K_t. Let v be a vertex of K_n. Then the degree of v in K_n is $n-1 = R(s-1,t) + R(s,t-1) - 1$. Let G be the spanning subgraph of K_n all of whose edges are colored red. Then $\overline{G}$ is the spanning subgraph of K_n all of whose edges are colored blue. We consider two cases, depending on the degree of v in G.

Case 1. $\deg_G v \ge R(s-1,t)$. Let A be the set of vertices in G that are adjacent to v. Thus, the order of the (complete) subgraph of K_n induced by A is $p = \deg_G v \ge R(s-1,t)$. Hence, this complete subgraph K_p contains either a red K_{s-1} or a blue K_t. If K_p contains a blue K_t, then K_n contains a blue K_t as well. On the other hand, if K_p contains a red K_{s-1}, then K_n contains a red K_s since v is joined to every vertex of A by a red edge.

Case 2. $\deg_G v \le R(s-1,t) - 1$. Then $\deg_{\overline{G}} v \ge R(s,t-1)$. Let B be the set of vertices in $\overline{G}$ that are adjacent to v. Therefore, the order of the (complete) subgraph of K_n induced by B is $q = \deg_{\overline{G}} v \ge R(s,t-1)$. Hence, this complete subgraph K_q contains either a red K_s or a blue K_{t-1}. If K_q contains a red K_s, then so does K_n. If K_q contains a blue K_{t-1}, then K_n contains a blue K_t since v is joined to every vertex of B by a blue edge.

Therefore,

$$R(s,t) \le R(s-1,t) + R(s,t-1) \le \binom{s+t-2}{s-1},$$

completing the proof. ∎

The proof of the preceding theorem provides an upper bound for $R(s,t)$, which is, in general, an improvement to that stated in Theorem 11.5.

Corollary 11.6 *For integers $s, t \ge 2$,*

$$R(s,t) \le R(s-1,t) + R(s,t-1). \tag{11.2}$$

Furthermore, if $R(s-1,t)$ and $R(s,t-1)$ are both even, then

$$R(s,t) \le R(s-1,t) + R(s,t-1) - 1.$$

Proof. The inequality in (11.2) follows from the proof of Theorem 11.5. Suppose that $R(s-1,t)$ and $R(s,t-1)$ are both even, and for

$$n = R(s - 1, t) + R(s, t - 1),$$

let there be given a red-blue coloring of K_{n-1}. Let G be the spanning subgraph of K_{n-1} all of whose edges are colored red. Then every edge of $\overline{G}$ is blue. Since G has odd order, some vertex v of G has even degree. If $\deg_G v \geq R(s - 1, t)$, then, proceeding as in the proof of Theorem 11.5, K_{n-1} contains a red K_s or a blue K_t. Otherwise, $\deg_G v \leq R(s - 1, t) - 2$ and so $\deg_{\overline{G}} v \geq R(s, t - 1)$. Again, proceeding as in the proof of Theorem 11.5, K_{n-1} contains a red K_s or a blue K_t.   ∎

Relatively few classical Ramsey numbers $R(s, t)$ are known for $s, t \geq 3$. The table in Figure 11.3, constructed by Stanislaw Radziszowski [160], gives the known values of $R(s, t)$ for integers s and t with $s, t \geq 3$.

s \ t	3	4	5	6	7	8	9
3	6	9	14	18	23	28	36
4	9	18	25	?	?	?	?

Figure 11.3: Some classical Ramsey numbers

While determining $R(F, H)$ is challenging in most instances, Vašek Chvátal [54] found the exact value of $R(F, H)$ whenever F is any tree of a fixed order and H is any complete graph of a fixed order.

Theorem 11.7 *Let T be a tree of order $p \geq 2$. For every integer $n \geq 2$,*

$$R(T, K_n) = (p - 1)(n - 1) + 1.$$

Proof.   First, we show that $R(T, K_n) \geq (p - 1)(n - 1) + 1$. Let there be given a red-blue coloring of the complete graph $K_{(p-1)(n-1)}$ such that the resulting red subgraph is $(n - 1)K_{p-1}$; that is, the red subgraph consists of $n - 1$ copies of K_{p-1}. Since each component of the red subgraph has order $p - 1$, it contains no connected subgraph of order greater than $p - 1$. In particular, there is no red tree of order p. The blue subgraph is then the complete $(n-1)$-partite graph $K_{p-1,p-1,\ldots,p-1}$, where every partite set contains exactly $p - 1$ vertices. Hence, there is no blue K_n either. Since this red-blue coloring avoids both a red tree T and a blue K_n, it follows that $R(T, K_n) \geq (p - 1)(n - 1) + 1$.

We now show that $R(T, K_n) \leq (p - 1)(n - 1) + 1$ for an arbitrary but fixed tree T of order $p \geq 2$ and an integer $n \geq 2$. We verify this inequality by induction on n. For $n = 2$, we show that $R(T, K_2) \leq (p - 1)(2 - 1) + 1 = p$. Let there be given a red-blue coloring of K_p. If any edge of K_p is colored blue, then a blue K_2 is produced. Otherwise, every edge of K_p is colored red and a red T is produced. Thus, $R(T, K_2) \leq p$. Therefore, the inequality $R(T, K_n) \leq (p - 1)(n - 1) + 1$ holds when $n = 2$. Assume for an integer $k \geq 2$ that $R(T, K_k) \leq (p - 1)(k - 1) + 1$. Consequently, every red-blue coloring of $K_{(p-1)(k-1)+1}$ contains either a red T or a blue K_k. We now show that $R(T, K_{k+1}) \leq (p - 1)k + 1$. Let there be given a red-blue coloring of $K_{(p-1)k+1}$. We show that there is either a red tree T or a blue K_{k+1}. We consider two cases.

Case 1. There exists a vertex v in $K_{(p-1)k+1}$ that is incident with at least $(p-1)(k-1)+1$ blue edges. Suppose that vv_i is a blue edge for $1 \le i \le (p-1)(k-1)+1$. Consider the subgraph H induced by the set $\{v_i : 1 \le i \le (p-1)(k-1)+1\}$. Thus, $H = K_{(p-1)(k-1)+1}$. By the induction hypothesis, H contains either a red T or a blue K_k. If H contains a red T, so does $K_{(p-1)k+1}$. On the other hand, if H contains a blue K_k, then, since v is joined to every vertex of H by a blue edge, there is a blue K_{k+1} in $K_{(p-1)k+1}$.

Case 2. Every vertex of $K_{(p-1)k+1}$ is incident with at most $(p-1)(k-1)$ blue edges. So, every vertex of $K_{(p-1)k+1}$ is incident with at least $p-1$ red edges. Thus, the red subgraph of $K_{(p-1)k+1}$ has minimum degree at least $p-1$. By Theorem 4.11, this red subgraph contains a red T. Therefore, $K_{(p-1)k+1}$ contains a red T as well. ∎

The Ramsey number $R(F, H)$ has been determined when F and H are both paths and when F and H are both cycles. In the case of two paths, the Ramsey number was determined by László Gerencsér and András Gyárfás [97].

Theorem 11.8 *For integers r and s with $2 \le r \le s$, $R(P_r, P_s) = s + \lfloor \frac{r}{2} \rfloor - 1$.*

When F and H are both cycles, the Ramsey number was determined by Ralph Faudree and Richard Schelp [83] and, independently, by Vera Rosta [173]. When one of F and H is a path and the other a cycle, the Ramsey number was determined in [82].

Theorem 11.9 *Let p and q be integers with $3 \le p \le q$.*

(i) $R(C_3, C_3) = R(C_4, C_4) = 6$;

(ii) *If p is odd and $(p, q) \ne (3, 3)$, then $R(C_p, C_q) = 2q - 1$;*

(iii) *If p and q are even and $(p, q) \ne (4, 4)$, then $R(C_p, C_q) = q + \frac{p}{2} - 1$;*

(iv) *If p is even and q is odd, then $R(C_p, C_q) = \max\{q + \frac{p}{2} - 1, 2p - 1\}$.*

Theorem 11.10 *Let m and n be integers with $m, n \ge 2$.*

$$R(P_n, C_m) = \begin{cases} 2n - 1 & \text{if } 3 \le m \le n \text{ and } m \text{ is odd} \\ n - 1 + \frac{m}{2} & \text{if } 4 \le m \le n \text{ and } m \text{ is even} \\ \max\left\{m - 1 + \lfloor \frac{n}{2} \rfloor, 2n - 1\right\} & \text{if } 2 \le n \le m \text{ and } m \text{ is odd} \\ m - 1 + \lfloor \frac{n}{2} \rfloor & \text{if } 2 \le n \le m \text{ and } m \text{ is even.} \end{cases}$$

Ramsey's theorem suggests that the Ramsey number $R(F, H)$ of two graphs F and H can be extended to more than two graphs. For $k \ge 2$ graphs $G_1, G_2, \ldots, G_k$, the **Ramsey number** $R(G_1, G_2, \ldots, G_k)$ is defined as the smallest positive integer n such that if every edge of K_n is colored with one of k given colors, say $c_1, c_2, \ldots, c_k$, a subgraph isomorphic to G_i results for some integer i $(1 \le i \le k)$, all of whose edges are colored c_i. While the existence of this more general Ramsey number is also a consequence of Ramsey's theorem, the existence of $R(F, H)$ for every two graphs F and H can also be used to show that $R(G_1, G_2, \ldots, G_k)$ exists for every $k \geq 2$ graphs $G_1, G_2, \ldots, G_k$ (see Exercise 9).

Theorem 11.11 *For every $k \geq 2$ graphs $G_1, G_2, \ldots, G_k$, the Ramsey number $R(G_1, G_2, \ldots, G_k)$ exists.*

If $G_i = K_{n_i}$ for $1 \leq i \leq k$, then we write $R(G_1, G_2, \ldots, G_k)$ as $R(n_1, n_2, \ldots, n_k)$. When the graphs G_i ($1 \leq i \leq k$) are all complete graphs of order at least 3 and $k \geq 3$, only the Ramsey numbers $R(3, 3, 3)$ and $R(3, 3, 4)$ are known. The first of these was obtained by Robert E. Greenwood and Andrew M. Gleason [100] in 1955.

Theorem 11.12 $R(3, 3, 3) = 17$.

Proof. Let there be given an edge coloring of K_{17} with the three colors red, blue, and green. Since K_{17} is 16-regular, some vertex of K_{17} must be incident with at least six edges that are colored the same, say vv_i ($1 \leq i \leq 6$) are six green edges of K_{17} incident with a vertex v of K_{17}. Let $H = K_6$ be the subgraph induced by $\{v_1, v_2, \ldots, v_6\}$. If any edge of H is colored green, then K_{17} has a green triangle. Thus, we may assume that no edge of H is colored green. Hence, every edge of H is colored red or blue. Since $H = K_6$ and $R(3, 3) = 6$ (Theorem 11.3), it follows that H and K_{17} as well contain either a red triangle or a blue triangle. Therefore, K_{17} contains a monochromatic triangle and so $R(3, 3, 3) \leq 17$.

To show that $R(3, 3, 3) = 17$, it remains to show that there is a 3-edge coloring of K_{16} for which there is no monochromatic triangle. In fact, there is an isomorphic factorization of K_{16} into a triangle-free graph that is commonly called the **Clebsch graph** or the **Greenwood-Gleason graph**. This graph can be constructed by beginning with the Petersen graph with vertices u_i and v_i ($1 \leq i \leq 5$), as illustrated in Figure 11.4 by solid vertices and bold edges. We then add six new vertices, namely x and w_i ($1 \leq i \leq 5$). The Clebsch graph CG (a 5-regular graph of order 16) is constructed as shown in Figure 11.4. This graph has the property that for every vertex v of CG, the subgraph $CG - N[v]$ is isomorphic to the Petersen graph. ∎

It wasn't until six decades later when another Ramsey number of this type was determined. In 2016, Michael Codish Michael Frank, Avraham Itzhakov, and Alice Miller [57] showed that $R(3, 3, 4) = 30$. The Ramsey number of $k \geq 2$ graphs, all of which are stars, has been determined by Stefan Burr and John Roberts [32].

Theorem 11.13 *Let $s_1, s_2, \ldots, s_k$ be $k \geq 2$ positive integers, t of which are even, and let $s = \sum_{i=1}^{k}(s_i - 1)$. Then*

$$R(K_{1,s_1}, K_{1,s_2}, \ldots, K_{1,s_k}) = \begin{cases} s + 1 & \text{if } t \text{ is positive and even} \\ s + 2 & \text{otherwise.} \end{cases}$$

Corollary 11.14 *For integers $s, t \geq 2$,*

$$R(K_{1,s}, K_{1,t}) = \begin{cases} s + t - 1 & \text{if } s \text{ and } t \text{ are both even} \\ s + t & \text{otherwise.} \end{cases}$$

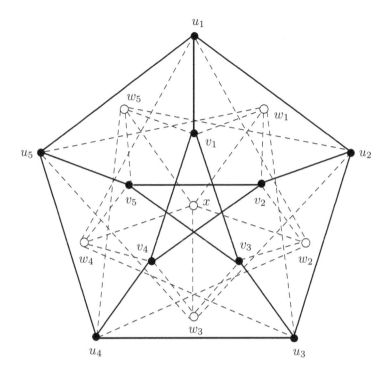

Figure 11.4: The Clebsch graph

11.2 Bipartite Ramsey Numbers

For the Ramsey numbers that were discussed in the preceding section, we often began with two graphs F and H and the two colors red and blue. We then sought the smallest order n of a complete graph K_n such that if each edge of K_n is colored red or blue, then we obtain either a red F or a blue H. There are also Ramsey numbers where complete graphs are replaced by other familiar graphs, primarily complete multipartite graphs. It is the complete bipartite graphs that has drawn the most attention, in which case this restricts the graphs F and H that can be considered.

In 1975 Lowell Beineke and Allen Schwenk [12] introduced the concept of bipartite Ramsey numbers. For two bipartite graphs F and H, the **bipartite Ramsey number** $BR(F, H)$ of F and H is the smallest positive integer r such that every red-blue coloring of the r-regular complete bipartite graph $K_{r,r}$ results in either a red F or a blue H. For example, the bipartite Ramsey number when $F = H = C_4$ is 5.

Example 11.15 $BR(C_4, C_4) = 5$.

Proof. Since $K_{4,4}$ can be decomposed into two copies of C_8, there is a red-blue coloring of $K_{4,4}$ resulting in a red C_8 and a blue C_8 (see Figure 11.5). That is, this red-blue coloring of $K_{4,4}$ avoids a monochromatic C_4. Thus, $BR(C_4, C_4) \geq 5$.

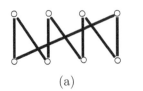 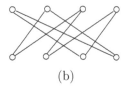

(a)   (b)

Figure 11.5: A red-blue coloring of $K_{4,4}$

To verify that $BR(C_4, C_4) \le 5$, it remains to show that every red-blue coloring of $K_{5,5}$ results in a monochromatic C_4. Let there be given a red-blue coloring of $G = K_{5,5}$ where U and W are the partite sets of $K_{5,5}$. Either U or W contains at least three vertices incident with at least three edges of the same color. Suppose that $u_1, u_2, u_3 \in U$ are three vertices incident with at least three red edges. Then two of these vertices, say u_1 and u_2, have at least two neighbors in common, producing a red C_4. ∎

For positive integers s and t, bipartite Ramsey numbers of the type $BR(K_{s,s}, K_{t,t})$ are referred to as **classical bipartite Ramsey numbers**, which is also denoted by $BR(s, t)$. In the next theorem, due to Johann Hattingh and Michael Henning [117], it is shown that $BR(s, t)$ exists by describing an upper bound for this number. If F and H are two bipartite graphs where the largest partite set of F has s vertices and the largest partite set of H has t vertices, then $BR(F, H) \le BR(s, t)$. That is, the following theorem also shows that the bipartite Ramsey number of every two bipartite graphs exists. This theorem is analogous to Corollary 11.6 for standard Ramsey numbers.

Theorem 11.16 *For every two integers $s, t \ge 2$,*

$$BR(s, t) \le BR(s - 1, t) + BR(s, t - 1) + 1.$$

Proof. Let $p = BR(s - 1, t) + BR(s, t - 1) + 1$ and let there be given a red-blue coloring of $G = K_{p,p}$, resulting in a red subgraph G_R and a blue subgraph G_B. We show that G contains either a red $K_{s,s}$ or a blue $K_{t,t}$. We now consider two cases.

Case 1. $\delta(G_R) \ge BR(s - 1, t) + 1$ or $\delta(G_B) \ge BR(s, t - 1) + 1$, say the former. Let $vw \in E(G_R)$. Then $\deg_{G_R} v \ge BR(s - 1, t) + 1$ and $\deg_{G_R} w \ge BR(s - 1, t) + 1$. Let $A = [N_{G_R}(v) \cup N_{G_R}(w)] - \{v, w\}$ and let $F = G[A]$. Since each partite set of F contains at least $BR(s - 1, t)$ vertices, it follows that F contains either a red $K_{s-1,s-1}$ or a blue $K_{t,t}$. If F contains a blue $K_{t,t}$, so does G. Thus, we may assume that F contains a red $K_{s-1,s-1}$. Since v is joined to every vertex in $N_{G_R}(v)$ by a red edge and w is joined to every vertex in $N_{G_R}(w)$ by a red edge, it follows that $G[N_{G_R}(v) \cup N_{G_R}(w)]$ contains a red $K_{s,s}$.

Case 2. $\delta(G_R) \le BR(s - 1, t)$ and $\delta(G_B) \le BR(s, t - 1)$. Let $v \in V(G_R)$ such that $\deg_{G_R} v = \delta(G_R)$. Thus, $\deg_{G_R} v \le BR(s - 1, t)$ and so $\deg_{G_B} v \ge BR(s, t - 1) + 1$. We consider two subcases.

Subcase 2.1. There is $w \in N_{G_B}(v)$ such that $\deg_{G_B} w \ge BR(s, t - 1) + 1$. Let $A = [N_{G_B}(v) \cup N_{G_R}(w)] - \{v, w\}$ and let $F = G[A]$. Since each partite set of F

contains at least $BR(s, t-1)$ vertices, it follows that F contains either a red $K_{s,s}$ or a blue $K_{t-1,t-1}$. If F contains a red $K_{s,s}$, so does G. Thus, we may assume that F contains a blue $K_{t-1,t-1}$. Since v is joined to every vertex in $N_{G_B}(v)$ by a blue edge and w is joined to every vertex in $N_{G_B}(w)$ by a blue edge, it follows that $G[N_{G_R}(v) \cup N_{G_R}(w)]$ contains a blue $K_{t,t}$.

Subcase 2.2. For every $x \in N_{G_B}(v)$, we have $\deg_{G_B} x \le BR(s, t-1)$. Let U and W be the partite sets of G, where $|U| = |W| = p$. We may assume that $v \in U$ and $x \in W$. We consider two subcases.

Subcase 2.2.1. For every $u \in U$, we have $\deg_{G_B} u \ge BR(s, t-1) + 1$. Thus, G contains at least $p(BR(s, t-1) + 1)$ blue edges that are incident with vertices of U. This implies that there is some $w \in W$ such that $\deg_{G_B} w \ge BR(s, t-1) + 1$. Let $x \in N_{G_B}(w) \subseteq U$. Then $\deg_{G_B} x \ge BR(s, t-1) + 1$. Let $A = [N_{G_B}(w) \cup N_{G_B}(x)] - \{w, x\}$ and let $F = G[A]$. Applying the same argument used in Subcase 2.1 to F, we see that G contains either a red $K_{s,s}$ or a blue $K_{t,t}$.

Subcase 2.2.2. There is $u \in U$ such that $\deg_{G_B} u \le BR(s, t-1)$. Thus, $\deg_{G_R} u \ge BR(s-1, t) + 1$. Since $v, u \in U$ with $\deg_{G_B} v \ge BR(s, t-1) + 1$ and $\deg_{G_R} u \ge BR(s-1, t) + 1$, it follows that there exists $w \in W$ such that $w \in N_{G_B}(v) \cap N_{G_R}(u)$. By the assumption of Subcase 2.2, $\deg_{G_B} w \le BR(s, t-1)$. Therefore, $\deg_{G_R} w \ge BR(s-1, t) + 1$. Let $A = [N_{G_R}(u) \cup N_{G_R}(w)] - \{u, w\}$ and let $F = G[A]$. Applying the same argument in Case 1 to F, we see that G contains either a red $K_{s,s}$ or a blue $K_{t,t}$. ∎

With the aid of Theorem 11.16, we can now present a numerical upper bound for the bipartite Ramsey numbers $BR(s, t)$, which is analogous to Theorem 11.5 for standard Ramsey numbers.

Corollary 11.17 *For every two positive integers s and t,*

$$BR(s,t) \le \binom{s+t}{s} - 1.$$

Proof. We proceed by induction on the integer $k = s + t \ge 2$. Since $BR(1, t) = t$ for $t \ge 1$ and $BR(s, 1) = s$ for $s \ge 1$, the inequality holds when $k = 2$ and $k = 3$. Suppose that

$$BR(s', t') \le \binom{s'+t'}{s'} - 1$$

for positive integers s' and t' such that $s' + t' = k - 1$ where $k \ge 4$. Now, let s and t be integers such that $s, t \ge 2$ and $s + t = k$. By the induction hypothesis,

$$BR(s-1, t) \le \binom{s+t-1}{s-1} - 1 \text{ and } BR(s, t-1) \le \binom{s+t-1}{s} - 1.$$

Since $\binom{s+t-1}{s-1} + \binom{s+t-1}{s} = \binom{s+t}{s}$, it follows that

$$BR(s-1, t) + BR(s, t-1) + 1 \le \binom{s+t}{s} - 1.$$

By Theorem 11.16, $BR(s, t) \leq \binom{s+t}{s} - 1$.   ∎

We have already seen that the upper bound for $BR(s, t)$ stated in Corollary 11.17 is attained when $s = 1$ or $t = 1$. Also, by Example 11.15 when $s = t = 2$, we have $BR(C_4, C_4) = BR(K_{2,2}, K_{2,2}) = BR(2, 2) = 5$ and the bound is attained here as well. It has been shown (see [12]) that $BR(2, 3) = 9$ and $BR(2, 4) = 14$ and so the bound is attained in these cases as well. On the other hand, $BR(3, 3) = 17$ (see [12]) and so the bound is not attained here.

This last bipartite Ramsey number provides an answer to the problem below.

> Suppose, for some positive integer r, that r girls and r boys are invited
> to a party where each girl-boy pair are either acquainted or are strangers.
> What is the smallest such r that guarantees that there exists a group of
> six people, three girls and three boys, such that either (1) every one of
> the three girls is acquainted with every one of the three boys or (2) every
> one of the three girls is a stranger of every one of the three boys?

The answer to this question is therefore $BR(3, 3) = 17$.

As is the case with standard Ramsey numbers, bipartite Ramsey numbers can also be extended to more than two (bipartite) graphs. For $k \geq 2$ bipartite graphs $G_1, G_2, \ldots, G_k$, the bipartite Ramsey number $BR(G_1, G_2, \ldots, G_k)$ is defined as the smallest positive integer r such that if every edge of $K_{r,r}$ is colored with one of k given colors $c_1, c_2, \ldots, c_k$, a subgraph isomorphic to G_i results for some i ($1 \leq i \leq k$), all of whose edges are colored c_i. That these numbers exist for all bipartite graphs is a consequence of a theorem of Paul Erdős and Richard Rado [74]. The only nontrivial example of a bipartite Ramsey number when $k \geq 3$ for which the exact value has been determined is $BR(2, 2, 2) = BR(C_4, C_4, C_4) = 11$, a result due to Wayne Goddard, Michael Henning, and Ortrud Oellermann [98].

As we have seen, if the bipartite Ramsey number of two bipartite graphs F and H is r, then every red-blue coloring of $K_{r,r}$ results in a red F or a blue H, while there exists a red-blue coloring of $K_{r-1,r-1}$ for which there is neither a red F nor a blue H. This brings up the question of what might occur for red-blue colorings of the intermediate graph $K_{r-1,r}$. This question leads to a more general concept.

For bipartite graphs F and H, the 2-**Ramsey number** $R_2(F, H)$ of F and H is the smallest positive integer n such that every red-blue coloring of the complete bipartite graph $K_{\lfloor n/2 \rfloor, \lceil n/2 \rceil}$ of order n results in a red F or a blue H. If the bipartite Ramsey number $BR(F, H)$ of two bipartite graphs F and H is r, then every red-blue coloring of $K_{r,r}$ produces a red F or a blue H, while there exists a red-blue coloring of $K_{r-1,r-1}$ that produces neither. Which of these two situations occurs for the graph $K_{r-1,r}$ depends on the graphs F and H. That is, either

$$R_2(F, H) = 2BR(F, H) \text{ or } R_2(F, H) = 2BR(F, H) - 1. \qquad (11.3)$$

To illustrate this concept, we show that $R_2(C_4, C_4) = 10$ (which was determined in [6]). We saw that $BR(C_4, C_4) = 5$. It then follows by (11.3) that $R_2(C_4, C_4) = 10$ or $R_2(C_4, C_4) = 9$. In fact, there is a red-blue coloring of $K_{4,5}$ that results in neither a red C_4 nor a blue C_4. To see this, consider the red-blue coloring of $K_{4,5}$

in which both the red subgraph shown in Figure 11.6(a) and the blue subgraph shown in Figure 11.6(b) are isomorphic to the graph in Figure 11.6(c). Since the graph in Figure 11.6(c) does not contain C_4 as a subgraph, this red-blue coloring of $K_{4,5}$ avoids both a red C_4 and a blue C_4. Therefore, $R_2(C_4, C_4) \geq 10$ and so $R_2(C_4, C_4) = 10$.

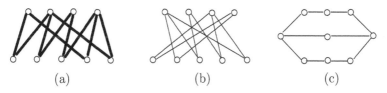

(a) (b) (c)

Figure 11.6: A red-blue coloring of $K_{4,5}$

The concept of the 2-Ramsey number of two bipartite graphs is a special case of a more general concept. For an integer $k \geq 2$, a **balanced complete k-partite graph** of order $n \geq k$ is the complete k-partite graph in which every partite set has either $\lfloor n/k \rfloor$ or $\lceil n/k \rceil$ vertices. So if $n = kq + r$ where $q \geq 1$ and $0 \leq r \leq k-1$, then the balanced complete k-partite graph G of order n has r partite sets with $q + 1$ vertices and the remaining $k - r$ partite sets have q vertices. If $r = 0$, then G is a $[(k-1)q]$-regular graph, which we denote by $K_{k(q)}$.

For bipartite graphs F and H and an integer k with $2 \leq k \leq R(F, H)$, the k-**Ramsey number** $R_k(F, H)$ is defined as the smallest positive integer n such that every red-blue coloring of a balanced complete k-partite graph of order n results in a red F or a blue H. That is, $R_k(F, H)$ is the minimum order of a balanced complete k-partite graph every red-blue coloring of which results in a red F or a blue H.

If F and H are two bipartite graphs for which $R(F, H) = n \geq 3$, then every red-blue coloring of K_n produces either a red F or a blue H. However, such is not the case for the smaller complete graphs $K_2, K_3, \ldots, K_{n-1}$. Equivalently, for every red-blue coloring of the complete n-partite graph K_n where each partite set consists of a single vertex, there is either a red F or a blue H. However, for each complete k-partite graph K_k, where $2 \leq k \leq n-1$ such that every partite set consists of a single vertex, there exists a red-blue coloring that produces neither a red F nor a blue H. On the other hand, for each of the graphs $K_2, K_3, \ldots, K_{n-1}$, we can continue to add vertices to each partite set, resulting in a balanced complete k-partite graph at each step where $2 \leq k \leq n - 1$ until eventually arriving at the balanced complete k-partite graph of smallest order $R_k(F, H)$ having the property that every red-blue coloring of this graph produces a red F or a blue H. Consequently, for every two bipartite graphs F and H and every integer k with $2 \leq k \leq R(F, H)$, the k-Ramsey number $R_k(F, H)$ exists. Furthermore, if $R(F, H) = n$, then

$$R_n(F, H) \leq R_{n-1}(F, H) \leq \cdots \leq R_3(F, H) \leq R_2(F, H).$$

We have seen that $R_2(C_4, C_4) = 10$. We now show that $R_3(C_4, C_4) = 9$.

Proposition 11.18 $R_3(C_4, C_4) = 9$.

Proof. Let H be the balanced complete 3-partite graph of order 8. Then $H = K_{2,3,3}$. Figure 11.7 shows a red-blue coloring of H having neither a red C_4 nor a blue C_4, where the bold edges represent red edges. Thus, $R_3(C_4, C_4) \geq 9$.

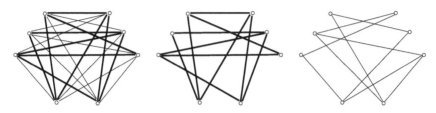

Figure 11.7: A red-blue coloring of $K_{2,3,3}$

To show that $R_3(C_4, C_4) = 9$, it remains to show that every red-blue coloring of $G = K_{3,3,3}$ results in a monochromatic C_4. Assume, to the contrary, that there is a red-blue coloring of G that produces neither a red C_4 nor a blue C_4. Let G_R and G_B denote the red and blue subgraphs of G, respectively, of sizes m_R and m_B. We may assume that $m_R \geq m_B$. Since $m_R + m_B = 27$, it follows that $m_R \geq 14$. Let V_1, V_2, and V_3 be the three partite sets of G and, for $1 \leq i < j \leq 3$, let $[V_i, V_j]$ denote the nine edges of G joining V_i and V_j. Let G'_R denote the subgraph of size m'_R in G_R with vertex set $V_1 \cup V_2$ such that $E(G'_R) \subseteq [V_1, V_2]$. The subgraphs G''_R and G'''_R with vertex sets $V_2 \cup V_3$ and $V_1 \cup V_3$ and sizes m''_R and m'''_R, respectively, are defined similarly. We may assume that $m'_R \geq m''_R \geq m'''_R$ and so $m'_R + m''_R \geq 10$. Let $V_1 = \{u_1, u_2, u_3\}$, $V_2 = \{v_1, v_2, v_3\}$, and $V_3 = \{w_1, w_2, w_3\}$. Observe that if any of u_1, u_2, and u_3 has degree 3 in G'_R, say u_1, then u_2 and u_3 have degree at most 1 in G'_R and each of w_1, w_2, and w_3 has degree at most 1 in G''_R, for otherwise a red C_4 is produced. However then, $m'_R + m''_R \leq 8$, a contradiction. Consequently, each of u_1, u_2, and u_3 has degree at most 2 in G'_R. Therefore, $m'_R = 6$ or $m'_R = 5$. We consider these two cases.

Case 1. $m'_R = 6$. Thus, $G'_R = C_6$, say $G'_R = (u_1, v_1, u_2, v_2, u_3, v_3, u_1)$. Hence, each of w_1, w_2, and w_3 has degree at most 1 in G''_R, for otherwise a red C_4 is produced. However then, $m'_R + m''_R \leq 9$, a contradiction.

Case 2. $m'_R = 5$. Hence, $m''_R = 5$ as well. We may assume that u_1 and u_2 have degree 2 in G'_R and w_1 and w_2 have degree 2 in G''_R. Neither u_1 and u_2 nor w_1 and w_2 have the same neighbors in G'_R and G''_R, respectively, for otherwise a red C_4 is produced. This, however, implies that two of the vertices v_1, v_2, and v_3 are neighbors of both a vertex u_i in G'_R and a vertex w_j in G''_R, producing a red C_4 and a contradiction. Therefore, $R_3(C_4, C_4) = 9$. ∎

Therefore, we now know that $R_2(C_4, C_4) = 10$ and $R_3(C_4, C_4) = 9$. In fact, the following was proved in [6].

Theorem 11.19 *For every integer k with $2 \leq k \leq 6$, $R_k(C_4, C_4) = 12 - k$.*

While the k-Ramsey number $R_k(F, H)$ exists for every two bipartite graphs F and H when $2 \leq k \leq R(F, H)$, such is not the case when F and H are not bipartite. For graphs F and H that are not bipartite, not only does $R_2(F, H)$ fail to exist but $R_3(F, H)$ and $R_4(F, H)$ also do not exist. To see this, let G be any balanced complete 3-partite graph with partite sets V_1, V_2, and V_3. Assigning the color red to every edge of $[V_1, V_2]$ and blue to all other edges of G results in G_R and G_B both being bipartite. Similarly, if G is a balanced complete 4-partite graph with partite sets V_1, V_2, V_3, and V_4 where the color red is assigned to every edge of $[V_1, V_2] \cup [V_2, V_3] \cup [V_3, V_4]$ and the color blue to all other edges of G, then G_R and G_B are both bipartite. Indeed, even if $\chi(F) = \chi(H) = 3$, $R_5(F, H)$ need not exist. For example, $R_5(K_3, K_3)$ does not exist. To see this, let G be a balanced complete 5-partite graph with partite sets V_i for $1 \le i \le 5$. If the edges in $[V_1, V_2] \cup [V_2, V_3] \cup [V_3, V_4] \cup [V_4, V_5] \cup [V_5, V_1]$ are colored red and all other edges are colored blue, then G does not contain a monochromatic K_3. Consequently, $R_k(K_3, K_3)$ exists only when $k = R(K_3, K_3) = 6$. Even if a red-blue coloring of K_5 does not contain a monochromatic K_3, there is another monochromatic graph it must contain.

Observation 11.20 *Every red-blue coloring of K_5 produces either a monochromatic C_3 or a monochromatic C_5.*

From our preceding discussion, the k-Ramsey number of two odd cycles does not exist when $k = 2, 3, 4$. Furthermore, the 5-Ramsey number of two triangles does not exist. However, if neither of the two odd cycles is a triangle, then the situation is different. The following was shown in [5].

Theorem 11.21 *For each pair k, ℓ of integers with $k, \ell \ge 2$, the 5-Ramsey number $R_5(C_{2\ell+1}, C_{2k+1})$ exists.*

Proof. We may assume that $k \ge \ell$. As we have seen, the bipartite Ramsey number $BR(K_{k,k}, K_{k,k})$ exists, say $BR(K_{k,k}, K_{k,k}) = p_1$. Proceeding recursively, we let

$$BR(K_{p_i,p_i}, K_{p_i,p_i}) = p_{i+1} \text{ for } i = 1, 2, 3, 4$$

and let $G = K_{5(p_5)}$ with partite sets U_5, W_5, X_5, Y_5, Z_5. Let there be given a red-blue coloring of G. We use the 10 pairs of elements of $S = \{U, W, X, Y, Z\}$ as a guide to prove the theorem. We consider the partition $\{S_1, S_2, S_3, S_4, S_5\}$ of the 2-element subsets of S as follows and use this to show that G contains a red $C_{2\ell+1}$ or a blue C_{2k+1}:

$$S_1 = \{\{U, W\}, \{X, Y\}\}, \ S_2 = \{\{U, X\}, \{Y, Z\}\},$$
$$S_3 = \{\{W, Y\}, \{X, Z\}\}, \ S_4 = \{\{U, Y\}, \{W, Z\}\},$$
$$S_5 = \{\{W, X\}, \{U, Z\}\}.$$

First, we employ the two pairs in S_1. Since $BR(K_{p_4,p_4}, K_{p_4,p_4}) = p_5$ and $|U_5| = |W_5| = p_5$ and $|X_5| = |Y_5| = p_5$, the complete bipartite graph K_{p_5,p_5} with partite sets U_5 and W_5 and the complete bipartite graph K_{p_5,p_5} with partite sets X_5 and Y_5 both contain a monochromatic subgraph K_{p_4,p_4} with partite sets $U_4 \subseteq U_5$ and

$W_4 \subseteq W_5$ in the first graph and $X_4 \subseteq X_5$ and $Y_4 \subseteq Y_5$ in the second graph. Furthermore, let $Z_4 \subseteq Z_5$ such that $|Z_4| = p_4$.

Next, we employ the two pairs in S_2. Since $BR(K_{p_3,p_3}, K_{p_3,p_3}) = p_4$ and $|U_4| = |X_4| = p_4$ and $|Y_4| = |Z_4| = p_4$, the complete bipartite graph K_{p_4,p_4} with partite sets U_4 and X_4 and the complete bipartite graph K_{p_4,p_4} with partite sets Y_4 and Z_4 both contain a monochromatic subgraph K_{p_3,p_3} with partite sets $U_3 \subseteq U_4$ and $X_3 \subseteq X_4$ in the first graph and $Y_3 \subseteq Y_4$ and $Z_3 \subseteq Z_4$ in the second graph. Furthermore, let $W_3 \subseteq W_4$ such that $|W_3| = p_3$.

Continuing in this manner, we arrive at a subgraph $H = K_{5(k)}$ of G with partite sets U_1, W_1, X_1, Y_1, Z_1 such that (i) $U_1 \subseteq U_5, W_1 \subseteq W_5, X_1 \subseteq X_5, Y_1 \subseteq Y_5, Z_1 \subseteq Z_5$ and $|U_1| = |W_1| = |X_1| = |Y_1| = |Z_1| = k$ and (ii) for each pair among the sets U_1, W_1, X_1, Y_1, Z_1, the complete bipartite graph $K_{k,k}$ with these partite sets is monochromatic.

By Observation 11.20, every red-blue coloring of K_5 has either a monochromatic C_3 or a monochromatic C_5. This implies that H has a monochromatic $K_{3(k)}$ or a monochromatic $C_{5(k)}$ (a graph obtained from the 5-cycle $(v_1, v_2, v_3, v_4, v_5, v_1)$ by replacing each vertex v_i with a set V_i of k vertices for $1 \le i \le 5$), where every vertex of V_i is adjacent to every vertex of V_{i+1} for $i = 1, 2, 3, 4$ and every vertex of V_5 is adjacent to every vertex of V_1. We consider these two cases.

Case 1. H has a monochromatic $K_{3(k)}$, say with partite sets X_1, Y_1, Z_1. Let $x_1 \in X_1$, $Y_1 = \{y_1, y_2, \ldots, y_k\}$ and $Z_1 = \{z_1, z_2, \ldots, z_k\}$. Then

$$C_{2k+1} = (x_1, y_1, z_1, y_2, z_2, \ldots, y_k, z_k, x_1)$$

is a monochromatic $(2k + 1)$-cycle and

$$C_{2\ell+1} = (x_1, y_1, z_1, y_2, z_2, \ldots, y_\ell, z_\ell, x_1)$$

is a monochromatic $(2\ell + 1)$-cycle in G.

Case 2. H has a monochromatic $C_{5(k)}$, say with edge set

$$[U_1, W_1] \cup [W_1, X_1] \cup [X_1, Y_1] \cup [Y_1, Z_1] \cup [Z_1, X_1].$$

Let $u_1 \in U_1, w_1 \in W_1$, and $x_1 \in X_1$. Furthermore, let $Y_1 = \{y_1, y_2, \ldots, y_k\}$ and $Z_1 = \{z_1, z_2, \ldots, z_k\}$. Then

$$C_{2k+1} = (u_1, w_1, x_1, y_1, z_1, y_2, z_2, \ldots, y_{k-1}, z_{k-1}, u_1)$$

is a monochromatic $(2k + 1)$-cycle and

$$C_{2\ell+1} = (u_1, w_1, x_1, y_1, z_1, y_2, z_2, \ldots, y_{\ell-1}, z_{\ell-1}, u_1)$$

is a monochromatic $(2\ell + 1)$-cycle in G.   ∎

We have seen that Ramsey numbers are defined for three or more graphs. In particular, for three graphs F_1, F_2, and F_3, the *Ramsey number* $R(F_1, F_2, F_3)$ of F_1, F_2, and F_3 is the smallest positive integer n for which every red-blue-green coloring (in which every edge is colored red, blue, or green) of the complete graph K_n results in a red F_1, a blue F_2, or a green F_3. This gives rise to the concept of the k-Ramsey number of three (or more) graphs. For three graphs F_1, F_2, and F_3 and an integer k with $2 \le k \le R(F_1, F_2, F_3)$, the k-Ramsey number $R_k(F_1, F_2, F_3)$ of F_1, F_2, and F_3, if it exists, is the smallest order of a balanced complete k-partite graph G for which every red-blue-green coloring of G results in a red F_1, a blue F_2, or a green F_3. In particular, if $k = 2$ and $F_i \cong F$ for some graph F where $i = 1, 2, 3$, then the 2-Ramsey number $R_2(F, F, F)$ is the smallest order of a balanced complete bipartite graph G for which every red-blue-green coloring of the edges of G results in a monochromatic F (all of whose edges are colored the same). For example, we mentioned that it was shown in [98] that $BR(C_4, C_4, C_4) = 11$. Furthermore, it was shown in [131] that $R_2(C_4, C_4, C_4) \le 21$. Consequently, $R_2(C_4, C_4, C_4) = 21$.

11.3 s-Bipartite Ramsey Numbers

Since $BR(C_4, C_4) = 5$ and $R_2(C_4, C_4) = 10$, it follows, as we have seen, that every red-blue coloring of $K_{5,5}$ contains a monochromatic C_4 and there is a red-blue coloring of $K_{4,5}$ that avoids a monochromatic C_4. This also brings up yet another question, namely whether every red-blue coloring of $K_{4,6}$ results in a monochromatic C_4 or whether there is some red-blue coloring of $K_{4,6}$ that avoids a monochromatic C_4. Since $K_{4,6}$ can be decomposed into two copies of the graph H shown in Figure 11.8 and H does not contain a 4-cycle, where the edges of one copy can be colored red and the edges of the other copy can be colored blue, it follows that there is a red-blue coloring of $K_{4,6}$ that results in no monochromatic C_4.

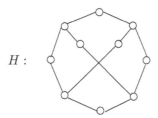

H :

Figure 11.8: A graph H for which $K_{4,6}$ is H-decomposable

On the other hand, every red-blue coloring of $K_{4,7}$ contains a monochromatic C_4. Indeed, every red-blue coloring of $K_{3,7}$ contains a monochromatic C_4. To see this, let there be given a red-blue coloring of $G = K_{3,7}$ resulting in a red subgraph G_R and a blue subgraph G_B. Let m_R and m_B denote the sizes of G_R and G_B, respectively. We may assume that $m_R > m_B$. Thus, $m_R \ge 11$. Let U and W be the partite sets of G where $|U| = 3$ and $|W| = 7$. If U contain vertices u_1 and u_2 such that $\deg_{G_R} u_1 + \deg_{G_R} u_2 \ge 9$, then u_1 and u_2 have two common neighbors in G_R and so G_R contains a 4-cycle. Otherwise, the degrees of the three vertices of U in G_R are either $5, 3, 3$, or $4, 4, 4$, or $4, 3, 3$. In any case, there are two vertices of U having two common neighbors in G_R and therefore G_R contains a 4-cycle.

This discussion brings up another concept, introduced in [18]. For two bipartite graphs F and H and a positive integer s, the s-**bipartite Ramsey number**

$BR_s(F, H)$ of F and H is the smallest integer t with $t \geq s$ such that every red-blue coloring of $K_{s,t}$ results in a red F or a blue H. From our discussion above, we then have the following result. Again, we write $BR_s(K_{p,p}, K_{q,q})$ as $BR_s(p, q)$.

Theorem 11.22 *For each integer $s \geq 2$,*

$$BR_s(2, 2) = \begin{cases} \text{does not exist} & \text{if } s = 2 \\ 7 & \text{if } s = 3, 4 \\ s & \text{if } s \geq 5. \end{cases}$$

Proof. First, let $t \geq 2$ be an integer and let $G = K_{2,t}$, where $\{u_1, u_2\}$ is one of the partite sets of G. If each edge of G incident with u_1 is colored red and each edge incident with u_2 is colored blue, then there is no monochromatic $K_{2,2}$. Thus, $BR_2(2, 2)$ does not exist.

We have seen that every red-blue coloring of $K_{3,7}$ has a monochromatic $C_4 = K_{2,2}$ and there exists a red-blue coloring of $K_{4,6}$ that avoids a monochromatic $K_{2,2}$. Therefore, $BR_3(2, 2) = BR_4(2, 2) = 7$.

Since $BR(2, 2) = 5$, it follows that $BR_s(2, 2) = s$ for each integer $s \geq 5$. ∎

Not only has $BR_s(K_{2,2}, H)$ been determined for $H = K_{2,2}$ and $s \geq 2$, it has also been determined when $H = K_{2,3}$ or $H = K_{3,3}$. For each of these two results, we give a proof for the smallest value of s in which this number exists.

Theorem 11.23 *For each integer $s \geq 2$,*

$$BR_s(K_{2,2}, K_{2,3}) = \begin{cases} \text{does not exist} & \text{if } s = 2 \\ 10 & \text{if } s = 3 \\ 8 & \text{if } 4 \leq s \leq 7 \\ s & \text{if } s \geq 8. \end{cases}$$

Proof for the cases $s = 2$ and $s = 3$. First, suppose that $s = 2$. For an arbitrary integer $t \geq 2$, the red-blue coloring of $K_{2,t}$, in which both red and blue subgraphs are $K_{1,t}$ produces neither a red $K_{2,2}$ nor a blue $K_{2,3}$.

Suppose next that $s = 3$. First, we show that there exists a red-blue coloring of $K_{3,9}$ that avoids both a red $K_{2,2}$ and a blue $K_{2,3}$. For $G = K_{3,9}$, let $U = \{u_1, u_2, u_3\}$ and $W = \{w_1, w_2, \ldots, w_9\}$ be the partite sets of G. Consider the following three 4-element subsets W_1, W_2, W_3 of W, where $\{w_a, w_b, w_c, w_d\}$ is denoted by $abcd$, and let $\overline{W}_i = W - W_i$ be the 5-element subset of W for $i = 1, 2, 3$.

W_i :	1378	5679	2489
$\overline{W}_i$:	24569	12348	13567

We now define a red-blue coloring of G by joining each vertex u_i $(1 \leq i \leq 3)$ to the four vertices in W_i by red edges and to the remaining five vertices in $\overline{W}_i$ by blue edges. This coloring is shown in Figure 11.9, where each solid line indicates a red edge and each dashed line indicates a blue edge. Since $|W_i \cap W_j| = 1$, $|\overline{W}_i \cap \overline{W}_j| = 2$ for $1 \leq i \neq j \leq 3$ and $|\overline{W}_1 \cap \overline{W}_2 \cap \overline{W}_3| = 0$, there is neither a red $K_{2,2}$ nor a blue $K_{2,3}$ in G. Hence, $BR_3(K_{2,2}, K_{2,3}) \geq 10$.

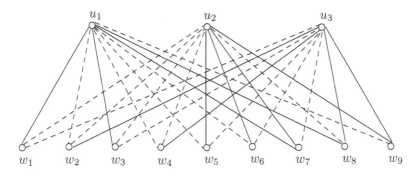

Figure 11.9: A red-blue coloring of $K_{3,9}$ avoiding both a red $K_{2,2}$ and a blue $K_{2,3}$

Next, we verify that $BR_3(K_{2,2}, K_{2,3}) \leq 10$ by showing that every red-blue coloring of $H = K_{3,10}$ results in a red $K_{2,2}$ or a blue $K_{2,3}$. Let there be given a red-blue coloring of H, where H_R denotes the red subgraph of H and H_B the blue subgraph. Let $U = \{u_1, u_2, u_3\}$ and $W = \{w_1, w_2, \ldots, w_{10}\}$ be the partite sets of H. First, suppose that W has at least four vertices of degree at most 1 in H_B; say $\deg_{H_B} w_i \leq 1$ for $1 \leq i \leq 4$. Thus, $\deg_{H_R} w_i \geq 2$ for $1 \leq i \leq 4$. Since there are $\binom{3}{2} = 3$ distinct 2-element subsets of U, at least two vertices in $\{w_1, w_2, w_3, w_4\}$ are joined to the same pair of vertices of U by red edges, producing a red $K_{2,2}$. On the other hand, if W has at most three vertices of degree at most 1 in H_B, then W has at least seven vertices of degree 2 or more in H_B. Since there are only three distinct 2-element subsets of U, at least three vertices of W are joined to the same pair of vertices in U by blue edges, producing a blue $K_{2,3}$. In any case, there is either a red $K_{2,2}$ or a blue $K_{2,3}$ in H. Therefore, $BR_3(K_{2,2}, K_{2,3}) = 10$. ∎

Theorem 11.24 *For each integer $s \geq 2$,*

$$BR_s(2,3) = BR_s(K_{2,2}, K_{3,3}) = \begin{cases} does\ not\ exist & if\ s = 2,3 \\ 15 & if\ s = 4 \\ 12 & if\ s = 5,6 \\ 9 & if\ s = 7,8 \\ s & if\ s \geq 9. \end{cases}$$

Proof for the cases $s = 2, 3, 4$. Suppose first that $s = 3$. For an arbitrary integer $t \geq s$, the red-blue coloring of $K_{3,t}$ in which the red subgraph is $K_{1,t}$ and the blue subgraph is $K_{2,t}$ produces neither a red $K_{2,2}$ nor a blue $K_{3,3}$. Therefore, neither $BR_2(2,3)$ or $BR_3(2,3)$ exists.

Suppose next that $s = 4$. First, we show that there exists a red-blue coloring of $K_{4,14}$ that avoids both a red $K_{2,2}$ and a blue $K_{3,3}$. For $G = K_{4,14}$, let

$$U = \{u_1, u_2, u_3, u_4\} \text{ and } W = \{w_1, w_2, \ldots, w_{14}\}$$

be the partite sets of G. Consider the following subsets $U_1, U_2, \ldots, U_{14}$ of U, where $\{u_a, u_b, \ldots\}$ is denoted by $ab\cdots$, and let $\overline{U}_i = U - U_i$ for $1 \leq i \leq 14$.

$U_i:$	1	1	2	2	3	3	4	4	12	13	14	23	24	34
$\overline{U}_i:$	234	234	134	134	124	124	123	123	34	24	23	14	13	12

We now define a red-blue coloring of G by joining each vertex w_i ($1 \le i \le 14$) to the vertices in U_i by red edges and to the remaining vertices in $\overline{U}_i$ by blue edges. The resulting red subgraph of this red-blue coloring is shown in Figure 11.10. Since $|U_i \cap U_j| \le 1$ and $|\overline{U}_i \cap \overline{U}_j \cap \overline{U}_k| \le 2$ for $1 \le i \ne j \ne k \le 14$ and $i \ne k$, there is neither a red $K_{2,2}$ nor a blue $K_{3,3}$ in G and so $BR_4(2,3) \ge 15$.

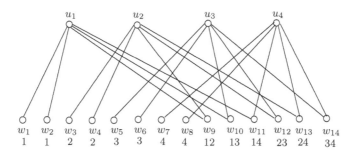

Figure 11.10: The red subgraph in a red-blue coloring of $K_{4,14}$

Next, we show that $BR_4(2,3) \le 15$. That is, we show that every red-blue coloring of $H = K_{4,15}$ results in a red $K_{2,2}$ or a blue $K_{3,3}$. Let there be given a red-blue coloring of H resulting in the red subgraph H_R and the blue subgraph H_B. Suppose that there is no red $K_{2,2}$. We show that there is a blue $K_{3,3}$. Let $U = \{u_1, u_2, u_3, u_4\}$ and $W = \{w_1, w_2, \ldots, w_{15}\}$ be the partite sets of H. First, we claim that W contains at least nine vertices of degree at least 3 in H_B. This is certainly true if the maximum degree of vertices of W in H_R is at most 1. Thus, we may assume that the maximum degree of vertices of W in H_R is 4, 3, or 2. We consider these three cases.

Case 1. The maximum degree of vertices of W in H_R is 4. Since there is no red $K_{2,2}$, it follows that W contains exactly one vertex of degree 4 in H_R, say $\deg_{H_R} w_1 = 4$. This implies that $\deg_{H_R} w_i \le 1$ for each integer $2 \le i \le 15$. Consequently, $\deg_{H_B} w_i \ge 3$ for $2 \le i \le 15$ and so W contains fourteen vertices of degree at least 3 in H_B.

Case 2. The maximum degree of vertices of W in H_R is 3. Since there is no red $K_{2,2}$, it follows that W contains exactly one vertex of degree 3, say $\deg_{H_R} w_1 = 3$ and $N_{H_R}(w_1) = \{u_1, u_2, u_3\}$. If $\deg_{H_R} w_i = 2$ for some integer i with $2 \le i \le 15$, then $N_{H_R}(w_i)$ cannot be a subset of $N_{H_R}(w_1)$, namely $N_{H_R}(w_i)$ cannot be any of the three sets $\{u_1, u_2\}, \{u_1, u_3\}, \{u_2, u_3\}$; for otherwise, there is a red $K_{2,2}$. Since there are only $\binom{4}{2} - 3 = 3$ available 2-element subsets of U for $N_{H_R}(w_i)$ for $2 \le i \le 15$, it follows that W has at most three vertices of degree 2 in H_R. Consequently, W has at least eleven vertices of degree at most 1 in H_R and so W has at least eleven vertices of degree at least 3 in H_B.

Case 3. The maximum degree of vertices of W in H_R is 2. Since there is no red $K_{2,2}$ and there are only six distinct 2-element subsets of U, it follows that W

contains at most six vertices of degree 2 in H_R. Consequently, W contains at least nine vertices of degree at most 1 in H_R and so W contains at least nine vertices of degree at least 3 in H_B.

In each case, W contains at least nine vertices of degree at least 3 in H_B, as claimed. Since there are only four distinct 3-element subsets of U, at least three vertices of W are joined to the same three vertices in U by blue edges, producing a blue $K_{3,3}$. Therefore, $BR_4(2,3) \leq 15$ and so $BR_4(2,3) = 15$. ∎

While $BR_s(F, H)$ has been determined when $F = H = K_{2,3}$ for all $s \geq 2$, there have been only partial results obtained when $H = K_{3,3}$ and $F = K_{2,3}$ or $F = K_{3,3}$. We state these partial results.

Theorem 11.25 *For each integer $s \geq 2$,*

$$BR_s(K_{2,3}, K_{2,3}) = \begin{cases} does\ not\ exist & if\ s = 2 \\ 13 & if\ s = 3,4 \\ 11 & if\ s = 5,6 \\ 9 & if\ s = 7,8 \\ s & if\ s \geq 9. \end{cases}$$

Theorem 11.26 *For each positive integer s,*

$$BR_s(K_{2,3}, K_{3,3}) = \begin{cases} does\ not\ exist & if\ 1 \leq s \leq 3 \\ 21 & if\ s = 4,5 \\ 15 & if\ s = 6,7 \\ 13\ or\ 14 & if\ s = 8,9 \\ 11,12,13\ or\ 14 & if\ 10 \leq s \leq BR(K_{2,3}, K_{3,3}). \end{cases}$$

Furthermore, $11 \leq BR(K_{2,3}, K_{3,3}) \leq 14$.

Theorem 11.27 *For each integer $s \geq 2$,*

$$BR_s(3,3) = \begin{cases} does\ not\ exist & if\ s = 2,3,4 \\ 41 & if\ s = 5,6 \\ 29 & if\ s = 7,8. \end{cases}$$

While $BR_s(3,3)$ is unknown for those integers s with $9 \leq s < BR(3,3)$, the results presented in [18] might suggest that $BR_9(3,3) = BR_{10}(3,3)$. It was shown in [18] that $17 \leq BR_{10}(3,3) \leq 23$.

The concept of the s-bipartite Ramsey number is also related to recreational problems, an example of which is the following:

> *There are five girls at a party. What is the minimum number of boys who must be invited to the party to guarantee that there exists a group of six people, three girls and three boys, such that either (1) every one of the three girls is acquainted with every one of the three boys or (2) every one of the three girls is a stranger of every one of the three boys?*

By Theorem 11.27, the answer to this question is $BR_5(3,3) = 41$.

As a final example of an s-bipartite Ramsey number, we determine the number $BR_s(F, H)$ when F and H are both stars.

Theorem 11.28 *For integers $p, q \geq 2$,*

$$BR_s(K_{1,p}, K_{1,q}) = \begin{cases} p + q - 1 & \text{if } 2 \leq s \leq p + q - 2 \\ s & \text{otherwise.} \end{cases}$$

Proof. We consider two cases, according to whether $2 \leq s \leq p + q - 2$ or $s \geq p + q - 1$.

Case 1. $2 \leq s \leq p + q - 2$. Let there be given a red-blue coloring of $H = K_{s,p+q-1}$ resulting in the red subgraph H_R and the blue subgraph H_B. Let U and W be the partite sets of H with $|U| = s$ and $|W| = p + q - 1$. Now let $u \in U$. If $\deg_{H_R} u \geq p$, then H contains a red $K_{1,p}$; while if $\deg_{H_R} u \leq p - 1$, then $\deg_{H_B} u \geq (p + q - 1) - (p - 1) = q$ and so H contains a blue $K_{1,q}$. Therefore,

$$BR_s(K_{1,p}, K_{1,q}) \leq p + q - 1.$$

Next, we show that there exists a red-blue coloring of $G = K_{s,p+q-2}$ that avoids both a red $K_{1,p}$ and a blue $K_{1,q}$. Let $U = \{u_1, u_2, \ldots, u_s\}$ and $W = \{w_1, w_2, \ldots, w_{p+q-2}\}$ be the partite sets of G. For each integer i with $1 \leq i \leq s$, assign the color red to the $p - 1$ edges $u_i w_i, u_i w_{i+1}, \ldots, u_i w_{i+p-2}$ incident with u_i, where the subscripts of vertices are expressed as integers modulo $p + q - 2$, and assign the color blue to the remaining edges of G.

Let G_R and G_B be the resulting red and blue subgraphs of G, respectively. If $u \in U$, then $\deg_{G_R} u = p - 1$ and $\deg_{G_B} u = q - 1$. Thus, this red-blue coloring of G produces neither a red $K_{1,p}$ nor a blue $K_{1,q}$ whose central vertex belongs to U. By this construction,

$$\max\{\deg_{G_R} w : w \in W\} = \deg_{G_R} w_s \leq p - 1 \tag{11.4}$$

$$\min\{\deg_{G_R} w : w \in W\} = \deg_{G_R} w_{p+q-2} \geq 0. \tag{11.5}$$

Hence, $0 \leq \delta(G_R) \leq \Delta(G_R) \leq p - 1$ and so there is no red $K_{1,p}$ whose central vertex belongs to W. Let $\deg_{G_R} w_{p+q-2} = k$. Since $\deg_{G_R} w + \deg_{G_B} w = s$ for each $w \in W$, it follows that

$$\deg_{G_B} w_{p+q-2} = s - \deg_{G_R} w_{p+q-2} = s - k. \tag{11.6}$$

First, suppose that $k \geq 1$. Since

$$N_{G_R}(u_{s-k+1}) = \{w_{s-k+1}, w_{s-k+2}, \ldots, w_{p+q-2}\}$$

and $\deg_{G_R} u_{s-k+1} = p - 1$, it follows that $(p + q - 2) - (s - k + 1) + 1 = p - 1$ and so $s = q + k - 1$. It then follows by (11.6) that $\deg_{G_B} w_{p+q-2} = q - 1$ and so $\deg_{G_B} w \leq q - 1$ for each $w \in W$ by (11.5). Next, suppose that $k = 0$. Since u_s is adjacent to $w_s, w_{s+1}, \ldots, w_{s+(p-2)}$, it follows that $s + (p - 2) < p + q - 2$ and so $s \leq q - 1$. Hence, $\deg_{G_B} w \leq \deg_G w = s \leq q - 1$ for each $w \in W$. Therefore, there is no blue $K_{1,q}$ whose central vertex belongs to W.

Hence, there is neither a red $K_{1,p}$ nor a blue $K_{1,q}$ in G. Therefore,

$$BR_s(K_{1,p}, K_{1,q}) \geq p + q - 1$$

and so $BR_s(K_{1,p}, K_{1,q}) = p + q - 1$ when $2 \leq s \leq p + q - 2$.

Case 2. $s \geq p + q - 1$. We show that every red-blue coloring of $H = K_{s,s}$ produces either a red $K_{1,p}$ or a blue $K_{1,q}$. Let there be given a red-blue coloring of H resulting in the red subgraph H_R and the blue subgraph H_B. Let U and W be the partite sets of H with $|U| = |W| = s$. Let v be any vertex of H, say $v \in U$. If $\deg_{H_R} v \geq p$, then H contains a red $K_{1,p}$; while if $\deg_{H_R} v \leq p - 1$, then

$$\deg_{H_B} v \geq s - (p - 1) \geq (p + q - 1) - (p - 1) = q$$

and so H contains a blue $K_{1,q}$. Therefore, $BR_s(K_{1,p}, K_{1,q}) = s$. ■

Exercises for Chapter 11

1. For a red-blue coloring c of K_6, let t_c denote the number of monochromatic triangles produced. By Theorem 11.3, $t_c \geq 1$ for every red-blue coloring c of K_6. Determine $\min\{t_c\}$ over all red-blue colorings c of K_6.

2. Determine $R(C_3, C_4)$.

3. Prove that $R(C_4, C_4) = 6$.

4. For the graphs F and H shown in Figure 11.2, it is shown in Theorem 11.4 that $R(F, H) = 7$. Determine $R(F, F)$ and $R(H, H)$.

5. Let F and H be two nontrivial graphs, where $x \in V(F)$ and $y \in V(H)$. Suppose that F' is isomorphic to $F - x$ and H' is isomorphic to $H - y$. Prove that $R(F, H) \leq R(F', H) + R(F, H')$.

6. Let F and H be two graphs and let maroon, scarlet, and blue be three colors. By the Ramsey number $R_{(2,1)}(F, H)$ is meant the minimum positive integer n such that if every edge of $G = K_n$ is colored with one of the three colors maroon, scarlet, and blue, there is either a subgraph of G isomorphic to F in which every edge is colored maroon or scarlet (a maroon-scarlet F) or a blue H. How is $R_{(2,1)}(F, H)$ related to $R(F, H)$?

7. Let F and H be two connected graphs, where F is a proper subgraph of H. Let $MR(F, H)$ denote the minimum positive integer n such that every red-blue coloring of K_n results in a red F, a red H, a blue F, or a blue H.

   (a) How is $MR(F, H)$ related to at least one of $R(F, F)$, $R(F, H)$ or $R(H, H)$?

   (b) The question in (a) should suggest another question to you.

8. For integers s and t with $1 \leq s \leq t$, define the Ramsey chromatic number $R_\chi(s, t)$ to be the smallest positive integer n such that for every red-blue coloring of K_n, either the red subgraph has chromatic number at least s or the blue subgraph has chromatic number at least t. Trivially, $R_\chi(1, t) = 1$ and $R_\chi(2, t) = t$.

(a) Does $R_\chi(s,t)$ exist for all integers s and t with $1 \leq s \leq t$?

(b) Prove that $R_\chi(s,t) \leq R_\chi(s-1,t) + R_\chi(s,t-1)$.

(c) Determine $R_\chi(3,3)$.

(d) For $3 \leq s \leq t$, determine a formula for $R_\chi(s,t)$ in terms of s and t.

9. Prove Theorem 11.11 *For every $k \geq 2$ graphs $G_1, G_2, \ldots, G_k$, the Ramsey number $R(G_1, G_2, \ldots, G_k)$ exists.*

10. Let G_1, G_2, G_3 be three nontrivial connected graphs. Define $TR(G_1, G_2, G_3)$ to be the minimum positive integer n such that every red-blue-green coloring of K_n results in a subgraph isomorphic to G_1 where no edge is colored green, a subgraph isomorphic to G_2 where no edge is colored red, or a subgraph isomorphic to G_3 where no edge is colored blue.

   (a) For which graphs G_1, G_2, G_3 is $TR(G_1, G_2, G_3)$ defined?

   (b) What question is suggested by your answer in (a)?

11. Determine $BR(K_{1,s}, K_{1,t})$ for positive integers s and t.

12. Determine $BR(P_4, P_4)$.

13. Determine $BR(P_5, P_5)$.

14. Determine $BR(K_{1,3}, C_4)$.

15. Give an example of bipartite graph F and H, neither of which is a star, such that $BR(F,H) \geq 6$. For your choice of F and H, find the exact value of $BR(F,H)$.

16. In the proof of Example 11.15, where it is shown that $BR(C_4, C_4) = 5$, we saw that there is a red-blue coloring of $K_{4,4}$ that results in a red C_8 and a blue C_8. In other words, $K_{4,4}$ is C_8-decomposable into two copies of C_8. This allows us to conclude that $BR(C_4, C_4) \geq 5$.

   (a) Use this observation to find other pairs F, H of bipartite graphs for which $BR(F,H) \geq 5$.

   (b) Recall that a graph G is *self-complementary* if $\overline{G} \cong G$. Let G be a bipartite graph of even order $2r$, each of whose partite sets consists of r vertices. The *bicomplement* $\overline{\overline{G}}$ of G is the graph with $V(\overline{\overline{G}}) = V(G)$ and $E(\overline{\overline{G}}) = E(K_{r,r}) - E(G)$. Then the graph G is *self-bicomplementary* if $\overline{\overline{G}} \cong G$. Give an example of a self-bicomplementary graph of order at least 8 that is distinct from C_8

   (c) Use the example given in (b) to determine two bipartite graphs F and H for which there is a lower bound for $BR(F,H)$.

17. Let F and H be two bipartite graphs. Prove for every integer k with $2 \leq k \leq R(F,H)$ that $R_k(F,H) \geq R(F,H)$.

18. Let F and H be two bipartite graphs. Prove for positive integers k and ℓ with $k \geq 2$ that $R_{\ell k}(F, H) \leq R_k(F, H)$.

19. Prove that $BR_s(K_{2,3}K_{2,3})$ does not exist when $s = 1, 2$.

20. Prove that $BR_s(K_{2,3}, K_{3,3})$ does not exist for $s = 1, 2, 3$.

21. Let $m, n, s \geq 2$ be integers. Prove the following.

    (a) If $2 \leq s \leq m + n - 2$, then $BR_s(mK_2, nK_2)$ does not exist.
    (b) If $s \geq m + n - 1$, then $BR_s(mK_2, nK_2) = s$.

22. Prove for integers m, n, s with $2 \leq n \leq s \leq m - 1$ that $BR_s(K_{1,m}, nK_2) = m + n - 1$.

23. Let n and s be integers with $n \geq 5$ and $s \geq 2$. Prove the following.

    (a) If n is odd, then $BR_s(P_n, P_n)$ exists only when $s \geq n - 2$.
    (b) If n is even, then $BR_s(P_n, P_n)$ exists only when $s \geq n - 1$.

24. Prove that if F is a bipartite graph such that $BR_k(F, F) = k + 1$ for some positive integer k, then $BR_s(F, F) = s$ for all integers $s \geq k + 1$.

25. Let $s \geq 2$ be an integer. Prove the following.

    (a) If $s = 3, 4$, then $BR_s(P_5, P_5) = 5$.
    (b) If $s \geq 5$, then $BR_s(P_5, P_5) = s$.

26. Prove that $BR_s(P_6, P_6) = s$ for each integer $s \geq 5$.

27. For integers b and s with $b \geq 2$ and $s \geq 2b - 1$, prove that $BR_s(S_{2,b}) = s$.

28. While the Ramsey number $R(K_6, K_6)$ is unknown, it is known that it exists, that is, $R(K_6, K_6) = k$ for some integer k. Prove that if every edge of K_k is colored with one of the three colors red, yellow, and green, then there is either a red K_6, a yellow K_3, or a green K_3.

29. Suppose that $R(5,5) = k$ and $R(k, k) = \ell$. Prove that if every edge of K_ℓ is colored with one of three colors red, yellow, and green, then there is a monochromatic K_5.

30. A sequence $G_1, G_2, G_3, \ldots$ of graphs is a **Ramsey sequence** if (1) G_i is isomorphic to a proper subgraph of G_{i+1} for every positive integer i and (2) for every positive integer r, there is an integer $s > r$ such that every red-blue coloring of G_s results in a monochromatic G_r. For example, the sequences $K_1, K_2, K_3, \ldots$ and $K_{1,1}, K_{2,2}, K_{3,3}, \ldots$ are both Ramsey sequences.

    (a) Give examples of other Ramsey sequences.
    (b) Give examples of sequences $G_1, G_2, G_3, \ldots$ of graphs that satisfy (1) but are not Ramsey sequences.

Chapter 12

Monochromatic Ramsey Theory

For two graphs F and H, we have seen that the Ramsey number $R(F, H)$ of F and H is the minimum positive integer n such that every red-blue coloring of K_n produces either a red F or a blue H. Consequently, each edge of a subgraph of K_n that is isomorphic to F must have a specific color, namely red, and each edge of a subgraph isomorphic to H must be blue. This also implies that for every red-blue coloring of K_n, there is a particular monochromatic F or a particular monochromatic H. There are other Ramsey numbers where the goal of a red-blue coloring is to require that either only one of F and H, say F, is monochromatic or H satisfies some other coloring condition. This is the primary topic of this chapter.

12.1 Monochromatic Ramsey Numbers

Recall that the Ramsey number $R(F, H)$ of two graphs F and H is the minimum positive integer n such that every red-blue coloring of K_n results in a red F or a blue H. Therefore, every red-blue coloring of K_n results in a monochromatic F or a monochromatic H. This observation corresponds to a related Ramsey number. The **monochromatic Ramsey number** $MR(F, H)$ of F and H is the minimum positive integer n such that every red-blue coloring of K_n produces either a monochromatic F or a monochromatic H. Here it doesn't matter what color the edges of F (or H) are, provided all colors are the same. Therefore, $MR(F, F) = R(F, F)$. Since the requirement of this coloring is weaker than that of the standard Ramsey number, it follows that

$$\min\{|V(F)|, |V(H)|\} \leq MR(F, H) \leq \min\{R(F, H), R(F, F), R(H, H)\}. \quad (12.1)$$

Another way to look at the monochromatic Ramsey number $MR(F, H)$ is the following. Suppose that $MR(F, H) = n$. Then, for every partition of $E(G)$ where $G = K_n$ into two sets E_1 and E_2, at least one of $G[E_1]$ and $G[E_2]$ contains a subgraph that is isomorphic to either F or H. On the other hand, there is some partition of $E(G')$ where $G' = K_{n-1}$ into two sets E'_1 and E'_2 such that neither $G'[E'_1]$ nor $G'[E'_2]$ contains a subgraph isomorphic to either F or H.

Let's look at an example of a monochromatic Ramsey number.

Example 12.1 $MR(C_3, P_4) = 4$.

Proof. Since K_3 can be decomposed into P_3 and K_2 (neither of which contains C_3 or P_4 as a subgraph), it follows that $MR(C_3, P_4) \geq 4$. Next, let there be given a red-blue of K_4. Let v be a vertex of K_4. At least two of the three edges incident with v are colored the same, say red. Let v_1, v_2, v_3 be the remaining vertices of K_4. We consider two cases.

Case 1. All three edges vv_1, vv_2, vv_3 are red. If any of v_1v_2, v_1v_3, v_2v_3 is red, then there is a red C_3; if not then there is a blue C_3.

Case 1. Exactly two of vv_1, vv_2, vv_3 are red, say vv_1 and vv_2 are red and vv_3 is blue. If v_1v_2 is red, then there is a red C_3. Thus, we may assume that v_1v_2 is blue. If v_1v_3 or v_2v_3 is blue, then there is a blue P_4; otherwise, both edges are red and there is a red P_4.   ∎

By Theorems 11.8–11.10, it follows that

$$R(P_4, P_4) = 5, \ R(C_3, C_3) = 6, \text{ and } R(C_3, P_4) = 7.$$

Therefore, the inequality in (12.1) can be strict. This brings up the following problem.

Problem 12.2 *For graphs F and H, let $M = \min\{R(F, H), R(F, F), R(H, H)\}$. We saw for $F = C_3$ and $H = P_4$ that $M - MR(F, H) = 1$. Does there exist an integer $k \geq 2$ such that $M - MR(F, H) = k$ for some pair F, H of graphs?*

Observation 12.3 *Let F and H be nonempty graphs.*

(a) *If $F \subseteq H$, then $MR(F, H) = MR(F, F) = R(F, F)$.*

(b) *If F' and H' are nonempty subgraphs of F and H, respectively, then*

$$MR(F', H') \leq MR(F, H).$$

Since $R(P_m, P_n) = \lfloor m/2 \rfloor + n - 1$ for integers m and n with $2 \leq m \leq n$ by Theorem 11.8, the following is an immediate consequence of Observation 12.3(a).

Theorem 12.4 *For integers m and n with $2 \leq m \leq n$,*

$$MR(P_m, P_n) = \lfloor 3m/2 \rfloor - 1.$$

The following two results give the monochromatic Ramsey numbers of every two cycles.

Theorem 12.5 $MR(C_3, C_5) = 5$, $MR(C_4, C_5) = 6$, *and*

$$MR(C_3, C_n) = MR(C_4, C_n) = 6 \text{ for } n \geq 6.$$

Theorem 12.6 *For $5 \leq m \leq n$,*

$$MR(C_m, C_n) = \begin{cases} \frac{3m}{2} - 1 & \text{if } m \text{ is even} \\ 2m - 1 & \text{if } m \text{ is odd and either } (i) \text{ } n \text{ is odd or} \\ & (ii) \text{ } n \text{ is even and } n \geq 2m. \end{cases}$$

If m is odd and n is even with $m + 1 \leq n \leq 2m$, then

$$m + \frac{n}{2} - 1 \leq MR(C_m, C_n) \leq \min\left\{2m - 1, \frac{3n}{2} - 1\right\}. \tag{12.2}$$

Proof. We consider the following two cases, according to the parity of m.

Case 1. m is even. Since the largest monochromatic cycle in the red-blue coloring of $K_{3m/2-2}$ produces a red $K_{m/2-1} + K_{m-1}$ and a blue $K_{m/2-1,m-1}$ is C_{m-1}, it follows that $MR(C_m, C_n) > 3m/2 - 2$. On the other hand, $R(C_m, C_m) = 3m/2 - 1$ for $m \geq 6$ and so $MR(C_m, C_n) = 3m/2 - 1$ by (12.1).

Case 2. m is odd. Then $MR(C_m, C_n) \leq R(C_m, C_m) = 2m - 1$. If n is odd or $n \geq 2m$, then the red-blue coloring of K_{2m-2} producing a red $2K_{m-1}$ and a blue $K_{m-1,m-1}$ shows that $MR(C_m, C_n) \geq 2m-1$. If n is even and $m+1 \leq n \leq 2m-2$, then the red-blue coloring of $K_{m+n/2-2}$ producing a red $K_{m-1} + K_{n/2-1}$ and a blue $K_{m-1,n/2-1}$ shows that $MR(C_m, C_n) \geq m+n/2-1$. Furthermore, $MR(C_m, C_n) \leq R(C_n, C_n) = 3n/2 - 1$. Thus, (12.2) holds. ∎

If F is a cycle and H is a path, then we have the following result [7].

Theorem 12.7 *For integers $m \geq 5$ and $n \geq 4$,*

$$MR(C_m, P_n) = \begin{cases} \lfloor 3n/2 \rfloor - 1 & \text{if } n \leq m \\ \frac{3m}{2} - 1 & \text{if } n \geq m + 1 \text{ and } m \text{ is even} \\ 2m - 1 & \text{if } n \geq 2m - 1 \text{ and } m \text{ is odd}. \end{cases}$$

If m is odd and $m + 1 \leq n \leq 2m - 2$, then

$$m + \lfloor n/2 \rfloor - 1 \leq MR(C_m, P_n) \leq \min\{2m - 1, \lfloor 3n/2 \rfloor - 1\}. \tag{12.3}$$

Proof. If $n \leq m$, then $MR(C_m, P_n) = R(P_n, P_n) = \lfloor 3n/2 \rfloor - 1$. Thus, we may assume that $5 \leq m < n$. If m is even, then the red-blue coloring of $K_{3m/2-2}$ producing a red $K_{m/2-1}+K_{m-1}$ and a blue $K_{m/2-1,m-1}$ shows that $MR(C_m, P_n) > 3m/2 - 2$. On the other hand, $MR(C_m, P_n) \leq MR(C_m, C_n) = 3m/2 - 1$. If m is odd, then $MR(C_m, P_n) \leq R(C_m, C_m) = 2m - 1$. In the red-blue coloring of K_{2m-2} producing a red $2K_{m-1}$ and a blue $K_{m-1,m-1}$, a longest monochromatic path is of order $2m - 2$ and a longest monochromatic odd cycle is of order $m - 2$. Therefore, if m is odd and $n \geq 2m - 1$, then $MR(C_m, P_n) = 2m - 1$.

It remains to consider the case where m is odd and $m + 1 \leq n \leq 2m - 2$. In the red-blue coloring of $K_{m+\lfloor n/2 \rfloor - 2}$ resulting in a red $K_{m-1} + K_{\lfloor n/2 \rfloor - 1}$ and a blue $K_{m-1,\lfloor n/2 \rfloor - 1}$, a longest monochromatic path is of order $n-1$ and a longest monochromatic odd cycle is of order $m-2$. Therefore, (12.3) follows by (12.1). ∎

There are many variations of the monochromatic Ramsey number. For example, for graphs $F_1, F_2, \ldots, F_k$ where $k \geq 2$, the k-**monochromatic Ramsey number** $MR_k(F_1, F_2, \ldots, F_k)$ of the graphs $F_1, F_2, \ldots, F_k$ is the minimum positive integer n such that every k-edge coloring of $G = K_n$ results in a monochromatic F_i for some integer i with $1 \leq i \leq k$. Equivalently, for every partition $\{E_1, E_2, \ldots, E_k\}$ of $E(G)$, at least one of these subgraphs $G[E_i]$ contains a subgraph isomorphic to at least one of the graphs $F_1, F_2, \ldots, F_k$; while for $H = K_{n-1}$, there exists a partition $\{E_1', E_2', \ldots, E_k'\}$ of $E(H)$ such that none of the subgraphs $H[E_i']$ $(1 \leq i \leq k)$ contains a subgraph isomorphic to any of the graphs $F_1, F_2, \ldots, F_k$.

There is also a bipartite version of monochromatic Ramsey numbers. For bipartite graphs F and H, the **monochromatic bipartite Ramsey number** $MBR(F, H)$ is the minimum positive integer r such that every red-blue coloring of $K_{r,r}$ produces either a monochromatic F or a monochromatic H. Similar to the inequality in (12.1), we have

$$MBR(F, H) \leq \min\{BR(F, H), BR(F, F), BR(H, H)\}. \tag{12.4}$$

An immediate problem from (12.4) is the following.

Problem 12.8 *Find bipartite graphs F and H such that*

$$MBR(F, H) < \min\{BR(F, H), BR(F, F), BR(H, H)\}.$$

A monochromatic edge coloring of a graph is only one of several possible edge colorings of a graph. Let H be a graph of size 2 or more and let $\pi = \{E_1, E_2\}$ be a partition of $E(H)$. A π-**coloring** of H is an edge coloring of H with two colors such that the resulting color classes are E_1 and E_2. Let F and H be two graphs where $\pi = \{E_1, E_2\}$ is a partition of $E(H)$. The π-**Ramsey number** $R_\pi(F, H)$ is the minimum positive integer n such that every red-blue coloring of K_n produces either a monochromatic F or a π-colored H, where the sets E_1 and E_2 are the two color classes (up to symmetry) of a subgraph isomorphic to H in K_n. In this case, a specific color, say red, cannot be specified for F since if all edges are colored blue for every complete graph, then the desired condition will never be satisfied. If either E_1 or E_2 is permitted to be empty, then $R_\pi(F, H) = MR(F, H)$. The concept of the π-Ramsey number can also be defined when $E(H)$ is partitioned into more than two sets and, consequently, when the edge coloring involves more than two colors. There is also a bipartite version for this type of Ramsey number.

Let H be a nonempty graph without isolated vertices. The monochromatic Ramsey number $MR(H, H)$ is the smallest integer n such that *every* red-blue coloring of K_n results in a monochromatic subgraph that is isomorphic to H. Suppose that $MR(H, H) = n$. Then for every red-blue coloring of K_n, there is a monochromatic H. Therefore, we refer H as a **monochromatic graph with respect to** n. A monochromatic graph with respect to n is **maximal** if it is not a proper subgraph of any other monochromatic graph with respect to n. The set of all maximal monochromatic graphs with respect to n is denoted by $\mathcal{M}_n$. There is a corresponding concept in the case of bipartite graphs and the complete bipartite graphs $K_{r,r}$ and a corresponding set $\mathcal{MB}_r$ for positive integers r.

12.2   Monochromatic-Bichromatic Ramsey Numbers

Let F and H be two graphs where the size of H is at least 2 and let $\pi = \{E_1, E_2\}$ be a partition of $E(H)$. A π-**coloring of** H is an edge coloring of H with two colors such that the resulting color classes are E_1 and E_2. The π-**Ramsey number** $R_\pi(F, H)$ of F and H is the minimum positive integer n such that every red-blue coloring of K_n produces either a monochromatic F or a π-colored H, where the sets E_1 and E_2 are the two color classes (up to symmetry) of a subgraph isomorphic to H in K_n. In this case, a specific color, say red, cannot be specified for F since if all edges are colored blue for every complete graph, then the desired condition will never be satisfied. If either E_1 or E_2 is permitted to be empty, then $R_\pi(F, H)$ is the monochromatic Ramsey number $MR(F, H)$.

In this section, we determine all π-Ramsey numbers $R_\pi(F, H)$, where $F, H \in \{C_3, C_4\}$. We begin with $F = H = C_3$. For a red-blue coloring of C_3, there is only one possible partition π of its edge set into two nonempty sets, namely the partition in which exactly two edges of C_3 are colored the same. Since every red-blue coloring of K_3 produces either a monochromatic C_3 or a π-colored C_3, it follows that $R_\pi(C_3) = 3$. Indeed, this is a special case of the following.

Observation 12.9   *If F is a graph of order n, then $R_\pi(F, C_3) = n$.*

We now consider the π-Ramsey numbers $R_\pi(F, H)$ where $H = C_4$. For a red-blue coloring of C_4, there are three possible partitions π_i $(1 \le i \le 3)$ of its edge set, all shown in Figure 12.1. For $1 \le i \le 3$, we compute the π_i-Ramsey numbers $R_{\pi_i}(C_3, C_4)$ and $R_{\pi_i}(C_4, C_4)$, beginning with the first of these.

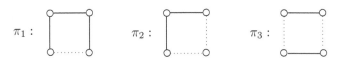

Figure 12.1: Three bichromatic graphs C_4

The following is a useful observation.

Observation 12.10   *A red-blue coloring of a complete graph of order at least 4 resulting in a vertex that is incident with three or more edges of the same color produces a monochromatic C_3.*

For a given red-blue coloring of a complete graph G, where we denote the subgraphs induced by the set of red edges and the set of blue edges by G_R and G_B, respectively.

Theorem 12.11   (a) $R_{\pi_1}(C_3, C_4) = 5$, (b) $R_{\pi_2}(C_3, C_4) = 6$, (c) $R_{\pi_3}(C_3, C_4) = 4$.

Proof.   (a)   The red-blue coloring of K_4 such that $G_R = C_4$ has neither a monochromatic C_3 nor a π_1-colored C_4; so $R_{\pi_1}(C_3, C_4) \geq 5$. We now show that $R_{\pi_1}(C_3, C_4) \leq 5$. Let there be given a red-blue coloring of K_5 avoiding monochromatic triangles. By Observation 12.10, we may assume that this coloring induces a red C_5 and a blue C_5. Then there is a π_1-colored C_4.

(b)   The red-blue coloring of K_5 that induces a red C_5 and a blue C_5 contains neither a monochromatic C_3 nor a π_2-colored C_4 and so $R_{\pi_2}(C_3, C_4) \geq 6$. Thus, $R_{\pi_2}(C_3, C_4) \leq 6$ by Observation 12.10.

(c)   Since the red-blue coloring of K_3 using both colors produces neither a monochromatic C_3 nor a π_3-colored C_4, it follows that $R_{\pi_3}(C_3, C_4) \geq 4$. Now consider a red-blue coloring of K_4 that avoids a monochromatic triangle. Then by Observation 12.10, either $G_R = G_B = P_4$ or $\{G_R, G_B\} = \{C_4, 2K_2\}$ and so there is a π_3-colored C_4.   ∎

Theorem 12.12   (a) $R_{\pi_1}(C_4, C_4) = 5$, (b) $R_{\pi_2}(C_4, C_4) = 6$, (c) $R_{\pi_3}(C_4, C_4) = 5$.

Proof.   (a)   First, the red-blue coloring of K_4 such that $G_R = K_{1,3}$ contains neither a monochromatic C_4 nor a π_1-colored C_4 and so $R_{\pi_1}(C_4, C_4) \geq 5$. It remains to show that $R_{\pi_1}(C_4, C_4) \leq 5$. Let there be given a red-blue coloring of K_5 where $V(K_5) = \{v_1, v_2, \ldots, v_5\}$ for which there is no π_1-colored C_4. By Theorem 12.11(a), there is a monochromatic C_3, say (v_1, v_2, v_3, v_1) is a red C_3. If any two edges of $A = [\{v_4\}, \{v_1, v_2, v_3\}]$ or $B = [\{v_5\}, \{v_1, v_2, v_3\}]$ are red, then there is a monochromatic C_4. So, at least two edges of A and B are blue. If one edge of A is red, there is a π_1-colored C_4; so all edges of A and B are blue, resulting a monochromatic C_4.

(b)   First, the red-blue coloring of K_5 producing a red C_5 and a blue C_5 contains neither a monochromatic C_4 nor a π_2-colored C_4 and so $R_{\pi_2}(C_4, C_4) \geq 6$. Next, let there be given a red-blue coloring of K_6 where $V(K_6) = \{v_1, v_2, \ldots, v_6\}$. Thus, each vertex is incident with three edges of the same color, say v_1v_2, v_1v_3 and v_1v_4 are colored red. If two of the three edges in $S = \{v_2v_3, v_2v_4, v_3v_4\}$ are colored red, then a monochromatic C_4 is produced; while if there are two edges in S that are colored blue, then a π_2-colored C_4 is produced.

(c)   First, the red-blue coloring of K_4 with $G_R = K_{1,3}$ contains neither a monochromatic C_4 nor a π_3-colored C_4 and so $R_{\pi_3}(C_4, C_4) \geq 5$. To show that $R_{\pi_3}(C_4, C_4) \leq 5$, consider an arbitrary red-blue coloring of K_5 where $V(K_5) = \{v_1, v_2, \ldots, v_5\}$. If every vertex is incident with two red edges and two blue edges, then $G_R = G_B = C_5$ and a π_3-colored C_4 is produced. Thus, we may assume that some vertex of K_5 is incident with three edges of the same color, say v_1v_2, v_1v_3 and v_1v_4 are colored red. If two of the three edges v_2v_3, v_2v_4 and v_3v_4 are also colored red, then there is a red C_4. Hence, assume that v_2v_3 and v_2v_4 are blue edges. If v_3v_5 and v_4v_5 are assigned the same color, then there is a monochromatic C_4. Otherwise, we may assume that v_3v_5 is a red edge while v_4v_5 is a blue edge. If v_2v_5 is a red edge, then $(v_1, v_2, v_5, v_3, v_1)$ is a red C_4. Hence, we may assume that v_2v_5 is colored blue. If v_3v_4 is a blue edge, then $(v_2, v_3, v_4, v_5, v_2)$ is a blue C_4.

Thus, we may assume that v_3v_4 is red. If v_1v_5 is red, then a monochromatic C_4 is produced; while if v_1v_5 is blue, then a π_3-colored C_4 is produced. This completes the proof.   ∎

The most studied Ramsey numbers are the classical Ramsey numbers, which deal with complete graphs. We thus consider the complete graph K_4. For an edge-colored K_4, each of whose edges is colored red or blue, there are five partitions $\pi_1, \pi_2, \ldots, \pi_5$ of the edge set, shown in Figure 12.2.

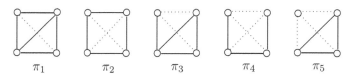

Figure 12.2: The five bichromatic partitions π_i $(1 \leq i \leq 5)$ of $E(K_4)$

We next determine the values of $R_{\pi_i}(K_4, K_4)$ for $1 \leq i \leq 5$. Here it is convenient to recall that $R(K_4, K_4 - e) = 10$ (see [160]). It turns out that $R_{\pi_i}(K_4, K_4) = 10$ for all bichromatic partitions π_i, with one exception. We verify this for $i = 1, 5$.

Theorem 12.13 $R_{\pi_1}(K_4, K_4) = R_{\pi_5}(K_4, K_4) = 10$.

Proof.   Since the complete graph K_9 can be decomposed into two copies of $C_3 \,\square\, C_3$, which has clique number 3 and does not contain a π_1-colored K_4, it follows that $R_{\pi_1}(K_4, K_4) \geq 10$. Let there be given a red-blue coloring of K_{10}. Since $R(K_4, K_4 - e) = 10$, there is either a red K_4 or a blue $K_4 - e$. If there is a red K_4, then it is a monochromatic K_4. If there is a blue $K_4 - e$, then there is a monochromatic K_4 if e is blue or a π_1-colored K_4 if e is red. In either case, $R_{\pi_1}(K_4, K_4) \leq 10$ and so $R_{\pi_1}(K_4, K_4) = 10$.

Next we verify that $R_{\pi_5}(K_4, K_4) \leq 10$. Suppose that there is a red-blue coloring of K_{10} that produces neither a monochromatic K_4 nor a π_5-colored K_4. We claim that no vertex in K_{10} has degree at least 6 in either G_R or G_B. If K_{10} should contain a vertex v_0 and six edges v_0v_i $(1 \leq i \leq 6)$, all of the same color, then the subgraph of K_{10} induced by $\{v_1, v_2, \ldots, v_6\}$ contains a monochromatic K_3 (since $R(K_3, K_3) = 6$), implying that K_{10} contains either a monochromatic K_4 or a π_5-colored K_4, a contradiction. Thus, every vertex of K_{10} has degree 4 or 5 in both G_R and G_B.

Since no graph has an odd number of odd vertices, there are at least six vertices of degree 5 in either G_R or G_B, say the former. Then there are two adjacent vertices v_1 and v_2 in G_R, each incident with five red edges. Since the only red-blue coloring of K_5 avoiding a monochromatic K_3 is the one resulting in a red C_5 and a blue C_5, it follows that the red subgraph induced by $N_{G_R}(v_i)$ is a 5-cycle for $i = 1, 2$. Suppose that $N_{G_R}(v_1) = \{v_2, v_3, v_4, v_5, v_6\}$ and the subgraph induced by $N_{G_R}(v_1)$ is $C_5 = (v_2, v_3, v_4, v_5, v_6, v_2)$. Thus, each of the remaining five edges joining two vertices in $N_{G_R}(v_1)$ is blue. In particular, v_3v_6 is blue. Let $N_{G_R}(v_2) = \{v_1, v_3, v_6, x, y\}$, where $x \neq y$ and $x, y \notin \{v_4, v_5\}$. Hence, v_1x and v_1y are blue edges. We may assume that the red subgraph induced by $N_{G_R}(v_2)$

is the 5-cycle $(v_1, v_6, x, y, v_3, v_1)$ and so each of the remaining five edges joining two vertices in $N_{G_R}(v_2)$ is blue. This implies that the red subgraph induced by $N_{G_R}[v_1] \cup N_{G_R}[v_2]$ is shown in Figure 12.3.

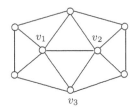

Figure 12.3: The red subgraph induced by
$N_{G_R}[v_1] \cup N_{G_R}[v_2]$ in the proof of Theorem 12.13

Let v_3 be one of the two vertices adjacent to both v_1 and v_2 in G_R. Since v_3 is incident with at most five red edges, there exists a vertex v_4 such that $v_i v_4$ is a blue edge for $1 \leq i \leq 3$. However then, the subgraph induced by $\{v_1, v_2, v_3, v_4\}$ is a π_5-colored K_4. This produces a contradiction. ∎

Theorem 12.14 $R_{\pi_2}(K_4, K_4) = 8$ and $R_{\pi_3}(K_4, K_4) = R_{\pi_4}(K_4, K_4) = 10$.

12.3 Proper Ramsey Numbers

Let F and H be two nontrivial connected graphs such that $\chi'(H) = t$. The **proper Ramsey number** $PR(F, H)$ of F and H is the smallest positive integer n such that every t-edge coloring of K_n results in either a monochromatic F or a properly colored H. This concept was introduced by Chartrand and first studied in [71, 72]. Since the Ramsey number $R(F_1, F_2, \ldots, F_t)$ where $F_i \cong F$ for $1 \leq i \leq t$ exists and $PR(F, H) \leq R(F_1, F_2, \ldots, F_t)$, it follows that the proper Ramsey number $PR(F, H)$ exists for every two graphs F and H. In fact, for each positive integer k, the k-**proper Ramsey number** $PR_k(F, H)$ of F and H is the minimum positive integer n such that every k-edge coloring of K_n results in either a monochromatic F or a properly colored H. The following observation is often useful.

Observation 12.15 *Let F and H be two nontrivial connected graphs and k a positive integer.*

(a) *If $k = 1 < \chi'(H)$, then $PR_1(F, H) = |V(F)|$.*

(b) *If $2 \leq k < \chi'(H)$, then $PR_k(F, H) = R(F_1, F_2, \ldots, F_k)$ where $F_i = F$ for $1 \leq i \leq k$.*

(c) *If $k = \chi'(H)$, then $PR_k(F, H) = PR(F, H)$.*

Here, we investigate the k-proper Ramsey number $PR_k(F, H)$ for several pairs F, H of connected graphs of order at least 3 where $\chi'(H) = 2$. For each such pair then,

$$|V(F)| \leq PR_k(F, H) \leq R(F, F). \tag{12.5}$$

First, we consider the numbers $PR_k(K_n, P)$ for integers $n \geq 3$ and positive integers k and for some paths P of small order. Of course, $\chi'(P) = 2$ for every path P of order at least 3. We begin with $PR_k(K_n, P_3)$.

Proposition 12.16 *Let $n \geq 3$ be an integer. Then $PR_k(K_n, P_3) = n$ for every positive integer k.*

Proof. Since $PR_1(K_n, P_3) = n$ by Observation 12.15, we may assume that $k \geq 2$. By (12.5), $PR_k(K_n, P_3) \geq n$. If a k-edge coloring of K_n assigns a single color to each edge, then a monochromatic K_n is produced. If a k-edge coloring of K_n assigns at least two colors to the edges of K_n, then two adjacent edges of K_n will be colored differently, resulting a properly colored P_3. Hence, $PR_k(K_n, P_3) = n$ for every integer $k \geq 2$. ∎

Theorem 12.17 *Let $n \geq 3$ be an integer. Then*

$$PR_k(K_n, P_4) = \begin{cases} n & \text{if } k = 1 \\ n+1 & \text{if } k \geq 2. \end{cases}$$

Proof. Since $PR_1(K_n, P_4) = n$ by Observation 12.15, we now assume that $k \geq 2$. First, suppose that $n = 3$. A k-edge coloring of K_3 in which at least two colors are used produces neither a monochromatic K_3 nor a properly colored P_4 and so $PR_k(K_3, P_4) \geq 4$. Let there be given a k-edge coloring c of K_4 where $V(K_4) = \{v_1, v_2, v_3, v_4\}$ and for which there is no monochromatic K_3. We may assume that $c(v_1v_2) = 1$, $c(v_1v_3) \neq 1$, and $c(v_2v_3) \neq 1$. Since at least one of v_1v_4 and v_2v_4 is not colored 1, there is a properly colored P_3. Hence, $PR_k(K_3, P_4) = 4$ for every integer $k \geq 2$.

We now assume that $n \geq 4$. Let v be a vertex of the graph K_n. A k-edge coloring of K_n in which each edge incident with v is colored 1 and all other edges of K_n are colored 2 has neither a monochromatic K_n nor a properly colored P_4. Hence, $PR_k(K_n, P_4) \geq n + 1$.

It remains to show that $PR_k(K_n, P_4) \leq n + 1$. Assume, to the contrary, that there is a k-edge coloring of $G = K_{n+1}$ which avoids both a monochromatic K_n and a properly colored P_4. By Proposition 12.16, there is a properly colored P_3, say (u, v, w), where uv is colored 1 and vw is colored 2. Let X be the set consisting of the remaining $n - 2$ vertices of G. Since there is no properly colored P_4 in G, the edge xu is colored 1 for each $x \in X$ and xw is colored 2 for each $x \in X$. Assume first that uw is colored either 1 or 2, say the former. Hence, xv is not colored 1 for each $x \in X$. Since there is no properly colored P_4 in G, all edges xv are colored 2 for each $x \in X$. This is illustrated in Figure 12.4, where an edge colored 1 is indicated by a solid line and an edge colored 2 is indicated by a dashed line.

If some edge of $G[X]$ is not colored 2, then there is a properly colored P_4. If all edges of $G[X]$ are colored 2, then $G[X \cup \{v, w\}]$ is a monochromatic K_n, a contradiction. Thus, $PR_k(K_n, P_4) = n + 1$. ∎

The proper Ramsey number of K_n and P_5 (and P_6) was determined in [72].

Theorem 12.18 *For every integer $n \geq 4$, $PR(K_n, P_5) = PR(K_n, P_6) = 2n - 2$.*

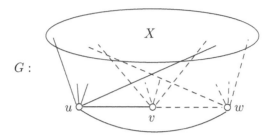

Figure 12.4: A k-edge coloring of $G = K_{n+1}$

We have seen that $PR(K_n, H) = 2n - 2$ for $n \geq 4$ when $H = P_5$ or $H = P_6$. In fact, this proper Ramsey number has the same value when H is the 2-chromatic graph C_4. First, we verify this for the case where $n = 3$.

Example 12.19 $PR(K_3, C_4) = 4$.

Proof. Since a red-blue coloring of K_3 in which not all edges are colored the same avoids both a monochromatic K_3 and a properly colored C_4, it follows that $PR(K_3, C_4) \geq 4$. Next, let there be given a red-blue coloring of $G = K_4$ that does not contain a monochromatic K_3. We may assume that the size of the red subgraph G_R is at least 3. Thus, G_R either contains $K_{1,3}$ or P_4. If G_R contains $K_{1,3}$, then G has a monochromatic K_3, a contradiction; while if G_R contains $P_4 = (v_1, v_2, v_3, v_4)$ but not K_3, then $(v_1, v_2, v_4, v_3, v_1)$ is a properly colored C_4. Therefore, $PR(K_3, C_4) = 4$. ∎

Example 12.19 is a special case of a more general result, which we now show.

Theorem 12.20 *For each integer $k \geq 2$, $PR_k(K_3, C_4) = k + 2$.*

Proof. We proceed by induction on k. Since $PR(K_3, C_4) = PR_2(K_3, C_4) = 4$ by Example 12.19, the statement is true for $k = 2$. Assume that $PR_k(K_3, C_4) = k + 2$ for an integer $k \geq 2$. We show that $PR_{k+1}(K_3, C_4) = k + 3$. Since $PR_k(K_3, C_4) = k + 2$, there exists a k-edge coloring c_0 of K_{k+1} using the colors $1, 2, \ldots, k$ that avoids both a monochromatic K_3 and a properly colored C_4. Define a $(k + 1)$-edge coloring of K_{k+2} such that (i) every edge incident with a vertex v of K_{k+2} is colored $k + 1$ and (ii) the subgraph $K_{k+2} - v = K_{k+1}$ has the k-edge coloring c_0. Then there is neither a monochromatic K_3 nor a properly colored C_4. Therefore, $PR_{k+1}(K_3, C_4) > k + 2 = PR_k(K_3, C_4)$. Therefore, $PR_{k+1}(K_3, C_4) \geq k + 3$.

Next, we show that $PR_{k+1}(K_3, C_4) \leq k+3$. Suppose that there is a $(k+1)$-edge coloring c of $G = K_{k+3}$ using the colors $1, 2, \ldots, k + 1$ for which there is neither a monochromatic K_3 nor a properly colored C_4. For each integer i with $1 \leq i \leq k+1$, let G_i denote the subgraph of G without isolated vertices whose edges are colored i. We consider the following three cases.

Case 1. At least one of the subgraphs G_i $(1 \leq i \leq k + 1)$ contains cycles. We may assume that G_1 contains cycles. Let C be a cycle of minimum length in

G_1. Since G has no monochromatic K_3, the length of C is at least 4. Because C is a cycle of minimum length in G_1, it follows that C has no chords in G_1. Let $C = (u_1, u_2, \ldots, u_r, u_1)$, where $r \geq 4$. Since $c(u_1 u_3) \neq 1$ and $c(u_2 u_4) \neq 1$, the cycle $(u_1, u_2, u_4, u_3, u_1)$ is a properly colored C_4, which is a contradiction. Thus, Case 1 cannot occur and so every subgraph G_i $(1 \leq i \leq k+1)$ is a forest.

Case 2. At least one of the subgraphs G_i $(1 \leq i \leq k+1)$ is a forest that is not a tree. We may assume that G_1 is a forest with at least two (nontrivial) components of G_1. Thus, G_1 contains edges $u_1 u_2$ and $v_1 v_2$ belonging to different components. Since $u_1 v_1$ and $u_2 v_2$ are not edges of G_1, it follows that $c(u_1 v_1) \neq 1$ and $c(u_2 v_2) \neq 1$. Hence, $(u_1, u_2, v_2, v_1, u_1)$ is a properly colored C_4, which is a contradiction. By Cases 1 and 2, each subgraph G_i $(1 \leq i \leq k+1)$ is a tree.

Case 3. Each subgraph G_i $(1 \leq i \leq k+1)$ is a tree. Suppose first that each tree G_i $(1 \leq i \leq k+1)$ is a star. Therefore, each star has a center and a vertex of G may be the center of more than one star. However, at least two vertices of G are not centers of any star. Let u be one of these vertices. Since $\deg_G u = k+2$, at least two edges of G incident with u are colored the same. Since u is not the center of a star, this is impossible. Therefore, at least one of the subgraphs G_i $(1 \leq i \leq k+1)$ is a tree of diameter at least 3, say G_1 is one such subgraph. Thus, G_1 contains a path (u_1, u_2, u_3, u_4). Since $u_1 u_3$ and $u_2 u_4$ are not edges of G_1, it follows that $c(u_1 u_3) \neq 1$ and $c(u_2 u_4) \neq 1$. Hence, $(u_1, u_2, u_4, u_3, u_1)$ is a properly colored C_4, which is a contradiction. ∎

While $PR_k(K_n, C_4) = k + 2$ when $k = 2$ and $n = 3$, it is also the case that $PR_k(K_n, C_4) = 2n - 2$ when $k = 2$ and $n = 3$. We now show that $PR_k(K_n, C_4) = 2n - 2$ for $k = 2$ and every integer $n \geq 3$.

Theorem 12.21 *For each integer $n \geq 3$, $PR(K_n, C_4) = 2n - 2$.*

Proof. We proceed by induction on $n \geq 3$. By Example 12.19, the statement holds for $n = 3$. Assume that $PR(K_{n-1}, C_4) = 2n - 4$ for some integer $n \geq 4$. We show that $PR(K_n, C_4) = 2n - 2$.

Since the red-blue coloring of K_{2n-3} in which every edge of some $(n-1)$-clique is colored red and all other edges are blue, contains neither a monochromatic K_n nor a properly colored C_4, it follows that $PR(K_n, C_4) \geq 2n - 2$. It remains to show that $PR(K_n, C_4) \leq 2n-2$. Assume to the contrary that there is a red-blue coloring of $G = K_{2n-2}$ that avoids a monochromatic K_n and a properly colored C_4. By the induction hypothesis, G contains a monochromatic K_{n-1}. We may assume that G contains a red K_{n-1} with vertex set $X = \{x_1, x_2, \ldots, x_{n-1}\}$. Let

$$Y = V(G) - X = \{y_1, y_2, \ldots, y_{n-1}\}.$$

We claim that $G[Y]$ is a blue K_{n-1}. If this were not the case, then $G[Y]$ contains a red edge, say $y_1 y_2$ is red. Since there is no red K_n, it follows that each vertex in Y is joined to at least one vertex in X by a blue edge. We may assume that $x_1 y_1$ is blue where $x_1 \in X$. If $x_i y_2$ is blue for some $i \in \{2, 3, \ldots, n-1\}$, then $(x_1, y_1, y_2, x_i, x_1)$ is a properly colored C_4. Thus, $x_i y_2$ is red for each $i \in \{2, 3, \ldots, n-1\}$. Since there is no red K_n, it follows that x_1y_2 is blue. Furthermore, y_1x_i is red for $2 \le i \le n-1$; for otherwise, $(y_1, x_i, x_1, y_2, y_1)$ is a properly colored C_4. So, each edge in $[\{y_1, y_2\}, \{x_2, x_3, \ldots, x_{n-1}\}]$ is red. However then, $G[\{x_2, x_3, \ldots, x_{n-1}, y_1, y_2\}]$ is a red K_n, a contradiction. Thus, as claimed, $G[Y]$ is a blue K_{n-1}.

Next, we claim that the vertices of X can be labeled as $u_1, u_2, \ldots, u_{n-1}$ and the vertices of Y can be labeled as $v_1, v_2, \ldots, v_{n-1}$ in such a way that for each integer k with $1 \le k \le n-1$, the edge u_iv_j $(1 \le i, j \le k)$ is red if and only if $1 \le i \le j$. We verify this statement by induction on k.

Since $G[Y]$ is a blue K_{n-1}, every vertex in X must be joined to some vertex in Y by a red edge. Let u_1v_1 be a red edge where $u_1 \in X$ and $v_1 \in Y$. Hence, the statement holds for $k = 1$. Assume for some integer k with $1 \le k < n-1$ that X contains k vertices $u_1, u_2, \ldots, u_k$ and Y contains k vertices $v_1, v_2, \ldots, v_k$ such that for each integer i with $1 \le i \le k$, the edge u_iv_j is red if $i \le j \le k$ and u_iv_j is blue if $1 \le j < i$.

We now show that the statement is true for $k + 1$. By assumption, v_k is joined to $u_1, u_2, \ldots, u_k$ by red edges. Since v_k cannot be joined to each vertex of X by a red edge, there must be a vertex $u_{k+1} \in X$ such that $u_{k+1}v_k$ is blue. If $u_{k+1}v_i$ were red for some i with $1 \le i < k$, then $(v_i, u_{k+1}, v_k, u_k, v_i)$ would be a properly colored C_4, which is impossible. Thus, $u_{k+1}v_i$ is blue for all i with $1 \le i < k$. However, u_{k+1} must be joined to some vertex of Y by a red edge, say $u_{k+1}v_{k+1}$ is red, where $v_{k+1} \in Y$. If u_iv_{k+1} were blue for some i with $1 \le i \le k$, then $(v_{k+1}, u_i, v_i, u_{k+1}, v_{k+1})$ would be a properly colored C_4, again impossible. Thus, u_iv_{k+1} is red for all i with $1 \le i \le k$ (see Figure 12.5). This verifies the claim. In particular then, v_{n-1} is joined to every vertex of X by a red edge. However then, $G[X \cup \{v_{n-1}\}]$ is a red K_n, a contradiction. Therefore, $PR(K_n, C_4) = 2n - 2$. ∎

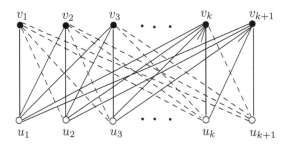

Figure 12.5: A step in the proof of Theorem 12.21

Here too, there is a bipartite version of this Ramsey number. For bipartite graphs F and H, and a positive integer k, the k-**proper bipartite Ramsey number** $PBR_k(F, H)$ is the minimum positive integer r such that for every k-edge coloring of $K_{r,r}$, there is either a monochromatic F or a properly colored H.

12.4 Rainbow Ramsey Numbers

While in an edge-colored graph G, a subgraph F of G is referred to as a monochromatic F if the edges of F are colored the same, F is referred to as a **rainbow** F if all edges of F are colored differently. Arie Bialostocki and William Voxman [20] defined, for a nonempty graph F, the **rainbow Ramsey number** $RR(F)$ of F as the smallest positive integer n such that if each edge of the complete graph K_n is colored from any set of colors, then either a monochromatic F or a rainbow F is produced. Unlike the situation for Ramsey numbers, however, rainbow Ramsey numbers are not defined for every graph F. For example, for the complete graph K_n ($n \geq 3$) with vertex set $\{v_1, v_2, \ldots, v_n\}$ and an edge coloring from the set $\{1, 2, \ldots, n-1\}$ of colors that assigns the color 1 to every edge incident with v_1, the color 2 to any uncolored edge incident with v_2, and so on has the property that no K_3 in K_n is monochromatic or rainbow. Consequently, $RR(K_3)$ is not defined. A natural question therefore arises: For which graphs F is $RR(F)$ defined?

Once again, let $\{v_1, v_2, \ldots, v_n\}$ be the vertex set of a complete graph K_n. An edge coloring of K_n using positive integers for colors is called a **minimum coloring** if two edges $v_i v_j$ and $v_k v_\ell$ are colored the same if and only if

$$\min\{i, j\} = \min\{k, \ell\};$$

while an edge coloring of K_n is called a **maximum coloring** if two edges $v_i v_j$ and $v_k v_\ell$ are colored the same if and only if

$$\max\{i, j\} = \max\{k, \ell\}.$$

In 1950 Paul Erdös and Richard Rado [73] showed that if the edges of a sufficiently large complete graph are colored from a set of positive integers, then there must be a complete subgraph of prescribed order that is either (i) monochromatic, (ii) rainbow, or (iii) has a minimum or maximum coloring.

Theorem 12.22 *For every positive integer p, there exists a positive integer n such that if each edge of the complete graph K_n is colored from a set of positive integers, then there is a complete subgraph of order p that is either monochromatic, rainbow, or has a minimum or maximum coloring.*

With the aid of Theorem 12.22, Bialostocki and Voxman [20] determined all those graphs F for which $RR(F)$ is defined.

Theorem 12.23 *Let F be a graph without isolated vertices. The rainbow Ramsey number $RR(F)$ is defined if and only if F is a forest.*

Proof. Let F be a graph of order $p \geq 2$. First, we show that $RR(F)$ is defined only if F is a forest. Suppose that F is not a forest. Then F contains a cycle C, of length $k \geq 3$ say. Let n be an integer with $n \geq p$ and let $\{v_1, v_2, \ldots, v_n\}$ be the vertex set of a complete graph K_n. Define an edge coloring c of K_n by $c(v_i v_j) = i$ if $i < j$. Hence, c is a minimum edge coloring of K_n. If k is the minimum positive integer such that v_k belongs to C, then two edges of C are colored k, implying that there is no rainbow F in K_n. Since any other edge in C is not colored k, it follows that F is not monochromatic either. Thus, $RR(F)$ is not defined.

For the converse, suppose that F is a forest of order $p \geq 2$. By Theorem 12.22, there exists an integer $n \geq p$ such that for any edge coloring of K_n with positive integers, there is a complete subgraph G of order p in K_n that is either monochromatic or rainbow or has a minimum or maximum coloring. If G is monochromatic or rainbow, then K_n contains a monochromatic or rainbow F. Hence, we may assume that the edge coloring of G is minimum or maximum, say the former. We show in this case that G contains a rainbow F. If F is not a tree, then we can add edges to F to produce a tree T of order p. Let

$$V(G) = \{v_{i_1}, v_{i_2}, \ldots, v_{i_p}\},$$

where $i_1 < i_2 < \ldots < i_p$. Select some vertex $v = v_{i_p}$ of T and label the vertices of T in the order

$$v = v_{i_p}, v_{i_{p-1}}, \ldots, v_{i_2}, v_{i_1}$$

of nondecreasing distances from v; that is,

$$d(v_{i_j}, v) \geq d(v_{i_{j+1}}, v)$$

for every integer j with $1 \leq j \leq p-1$. Hence, there exists exactly one edge of T having color i_j for each j with $1 \leq j \leq p-1$. Thus, T and hence F is rainbow. The rainbow Ramsey number $RR(F)$ is therefore defined.   ∎

By Theorem 12.23, $RR(T)$ is therefore defined for every tree T. We determine the exact value of the rainbow Ramsey number when T is a star.

Example 12.24 *For each integer $k \geq 2$, $RR(K_{1,k}) = (k-1)^2 + 2$.*

Proof. We first show that $RR(K_{1,k}) \geq (k-1)^2 + 2$. Let

$$n = (k-1)^2 + 1.$$

We consider two cases.

Case 1. k is odd. Then n is odd and K_n can be factored into $\frac{n-1}{2} = \frac{(k-1)^2}{2}$ Hamiltonian cycles. These cycles are now partitioned into $k-1$ sets S_i ($1 \leq i \leq k-1$) of $\frac{k-1}{2}$ Hamiltonian cycles. Each edge of every cycle in S_i is colored i. Then there is neither a monochromatic $K_{1,k}$ nor a rainbow $K_{1,k}$.

Case 2. k is even. Since n is even, K_n can be factored into $n-1 = (k-1)^2$ 1-factors. These 1-factors are partitioned into $k-1$ sets S_i ($1 \leq i \leq k-1$) of $k-1$ 1-factors. Each edge of every 1-factor in S_i is colored i. Then there is neither a monochromatic $K_{1,k}$ nor a rainbow $K_{1,k}$.

Therefore, $RR(K_{1,k}) \geq (k-1)^2 + 2$. It remains to show that $RR(K_{1,k}) \leq (k-1)^2 + 2$. Let $N = (k-1)^2 + 2$. Suppose that there is an edge coloring of K_N from any set of colors in which no monochromatic $K_{1,k}$ results. Let v be a vertex of K_N. Since $\deg v = N-1$ and there is no monochromatic $K_{1,k}$, at most $k-1$

edges incident with v can be colored the same. Thus, there are at least $\lceil \frac{N}{k-1} \rceil = k$ edges incident with v that are colored differently, producing a rainbow $K_{1,k}$. ∎

More generally, for two nonempty graphs F_1 and F_2, the **rainbow Ramsey number** $RR(F_1, F_2)$ is defined as the smallest positive integer n such that if each edge of K_n is colored from any set of colors, then there is either a monochromatic F_1 or a rainbow F_2. In view of Theorem 12.23, one wouldn't expect $RR(F_1, F_2)$ to be defined for every pair F_1, F_2 of nonempty graphs; in fact, this is the case. While it is a consequence of a result of Erdös and Rado [73] for which graphs F_1 and F_2 the rainbow Ramsey number $RR(F_1, F_2)$ exists, the proof of the following theorem is due to Linda Eroh [79].

Theorem 12.25 *Let F_1 and F_2 be two graphs without isolated vertices. The rainbow Ramsey number $RR(F_1, F_2)$ is defined if and only if F_1 is a star or F_2 is a forest.*

Proof. First, we show that $RR(F_1, F_2)$ exists only if F_1 is a star or F_2 is a forest. Suppose that F_1 is not a star and F_2 is not a forest. Let G be a complete graph of order n with $V(G) = \{v_1, v_2, \ldots, v_n\}$ such that both F_1 and F_2 are subgraphs of G. Define an $(n-1)$-edge coloring on G such that the edge $v_i v_j$ is assigned the color i if $i < j$. Hence, this coloring is a minimum edge coloring of G.

Let G_1 be a copy of F_1 in G and let a be the minimum integer such that v_a is a vertex of G_1. Then every edge incident with v_a is colored a. Since G_1 is not a star, some edge of G_1 is not incident with v_a and is therefore not colored a. Hence, G_1 is not monochromatic. Next, let G_2 be a copy of F_2 in G. Since G_2 is not a forest, G_2 contains a cycle C. Let b be the minimum integer such that v_b is a vertex of G_2 belonging to C. Since the two edges of C incident with v_b are colored b (and G_2 contains at least two edges colored b), G_2 is not a rainbow subgraph of G. Hence, $RR(F_1, F_2)$ is not defined.

We now verify the converse. Let F_1 and F_2 be two graphs without isolated vertices such that either F_1 is a star or F_2 is a forest. We show that there exists a positive integer n such that for every edge coloring of K_n, either a monochromatic F_1 or a rainbow F_2 results. Suppose that the order of F_1 is $s+1$ and the order of F_2 is $t+1$ for positive integers s and t. Hence, $F_1 = K_{1,s}$. We now consider two cases, depending on whether F_1 is a star or F_2 is a forest. It is convenient to begin with the case where F_2 is a forest.

Case 1. F_2 is a forest. If F_2 is not a tree, then we may add edges to F_2 so that a tree G_2 results. If F_2 is a tree, then let $G_2 = F_2$. Furthermore, if F_1 is not complete, then we may add edges to F_1 so that a complete graph $G_1 = K_{s+1}$ results. If F_1 is complete, then let $G_1 = F_1$. Hence, $G_1 = K_{s+1}$ and G_2 is a tree of order $t+1$. We now show that $RR(G_1, G_2)$ is defined by establishing the existence of a positive integer n such that any edge coloring of K_n from any set of colors results in either a monochromatic G_1 or a rainbow G_2. This, in turn, implies the existence of a monochromatic F_1 or a rainbow F_2. We now consider two subcases, depending on whether G_2 is a star.

Subcase 1.1. G_2 is a star of order $t+1$, that is, $G_2 = K_{1,t}$. Therefore, in this subcase, $G_1 = K_{s+1}$ and $G_2 = K_{1,t}$. (This subcase will aid us later in the proof.) In this subcase, let

$$n = \sum_{i=0}^{(s-1)(t-1)+1} (t-1)^i$$

and let an edge coloring of K_n be given from any set of colors. If K_n contains a vertex incident with t or more edges assigned distinct colors, then K_n contains a rainbow G_2. Hence, we may assume that every vertex of K_n is incident with at most $t-1$ edges assigned distinct colors. Let v_1 be a vertex of K_n. Since the degree of v_1 in K_n is $n-1$, there are at least

$$\frac{n-1}{t-1} = \sum_{i=0}^{(s-1)(t-1)} (t-1)^i$$

edges incident with v_1 that are assigned the same color, say color c_1.

Let S_1 be the set of vertices joined to v_1 by edges colored c_1 and let $v_2 \in S_1$. There are at least

$$\frac{|S_1|-1}{t-1} \geq \sum_{i=0}^{(s-1)(t-1)-1} (t-1)^i$$

edges of the same color, say color c_2, joining v_2 and vertices of S_1, where possibly $c_2 = c_1$. Let S_2 be the set of vertices in S_1 joined to v_2 by edges colored c_2. Continuing in this manner, we construct sets $S_1, S_2, \ldots, S_{(s-1)(t-1)}$ and vertices $v_1, v_2, \ldots, v_{(s-1)(t-1)+1}$ such that for $2 \leq i \leq (s-1)(t-1)+1$, the vertex v_i belongs to S_{i-1} and is joined to at least

$$\sum_{i=0}^{(s-1)(t-1)-(i-1)} (t-1)^i$$

vertices of S_{i-1} by edges colored c_i. Finally, in the set $S_{(s-1)(t-1)}$, the vertex $v_{(s-1)(t-1)+1}$ is joined to a vertex $v_{(s-1)(t-1)+2}$ in $S_{(s-1)(t-1)}$ by an edge colored $c_{(s-1)(t-1)+1}$. Thus, we have a sequence

$$v_1, v_2, \ldots, v_{(s-1)(t-1)+2} \tag{12.6}$$

of vertices such that every edge $v_i v_j$ for $1 \leq i < j \leq (s-1)(t-1)+2$ is colored c_i and where the colors $c_1, c_2, \ldots, c_{(s-1)(t-1)+1}$ are not necessarily distinct. In the complete subgraph H of order $(s-1)(t-1)+2$ induced by the vertices listed in (12.6), the vertex $v_{(s-1)(t-1)+2}$ is incident with at most $t-1$ edges having distinct colors. Hence, there is a set of at least

$$\left\lceil \frac{(s-1)(t-1)+1}{t-1} \right\rceil = s$$

vertices in H joined to $v_{(s-1)(t-1)+2}$ by edges of the same color. Let $v_{i_1}, v_{i_2}, \ldots, v_{i_s}$ be s of these vertices, where $i_1 < i_2 < \cdots < i_s$. Then $c_{i_1} = c_{i_2} = \cdots = c_{i_s}$ and the complete subgraph of order $s+1$ induced by

$$\{v_{i_1}, v_{i_2}, \ldots, v_{i_s}, v_{(s-1)(t-1)+2}\}$$

is monochromatic.

Subcase 1.2. G_2 *is a tree of order* $t + 1$ *that is not necessarily a star.* Recall that $G_1 = K_{s+1}$. We proceed by induction on the positive integer t. If $t = 1$ or $t = 2$, then G_2 is a star and the base case of the induction follows by Subcase 1.1. Suppose that $RR(G_1, G_2)$ exists for $G_1 = K_{s+1}$ and for every tree G_2 of order $t+1$ where $t \geq 2$. Let T be a tree of order $t + 2$. We show that $RR(G_1, T)$ exists. Let v be an end-vertex of T and let u be the vertex of T that is adjacent to v. Let $T' = T - v$. Since T' is a tree of order $t+1$, it follows by the induction hypothesis that $RR(G_1, T')$ exists, say $RR(G_1, T') = p$. Hence, for any edge coloring of K_p from any set of colors, there is either a monochromatic $G_1 = K_{s+1}$ or a rainbow T'. From Subcase 1.1, we know that $RR(G_1, K_{1,t+1})$ exists. Suppose that $RR(G_1, K_{1,t+1}) = q$ and let $n = pq$ in this subcase.

Let there be given an edge coloring of K_n using any number of colors. Consider a partition of the vertex set of K_n into q mutually disjoint sets of p vertices each. By the induction hypothesis, the complete subgraph induced by each set of p vertices contains either a monochromatic K_{s+1} or a rainbow T'. If a monochromatic K_{s+1} occurs in any of these complete subgraphs K_p, then Subcase 1.2 is verified. Hence, we may assume that there are q pairwise mutually rainbow copies

$$T'_1, T'_2, \ldots, T'_q$$

of T', where u_i is the vertex in T'_i $(1 \leq i \leq q)$ corresponding to the vertex u in T'.

Let H be the complete subgraph of order q induced by $\{u_1, u_2, \ldots, u_q\}$. Since $RR(K_{s+1}, K_{1,t+1}) = q$, it follows that either H contains a monochromatic K_{s+1} or a rainbow $K_{1,t+1}$. If H contains a monochromatic K_{s+1}, then, once again, the proof of Subcase 1.2 is complete. So we may assume that H contains a rainbow $K_{1,t+1}$. Let u_j be the center of a rainbow star $K_{1,t+1}$ in H. At least one of the $t + 1$ colors of the edges of $K_{1,t+1}$ is different from the colors of the t edges of T'_j. Adding the edge having this color at u_j in T'_j produces a rainbow copy of T.

Case 2. F_1 *is a star.* Denote F_1 by G_1 as well and so $G_1 = K_{1,s}$. If F_2 is complete, then let $G_2 = F_2$. If F_2 is not complete, then we may add edges to F_2 so that a complete graph $G_2 = K_{t+1}$ results. We verify that $RR(G_1, G_2)$ exists by establishing the existence of a positive integer n such that for any edge coloring of K_n from any set of colors, either a monochromatic G_1 or a rainbow G_2 results. This then shows that K_n will have a monochromatic F_1 or a rainbow F_2. For positive integers p and r with $r < p$, let

$$p^{(r)} = \frac{p!}{(p-r)!} = p(p-1)\cdots(p-r+1).$$

Now let n be an integer such that $s - 1$ divides $n - 1$ and

$$n \geq 3 + \frac{(s-1)(t+2)^{(4)}}{8}. \tag{12.7}$$

Then $n-1 = (s-1)q$ for some positive integer q. Let there be given an edge coloring of K_n from any set of colors and suppose that no monochromatic $G_1 = K_{1,s}$ occurs. We show that there is a rainbow $G_2 = K_{t+1}$. Observe that the total number of different copies of K_{t+1} in K_n is $\binom{n}{t+1}$. We show that the number of copies of K_{t+1} that are not rainbow is less than $\binom{n}{t+1}$, implying the existence of at least one rainbow K_{t+1}.

First consider the number of copies of K_{t+1} containing *adjacent* edges uv and uw that are colored the same. There are n possible choices for the vertex u. Suppose that there are a_i edges incident with u that are colored i for $1 \le i \le k$. Then

$$\sum_{i=1}^{k} a_i = n - 1,$$

where, by assumption, $1 \le a_i \le s - 1$ for each i. For each color i ($1 \le i \le k$), the number of different choices for v and w where uv and uw are colored i is $\binom{a_i}{2}$. Hence, the number of different choices for v and w where uv and uw are colored the same is

$$\sum_{i=1}^{k} \binom{a_i}{2}.$$

Since the maximum value of this sum occurs when each a_i is as large as possible, the largest value of this sum is when each a_i is $s - 1$ and when $k = q$, that is, there are at most

$$\sum_{i=1}^{q} \binom{s-1}{2} = q\binom{s-1}{2}$$

choices for v and w such that uv and uw are colored the same. Since there are $\binom{n-3}{t-2}$ choices for the remaining $t - 2$ vertices of K_{t+1}, it follows that there are at most

$$nq\binom{s-1}{2}\binom{n-3}{t-2}$$

copies of K_{t+1} containing two adjacent edges that are colored the same.

We now consider copies of K_{t+1} in which there are two *nonadjacent* edges colored the same, say $e = xy$ and $f = wz$. There are $\binom{n}{2}$ choices for e and $n - 2$ choices for one vertex, say w, that is incident with f. The vertex w is incident with at most $s - 1$ edges having the same color as e and not adjacent to e. Since there are four ways of counting such a pair of edges in this way (namely e and either w or z, or f and either x or y), there are at most

$$\frac{\binom{n}{2}(n-2)(s-1)}{4} = \frac{n(n-1)(n-2)(s-1)}{8}$$

ways to choose nonadjacent edges of the same color and $\binom{n-4}{t-3}$ ways to choose the remaining $t - 3$ vertices of K_{t+1}. Hence, there are at most

$$\frac{n(n-1)(n-2)(s-1)}{8}\binom{n-4}{t-3}$$

copies of K_{t+1} containing two nonadjacent edges that are colored the same. Therefore, the number of non-rainbow copies of K_{t+1} is at most

$$nq\binom{s-1}{2}\binom{n-3}{t-2} + \frac{n(n-1)(n-2)(s-1)}{8}\binom{n-4}{t-3}$$

$$= n\left(\frac{n-1}{s-1}\right)\frac{(s-1)(s-2)}{2}\binom{n-2}{n-2}\binom{n-3}{t-2}$$

$$\quad + \frac{n(n-1)(n-2)(s-1)}{8}\binom{n-3}{n-3}\binom{n-4}{t-3}$$

$$= \binom{n}{t+1}\left[\frac{(s-2)(t+1)^{(3)}}{2(n-2)} + \frac{(s-1)(t+1)^{(4)}}{8(n-3)}\right]$$

$$< \binom{n}{t+1}\left[\frac{(s-1)(t+1)^{(3)}}{2(n-3)} + \frac{(s-1)(t+1)^{(4)}}{8(n-3)}\right]$$

$$= \binom{n}{t+1}\left[\frac{(s-1)(t+1)^{(3)}(t+2)}{8(n-3)}\right]$$

$$= \binom{n}{t+1}\left[\frac{(s-1)(t+2)^{(4)}}{8(n-3)}\right] \le \binom{n}{t+1},$$

where the final inequality follows from (12.7). Hence, there is a rainbow K_{t+1} in K_n. ∎

We now determine the value of a rainbow Ramsey number.

Example 12.26 $RR(K_{1,3}, K_3) = 6$.

Proof. Since the 2-edge coloring of K_5 producing two monochromatic 5-cycles results in neither a monochromatic $K_{1,3}$ nor a rainbow K_3, it follows that

$$RR(K_{1,3}, K_3) \ge 6.$$

Next, suppose that there exists an edge coloring c of K_6 that produces neither a monochromatic $K_{1,3}$ nor a rainbow K_3. Let $V(K_6) = \{v_0, v_1, \ldots, v_5\}$. Since at most two edges incident with v_0 are assigned the same color, there are three edges incident with v_0 that are assigned distinct colors. We may assume that $c(v_0 v_i) = i$ for $i = 1, 2, 3$ (see Figure 12.6(a)).

Since K_6 has no rainbow K_3, we may assume that $c(v_1 v_2) = 1$. Since there is no monochromatic $K_{1,3}$ and no rainbow K_3, it follows that $c(v_1 v_3) = 3$. Since the triangle with vertices v_1, v_2, v_3 is not a rainbow triangle and $c(v_2 v_3) \ne 3$, it follows that $c(v_2 v_3) = 1$. However then, the triangle with vertices v_0, v_2, v_3 is a rainbow triangle, producing a contradiction (see Figure 12.6(b)). ∎

12.5 Gallai-Ramsey Numbers

In the preceding section, we saw for graphs F and H that the rainbow Ramsey number $RR(F, H)$ of F and H is the minimum positive integer n such that for

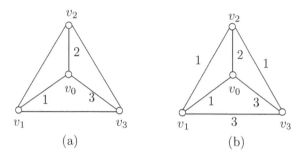

Figure 12.6: A step in proving $RR(K_{1,3}, K_3) = 6$

every edge coloring of K_n using any number of colors, either a monochromatic F or a rainbow H is produced. We also saw that this number is defined only for certain graphs F and H. If, however, a specific number k of edge colors is used, then there are always positive integers n such that every coloring of the edges of the complete graph K_n with k or fewer colors results in a monochromatic F or a rainbow H. Specifically, for graphs F and H, the k-**rainbow Ramsey number** $RR_k(F, H)$ of F and H is the minimum positive integer n such that every k-edge coloring of K_n produces either a monochromatic F or a rainbow H. In particular, $RR_1(F, H) = |V(F)|$ and $RR_2(F, H) = RR(F, H)$. The special case where $H = K_3$ has attracted special attention and a special name.

Dénes König, who wrote the first book on graph theory [137] (published in 1936), was one of the early teachers of graph theory. He had several students who made important contributions to graph theory. Two of his best-known students were Paul Erdős and Paul Turán. Another student, who was a doctoral student of König, was Tibor Gallai (1912-1992). Among the results on graph theory obtained by Gallai was the following [92].

Theorem 12.27 (Gallai's Theorem) *Let k be a positive integer. In any k-edge coloring of the complete graph $G = K_n$ where there is no rainbow triangle, there exists a partition of $V(G)$ into subsets $V_1, V_2, \ldots, V_t$ $(t \geq 2)$ such that*

(1) *for each pair i, j of integers with $1 \leq i < j \leq t$, all edges in $[V_i, V_j]$ are colored the same and*

(2) *the number of colors of the edges in the set $\displaystyle\bigcup_{1 \leq i < j \leq t} [V_i, V_j]$ is at most 2.*

Any partition of an edge-colored complete graph described in Theorem 12.27 is often referred as a **Gallai partition**. From Theorem 12.27, the following concept was derived.

Let F be a graph and let k be a positive integer. The k-**Gallai-Ramsey number** $GR_k(F)$ of F is the minimum positive integer n such that every k-edge coloring of K_n produces either a monochromatic F or a rainbow triangle.

Therefore, $GR_1(F) = |V(F)|$; while for $k \geq 2$, $GR_k(F) \leq MR(F_1, F_2, \ldots, F_k)$, where $F_i \cong F$ for $1 \leq i \leq k$. In fact, if $|E(F)| = m$ and $k < m$, then $GR_k(F) = R(F_1, F_2, \ldots, F_k)$, where $F_i \cong F$ for $1 \leq i \leq k$.

For a k-edge coloring of a complete graph $G = K_n$ for positive integers k and n, where there is no rainbow triangle, let there be given a Gallai partition $\{V_1, V_2, \ldots, V_t\}$ of $V(G)$ into $t \geq 2$ subsets. Let $u_i \in V_i$ for $1 \leq i \leq t$ and let $U = \{u_1, u_2, \ldots, u_t\}$. Then $G[U] = K_t$, where the edge coloring of $G[U]$ is a 2-edge coloring. Consequently, $G[U]$ is a 2-edge-colored subgraph of G. Because much more is known about Ramsey numbers dealing with 2-edge colorings of complete graphs, it is often useful to make use of $G[U]$ in results involving Gallai-Ramsey numbers.

In the case of $GR_k(C_3)$, we have $GR_1(C_3) = 3$ and $GR_2(C_3) = R(3,3) = 6$. The following result establishes the value of $GR_3(C_3)$.

Theorem 12.28 $GR_3(C_3) = 11$.

Proof. First, we show that there is a 3-edge coloring of K_{10} that avoids both a monochromatic C_3 and a rainbow K_3. Let $V(K_{10}) = \{v_1, v_2, \cdots, v_{10}\}$. Define a 3-edge coloring of K_{10} using the three colors $1, 2, 3$ as follows:

(i) The subgraph K_5 of K_{10} induced by $\{v_1, v_2, \ldots, v_5\}$ is decomposed into two copies H_1 and H_2 of C_5 such that all edges of H_i are colored i for $i = 1, 2$. Similarly, the subgraph K_5 induced by $\{v_6, v_7, \ldots, v_{10}\}$ is decomposed into two copies F_1 and F_2 of C_5 such that all edges of F_i are colored i for $i = 1, 2$.

(ii) Each edge $v_i v_j$ where $1 \leq i \leq 5$ and $6 \leq j \leq 10$ is colored 3.

Since this 3-edge coloring of K_{10} produces neither a monochromatic C_3 nor a rainbow K_3, it follows that $GR_3(C_3) \geq 11$.

To verify that $GR_3(C_3) \leq 11$, we show that every 3-edge coloring of K_{11} produces a monochromatic C_3 or a rainbow K_3. Let there be given a 3-edge coloring of $G = K_{11}$ and assume, to the contrary, that there is neither a monochromatic C_3 nor a rainbow K_3. By Theorem 12.27, there exists Gallai partition $\{V_1, V_2, \ldots, V_t\}$, $t \geq 2$, of $V(G)$, where $|V_1| \geq |V_2| \geq \cdots \geq |V_t| \geq 1$. Thus, the edges of $[V_i, V_j]$ are colored the same for each pair i, j of integers with $1 \leq i < j \leq t$ and the number of colors of the edges in the set $\bigcup_{1 \leq i < j \leq t}[V_i, V_j]$ is at most 2.

We may assume that all edges in $[V_1, V_2]$ are colored 1. If some edge in $G[V_1]$ is colored 1, then there is a monochromatic C_3, a contradiction. Thus, all edges in $G[V_1]$ are colored 2 or 3. Since $R(3,3) = 6$ and there is no monochromatic C_3, it follows that $|V_1| \leq 5$. Consequently, $t \geq 3$. If $t \geq 6$, then let $u_i \in V_i$ for $1 \leq i \leq t$ and let $U = \{u_1, u_2, \ldots, u_t\}$. By Theorem 12.27, the number of colors of the edges in $G[U]$ is at most 2. Since $R(3,3) = 6$, it follows that $G[U]$ contains a monochromatic C_3, producing a contradiction. Therefore, $3 \leq t \leq 5$ and so $3 \leq |V_1| \leq 5$. We consider two cases.

Case 1. For each integer i with $3 \leq i \leq t$, all edges in $[V_1, V_i]$ all colored 1. Let $W = V(G) - V_1$. Since $|V_1| \leq 5$, it follows that $|W| \geq 6$. First, suppose that some edge in $G[W]$ is colored 1, say ww' is colored 1. For each $v \in V_1$, it follows that

(v, w, w', v) is a monochromatic C_3, a contradiction. Next, suppose that no edge in $G[W]$ is colored 1. Then all edges in $G[W]$ are colored 2 or 3. Since $|W| \geq 6$ and $R(3,3) = 6$, the graph G contains a monochromatic C_3 in $G[W]$, a contradiction.

Case 2. For some integer i with $3 \leq i \leq t$, all edges in $[V_1, V_i]$ are colored by the same color that is not 1. We may assume that all edges in $[V_1, V_3]$ are colored 2. Since there is no monochromatic C_3, it follows that no edge in $G[V_1]$ can be colored 1 or 2. Hence, all edges in $G[V_1]$ are colored 3. Since $|V_1| \geq 3$, there is a monochromatic C_3 in G. ∎

In the case of a general positive integer k, the following is due to Fan Chung and Ronald Graham [53].

Theorem 12.29 *For each positive integer k,*

$$GR_k(C_3) = \begin{cases} 5^{\frac{k}{2}} + 1 & \text{if } k \text{ is even} \\ 2 \cdot 5^{\frac{k-1}{2}} + 1 & \text{if } k \text{ is odd.} \end{cases}$$

For some larger odd cycles, the following Gallai-Ramsey numbers are known (see [30, 89, 111]).

Theorem 12.30 *For positive integers k and ℓ with $2 \leq \ell \leq 7$,*

$$GR_k(C_{2\ell+1}) = \ell \cdot 2^k + 1.$$

In case of the 4-cycle C_4, the following result, due to Ralph Faudree, Ronald Gould, Michael Jacobson, and Colton Magnant [81], establishes the value of $GR_k(C_4)$ for every positive integer k.

Theorem 12.31 *For integers $k \geq 2$, $GR_k(C_4) = k + 4$.*

Proof. First, we show that there is a k-edge coloring of K_{k+3} that avoids both a monochromatic C_4 and a rainbow K_3. Let $V(K_{k+3}) = \{v_1, v_2, \cdots, v_{k+3}\}$. Define a k-edge coloring of K_{k+3} using k colors $1, 2, \ldots, k$ as follows:

(i) the subgraph K_5 induced by $\{v_1, v_2, \ldots, v_5\}$ is decomposed into two copies H_1 and H_2 of C_5 such that all edges of H_i are colored i for $i = 1, 2$ and

(ii) for each pair i, j of integers with $1 \leq i < j$ and $6 \leq j \leq k+3$, assign the color $j - 3$ to the edge $v_i v_j$.

Since this k-edge coloring of K_{k+3} produces neither a monochromatic C_4 nor a rainbow K_3, it follows that $GR_k(C_4) \geq k + 4$.

Next, we show that $GR_k(C_4) \leq k+4$. Since $GR_2(C_4) = R(C_4, C_4) = 6$, we may assume that $k \geq 3$. We proceed by induction on $k \geq 3$ to show that every k-edge coloring c of K_{k+4} produces either a monochromatic C_4 or a rainbow K_3.

First, let $k = 3$ and let there be given a 3-edge coloring c of $G = K_7$ using the three colors $1, 2, 3$. Assume, to the contrary, that c produces neither a monochromatic C_4 nor a rainbow K_3. Let $\{V_1, V_2, \ldots, V_t\}$, $t \geq 2$, be a Gallai partition of

$V(G)$, where $|V_1| \geq |V_2| \geq \cdots \geq |V_t| \geq 1$, as described in Theorem 12.27. Thus, the edges of $[V_i, V_j]$, $1 \leq i < j \leq t$, are colored the same and the number of colors of the edges in the set $\bigcup_{1 \leq i < j \leq t}[V_i, V_j]$ is at most 2. If $|V_1| = 1$ for $1 \leq i \leq t$, then G has a rainbow K_3. Thus, $|V_1| \geq 2$. This implies that $|V_i| = 1$ for $2 \leq i \leq t$, for otherwise, there is a monochromatic C_4. If $t \geq 4$, then we may assume that the edges in $[V_1, V_2]$ and $[V_1, V_3]$ are colored the same, producing a monochromatic C_4. Hence, $t = 2$ or $t = 3$.

⋆ First, assume that $t = 2$. Therefore, $|V_1| = 6$ and $|V_2| = 1$. We may assume that each edge in $[V_1, V_2]$ is colored 1. Since there is no monochromatic C_4, no two edges colored 1 in $G[V_1]$ are adjacent. Suppose that there are two edges in $G[V_1]$ that are colored 1, say $c(uv) = c(xy) = 1$. Since there is no rainbow K_3, all edges in $\{ux, uy, vx, vy\}$ are colored the same, producing a monochromatic C_4. Next, suppose that exactly one edge in $G[V_1]$ is colored 1, say $c(uv) = 1$. Hence, $c(xu) = c(xv) \neq 1$ for each vertex $x \in V(G_1) - \{u, v\}$. Since $|V(G_1) - \{u, v\}| \geq 4$, there are $x, y \in V(G) - \{u, v\}$ and $x \neq y$ for which $c(xu) = c(xv) = c(yu) = c(yv)$, producing a monochromatic C_4. Finally, suppose that no edge in $G[V_1]$ is colored 1. Thus, the restriction of c to $G[V_1]$ is a 2-edge coloring of K_6. Since $R(C_4, C_4) = 6$, there is a monochromatic C_4.

⋆ Next, assume that $t = 3$. So, $|V_1| = 5$ and $|V_2| = |V_3| = 1$. We may assume that each edge in $[V_1, V_2]$ is colored 1 and each edge in $[V_1, V_3]$ is colored 2. An argument similar to the one in the case when $t = 2$ shows that no edge in $G[V_1]$ can be colored 1 or 2. Thus, $G[V_1]$ is a monochromatic K_5 and so there is a monochromatic C_4.

This verifies the base step of the induction.

Suppose for some integer $k \geq 4$ that every $(k-1)$-edge coloring of K_{k+3} produces either a monochromatic C_4 or a rainbow K_3. Let c be a k-edge coloring of $G = K_{k+4}$. Assume, to the contrary, that c produces neither a monochromatic C_4 nor a rainbow K_3. Let $\{V_1, V_2, \ldots, V_t\}$, $t \geq 2$, be a Gallai partition of $V(G)$, where $|V_1| \geq |V_2| \geq \cdots \geq |V_t| \geq 1$. Since G has no rainbow triangle, it follows that $|V_1| \geq 2$. Because there is no monochromatic C_4, it follows that $|V_i| = 1$ for $2 \leq i \leq t$. If $t \geq 4$, then we may assume that the edges in $[V_1, V_2]$ and $[V_1, V_3]$ are colored the same, producing a monochromatic C_4. Hence, $t = 2$ or $t = 3$. An argument similar to the one in the case when $k = 3$ and $t \in \{2, 3\}$ shows that no edge in $G[V_1]$ can be colored 1 or 2. Let $V_2 = \{v\}$ and let $G' = G - v = K_{k+3}$. Then the restriction c' of the coloring c to G' is a $(k-1)$-edge coloring of K_{k+3}. By the induction hypothesis, c' produces either a monochromatic C_4 or a rainbow K_3. Hence, there is a monochromatic C_4 or a rainbow K_3 in G as well. Therefore, $GR_k(C_4) = k + 4$. ∎

For the even cycles, C_4, C_6, and C_8, we have the following result [89, 101, 90].

Theorem 12.32 *For positive integers k and ℓ with $2 \leq \ell \leq 4$,*

$$GR_k(C_{2\ell}) = (\ell - 1)k + \left\lceil \frac{\ell+5}{2} \right\rceil.$$

For cycles in general, the following bounds were determined [89, 111].

Theorem 12.33 *For integers $k \geq 1$ and $\ell \geq 2$,*

$$(\ell - 1)k + \ell + 1 \leq GR_k(C_{2\ell}) \leq (\ell - 1)k + 3\ell$$
$$\text{and}$$
$$2^k \ell + 1 \leq GR_k(C_{2\ell+1}) \leq (2^{k+3} - 1)\ell \log_2 \ell.$$

For paths P_n, $4 \leq n \leq 8$, the following results are known [81, 90].

Theorem 12.34 *For integers k and n with $4 \leq n \leq 8$,*

$$GR_k(P_n) = \left\lceil \frac{n-3}{2} \right\rceil k + \left\lceil \frac{n+2}{2} \right\rceil.$$

For general paths, the following bounds were established [81, 111].

Theorem 12.35 *For integers $k \geq 1$ and $n \geq 3$,*

$$\left\lfloor \frac{n-2}{2} \right\rfloor k + \left\lceil \frac{n}{2} \right\rceil + 1 \leq GR_k(P_n) \leq \left\lfloor \frac{n-2}{2} \right\rfloor k + 3 \left\lceil \frac{n}{2} \right\rceil.$$

The Gallai-Ramsey number of a graph is a special case of more general Ramsey numbers. For graphs F and H and a positive integer k, the k-**rainbow Ramsey number** $RR_k(F, H)$ of F and H is the minimum positive integer n such that every k-edge coloring of K_n results in a monochromatic F or a rainbow H. If $H = K_3$, then $RR_k(F, H)$ is therefore the Gallai-Ramsey number $GR_k(F)$. That is, the Gallai-Ramsey number is a special case of k-rainbow Ramsey numbers. We now give some examples of k-rainbow Ramsey numbers that are not Gallai-Ramsey numbers.

Theorem 12.36 *For each integer $k \geq 3$, $RR_k(K_{1,3}, P_4) = 6$.*

Proof. For each integer $k \geq 3$, the edge coloring of K_5 that induces two copies of monochromatic 5-cycle using two colors from a set of k colors contains neither a monochromatic $K_{1,3}$ nor a rainbow P_4 and so $RR_k(K_{1,3}, P_4) \geq 6$.

It remains to show that $RR_k(K_{1,3}, P_4) \leq 6$. Let $V(K_6) = \{v_0, v_1, \ldots, v_5\}$ and let $c : E(K_6) \rightarrow \{1, 2, \ldots, k\}$ be an edge coloring that avoids a monochromatic $K_{1,3}$. We show that there is a rainbow P_4. If no vertex of K_6 is incident with two or more edges of the same color, then we may assume that $c(v_0 v_i) = i$ for $1 \leq i \leq 5$. Thus, $c(v_1 v_2) \notin \{1, 2\}$ and a rainbow P_4 is produced. Otherwise, we may assume that v_0 is incident with two edges of the same color, say $c(v_0 v_1) = c(v_0 v_2) = 1$. Since there is no monochromatic $K_{1,3}$, we may further assume that $c(v_0 v_3) = 2$ and $c(v_0 v_4) = 3$. Then observe that there is a rainbow P_4 regardless of the color of $v_1 v_3$. ∎

We next determine four k-rainbow Ramsey numbers $RR_k(F, H)$ where $F \in \{P_4, K_3\}$, $H \in \{K_{1,3}, P_4\}$, and $k \geq 3$, beginning with $F = P_4$.

Theorem 12.37 *For each integer $k \geq 3$, $RR_k(P_4, K_{1,3}) = RR_k(P_4, P_4) = 5$.*

Proof. The edge coloring of K_4 that produces a monochromatic $K_{1,3}$ of one color and a monochromatic K_3 of another color shows that $RR_k(P_4, H) \geq 5$ for $H \in \{K_{1,3}, P_4\}$. To complete the proof, let $V(K_5) = \{v_0, v_1, \ldots, v_4\}$ and consider an edge coloring $c : E(K_5) \to \{1, 2, \ldots, k\}$ that avoids a monochromatic P_4. We show that there must be both a rainbow $K_{1,3}$ and a rainbow P_4.

If there exists a vertex that is incident with at least three edges of the same color, then we may assume that $c(v_0v_i) = 1$ for $1 \leq i \leq 3$. Since there is no monochromatic P_4, we may further assume that $1 \notin \{c(v_1v_2), c(v_1v_3), v(v_2v_3)\}$. If $c(v_1v_2) \neq c(v_1v_3)$, say, then both a rainbow $K_{1,3}$ and a rainbow P_4 are produced. Otherwise, we may assume that $c(v_1v_2) = c(v_1v_3) = c(v_2v_3) = 2$. Since there is no monochromatic P_4, it follows that $c(v_1v_4) \geq 3$, resulting in a rainbow $K_{1,3}$ and a rainbow P_4.

Hence, we next assume that every vertex is incident with at most two edges of the same color. If there exists a vertex that is incident with three edges of distinct colors, then there is a rainbow $K_{1,3}$. To verify that there is also a rainbow P_4, assume, without loss of generality, that $c(v_0v_i) = i$ for $1 \leq i \leq 3$. If $c(v_iv_4) \neq i$ for some $i \in \{1, 2, 3\}$, then there is a rainbow P_4. Otherwise, assume that $c(v_iv_4) = i$ for $1 \leq i \leq 3$ and observe that there is still a rainbow P_4 regardless of the color of v_0v_4.

Finally, assume that each vertex is incident with two edges of one color and two edges of another color. We show that this is impossible. Suppose that $c(v_0v_1) = c(v_0v_2) = 1$ and $c(v_0v_3) = c(v_0v_4) = 2$. Since there is no monochromatic P_4, it follows that $c(v_1v_3) \notin \{1, 2\}$, say $c(v_1v_3) = 3$. Then $c(v_1v_2) = 1$ and $c(v_1v_4) = 3$. Now observe that $c(v_2v_3) \in \{c(v_0v_3), c(v_1v_3)\} = \{2, 3\}$, producing a monochromatic P_4. Thus, this case does not occur, as claimed. ∎

Theorem 12.38 *For each integer $k \geq 3$, $RR_k(K_3, K_{1,3}) = RR_k(K_3, P_4) = 6$.*

Proof. The edge coloring of K_5 that produces two copies of monochromatic 5-cycles of different colors has no monochromatic K_3, no rainbow $K_{1,3}$, and no rainbow P_4. Thus, $PR_k(K_3, K_{1,3}) \geq 6$ and $RR_k(K_3, P_4) \geq 6$. It suffices to show that $RR_k(K_3, H) \leq 6$ for $H \in \{K_{1,3}, P_4\}$. Let $V(K_6) = \{v_0, v_1, \ldots, v_5\}$ and consider an edge coloring $c : E(K_6) \to \{1, 2, \ldots, k\}$ that avoids a monochromatic K_3. We show that there must be both a rainbow $K_{1,3}$ and a rainbow P_4.

If there exists a vertex that is incident with at least three edges of the same color, then we may assume that $c(v_0v_i) = 1$ for $1 \leq i \leq 3$. Since there is no monochromatic K_3, we may further assume that $c(v_1v_2), c(v_1v_3) \neq 1$ and $c(v_1v_2) \neq c(v_1v_3)$. Then there exists a rainbow $K_{1,3}$ as well as a rainbow P_4.

Hence, suppose next that no vertex is incident with three or more edges of the same color. This implies that every vertex is incident with three edges of distinct colors, which guarantees the existence of a rainbow $K_{1,3}$. It remains to show that there exists a rainbow P_4. If v_0 is incident with two edges of the same color, say $c(v_0v_1) = c(v_0v_2) = 1$, then we may further assume that $c(v_0v_3) = 2$ and $c(v_0v_4) = 3$. Then there is a rainbow P_4 since $c(v_1v_2) \neq 1$. Finally, if every vertex is incident with five edges of distinct colors, then there is a rainbow P_4. ∎

Once again, there is a bipartite version of this Ramsey number as well. For bipartite graphs F and H, and a positive integer k, the **k-rainbow bipartite Ramsey number** $RBR_k(F, H)$ is the minimum positive integer r such that every k-edge coloring of $K_{r,r}$ results in either a monochromatic F or a rainbow H.

Exercises for Chapter 12

1. By (12.1), $MR(K_{1,3}, C_4) \leq \min\{R(K_{1,3}, C_4), R(K_{1,3}, K_{1,3}), R(C_4, C_4)\}$.
   By Theorem 11.9 and Corollary 11.14, $R(K_{1,3}, K_{1,3}) = R(C_4, C_4) = 6$.

   (a) Determine $R(K_{1,3}, C_4)$.
   (b) Determine $MR(K_{1,3}, C_4)$.

2. Let F and H be graphs without isolated vertices.

   (a) Prove that $MR(F, H) = 2$ if and only if $F = K_2$ or $H = K_2$.
   (b) Prove that if the order of F is at least 3, then $MR(F, P_3) = 3$.

3. Let F be nonempty graphs without isolated vertices. Prove the following:

   (a) If F is not a subgraph of C_5, then $MR(F, C_3) = 6$.
   (b) If F is a subgraph of C_5, then $MR(F, C_3) = |V(F)|$.

4. (a) Determine $MR(C_3, C_5)$.
   (b) Determine $MR(C_4, C_5)$.
   (c) For each integer $n \geq 3$ and $n \neq 5$, prove that
   $$MR(C_3, C_n) = MR(C_4, C_n) = 6.$$

5. For the bichromatic partitions π_2, π_3, and π_4 of $E(K_4)$ in Figure 12.2, prove that $R_{\pi_2}(K_4, K_4) \geq 8$ and $R_{\pi_i}(K_4, K_4) \geq 10$ for $i = 3, 4$.

6. Let F and H be two graphs with $\chi'(H) = t$ and let k be an integer with $k \geq t$. Prove that $PR_{k+1}(F, H) \geq PR_k(F, H)$.

7. By Theorem 12.20, $PR_3(K_3, C_4) = 5$. Give an independent proof of this.

8. (a) Determine $PR(C_4, C_4)$.
   (b) Prove for every integer $k \geq 2$ that $PR_{k+1}(C_4, C_4) > PR_k(C_4, C_4)$.
   (c) Provide a lower bound for $PR_k(C_4, C_4)$ in terms of k.

9. Show that $PR(K_3, P_5) = 5$.

10. Show that $PR(K_3, P_6) = PR(K_4, P_6) = 6$.

11. Determine $PR(K_3, C_6)$.

12. Prove for every integer $n \geq 3$ that $PR(K_{1,n}, C_4) = n + 1$.

13. Prove for each integer $n \geq 3$ that $PR(K_{1,n}, P_3) = n + 1$.

14. Prove for each integer $n \geq 3$ that $PR(K_{1,n}, P_4) = n + 1$.

15. Let k be an integer with $k \geq 3$. Prove that there exists a positive integer n such that for every edge coloring of K_n from a set of positive integers, there exists some subgraph K_k such that either (1) K_k is monochromatic, (2) K_k is rainbow, or (3) K_k is neither monochromatic nor rainbow and for every two distinct colors a and b used in the edge coloring of K_k, the number of edges colored a is distinct from the number of edges colored b.

16. Give an example of two graphs F_1 and F_2, each of order at least 4, such that both $RR(F_1, F_2)$ and $RR(F_2, F_1)$ are defined but $RR(F_1, F_2) \neq RR(F_2, F_1)$.

17. Show that $RR(K_3, 3K_2) \geq 7$.

18. Let $k \geq 2$ be an integer. For $1 \leq i \leq k - 1$, let G_i be a copy of a graph G and $R(G; k - 1) = R(G_1, G_2, \ldots, G_{k-1})$. Prove that

$$R(G; k - 1) \leq RR(G, kK_2) \leq R(G; k - 1) + 2(k - 1)$$

for each integer $k \geq 2$ and every graph G.

19. Let $k \geq 3$ be an integer.

   (a) Prove for each odd integer $2\ell + 1 \geq 3$ that $GR_k(K_{1,2\ell+1}) \geq 5\ell + 1$.

   (a) Prove for each even integer $2\ell \geq 4$ that $GR_k(K_{1,2\ell}) \geq 4\ell - 1$.

20. Prove that $GR_k(K_{1,3}) = 6$ for each integer $k \geq 3$.

21. Prove that $RR_k(K_{1,4}, K_3) = 7$ for each integer $k \geq 3$.

22. Prove for each integer $k \geq 3$ that $RR_k(P_4) = k + 3$.

23. Prove for integers $r \geq 2$ and $k \geq 3$ that $GR_k(rK_2) \geq (r - 1)k + r + 1$.

24. Prove that $GR_k(2K_2) = k + 3$ for each integer $k \geq 3$.

25. Prove that $GR_3(rK_2) = 4r - 2$ for each integer $r \geq 2$.

26. Prove that $RR_k(2K_2, 2K_2) = 4$ for each integer $k \geq 2$.

Chapter 13

Color Connection

It was stated in [48] that the Department of Homeland Security in the United States was created in 2003 in response to weaknesses discovered in the transfer of classified information after the September 11, 2001 terrorist attacks. In [78], Anne Baye Ericksen made the following observation:

> An unanticipated aftermath of those deadly attacks was the realization that law enforcement and intelligence agencies couldn't communicate with each other through their regular channels from radio systems to databases. The technologies utilized were separate entities and prohibited shared access, meaning there was no way for officers and agents to cross check information between various organizations.

While information related to national security needs to be protected, there must be procedures in place that permit access between appropriate parties. This two-fold issue can be addressed by assigning information-transfer paths between agencies which may have other agencies as intermediaries while requiring a large enough number of passwords and firewalls that is prohibitive to intruders, yet small enough to manage. An immediate question arises:

> What is the minimum number of passwords or firewalls needed that permits the existence of one or more secure paths between every two agencies?

To answer this question, we need an understanding of what is meant by a secure information-transfer path. There are many possible interpretations of what might be meant by such a secure path. At one extreme, a path may be considered secure only if the passwords along the path are distinct. A considerably less stringent interpretation might require only that every pair of consecutive passwords along the path be distinct. As described in [48], situations such as this can be represented by a graph whose vertices are the agencies and where two vertices are adjacent if there is direct access between them. Such graphs can then be studied by means of certain edge colorings of the graphs, where colors here refer to passwords. It is edge colorings possessing such properties that is the primary topic of this chapter.

13.1   Rainbow Connection

For a nontrivial connected graph G and a positive integer k, let $c : E(G) \to \{1, 2, \ldots, k\}$ be an edge coloring of G, where adjacent edges of G are permitted to be colored the same. A path in this edge-colored graph G is called a **rainbow path** if no two of its edges are assigned the same color. The graph G is **rainbow-connected** (with respect to c) if G contains a rainbow $u - v$ path for every pair u, v of vertices G. In this context, the coloring c is called a **rainbow edge coloring** or, more simply, a **rainbow coloring**. If k colors are used in a rainbow coloring of G, then c is a **rainbow k-coloring**. The edge coloring of the Petersen graph shown in Figure 13.1 is a rainbow 3-coloring.

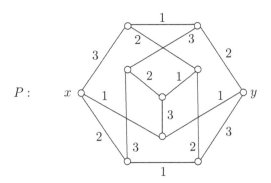

Figure 13.1: A rainbow 3-coloring of the Petersen graph

The minimum positive integer k for which there exists a rainbow k-coloring of a connected graph G is the **rainbow connection number** $\mathrm{rc}(G)$ of G. If $\mathrm{diam}(G) = k$, then necessarily $\mathrm{rc}(G) \geq k$. The rainbow connection number is defined for every nontrivial connected graph G since every edge coloring of G in which distinct edges are assigned distinct colors is a rainbow coloring.

Since there exists a rainbow 3-coloring of the Petersen graph P, it follows that $\mathrm{rc}(P) \leq 3$. Furthermore, since $\mathrm{diam}(P) = 2$, it follows that $\mathrm{rc}(P) \geq 2$. There is no rainbow 2-coloring of the Petersen graph, however, for suppose that such an edge coloring c exists. Because P is cubic, there are two adjacent edges, say uv and vw, that must be assigned the same color by c. Since the girth of P is 5, (u, v, w) is the only $u - w$ path of length 2 in P. Because this path is not a rainbow path, c is not a rainbow coloring and so $\mathrm{rc}(P) = 3$.

Let c be an edge coloring of a nontrivial connected graph G. For two vertices u and v of G, a **rainbow $u - v$ geodesic** in G is a $u - v$ rainbow path of length $d(u, v)$. The graph G is said to be **strongly rainbow-connected** if G contains a rainbow $u - v$ geodesic for every two vertices u and v of G. In this case, the coloring c is called a **strong rainbow coloring** of G. The minimum positive integer k for which G has a strong rainbow k-coloring is the **strong rainbow connection number** $\mathrm{src}(G)$ of G. In general, if G is a nontrivial connected graph of size m, then

$$\mathrm{diam}(G) \leq \mathrm{rc}(G) \leq \mathrm{src}(G) \leq m. \tag{13.1}$$

These concepts and the results that follow were introduced in [47, 48].

Since the rainbow connection number of the Petersen graph P is 3, it follows by (13.1) that $\mathrm{src}(P) \geq 3$. The rainbow 3-coloring of the Petersen graph shown in Figure 13.1 is not a strong rainbow 3-coloring, however, since the unique $x - y$ geodesic in P is not a rainbow $x - y$ geodesic. Indeed, any strong rainbow coloring of P must not assign the same color to adjacent edges, implying that the coloring is a proper edge coloring. Because the chromatic index $\chi'(P)$ of P is 4, it follows that $\mathrm{src}(P) \geq 4$. Since the edge coloring of P shown in Figure 13.2 is a strong rainbow 4-coloring, $\mathrm{src}(P) = 4$.

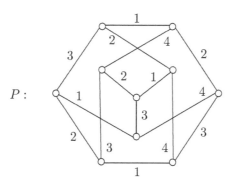

Figure 13.2: A strong rainbow 4-coloring of the Petersen graph

While the rainbow connection number and strong rainbow connection number have different values for the Petersen graph, this situation does not occur for graphs whose rainbow connection number is less than 3.

Theorem 13.1 *For $k \in \{1, 2\}$ and a nontrivial connected graph G,*

$$\mathrm{rc}(G) = k \text{ if and only if } \mathrm{src}(G) = k.$$

Proof. Suppose that $\mathrm{src}(G) = 1$. Since $\mathrm{rc}(G) \leq \mathrm{src}(G)$, it follows that $\mathrm{rc}(G) = 1$. If $\mathrm{rc}(G) = 1$, then $\mathrm{diam}(G) = 1$ and so G is complete. The coloring that assigns the color 1 to every edge of G is a strong rainbow 1-coloring of G and so $\mathrm{src}(G) = 1$.

If $\mathrm{rc}(G) = 2$, then $\mathrm{diam}(G) = 2$ and there exists a rainbow 2-coloring of G. Hence, every two nonadjacent vertices of G are connected by a rainbow path of length 2, which is necessarily a geodesic. Thus, $\mathrm{src}(G) = 2$. If G is a graph with $\mathrm{src}(G) = 2$, then $\mathrm{rc}(G) \leq 2$. However, $\mathrm{rc}(G) \neq 1$, for otherwise, $\mathrm{src}(G) = 1$. Thus, $\mathrm{rc}(G) = 2$. ∎

In (13.1), we saw that the rainbow connection number of a connected graph G can never exceed its size by (13.1). The graphs G for which $\mathrm{rc}(G) = |E(G)|$ belong to a familiar class.

Theorem 13.2 *Let G be a nontrivial connected graph of size m. Then*

$$\mathrm{rc}(G) = \mathrm{src}(G) = m \text{ if and only if } G \text{ is a tree.}$$

Proof. First, suppose that G is not a tree. Then G contains a cycle

$$C = (v_1, v_2, \ldots, v_k, v_1),$$

where $k \geq 3$. Then the $(m-1)$-edge coloring that assigns the color 1 to the edges v_1v_2 and v_2v_3 and distinct colors from $\{2, 3, \ldots, m-1\}$ to the remaining $m-2$ edges of G is a rainbow $(m-1)$-coloring. Thus, $\text{rc}(G) \leq m-1$.

For the converse, suppose that G is a tree of size m and assume, to the contrary, that there is a rainbow $(m-1)$-coloring of G. Then there exist edges $e = uv$ and $f = xy$ that are assigned the same color. For either u or v and for either x or y, say u and x, there exists a $u-x$ path in G containing both e and f. Since this is the unique $u-x$ path in G, there is no rainbow $u-x$ path in G, which is a contradiction.   ∎

For the complete graph K_n, $n \geq 2$, $\text{src}(K_n) = \text{rc}(K_n) = \text{diam}(K_n) = 1$; while for the Petersen graph P, $\text{src}(P) = 4$, $\text{rc}(P) = 3$ and $\text{diam}(P) = 2$. Furthermore, $\text{src}(K_{1,t}) = \text{rc}(K_{1,t}) = t$ and $\text{diam}(K_{1,t}) = 2$. Thus, both the strong rainbow connection number and rainbow connection number of a graph can be considerably larger than its diameter. Another well-known class of graphs having diameter 2, containing the stars as a subclass, is that of the complete bipartite graphs.

Theorem 13.3 *For integers s and t with $1 \leq s \leq t$,*

$$\text{src}(K_{s,t}) = \left\lceil \sqrt[s]{t} \, \right\rceil.$$

Proof. Since $\text{src}(K_{1,t}) = t$, the result is true for $s = 1$. So we may assume that $s \geq 2$. Let $\left\lceil \sqrt[s]{t} \, \right\rceil = k$. Hence,

$$1 \leq k - 1 < \sqrt[s]{t} \leq k.$$

Therefore, $(k-1)^s < t \leq k^s$.

First, we show that $\text{src}(K_{s,t}) \geq k$. Assume, to the contrary, that

$$\text{src}(K_{s,t}) \leq k - 1.$$

Then there exists a strong rainbow $(k-1)$-coloring c of $K_{s,t}$. Let U and W be the partite sets of $K_{s,t}$, where $|U| = s$ and $|W| = t$ with $U = \{u_1, u_2, \ldots, u_s\}$. For each vertex $w \in W$, define its color code, denoted by $\text{code}(w)$, as the ordered s-tuple

$$(a_1, a_2, \ldots, a_s),$$

where $a_i = c(u_i w)$ for $1 \leq i \leq s$. Since $1 \leq a_i \leq k-1$ for each i ($1 \leq i \leq s$), the number of distinct color codes of the vertices of W is at most $(k-1)^s$. Since $t > (k-1)^s$, there exist two distinct vertices w' and w'' of W such that $\text{code}(w') = \text{code}(w'')$, which implies that $c(u_i w') = c(u_i w'')$ for all i ($1 \leq i \leq s$). Consequently, there is no rainbow $w'-w''$ geodesic in $K_{s,t}$, contradicting our assumption that c is a strong rainbow $(k-1)$-coloring of $K_{s,t}$. Therefore, $\text{src}(K_{s,t}) \geq k$.

Next, we show that $\text{src}(K_{s,t}) \leq k$. Let

$$A = \{1, 2, \ldots, k\} \text{ and } B = \{1, 2, \ldots, k-1\}.$$

Furthermore, let A^s and B^s be Cartesian products of s sets A and s sets B, respectively. Thus,

$$|A^s| = k^s \text{ and } |B^s| = (k-1)^s.$$

Hence, $|B^s| < t \le |A^s|$. Let

$$W = \{w_1, w_2, \ldots, w_t\},$$

where the t vertices of W are labeled with t elements of A^s such that the vertices $w_1, w_2, \ldots, w_{(k-1)^s}$ are labeled with the $(k-1)^s$ elements of B^s. For $1 \le i \le t$, denote the label of w_i by

$$\ell(w_i) = (a_{i,1}, a_{i,2}, \ldots, a_{i,s}).$$

Thus, for each i with $1 \le i \le (k-1)^s$ and each j with $1 \le j \le s$, it follows that $1 \le a_{i,j} \le k-1$.

We now define an edge coloring c of $K_{s,t}$ by

$$c(w_i u_j) = a_{i,j} \text{ where } 1 \le i \le t \text{ and } 1 \le j \le s.$$

Hence, the color code of w_i ($1 \le i \le t$) is $\text{code}(w_i) = \ell(w_i)$ and so distinct vertices in W have distinct color codes.

We show that c is a strong rainbow k-coloring of $K_{s,t}$. Certainly, for $w_i \in W$ and $u_j \in U$, the $w_i - u_j$ path (w_i, u_j) is a rainbow geodesic. Now let w_a and w_b be two vertices of W. Because w_a and w_b have distinct color codes, there exists an integer r with $1 \le r \le s$ such that the r-th coordinates of $\text{code}(w_a)$ and $\text{code}(w_b)$ are different. Thus, $c(w_a u_r) \ne c(w_b u_r)$ and (w_a, u_r, w_b) is a rainbow $w_a - w_b$ geodesic in $K_{s,t}$. Next, let u_p and u_q be two vertices in U. Since there exists a vertex $w_i \in W$ with $1 \le i \le (k-1)^s$ such that $a_{i,p} \ne a_{i,q}$, it follows that (u_p, w_i, u_q) is a rainbow $u_p - u_q$ geodesic in $K_{s,t}$. Thus, as claimed, c is a strong rainbow k-coloring of $K_{s,t}$ and so $\text{src}(K_{s,t}) \le k$. ∎

Of course, for integers s and t with $1 \le s \le t$, $\text{rc}(K_{s,t}) \le \text{src}(K_{s,t})$. The following result gives the value of $\text{rc}(K_{s,t})$.

Theorem 13.4 *For integers s and t with $2 \le s \le t$,*

$$\text{rc}(K_{s,t}) = \min\left\{\left\lceil \sqrt[s]{t} \right\rceil, 4\right\}.$$

By Theorem 2.19 for every two distinct vertices u and v of a graph G with connectivity k, there exist k internally disjoint $u - v$ paths. The **rainbow connectivity** $\kappa_r(G)$ of G is the minimum number of colors required of an edge coloring of G such that every two vertices are connected by k internally disjoint rainbow paths.

For example, since the connectivity of the graph $H = K_3 \,\square\, K_2$ is 3, it follows that $\kappa_r(H)$ is the minimum number of colors in an edge coloring of H such that every two vertices of H are connected by three internally disjoint rainbow paths. Since the 6-edge coloring of H shown in Figure 13.3 has this property, $\kappa_r(H) \le 6$.

We now show that $\kappa_r(H) \ge 6$. A useful observation about the graph $K_3 \,\square\, K_2$ is that for every two vertices x and y belonging to different triangles, there exists

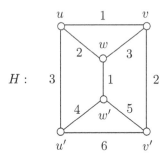

Figure 13.3: A 6-edge coloring of $K_3 \,\square\, K_2$

a unique set of three internally disjoint $x - y$ paths. Suppose that $\kappa_r(H) = k$ and let a k-edge coloring c of H be given such that for every two vertices x and y of H, there are three internally disjoint rainbow $x - y$ paths. Suppose first that x and y belong to a common triangle of H. Since two paths in any set of three internally disjoint $x - y$ paths have lengths 1 and 2 and the edges of these paths are the edges of the triangle, all three edges of each triangle must be assigned different colors by c. Assume that uv is colored 1, uw is colored 2, and vw is colored 3. See Figure 13.4(a).

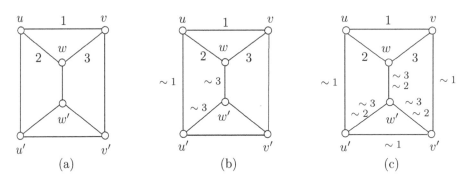

Figure 13.4: Steps in determining $\kappa_r(H)$

By considering the three internally disjoint $u' - v$ paths in H, we see that uu' is not colored by 1 (denoted by ~ 1) and neither ww' nor $u'w'$ is colored 3 (see Figure 13.4(b)). Similarly, by considering the pairs $\{u, v'\}$, $\{u, u'\}$, and $\{v, v'\}$ of vertices of H, we have the conditions on the coloring c shown in Figure 13.4(c).

Then, by considering the pairs $\{u, w'\}$, $\{v, w'\}$, $\{w, u'\}$, and $\{w, v'\}$ of vertices of H, we have the added conditions on c shown in Figure 13.5. This shows that none of the edges of the triangle with vertices u', v', and w' can be assigned any of the colors 1, 2, 3. Since no two edges belonging to a triangle can be colored the same, at least six colors are required to color the edges of H in order for every two distinct vertices of H to be connected by three internally disjoint rainbow paths. Thus, $\kappa_r(H) \geq 6$ and so $\kappa_r(H) = 6$.

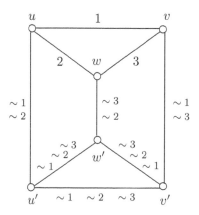

Figure 13.5: A step in determining $\kappa_r(H)$

We first determine the rainbow connectivity of complete graphs. Recall (from Section 10.2) that the chromatic index of K_n is

$$\chi'(K_n) = 2\lceil n/2 \rceil - 1.$$

Theorem 13.5 *For every integer $n \geq 2$,*

$$\kappa_r(K_n) = \chi'(K_n).$$

Proof. Since $\kappa_r(K_2) = 1$, the result holds for $n = 2$. Hence, we may assume that $n \geq 3$. Let $\chi'(K_n) = k$ and let there be given a proper k-edge coloring of K_n. Consider two vertices u and v of K_n. Suppose that $v_1, v_2, \ldots, v_{n-2}$ are the remaining vertices of K_n. For each i with $1 \leq i \leq n-2$, the colors of uv_i and v_iv are different. Therefore, the path (u, v) together with the paths (u, v_i, v), $1 \leq i \leq n-2$, are $n - 1$ internally disjoint $u - v$ rainbow paths. Hence, $\kappa_r(K_n) \leq \chi'(K_n)$.

We now show that $\kappa_r(K_n) \geq \chi'(K_n)$ for each $n \geq 3$. Assume, to the contrary, that $\kappa_r(K_n) = \ell < \chi'(K_n)$ for some integer $n \geq 3$. Then there exists an ℓ-edge coloring c of K_n such that every two vertices of K_n are connected by $n-1$ internally disjoint rainbow paths. Since $\chi'(K_n) > \ell$, there exist two adjacent edges of K_n, say xy and yz, that are assigned the same color. Since (x, y, z) is one of the $n - 1$ internally disjoint $x - z$ rainbow paths, a contradiction is produced. Therefore, $\kappa_r(K_n) \geq \chi'(K_n)$ and so $\kappa_r(K_n) = \chi'(K_n)$. ∎

We have seen that the rainbow connection number of the Petersen graph is 3 and its strong rainbow connection number is 4. The connectivity of the Petersen graph is 3. That $\kappa_r(P) \leq 5$ is shown in Figure 13.6. Showing that $\kappa_r(P) \geq 5$ is more complicated, but, nevertheless, such is the case. Hence, $\kappa_r(P) = 5$.

While, in a rainbow-connected graph G, every two vertices u and v of G are connected by a $u - v$ rainbow path, there is no condition on what the length of such a path must be. For certain highly Hamiltonian graphs G, however, it is natural to ask whether there exists an edge coloring of G using a certain number of colors such

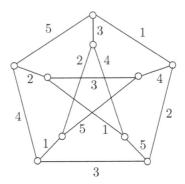

Figure 13.6: $\kappa_r(P) = 5$

that every two vertices of G can be connected by a rainbow path of a prescribed length.

For a Hamiltonian-connected graph G, an edge coloring $c : E(G) \rightarrow [k] = \{1, 2, \ldots, k\}$ is called a **Hamiltonian-connected rainbow k-coloring** if every two vertices of G are connected by a rainbow Hamiltonian path in G. An edge coloring c is a **Hamiltonian-connected rainbow coloring** if c is a Hamiltonian-connected rainbow k-coloring for a positive integer k. The minimum k for which G has a Hamiltonian-connected rainbow k-coloring is the **rainbow Hamiltonian-connection number** of G, denoted by $\mathrm{hrc}(G)$.

If H is a Hamiltonian-connected spanning subgraph of a graph G and c is a Hamiltonian-connected rainbow coloring of H, then the coloring c can be extended to a Hamiltonian-connected rainbow coloring of G by assigning any color used by c to each edge in $E(G) - E(H)$. Thus, we have the following observation.

Observation 13.6 *If H is a Hamiltonian-connected spanning subgraph of a graph G, then*

$$\mathrm{hrc}(G) \leq \mathrm{hrc}(H).$$

Let G be a Hamiltonian-connected graph of order $n \geq 4$. Note that (1) there is no Hamiltonian-connected rainbow coloring of G using less than $n - 1$ colors and (2) the edge coloring that assigns distinct colors to distinct edges of G is a Hamiltonian-connected rainbow coloring. This gives us the following observation.

Observation 13.7 *If G is a Hamiltonian-connected graph of order $n \geq 3$ and size m, then*

$$n - 1 \leq \mathrm{hrc}(G) \leq m.$$

If G is a Hamiltonian-connected graph of order $n \geq 3$ and size at most $2n - 3$, then $\mathrm{hrc}(G) \neq n - 1$, for otherwise, each $(n - 1)$-edge coloring of G results in some edge $e = uv$ of G having the property that e is the only edge possessing the color assigned to it. However then, there is no rainbow Hamiltonian $u - v$ path P in G. This gives rise to the following observation.

Observation 13.8 *If G is a Hamiltonian-connected graph of order $n \geq 3$ and size at most $2n - 3$, then $\mathrm{hrc}(G) \geq n$.*

We now present infinite classes of Hamiltonian-connected graphs G such that $\mathrm{hrc}(G) = |V(G)| - 1$. For an integer $n \geq 3$, the wheel $W_n = C_n \vee K_1$ of order $n+1$ is Hamiltonian-connected.

Theorem 13.9 *For each integer $n \geq 3$, $\mathrm{hrc}(W_n) = n$.*

Proof. Let $W_n = C_n \vee K_1$, where $C_n = (v_1, v_2, \ldots, v_n, v_1)$ and the vertex $v \in V(K_1)$ is adjacent to each vertex of C_n. Since $\mathrm{hrc}(W_n) \geq n$ by Observation 13.7, it remains to show that $\mathrm{hrc}(W_n) \leq n$. Define an n-edge coloring $c : E(W_n) \to [n]$ by $c(v_i v_{i+1}) = c(v_i v) = i$ for $1 \leq i \leq n$, where $v_{n+1} = v_1$. We show that every two vertices x and y in W_n are connected by a rainbow Hamiltonian path in W_n. We may assume, without loss of generality, that $x = v_1$. If $y = v$, then $(v_1, v_2, \ldots, v_n, v)$ is a rainbow Hamiltonian $v_1 - v$ path. If $y = v_n$, then $(v_1, v_2, v_3, \ldots, v_{n-1}, v, v_n)$ is a rainbow Hamiltonian $v_1 - v_n$ path. If $y = v_i$ where $2 \leq i \leq n - 1$, then $(v_1, v_2, \ldots, v_{i-1}, v, v_n, v_{n-1}, \ldots, v_i)$ is a rainbow Hamiltonian $v_1 - v_i$ path. Thus, c is a Hamiltonian-connected rainbow coloring of W_n and so $\mathrm{hrc}(W_n) \leq n$. Therefore, $\mathrm{hrc}(W_n) = n$. ∎

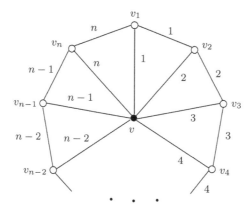

Figure 13.7: An n-edge coloring c of $W_n = C_n \vee K_1$

The following results are consequences of Observations 13.6 and 13.7 and Theorem 13.9.

Corollary 13.10 *If G is a Hamiltonian graph of order $n \geq 3$, then*
$$\mathrm{hrc}(G \vee K_1) = n.$$

Corollary 13.11 *For each integer $n \geq 4$, $\mathrm{hrc}(K_n) = n - 1$.*

Not only are there Hamiltonian-connected graphs G of order $n \geq 4$ and size m for which $\mathrm{hrc}(G)$ attains the lower bound $n - 1$ in Observation 13.7, there are Hamiltonian-connected graphs G of order $n \geq 3$ and size m for which $\mathrm{hrc}(G) = m$. Indeed, it was shown in [16] that if $n \geq 5$ is odd and $G = C_n \square K_2$, then $\mathrm{hrc}(G) = |E(G)|$.

13.2   Proper Connection

One property that a connected graph G with a proper edge coloring has is that for every two vertices u and v of G, there is a properly colored $u - v$ path in G. In fact, every $u - v$ path in such a graph G is properly colored. Therefore, there is an edge coloring of G with $\chi'(G)$ colors such that every two vertices of G are connected by a properly colored path. However, if our primary interest concerns edge colorings of G with the property that for every two vertices u and v of G, there exists a properly colored $u - v$ path in G, then this may very well be possible using fewer than $\chi'(G)$ colors.

Inspired by rainbow colorings and proper colorings of graphs, Chartrand [8] and, independently, Borozan, Fujita, Gerek, Magnant, Manoussakis, Montero, and Tuza [26] introduced the concept of proper-path colorings of graphs. This concept has now been studied by many; in fact, there is a dynamic survey on this topic due to Li and Magnant [141].

Let G be an edge-colored connected graph, where adjacent edges may be colored the same. A path P in G is **properly colored** or, more simply, P is a **proper path** in G if no two adjacent edges of P are colored the same. An edge coloring c is a **proper-path coloring** of a connected graph G if every pair u, v of distinct vertices of G are connected by a proper $u - v$ path in G. If k colors are used, then c is referred to as a **proper-path k-coloring**. The minimum k for which G has a proper-path k-coloring is called the **proper connection number** $\mathrm{pc}(G)$ of G. A proper-path coloring using $\mathrm{pc}(G)$ colors is referred to as a **minimum proper-path coloring**.

To illustrate this concept, we consider the 3-regular graph G of Figure 13.8 and a proper-path 3-coloring of G using the colors 1, 2, 3 where the uncolored edges can be colored with any of these three colors. Since the three bridges in G must be assigned distinct colors, $\mathrm{pc}(G) = 3$. Note that this 3-regular graph G is not 1-factorable and so $\chi'(G) = 4$. That is, G has a proper-path coloring using fewer than $\chi'(G)$ colors. Indeed, if we were to replace a block of order 5 in the graph G of Figure 13.8 by a complete graph K_n of arbitrarily large odd order n, say, then the resulting graph H has the property that $\chi'(H) = n$ but $\mathrm{pc}(H) = 3$ and so $\chi'(H) - \mathrm{pc}(H)$ can be arbitrarily large. In fact, $\mathrm{pc}(K_n) = 1$ and $\chi'(K_n) = n$ if $n \geq 3$ is odd.

Let G be a nontrivial connected graph of order n and size m. Since every rainbow coloring is a proper-path coloring, it follows that $\mathrm{pc}(G)$ exists and

$$1 \leq \mathrm{pc}(G) \leq \min\{\chi'(G), \mathrm{rc}(G)\} \leq m. \tag{13.2}$$

Furthermore, not only is $\mathrm{pc}(K_n) = 1$ but $\mathrm{pc}(G) = 1$ if and only if $G = K_n$. Also, $\mathrm{pc}(G) = m$ if and only if $G = K_{1,m}$ is a star of size m (see Exercise 10).

While these concepts were introduced to parallel concepts dealing with rainbow colorings for the purpose of studying connected graphs by means of properly colored paths in edge-colored graphs, there is also corresponding motivation to that introduced for rainbow colorings of graphs. With regard to the national security discussion mentioned earlier, we are also interested in the answer to the following question:

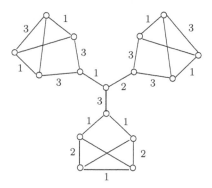

Figure 13.8: A proper-path 3-coloring of a graph G

What is the minimum number of passwords or firewalls that allow one or more secure paths between every two agencies where as we progress from one step to another along such a path, we are required to change passwords?

Furthermore, Li and Magnant made the following observations in [141]:

When building a communication network between wireless signal towers, one fundamental requirement is that the network is connected. If there cannot be a direct connection between two towers A and B, say for example if there is a mountain in between, there must be a route through other towers to get from A to B. As a wireless transmission passes through a signal tower, to avoid interference, it would help if the incoming signal and the outgoing signal do not share the same frequency. Suppose we assign a vertex to each signal tower, an edge between two vertices if the corresponding signal towers are directly connected by a signal and assign a color to each edge based on the assigned frequency used for the communication. Then the number of frequencies needed to assign the connections between towers so that there is always a path avoiding interference between each pair of towers is precisely the proper connection number of the corresponding graph. . . .

Other motivation has been given to study proper-path colorings of graphs. The Knight's Tour Problem is a famous problem that asks whether it's possible for a knight to tour an 8×8 chessboard where each square of the chessboard is visited exactly once (except that the final square visited is the initial square of the tour) and each step along the tour is a single legal move of a knight. It is well known that such a tour is, in fact, possible (see [13], for example). Since a single move of a knight moves it from a square of one color to a square of the other color, it follows that if two knights were placed on any two squares of a chessboard, then there exists a path of legal moves from one knight to the other on a chessboard, using legal knight moves, such that the squares visited along the path alternate in color.

Because the Knight's Tour Problem is equivalent to that of finding a particular type of Hamiltonian cycle on the grid $P_8 \,\square\, P_8$ (the Cartesian product of P_8 and P_8), this brings up two items.

(1) Assign colors to the edges of a connected graph G so that for every two vertices u and v of G, there exists a $u - v$ path P in G such that every two adjacent edges on P have distinct colors.

(2) Assign colors to the vertices of a connected graph G so that for every two vertices u and v of G, there exists a $u - v$ path P in G such that every two adjacent vertices on P have distinct colors.

It is item (1), of course, that leads us once again to proper-path colorings of connected graphs (see [37]).

Although the following result is straightforward and easy to verify, it is very useful.

Proposition 13.12 *If G is a nontrivial connected graph and H is a connected spanning subgraph of G, then $\mathrm{pc}(G) \le \mathrm{pc}(H)$. In particular, $\mathrm{pc}(G) \le \mathrm{pc}(T)$ for every spanning tree T of G.*

Proof. Let H be a spanning subgraph of G and c_H a proper-path coloring of H using colors from the set $\{1, 2, \ldots, k\}$ where $k = \mathrm{pc}(H)$. Define a coloring c of G by $c(e) = c_H(e)$ if $e \in E(H)$ and $c(e) = 1$ for the remaining edges of G. Then c is a proper-path coloring of G using $\mathrm{pc}(H)$ colors and so $\mathrm{pc}(G) \le \mathrm{pc}(H)$.   ∎

While the preceding result provides an upper bound for the path connection number of a graph G, the following result gives a lower bound for $\mathrm{pc}(G)$ when G possesses bridges.

Proposition 13.13 *Let G be a nontrivial connected graph that contains bridges. If b is the maximum number of bridges incident with a single vertex in G, then $\mathrm{pc}(G) \ge b$.*

Proof. Since $\mathrm{pc}(G) \ge 1$, the result is trivial when $b = 1$. Thus, we may assume that $b \ge 2$. Let v be a vertex of G that is incident with b bridges and let vw_1 and vw_2 be two bridges incident with v. Since (w_1, v, w_2) is the only $w_1 - w_2$ path in G, it follows that every proper-path coloring of G must assign distinct colors to vw_1 and vw_2. Hence, all b bridges incident with v are colored differently and so $\mathrm{pc}(G) \ge b$.   ∎

Since every edge is a bridge in a nontrivial tree T, it follows that $\mathrm{pc}(T) \ge \Delta(T)$ by Proposition 13.13. By Theorem 10.8 (König's theorem), $\chi'(G) = \Delta(G)$ for every nonempty bipartite graph G. Hence, the following is a consequence of (13.2), Proposition 13.13 and Theorem 10.8.

Proposition 13.14 *If T is a nontrivial tree, then $\mathrm{pc}(T) = \chi'(T) = \Delta(T)$.*

Propositions 13.12 and 13.14 provide an upper bound for the proper connection number of a graph.

Proposition 13.15 *For a nontrivial connected graph G,*

$$\mathrm{pc}(G) \leq \min\{\Delta(T) : \ T \text{ is a spanning tree of } G\}.$$

The following is an immediate consequence of Proposition 13.15.

Corollary 13.16 *If G is a graph with a Hamiltonian path that is not complete, then $\mathrm{pc}(G) = 2$.*

We saw in (13.2) that if G is a nontrivial connected graph that is not complete such that $\mathrm{pc}(G) = a$ and $\mathrm{rc}(G) = b$, then $2 \leq a \leq b$. In fact, this is the only restriction on these two parameters.

Theorem 13.17 *For every pair a, b of positive integers with $2 \leq a \leq b$, there is a connected graph G such that*

$$\mathrm{pc}(G) = a \ and \ \mathrm{rc}(G) = b.$$

Proof. By Proposition 13.14, if T is a tree of order at least 3, then $\mathrm{pc}(T) = \Delta(T) \geq 2$. Since $\mathrm{rc}(T)$ is the size of T, it follows that for integers a and b with $2 \leq a \leq b$, a tree T of size b and $\Delta(T) = a$ has $\mathrm{pc}(T) = a$ and $\mathrm{rc}(T) = b$. ∎

By (13.2), if G is a nontrivial connected graph that is not complete such that $\mathrm{pc}(G) = a$ and $\chi'(G) = b$, then $2 \leq a \leq b$. This restriction is all that is required for these two parameters.

Proposition 13.18 *For every pair a, b of integers with $2 \leq a \leq b$, there exists a connected graph G such that*

$$\mathrm{pc}(G) = a \ and \ \chi'(G) = b.$$

Proof. If $b = 2$, then $a = b = 2$ and any path of order at least 3 has the desired property by Corollary 13.16. Thus, we may assume that $b \geq 3$. Let G be the graph obtained from the path $(x_1, x_2, x_3, x_4, x_5)$ of order 5 by (i) adding the $b - 2$ new vertices $v_1, v_2, \ldots, v_{b-2}$ and joining each v_i ($1 \leq i \leq b-2$) to both x_2 and x_4 and (ii) adding the $a - 1$ new vertices $w_1, w_2, \ldots, w_{a-1}$ and joining each w_i ($1 \leq i \leq a - 1$) to x_5. Since G is a bipartite graph and $\Delta(G) = b$, it follows that $\chi'(G) = b$. It remains to show that $\mathrm{pc}(G) = a$. Define an edge coloring c by assigning (1) the color 1 to each of $x_1 x_2, x_2 v_i$ ($1 \leq i \leq b - 2$), $x_3 x_4$ and $x_5 w_1$, (2) the color 2 to each of $x_2 x_3, x_4 v_i$ ($1 \leq i \leq b - 2$) and $x_5 w_2$ (3) the color i to $x_5 w_i$ ($3 \leq i \leq a - 1$ if $a \geq 4$) and (4) the color a to $x_4 x_5$. Then every two vertices u and v are connected by a proper $u - v$ path. For example, v_1 and w_1 are connected by the proper path $(v_1, x_2, x_3, x_4, x_5, w_1)$. Hence, c is a proper-path coloring of G using a colors and so $\mathrm{pc}(G) \leq a$. Assume, to the contrary, that $\mathrm{pc}(G) \leq a - 1$. Let c^* be a minimum proper-path coloring of G. Since $\deg x_5 = a$ and at most $a - 1$ colors are used by c^*, there are two edges e and f incident with x_5 that are colored the same, say $e = u x_5$ and $f = x_5 v$. However then, there is no proper $u - v$ path in G, which is impossible. Thus, $\mathrm{pc}(G) \geq a$ and so $\mathrm{pc}(G) = a$. ∎

While many of results mentioned indicate that the proper connection number of a graph can be arbitrarily large, all of the graphs G described above for which $pc(G)$ is large contain cut-vertices. If a graph does not contain a cut-vertex, the situation is quite different. Borozan, Fujita, Gerek, Magnant, Manoussakis, Montero and Tuza [26] obtained the following three results.

Theorem 13.19 *If G is a 2-connected graph that is not complete, then $pc(G) \leq 3$.*

Theorem 13.20 *If G is a connected non-complete graph of order $n \geq 68$ and $\delta(G) \geq n/4$, then $pc(G) = 2$.*

Theorem 13.21 *If G is a 3-connected graph that is not complete, then $pc(G) = 2$.*

Just as there is a "strong" version of rainbow colorings, there is a "strong" version of proper-path colorings. Let c be a proper-path coloring of a nontrivial connected graph G. For two vertices u and v of G, a **proper $u - v$ geodesic** in G is a proper $u - v$ path of length $d(u, v)$. If there is a proper $u - v$ geodesic for every two vertices u and v of G, then c is called a **strong proper-path coloring** of G or a **strong proper-path k-coloring** if k colors are used. The minimum number of colors needed to produce a strong proper-path coloring of G is called the **strong proper connection number** $spc(G)$ of G. A strong proper-path coloring using $spc(G)$ colors is a **minimum strong proper-path coloring**. The concept of strong proper-path colorings of graphs was introduced by Chartrand and first studied in [8]. In general, if G is a nontrivial connected graph, then

$$1 \leq pc(G) \leq spc(G) \leq \chi'(G).$$

The following result was obtained in [8].

Theorem 13.22 *For every three integers $a, b,$ and n with $2 \leq a \leq b < n$, there exists a connected graph G of order n such that*

$$spc(G) = a \text{ and } \chi'(G) = b.$$

Since every strong rainbow coloring of G is a strong proper-path coloring of G, it follows that $spc(G) \leq src(G)$. Therefore, if G is a nontrivial connected graph of order n, and size m, then

$$spc(G) \leq src(G). \tag{13.3}$$

Analogous to Theorem 13.17, the following result was obtained in [8].

Theorem 13.23 *For every three integers $a, b,$ and n with $2 \leq a \leq b < n$, there exists a connected graph G of order n such that*

$$spc(G) = a \text{ and } src(G) = b.$$

To illustrate these concepts, consider the two proper-path colorings of the 5-cycle C_5 and the proper-path coloring of the 3-regular graph G shown in Figure 13.9, where the graph G is the same graph in Figure 13.8. The coloring in Figure 13.9(a) is a minimum proper-path coloring of C_5 and so $\text{pc}(C_5) = 2$. The coloring in Figure 13.9(b) is a minimum strong proper-path coloring of C_5 and so $\text{spc}(C_5) = 3$. The coloring in Figure 13.9(c) is both a minimum proper-path coloring and a minimum strong proper-path coloring of G and so $\text{pc}(G) = \text{spc}(G) = 3$. As we saw, this 3-regular graph G is not 1-factorable and so $\chi'(G) = 4$. Notice that the coloring of the graph G in Figure 13.8 is a minimum proper-path coloring but whether it's a minimum strong proper-path coloring of G depends on how the six uncolored edges are colored. For example, if all six edges are colored 1, then the resulting coloring is not a minimum strong proper-path coloring of G.

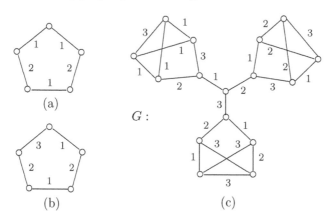

Figure 13.9: Illustrating proper-path colorings
and strong proper-path colorings

For a Hamiltonian-connected graph G, an edge coloring $c : E(G) \to [k]$ is a **proper Hamiltonian-path k-coloring** if every two vertices of G are connected by a proper Hamiltonian path in G. An edge coloring c is a **proper Hamiltonian-path coloring** if c is a proper Hamiltonian-path k-coloring for some positive integer k. The minimum number of colors required of a proper Hamiltonian-path coloring of G is the **proper Hamiltonian-connection number** of G, denoted by $\text{hpc}(G)$. These concepts were introduced by Chartrand and first studied in [17]. Since every proper edge coloring of a Hamiltonian-connected graph G is a proper Hamiltonian-path coloring of G and there is no proper Hamiltonian-path 1-coloring of G, it follows that

$$2 \le \text{hpc}(G) \le \chi'(G). \tag{13.4}$$

To illustrate these concepts, consider the graph $G = C_6^2$. Since $\Delta(G) = 4$ and the edge coloring of G in Figure 13.10(a) is a proper 4-edge coloring, it follows that $\chi'(G) = \Delta(G) = 4$. Next, consider the 2-edge coloring c of G shown in Figure 13.10(b). We show that c is a proper Hamiltonian-path coloring of G; that is, every two vertices u and v of G are connected by a proper Hamiltonian $u - v$ path P in G. If $\{u, v\} = \{v_1, v_2\}$ or $\{u, v\} = \{v_1, v_6\}$, say the former, let $P = (v_1, v_6, v_5, v_4, v_3, v_2)$; if $\{u, v\} = \{v_1, v_3\}$ or $\{u, v\} = \{v_1, v_5\}$, say the former, let $P = (v_1, v_2, v_6, v_5, v_4, v_3)$; while if $\{u, v\} = \{v_1, v_4\}$, let $P = (v_1, v_2, v_6, v_5, v_3, v_4)$.

By the symmetry of this edge coloring, c is a proper Hamiltonian-path 2-coloring and so $\text{hpc}(G) = 2$. Therefore, $\text{hpc}(G) < \chi'(G)$.

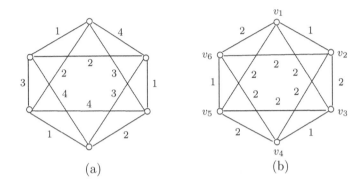

Figure 13.10: A proper 4-edge coloring and a proper Hamiltonian-path 2-coloring of C_6^2

Next, we give an example of a graph G with $\text{hpc}(G) = \chi'(G)$. Let $G = K_3 \,\square\, K_2$, where the two triangles K_3 in G are (u, x, w, u) and (v, y, z, v) and $uv, xy, wz \in E(G)$. Since there is a proper 3-edge coloring of G shown in Figure 13.11 and $\Delta(G) = 3$, it follows that $\chi'(G) = 3$. Hence, $\text{hpc}(G) \le 3$.

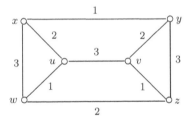

Figure 13.11: A proper 3-edge coloring of $K_3 \,\square\, K_2$

We now show that $\text{hpc}(G) \ge 3$. Assume, to the contrary, that there is a proper Hamiltonian-path 2-coloring c of G using the colors red (color 1) and blue (color 2). There are only two Hamiltonian $u - v$ paths, namely (u, w, x, y, z, v) and (u, x, w, z, y, v). Because of the symmetry of these paths, we may assume that the first path is a proper Hamiltonian $u - v$ path and its edges are colored as $c(uw) = c(xy) = c(zv) = 1$ and $c(wx) = c(yz) = 2$. Next, we consider a proper Hamiltonian $x - z$ path. There are only two Hamiltonian $x - z$ paths in G, namely, $Q_1 = (x, w, u, v, y, z)$ and $Q_2 = (x, y, v, u, w, z)$. Since the path $Q = (w, u, v, y)$ lies on both Q_1 and Q_2, it follows that Q must be proper. This implies that $c(uv) = 2$ and $c(vy) = 1$. Similarly, there are only two Hamiltonian $w - y$ paths in G, each of which contains the path (x, u, v, z), and so this path must be proper. This implies that $c(ux) = 1$. We now consider a proper Hamiltonian $x - v$ path. There are only two Hamiltonian $x - v$ paths in G, namely, $R_1 = (x, u, w, z, y, v)$ and

$R_2 = (x, y, z, w, u, v)$. Since the path $R = (y, z, w, u)$ lies on both R_1 and R_2, it follows that R must be properly colored by the colors 1 and 2. Since $c(yz) = 2$ and $c(wu) = 1$, this is impossible. Thus, there is no proper Hamiltonian $x - v$ path in G, which is a contradiction. Therefore, $\text{hpc}(G) \geq 3$ and so $\text{hpc}(G) = 3$.

We now consider some well-known Hamiltonian-connected graphs, beginning with complete graphs, which are supergraphs of all Hamiltonian-connected graphs. It is easy to see that $\text{hpc}(K_3) = 3$. When $n \geq 4$, $\text{hpc}(K_n) = 2$, however, which we verify next.

Theorem 13.24 *For every integer $n \geq 4$, $\text{hpc}(K_n) = 2$.*

Proof. We consider two cases, according to whether n is even or n is odd.

Case 1. n is even. The complete graph $G = K_n$ contains a 1-factor F. Define an edge coloring c of G by assigning the color red to each edge of F and the color blue to the remaining edges of G. We show that c is a proper Hamiltonian-path 2-coloring of G; that is, for every two vertices u and v of G, there is a proper Hamiltonian $u - v$ path in G. Let $n = 2k$ and let $V(G) = \{v_1, v_2, \ldots, v_{2k}\}$. Suppose that $E(F) = \{v_{2i-1}v_{2i} : 1 \leq i \leq k\}$. There are two possibilities, depending on whether uv is a blue edge or uv is a red edge. Thus, we may assume that either (1) $u = v_1$ and $v = v_{2k}$ or (2) $u = v_2$ and $v = v_1$. Consider the properly colored Hamiltonian cycle $C = (v_1, v_2, \ldots, v_{2k}, v_1)$ of G. If (1) occurs, then $(u = v_1, v_2, \ldots, v_{2k} = v)$ is a proper Hamiltonian $u - v$ path in G; while if (2) occurs, then $(u = v_2, v_3, \ldots, v_{2k}, v_1 = v)$ is a proper Hamiltonian $u - v$ path in G. Therefore, $\text{hpc}(K_n) = 2$.

Case 2. $n \geq 5$ is odd. Let $C = (v_1, v_2, \ldots, v_n, v_1)$ be a Hamiltonian cycle in $G = K_n$. Define a coloring c of G by assigning the color red to each edge of C and the color blue to the remaining edges of G. We show that c is a proper Hamiltonian-path 2-coloring of G; that is, for every two vertices u and v of G, there is a proper Hamiltonian $u - v$ path in G. We may assume that $v = v_n$ and $u = v_i$ for some integer i with $1 \leq i \leq (n-1)/2$.

First, suppose that $u = v_1$. If $n \equiv 1 \,(\text{mod } 4)$, then

$$(u = v_1, v_2, v_4, v_3, v_5, v_6, v_8, v_7, v_9, \ldots, v_{n-3}, v_{n-1}, v_{n-2}, v_n = v)$$

is a proper Hamiltonian $u - v$ path in G; while if $n \equiv 3 \pmod 4$, then

$$(u = v_1, v_2, v_4, v_3, v_5, v_6, v_8, v_7, v_9, \ldots, v_{n-5}, v_{n-3}, v_{n-4}, v_{n-1}, v_{n-2}, v_n = v)$$

is a proper Hamiltonian $u - v$ path in G.

Next, suppose that $u = v_j$ where $2 \leq j \leq (n-1)/2$. If $n = 5$, then $u = v_2$ and $(v_5, v_3, v_4, v_1, v_2)$ is a proper Hamiltonian $u - v$ path in G. Thus, we may assume that $n \geq 7$ is odd. Let $A = \{v_1, v_2, \ldots, v_{j-1}\}$ and $B = \{v_{j+1}, v_{j+2}, \ldots, v_{n-1}\}$. Let $|A| = a$ and $|B| = b$. Since $n \geq 7$ is odd, it follows that (1) $b \geq 3$ and (2) $a + b = n - 2$ is odd and so a and b are of opposite parity. We consider two subcases, according to whether a is even or a is odd.

Subcase 2.1. a is even. Then

$$Q = (u = v_j, v_{j-2}, v_{j-1}, v_{j-4}, u_{j-3}, v_{j-6}, v_{j-5}, \ldots, v_1, v_2, v_{j+2})$$

is a proper $u - v_{j+2}$ path in G with $V(Q) = \{v_1, v_2, \ldots, v_j\} \cup \{v_{j+2}\}$ and

$$Q' = (v_{j+2}, v_{j+1}, v_{j+4}, v_{j+3}, u_{j+6}, v_{j+5}, v_{j+8}, u_{j+7}, \ldots, v_{n-2}, v_{n-3}, v_{n-1}, v_n = v)$$

is a proper $v_{j+2} - v$ path in G with $V(Q') = \{v_{j+1}, v_{j+2}, \ldots, v_n\}$. Thus, $V(Q) \cup V(Q') = V(G)$, $V(Q) \cap V(Q') = \{v_{j+1}\}$ and $v_2 v_{j+2}$ and $v_{j+1} v_{j+2}$ have distinct colors (namely, $v_2 v_{j+2}$ is blue and $v_{j+1} v_{j+2}$ is red). Therefore, the path Q followed by Q' produces a proper Hamiltonian $u - v$ path in G.

Subcase 2.2. a is odd. If $a \equiv 3 \pmod 4$, then

$$Q = (u = v_j, v_{j-1}, v_{j-3}, v_{j-2}, u_{j-4}, v_{j-5}, v_{j-7}, u_{j-6}, \ldots, v_1, v_2, v_{j+1})$$

is a proper $u - v_{j+1}$ path in G; while if $a \equiv 1 \pmod 4$, then

$$Q = (u = v_j, v_{j-1}, v_{j-3}, v_{j-2}, u_{j-4}, v_{j-5}, v_{j-7}, u_{j-6}, \ldots, v_3, v_4, v_1, v_2, v_{j+1})$$

is a proper $u - v_{j+1}$ path in G. We now show that Q can be extended to a proper Hamiltonian $u - v$ path in G. If $b \equiv 0 \pmod 4$, then

$$Q' = (v_{j+1}, v_{j+2}, v_{j+4}, v_{j+3}, u_{j+5}, v_{j+6}, v_{j+8}, u_{j+7}, \ldots, v_{n-3}, v_{n-1}, v_{n-2}, v_n = v)$$

is a proper $v_{j+1} - v$ path in G; while if $b \equiv 2 \pmod 4$, then $b \geq 6$ (since $b \geq 3$) and

$$Q' = (v_{j+1}, v_{j+2}, v_{j+4}, v_{j+3}, u_{j+5}, v_{j+6}, v_{j+8}, u_{j+7}, \ldots, v_{n-4}, v_{n-1}, v_{n-2}, v_n = v)$$

is a proper $v_{j+1} - v$ path in G. Thus, as in Case 1, the path Q followed by Q' produces a proper Hamiltonian $u - v$ path in G. $\blacksquare$

It was shown in [26] that if G is a 2-connected graph, then the proper connection number of G is at most 3. Since every Hamiltonian-connected graph G of order at least 4 is 2-connected (in fact, 3-connected), $\mathrm{pc}(G) \leq 3$. Since we have seen no Hamiltonian-connected graph G for which $\mathrm{hpc}(G) > 3$, we are led to the following conjecture.

Conjecture 13.25 *If G is a Hamiltonian-connected graph, then $\mathrm{hpc}(G) \leq 3$.*

Let us review the various edge colorings we have discussed. First, if G is an edge-colored graph such that for every two vertices u and v, there exists a $u - v$ path P having the property that every subpath of P is a rainbow path, then this edge coloring is a rainbow coloring. On the other hand, if for every two vertices u and v, there exists a $u - v$ path Q having the property that every subpath of Q of length (at most) 2 is a rainbow path, then this edge coloring is a proper-path coloring. However, what if we require this for subpaths of some specific length greater than 2? Looking at rainbow colorings and proper-path colorings in this way brings up, quite naturally, other edge colorings that are intermediate to rainbow and proper-path colorings.

More formally, let G be an edge-colored nontrivial connected graph, where adjacent edges may be colored the same. For an integer $k \geq 2$, a path P in G is a **k-rainbow path** (*with respect to the edge coloring*) if every subpath of P having length k or less is a rainbow path. Thus, every proper path is a 2-rainbow path and for each $k \geq 3$, a k-rainbow path is also an ℓ-rainbow path for every integer ℓ with $2 \leq \ell \leq k$. In particular, every k-rainbow path is a proper path for each integer $k \geq 2$. Also, if k is the length of a longest path $u - v$ path in G, then a k-rainbow $u - v$ path is a rainbow $u - v$ path.

For an integer $k \geq 2$, an edge coloring c is a k-**rainbow coloring** of a connected graph G if every pair of distinct vertices of G are connected by a k-rainbow path in G. In this case, the graph G is k-**rainbow connected** (*with respect to c*). If j colors are used to produce a k-rainbow coloring of G, then c is referred to as a k-**rainbow j-edge coloring** (or simply a k-rainbow j-coloring). The minimum j for which G has a k-rainbow j-coloring is called the k-**rainbow connection number** $\mathrm{rc}_k(G)$ of G. Hence, $\mathrm{rc}_2(G) = \mathrm{pc}(G)$ and $r_\ell(G) = \mathrm{rc}(G)$ if ℓ is the length of a longest path in G. In general, for every nontrivial connected graph G whose longest paths have length ℓ,

$$\mathrm{pc}(G) = \mathrm{rc}_2(G) \leq \mathrm{rc}_3(G) \leq \cdots \leq \mathrm{rc}_\ell(G) = \mathrm{rc}(G). \tag{13.5}$$

This concept was introduced and studied in [37].

To illustrate these concepts, a proper-path 2-coloring, a 3-rainbow 3-coloring and a rainbow 4-coloring are shown for the graph G in Figure 13.12. In fact, for this graph G, we have $\mathrm{pc}(G) = 2$, $\mathrm{rc}_3(G) = 3$ and $\mathrm{rc}(G) = 4$.

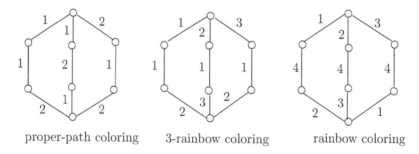

proper-path coloring   3-rainbow coloring   rainbow coloring

Figure 13.12: Three edge colorings of a graph G

First, we state some observations concerning k-rainbow connection numbers in general.

Observation 13.26 *If H is a connected spanning subgraph of a nontrivial connected graph G, then $\mathrm{rc}_k(G) \leq \mathrm{rc}_k(H)$ for each integer $k \geq 3$. In particular, if T is a spanning tree of G, then $\mathrm{rc}_k(G) \leq \mathrm{rc}_k(T)$.*

Proposition 13.27 *Let G be a connected graph of diameter $d \geq 2$ whose longest paths have length ℓ.*

(a) If $2 \leq k \leq d$, then $\mathrm{rc}_k(G) \geq k$.

(b) If $d + 1 \leq k \leq \ell$, then $\mathrm{rc}_k(G) \geq d$.

Proof. Let u and v be two antipodal vertices of G such that $d(u, v) = d$. If $2 \leq k \leq d$, then every k-rainbow coloring of G must assign at least k distinct colors to the edges of any k-rainbow $u - v$ path in G. Hence, $\mathrm{rc}_k(G) \geq k$ and so (a) holds. If $d + 1 \leq k \leq \ell$, then every k-rainbow coloring of G must assign at least d distinct colors to the edges of any k-rainbow $u - v$ path in G. Hence, $\mathrm{rc}_k(G) \geq d$ and so (b) holds.   ∎

Observation 13.28 *For integers k and n with $2 \leq k \leq n - 1$, $\mathrm{rc}_k(P_n) = k$.*

We now turn to k-rainbow connection numbers of cycles.

Theorem 13.29 *For integers k and n with $3 \leq k \leq n - 1$ and $n \geq 5$,*

$$\mathrm{rc}_k(C_n) = \min \left\{ \lceil n/2 \rceil,\ k \right\}.$$

Proof. Let $C_n = (v_1, v_2, \ldots, v_n, v_{n+1} = v_1)$ where $e_i = v_i v_{i+1}$ for $1 \leq i \leq n$. The diameter of C_n is $\mathrm{diam}(C_n) = \lfloor n/2 \rfloor$. We consider two cases, according to whether $\lceil n/2 \rceil \leq k$ or $\lceil n/2 \rceil > k$.

Case 1. $\lceil n/2 \rceil \leq k$. Here, we show that $\mathrm{rc}_k(C_n) = \min \left\{ \lceil n/2 \rceil,\ k \right\} = \lceil n/2 \rceil$. First, define an edge coloring c of C_n by

$$c(e_i) = \begin{cases} i & \text{if } 1 \leq i \leq \lceil n/2 \rceil \\ i - \lceil n/2 \rceil & \text{if } \lceil n/2 \rceil + 1 \leq i \leq n. \end{cases}$$

Thus, the *color sequence of the edges of C_n* with respect to c is

$$S_c = (c(e_1), c(e_2), \ldots, c(e_n)) = (1, 2, \ldots, \lceil n/2 \rceil, 1, 2, \ldots, \lfloor n/2 \rfloor).$$

Note that every subsequence of length at most $\lfloor n/2 \rfloor$ in S_c has distinct terms. For two vertices v_i and v_j of C_n where $1 \leq i < j \leq n$, there are exactly two $v_i - v_j$ paths P and Q in C_n. We may assume that $|E(P)| \leq |E(Q)|$. Then P is a rainbow path of length at most $\lfloor n/2 \rfloor$. Since $\lfloor n/2 \rfloor \leq \lceil n/2 \rceil \leq k$, it follows that P is a k-rainbow $v_i - v_j$ path. Therefore, c is a k-rainbow $\lceil n/2 \rceil$-coloring of C_n and so $\mathrm{rc}_k(C_n) \leq \lceil n/2 \rceil$.

Next, we show that $\mathrm{rc}_k(C_n) \geq \lceil n/2 \rceil$. Assume, to the contrary, that C_n has a k-rainbow coloring c^* using the colors $1, 2, \ldots, \lceil n/2 \rceil - 1$. Of the two $v_1 - v_{\lfloor n/2 \rfloor + 1}$ paths on C_n, one has length $\lceil n/2 \rceil$ and the other $\lfloor n/2 \rfloor$. If n is even, then neither path is a k-rainbow path; while if n is odd, the path of length $\lceil n/2 \rceil$ cannot be a k-rainbow path. In this case, let $n = 2t + 1$ for some integer $t \geq 2$ and consider the path $(v_1, v_2, \ldots, v_{t+1})$ of length $t = \lfloor n/2 \rfloor$, which is necessarily a k-rainbow path. Hence, we may assume that $c^*(v_i v_{i+1}) = i$ for $1 \leq i \leq t$. The path $(v_2, v_3, \ldots, v_{t+s})$ also has length t and so is a k-rainbow path, implying that $c^*(v_{t+1} v_{t+2}) = 1$. Continuing in this manner, we see that

$$c^*(v_i v_{i+1}) = \begin{cases} i & \text{if } 1 \leq i \leq t \\ t - i & \text{if } t + 1 \leq i \leq 2t \\ 1 & \text{If } i = 2t + 1. \end{cases}$$

The $v_{2t+1} - v_t$ path $P = (v_{2t+1}, v_1, v_2, \ldots, v_t)$ of length t has $c^*(v_{2t+1}v_1) = v^*(v_1v_2) = 1$, implying that neither P nor the $v_{2t+1} - v_t$ path of length $t+1$ is a k-rainbow path, producing a contradiction.

Case 2. $\lceil n/2 \rceil > k$. We show that $rc_k(C_n) = \min\{\lceil n/2 \rceil, k\} = k$. First, by Observations 13.26 and 13.28, $rc_k(C_n) \leq rc_k(P_n) \leq k$. Next, we show that $rc_k(C_n) \geq k$. Assume, to the contrary, that C_n has a k-rainbow coloring c^* using the $k - 1$ colors $1, 2, \ldots, k - 1$. There are two $v_1 - v_{\lceil n/2 \rceil+1}$ paths in C_n, one of which has length $\lfloor n/2 \rfloor$ and the other has length $\lceil n/2 \rceil$. Since the coloring c^* only uses $k - 1$ distinct colors, neither path can be a k-rainbow $v_1 - v_{t+1}$ path in C_n, producing a contradiction. ∎

The size of a graph H is denoted by $m(H)$. The k-rainbow connection number of a tree has been determined in [65].

Theorem 13.30 *If T is a tree of diameter at least $k \geq 2$ for some integer k, then*

$$rc_k(T) = \max\{m(T') : T' \text{ is a subtree of } T \text{ with } \text{diam}(T') = k\}.$$

13.3 Rainbow Disconnection

In this section, the rainbow disconnection number of a graph is discussed, which is a concept that is somewhat the reverse of rainbow connection. While rainbow connection concerns connecting each pair of vertices by a rainbow set of edges, the concept described here concerns disconnecting each pair of vertices by a rainbow set of edges.

First, we review some concepts and results discussed in Chapter 2. An **edge-cut** of a nontrivial connected graph G is a set R of edges of G such that $G - R$ is disconnected. The minimum number of edges in an edge-cut of G is its **edge-connectivity** $\lambda(G)$. We then have the inequality $\lambda(G) \leq \delta(G)$. For two distinct vertices u and v of G, let $\lambda(u, v)$ denote the minimum number of edges in an edge-cut R of G such that u and v lie in different components of $G - R$. Thus,

$$\lambda(G) = \min\{\lambda(u, v) : u, v \in V(G)\}.$$

The following result of Elias, Feinstein, and Shannon [70] and Ford and Fulkerson [86] presents an alternate interpretation of $\lambda(u, v)$.

Theorem 13.31 *For every two vertices u and v in a graph G, $\lambda(u, v)$ is the maximum number of pairwise edge-disjoint $u - v$ paths in G.*

The **upper edge-connectivity** $\lambda^+(G)$ is defined by

$$\lambda^+(G) = \max\{\lambda(u, v) : u, v \in V(G)\}.$$

Consider, for example, the graph $K_n + v$, which is obtained from the complete graph K_n by adding a new vertex v and joining it to a vertex of K_n. Therefore, $\lambda(K_n + v) =$

1 while $\lambda^+(K_n + v) = n - 1$. Thus, $\lambda(G)$ denotes the global minimum edge-connectivity of a graph, while $\lambda^+(G)$ denotes the local maximum edge-connectivity of a graph.

A set R of edges in a connected edge-colored graph G is a **rainbow set** if no two edges in R are colored the same. A set R of edges in a nontrivial connected, edge-colored graph G is a **rainbow cut** of G if R is both a rainbow set and an edge-cut. A rainbow cut R is said to **separate** two vertices u and v of G if u and v belong to different components of $G - R$. Any such rainbow cut in G is called a $u-v$ **rainbow cut** in G. An edge-coloring of G is a **rainbow disconnection coloring** if for every two distinct vertices u and v of G, there exists a $u-v$ rainbow cut in G. The **rainbow disconnection number** $\mathrm{rd}(G)$ of G is the minimum number of colors required of a rainbow disconnection coloring of G. A rainbow disconnection coloring with $\mathrm{rd}(G)$ colors is called an rd-**coloring** of G. These concepts were introduced and studied by Gary Chartrand, Teresa Haynes, Stephen Hedetniemi, and Ping Zhang in [46]. We now present bounds for the rainbow disconnection number of a graph.

Proposition 13.32 *If G is a nontrivial connected graph, then*

$$\lambda(G) \le \lambda^+(G) \le \mathrm{rd}(G) \le \chi'(G) \le \Delta(G) + 1.$$

Proof. First, by Vizing's theorem (Theorem 1.5.6), $\chi'(G) \le \Delta(G) + 1$. Now, let there be given a proper edge-coloring of G using $\chi'(G)$ colors. Then, for each vertex x of G, the set E_x of edges incident with x is a rainbow set and $|E_x| = \deg x \le \Delta(G) \le \chi'(G)$. Furthermore, E_x is a rainbow cut in G and so $\mathrm{rd}(G) \le \chi'(G)$.

Next, let there be given an rd-coloring of G. Let u and v be two vertices of G such that $\lambda^+(G) = \lambda(u, v)$ and let R be a $u - v$ rainbow cut with $|R| = \lambda(u, v)$. Then $|R| \le \mathrm{rd}(G)$. Thus, $\lambda(G) \le \lambda^+(G) = |R| \le \mathrm{rd}(G)$. ∎

We now present examples of two classes of connected graphs G for which $\lambda(G) = \mathrm{rd}(G)$, namely cycles and wheels.

Proposition 13.33 *If C_n is a cycle of order $n \ge 3$, then $\mathrm{rd}(C_n) = 2$.*

Proof. Since $\lambda(C_n) = 2$, it follows by Proposition 13.32 that $\mathrm{rd}(C_n) \ge 2$. To show that $\mathrm{rd}(C_n) \le 2$, let c be an edge-coloring of C_n that assigns the color 1 to exactly $n - 1$ edges of C_n and the color 2 to the remaining edge e of C_n. Let u and v be two vertices of C_n. There are two $u - v$ paths P and Q in C_n, exactly one of which contains the edge e, say $e \in E(P)$. Then any set $\{e, f\}$, where $f \in E(Q)$, is a $u - v$ rainbow cut. Thus, c is a rainbow disconnection coloring of C_n using two colors. Hence, $\mathrm{rd}(C_n) = 2$. ∎

Proposition 13.34 *If $W_n = C_n \vee K_1$ is the wheel of order $n + 1 \ge 4$, then $\mathrm{rd}(W_n) = 3$.*

Proof. Since $\lambda(W_n) = 3$, it follows by Proposition 13.32 that $\mathrm{rd}(W_n) \ge 3$. It remains to show that there is a rainbow disconnection coloring of W_n using only the colors $1, 2, 3$. Suppose that $C_n = (v_1, v_2, \ldots, v_n, v_1)$ and that v is the center of W_n. Define an edge-coloring $c : E(W_n) \to \{1, 2, 3\}$ of W_n as follows. First, let c be a proper edge-coloring of C_n using the colors $1, 2$ when n is even and the colors $1, 2, 3$ when n is odd. For each integer i with $1 \le i \le n$, let

$$a_i \in \{1, 2, 3\} - \{c(v_{i-1}v_i), c(v_iv_{i+1})\}$$

where each subscript is expressed as an integer $1, 2, \ldots, n$ modulo n, and let $c(vv_i) = a_i$. This coloring is illustrated for W_6 and W_7 in Figure 13.13.

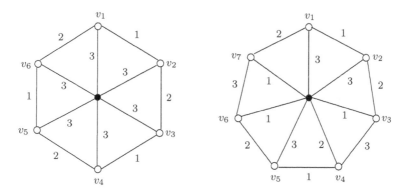

Figure 13.13: Rainbow disconnection colorings of W_6 and W_7

Thus, the set E_{v_i} of the three edges incident with v_i is a rainbow set for $1 \le i \le n$. Let x and y be two distinct vertices of W_n. Then at least one of x and y belongs to C_n, say $x \in V(C_n)$. Since E_x separates x and y, it follows that c is a rainbow disconnection coloring of W_n using three colors. Hence, $\text{rd}(W_n) = 3$. ∎

Since $\chi'(C_n) = 3$ when $n \ge 3$ is odd and $\chi'(W_n) = n$ for each integer $n \ge 3$, it follows that $\text{rd}(G) < \chi'(G)$ if G is an odd cycle or if G is a wheel of order at least 4. Wheels therefore are examples of graphs G for which $\chi'(G) - \text{rd}(G)$ can be arbitrarily large. We now give an example of a graph G for which $\lambda^+(G) < \text{rd}(G) = \chi'(G)$.

Proposition 13.35 *The rainbow disconnection number of the Petersen graph is 4.*

Proof. Let P denote the Petersen graph where $V(P) = \{v_1, v_2, \ldots, v_{10}\}$. Since $\lambda(P) = 3$ and $\chi'(P) = 4$, it follows by Proposition 13.32 that $\text{rd}(P) = 3$ or $\text{rd}(P) = 4$. Assume, to the contrary, that $\text{rd}(P) = 3$ and let there be given a rainbow disconnection 3-coloring of P. Now, let u and v be two vertices of P and let R be a $u - v$ rainbow cut. Hence, $|R| \le 3$ and $P - R$ is disconnected, where u and v belong to different components of $P - R$. Let U be the vertex set of the component of $P - R$ containing u, where $|U| = k$. We may assume that $1 \le k \le 5$. First, suppose that $1 \le k \le 4$. Since the girth of P is 5, the subgraph $P[U]$ induced by U contains $k - 1$ edges. Therefore, $|R| = 3k - (2k - 2) = k + 2$, where then $3 \le k + 2 \le 6$. If $k = 5$, then $P[U]$ contains at most five edges and so $|R| \ge 5$, which is impossible. Since $\text{rd}(P) = 3$, it follows that $|R| \le 3$ and so $k = 1$. Hence, the only possible $u - v$ rainbow cut is the set consisting of the three edges incident with u (or with v).

Let the colors assigned to the edges of P be red, blue, and green. Since $\chi'(P) = 4$, there is at least one vertex of P that is incident with two edges of the same color. We claim, in fact, that there are at least two such vertices. Let E_R, E_B, and E_G denote the sets of edges of P colored red, blue, and green, respectively, and let P_R, P_B, and P_G be the spanning subgraphs of P with edge sets E_R, E_B, and E_G. We may assume that $|E_R| \geq |E_B| \geq |E_G|$ and so $|E_R| \geq 5$. If $|E_R| \geq 7$, then $\sum_{i=1}^{10} \deg_{P_R} v_i \geq 14$. Since $\deg_{P_R} v_i \leq 3$ for each i with $1 \leq i \leq 10$, at least two vertices are incident with two red edges, verifying the claim. If $|E_R| = 6$, then $\sum_{i=1}^{10} \deg_{P_R} v_i = 12$. Then either (i) at least two vertices are incident with two red edges or (ii) there is a vertex, say v_{10}, incident with three red edges and each of $v_1, v_2, \ldots, v_9$ is incident with exactly one red edge. If (ii) occurs, then either $|E_B| = 6$ or $|E_B| = 5$ and so $\sum_{i=1}^{9} \deg_{P_B} v_i \geq 10$, which implies that at least one of the vertices $v_1, v_2, \ldots, v_9$ is incident with two blue edges, again verifying the claim. The only remaining possibility is therefore $|E_R| = |E_B| = |E_G| = 5$. If E_R is an independent set of five edges, then $P - E_R$ is a 2-regular graph. Since the girth of P is 5 and P is not Hamiltonian, it follows that $P - E_R$ consists of two vertex-disjoint 5-cycles. Thus, there is a vertex of P in each cycle incident with two blue edges or with two green edges, verifying the claim. Hence, none of E_R, E_B, or E_G is an independent set. This implies that for each of these colors, there is a vertex of P incident with two edges of this color, verifying the claim in general.

Thus, P contains two vertices u and v, each of which is incident with two edges of the same color. Since the only $u - v$ rainbow cut is the set of edges incident with u or v, this is a contradiction. ∎

The following two results are useful.

Proposition 13.36 *If H is a connected subgraph of a graph G, then*

$$\mathrm{rd}(H) \leq \mathrm{rd}(G).$$

Proof. Let c be an rd-coloring of G and let u and v be two vertices of G. Suppose that R is a $u - v$ rainbow cut. Then $R \cap E(H)$ is a $u - v$ rainbow cut in H. Hence, c restricted to H is a rainbow disconnection coloring of H. Thus, $\mathrm{rd}(H) \leq \mathrm{rd}(G)$. ∎

A **block** of a graph is a maximal connected graph of G containing no cut-vertices. The **block decomposition** of G is the set of blocks of G.

Proposition 13.37 *Let G be a nontrivial connected graph, and let B be a block of G such that $\mathrm{rd}(B)$ is maximum among all blocks of G. Then $\mathrm{rd}(G) = \mathrm{rd}(B)$.*

Proof. Let G be a nontrivial connected graph. Let $\{B_1, B_2, \ldots, B_t\}$ be a block decomposition of G, and let $k = \max\{\mathrm{rd}(B_i) \,|\, 1 \leq i \leq t\}$. If G has no cut-vertices, then $G = B_1$ and the result follows. Hence, we may assume that G has at least one cut-vertex. By Proposition 13.36, $k \leq \mathrm{rd}(G)$.

Let c_i be an rd-coloring of B_i. We define the edge-coloring $c : E(G) \to [k]$ of G by $c(e) = c_i(e)$ if $e \in E(B_i)$.

Let $x, y \in V(G)$. If there exists a block, say B_i, that contains both x and y, then any $x - y$ rainbow cut in B_i is an $x - y$ rainbow cut in G. Hence, we can assume that no block of G contains both x and y, and that $x \in B_i$ and $y \in B_j$, where $i \neq j$. Now every $x - y$ path contains a cut-vertex, say v, of G in B_i and a cut-vertex, say w, of G in B_j. Note that v could equal w. If $x \neq v$, then any $x - v$ rainbow cut of B_i is an $x - y$ rainbow cut in G. Similarly, if $y \neq w$, then any $y - w$ rainbow cut of B_j is an $x - y$ rainbow cut in G. Thus, we may assume that $x = v$ and $y = w$. It follows that $v \neq w$. Consider the $x - y$ path $P = (x = v_1, v_2, ..., v_p = y)$. Since x and y are cut-vertices in different blocks and no block contains both x and y, P contains a cut-vertex z of G in B_i, that is, $z = v_k$ for some k $(2 \leq k \leq p-1)$. Then any $x - z$ rainbow cut of B_i is an $x - y$ rainbow cut of G. Hence, $\mathrm{rd}(G) \leq k$, and so $\mathrm{rd}(G) = k$. ∎

As a consequence of Proposition 13.37, the study of rainbow disconnection numbers can be restricted to 2-connected graphs. We now present several corollaries of Proposition 13.37.

Corollary 13.38 *Let G and H be any two nontrivial connected graphs, and let $G \cdot H$ be a graph formed by identifying a vertex in G with a vertex in H. Then*

$$\mathrm{rd}(G \cdot H) = \max\{\mathrm{rd}(G), \mathrm{rd}(H)\}.$$

Corollary 13.39 *Let G and H be any two nontrivial connected graphs, and let F be a graph formed by adding an edge between any vertex u in G and any vertex v in H. Then*

$$\mathrm{rd}(F) = \max\{\mathrm{rd}(G), \mathrm{rd}(H)\}.$$

Corollary 13.40 *Let G be a nontrivial connected graph and G' the graph obtained by attaching a pendant edge uv to some vertex u of G. Then $\mathrm{rd}(G') = \mathrm{rd}(G)$.*

Recall that the **corona** $\mathrm{cor}(G)$ is the graph obtained from G by attaching a leaf to each vertex of G. Thus, if G has order n, then $\mathrm{cor}(G)$ has order $2n$ and has precisely n leaves.

Corollary 13.41 *If G is a nontrivial connected graph, then $\mathrm{rd}(\mathrm{cor}(G)) = \mathrm{rd}(G)$.*

Corollary 13.42 *Let G be a nontrivial connected graph, let T be a nontrivial tree and let u and v be vertices of G and T, respectively. If H is the graph obtained from G and T by identifying u and v, then $\mathrm{rd}(H) = \mathrm{rd}(G)$.*

Corollary 13.43 *If G is a unicyclic graph G, then $\mathrm{rd}(G) = 2$.*

We now characterize all those nontrivial connected graphs of order n with rainbow disconnection number k for each $k \in \{1, 2, n-1\}$. The result for graphs having rainbow disconnection number 1 follows directly from Propositions 13.36 and 13.37.

Proposition 13.44 *Let G be a nontrivial connected graph. Then $\mathrm{rd}(G) = 1$ if and only if G is a tree.*

The following result is useful.

Proposition 13.45 *A 2-connected graph G is a cycle if and only if for every two vertices u and v of G, there are exactly two internally disjoint $u - v$ paths in G.*

Theorem 13.46 *Let G be a nontrivial connected graph. Then $\mathrm{rd}(G) = 2$ if and only if each block of G is either K_2 or a cycle and at least one block of G is a cycle.*

Proof. If G a nontrivial connected graph, each block of which is either K_2 or a cycle and at least one block of G is a cycle, then Propositions 13.33 and 13.37 imply that $\mathrm{rd}(G) = 2$.

We now verify the converse. Assume, to the contrary, that there is a connected graph G with $\mathrm{rd}(G) = 2$ that does not have the property that each block of G is either K_2 or a cycle and at least one block of G is a cycle. First, not all blocks can be K_2, for otherwise, G is a tree and so $\mathrm{rd}(G) = 1$ by Proposition 13.44. Hence, G contains a block that is neither K_2 nor a cycle. By Proposition 13.45, there exist two distinct vertices u and v of G for which G contains at least three internally disjoint $u - v$ paths P_1, P_2 and P_3. Thus, any $u - v$ rainbow cut R must contain at least one edge from each of P_1, P_2 and P_3 and so $|R| \geq 3$, which is impossible. ■

We now consider those graphs that are, in a sense, opposite to trees.

Proposition 13.47 *For each integer $n \geq 4$, $\mathrm{rd}(K_n) = n - 1$.*

Proof. Suppose first that $n \geq 4$ is even. Then $\lambda(K_n) = \chi'(K_n) = n - 1$. It then follows by Proposition 13.32 that $\mathrm{rd}(K_n) = n - 1$. Next, suppose that $n \geq 5$ is odd. Then

$$n - 1 = \lambda(K_n) \leq \mathrm{rd}(K_n) \leq \chi'(K_n) = n$$

by Proposition 13.32. To show that $\mathrm{rd}(K_n) = n - 1$, it remains to show that there is a rainbow disconnection coloring of K_n using $n - 1$ colors. Let $x \in V(K_n)$. Then $K_n - x = K_{n-1}$. Since $n - 1$ is even, it follows that $\chi'(K_{n-1}) = n - 2$. Thus, there is a proper edge-coloring c_0 of K_{n-1} using the colors $1, 2, \ldots, n-2$. We now extend c_0 to an edge-coloring c of K_n by assigning the color $n - 1$ to each edge of K_n that is incident with x. We show that c is a rainbow disconnection coloring of K_n. Let u and v be two vertices of K_n, where say $u \neq x$. Then the set E_u of edges incident with u is a $u - v$ rainbow cut. Thus, c is a rainbow disconnection coloring of K_n and so $\mathrm{rd}(K_n) \leq n - 1$ and so $\mathrm{rd}(K_n) = n - 1$. ■

By Propositions 13.32, 13.36, and 13.47, if G is a nontrivial connected graph of order n, then

$$1 \leq \mathrm{rd}(G) \leq n - 1. \tag{13.6}$$

Furthermore, $\mathrm{rd}(G) = 1$ if and only if G is a tree by Proposition 13.44. We have seen that the complete graphs K_n of order $n \geq 2$ have rainbow disconnection number $n - 1$. We now characterize all nontrivial connected graphs of order n having rainbow disconnection number $n - 1$.

Theorem 13.48 *Let G be a nontrivial connected graph of order n. Then $\mathrm{rd}(G) = n - 1$ if and only if G contains at least two vertices of degree $n - 1$.*

Proof. First, suppose that G is a nontrivial connected graph of order n containing at least two vertices of degree $n - 1$. Since $\mathrm{rd}(G) \le n - 1$ by (13.6), it remains to show that $\mathrm{rd}(G) \ge n - 1$. Let $u, v \in V(G)$ such that $\deg u = \deg v = n - 1$. Among all sets of edges that separate u and v in G, let S be one of minimum size. We show that $|S| \ge n - 1$. Let U be a component of $G - S$ that contains u and let $W = V(G) - U$. Thus, $v \in W$ and $S = [U, W]$ consists of those edges in $G - S$ joining a vertex of U and a vertex of W. Suppose that $|U| = k$ for some integer k with $1 \le k \le n - 1$ and then $|W| = n - k$. The vertex u is adjacent to each of the $n - k$ vertices of W and each of the remaining $k - 1$ vertices in U is adjacent to at least one vertex in W. Hence,

$$|S| \ge n - k + (k - 1) = n - 1.$$

This implies that every $u - v$ rainbow cut contains at least $n - 1$ edges of G and so $\mathrm{rd}(G) \ge n - 1$.

For the converse, suppose that G is a nontrivial connected graph of order n having at most one vertex of degree $n - 1$. We show that $\mathrm{rd}(G) \le n - 2$. We consider two cases.

Case 1. Exactly one vertex v of G has degree $n - 1$. Let $H = G - v$. Thus, $\Delta(H) \le n - 3$. Since

$$\chi'(H) \le \Delta(H) + 1 = n - 2,$$

there is a proper edge-coloring of H using $n - 2$ colors. We now define an edge-coloring $c : E(G) \to [n - 2]$ of G. First, let c be a proper $(n - 2)$-edge-coloring of H. For each vertex $x \in V(H)$, since $\deg_H x \le n - 3$, there is $a_x \in [n - 2]$ such that a_x is not assigned to any edge incident with x. Define $c(vx) = a_x$. Thus, the set E_x of edges incident with x is a rainbow set for each $x \in V(H)$. Let u and w be two distinct vertices of G. Then at least one of u and w belongs to H, say $u \in V(H)$. Since E_u separates u and w, it follows that c is a rainbow disconnection coloring of G using $n - 2$ colors. Hence, $\mathrm{rd}(G) \le n - 2$.

Case 2. No vertex of G has degree $n - 1$. Therefore $\Delta(G) \le n - 2$. If $\Delta(G) \le n - 3$, then $\mathrm{rd}(G) \le \chi'(G) \le n - 2$ by Proposition 13.32. Thus, we may assume that $\Delta(G) = n - 2$. Suppose first that G is not $(n - 2)$-regular. We claim that G is a connected spanning subgraph of some graph G^* of order n having exactly one vertex of degree $n - 1$. Let u be a vertex of degree $k \le n - 3$ in G. Let $N(u)$ be the neighborhood of u and $W = V(G) - N[u]$, where $N[u] = N(u) \cup \{u\}$ is the closed neighborhood of u. Then $|N(u)| = k$ and $|W| = n - k - 1 \ge 2$. If W contains a vertex v of degree $n - 2$ in G, then v is the only vertex of degree $n - 1$ in $G^* = G + uv$. If no vertex in W has degree $n - 2$ in G, then let G^* be the graph obtained from G by joining u to each vertex in W. In this case, u is the only vertex of degree $n - 1$ in G^*. It then follows by Case 1 that $\mathrm{rd}(G^*) \le n - 2$. Since G is a connected spanning subgraph of G^*, it follows by Proposition 13.36 that

$$\mathrm{rd}(G) \le \mathrm{rd}(G^*) \le n - 2.$$

Finally, suppose that G is $(n-2)$-regular. Thus, G is 1-factorable and so $\chi'(G) = \Delta(G) = n-2$. Therefore, $\text{rd}(G) \leq \chi'(G) = n-2$ by Proposition 13.32. ∎

We now consider the following question:

> For a given pair k, n of positive integers with $k \leq n-1$, what are the minimum possible size and maximum possible size of a connected graph G of order n such that the rainbow disconnection number of G is k?

We have seen in Proposition 13.44 that the only connected graphs of order n having rainbow disconnection number 1 are the trees of order n. That is, the connected graphs of order n having rainbow disconnection number 1 have size $n-1$. We have also seen in Theorem 13.46 that the minimum size of a connected graph of order $n \geq 3$ having rainbow disconnection number 2 is n. Furthermore, we have seen in Theorem 13.48 that the minimum size of a connected graph of order $n \geq 2$ having rainbow disconnection number $n-1$ is $2n-3$. In fact, these are special cases of a more general result. In order to show this, we first present the following observation.

Proposition 13.49 *Let H be a connected graph of order n that is not complete and let x and y be two nonadjacent vertices of H. Then $\text{rd}(H + xy) \leq \text{rd}(H) + 1$.*

Proof. Suppose that $\text{rd}(H) = k$ for some positive integer k and let c_0 be a rainbow disconnection coloring of H using the colors $1, 2, \ldots, k$. Extend the coloring c_0 to the edge-coloring c of $H + xy$ by assigning the color $k+1$ to the edge xy. Let u and v be two vertices of H and let R be a $u - v$ rainbow cut in H. Then $R \cup \{xy\}$ is a $u - v$ rainbow cut in $H + xy$. Hence, c is a rainbow disconnection $(k+1)$-coloring of $H + xy$. Therefore, $\text{rd}(H + xy) \leq k + 1 = \text{rd}(H) + 1$. ∎

Theorem 13.50 *For integers k and n with $1 \leq k \leq n-1$, the minimum size of a connected graph of order n having rainbow disconnection number k is $n + k - 2$.*

Proof. By Theorem 13.48, the result is true for $k = n-1$. Hence, we may assume that $1 \leq k \leq n-2$. First, we show that if the size of a connected graph G of order n is $n + k - 2$, then $\text{rd}(G) \leq k$. We proceed by induction on k. We have seen that the result is true for $k = 1, 2$ by Proposition 13.44 and Theorem 13.46. Suppose that if the size of a connected graph H of order n is $n + k - 2$ for some integer k with $2 \leq k \leq n-3$, then $\text{rd}(H) \leq k$. Let G be a connected graph of order n and size $n + (k+1) - 2 = n + k - 1$. We show that $\text{rd}(G) \leq k + 1$. Since G is not a tree, there is an edge e such that $H = G - e$ is a connected spanning subgraph of G. Since the size of H is $n + k - 2$, it follows by induction hypothesis that $\text{rd}(H) \leq k$. Hence, $\text{rd}(G) = \text{rd}(H + e) \leq k + 1$ by Proposition 13.49. Therefore, the minimum possible size for a connected graph G of order n to have $\text{rd}(G) = k$ is $n + k - 2$. [Note that if F is a connected graph of order n and size $m < n + k - 2$, then $m = n + k - a$ for some integer $a \geq 3$. Since $m = n + (k - a + 2) - 2$, it follows that $\text{rd}(F) \leq k - a + 2 \leq k - 1$.]

It remains to show that for each pair k, n of integers with $1 \leq k \leq n-1$ there is a connected graph G of order n and size $n + k - 2$ such that $\text{rd}(G) = k$. Since this is true for $k = 1, 2, n-1$, we now assume that $3 \leq k \leq n-2$. Let $H = K_{2,k}$ with partite set $U = \{u_1, u_2\}$ and $W = \{w_1, w_2, \ldots, w_k\}$. Now, let G be the graph of order n and size $n+k-2$ obtained from H by subdividing the edge u_1w_1 a total of $n-k-2$ times, producing the path $P = (u_1, v_1, v_2, \ldots, v_{n-k-2}, w_1)$ in G. Since $\chi'(H) = k$, there is a proper edge-coloring c_H of H using the colors $1, 2, \ldots, k$. We may assume that $c(u_1w_1) = 1$ and $c(u_2w_1) = 2$. Next, we extend the coloring c_H to a proper edge-coloring c_G of G using the colors $1, 2, \ldots, k$ by defining $c_G(u_1v_1) = 1$ and alternating the colors of the edges of P with 3 and 1 thereafter. Hence, $\chi'(G) = k$ and so $\mathrm{rd}(G) \leq \chi'(G) = k$ by Proposition 13.32. Furthermore, since $\lambda(u_1, u_2) = k$ and $\lambda(x, y) = 2$ for all other pairs x, y of vertices of G, it follows that $\lambda^+(G) = k$. Again, by Proposition 13.32, $\mathrm{rd}(G) \geq \lambda^+(G) = k$ and so $\mathrm{rd}(G) = k$. ∎

For given integers k and n with $1 \leq k \leq n-1$, we've determined the minimum size of a connected graph G of order n with $\mathrm{rd}(G) = k$. So, this brings up the question of determining the maximum size of a connected graph G of order n with $\mathrm{rd}(G) = k$. Of course, we know this size when $k = 1$; it's $n-1$. Also, we know this size when $k = n-1$; it's $\binom{n}{2}$. For odd integers n, we have the following conjecture.

Conjecture 13.51 *Let k and n be integers with $1 \leq k \leq n-1$ and $n \geq 5$ is odd. Then the maximum size of a connected graph G of order n with $\mathrm{rd}(G) = k$ is $\frac{(k+1)(n-1)}{2}$.*

Notice that when $k = 1$, then $\frac{(k+1)(n-1)}{2} = n-1$ and when $k = n-1$, then $\frac{(k+1)(n-1)}{2} = \binom{n}{2}$. Also, when $k = 2$, then $\frac{(k+1)(n-1)}{2} = \frac{3n-3}{2}$. This is the size of the so-called **friendship graph** $\left(\frac{k-1}{2}\right) K_2 \vee K_1$ of order n (every two vertices has a unique friend). Since each block of a friendship graph is a triangle, it follows by Theorem 13.46 that each such graph has rainbow disconnection number 2.

For given integers k and n with $1 \leq k \leq n-1$ and $n \geq 5$ is odd, let H_k be a $(k-1)$-regular graph of order $n-1$. Since $n-1$ is even, such graphs H_k exist. Now, let $G_k = H_k \vee K_1$ be the join of H_k and K_1. Thus, G_k is a connected graph of order n having one vertex of degree $n-1$ and $n-1$ vertices of degree k. The size m of G_k satisfies the equation:

$$2m = (n-1) + (n-1)k = (k+1)(n-1)$$

and so $m = \frac{(k+1)(n-1)}{2}$. The graph H_k can be selected so that it is 1-factorable and so $\chi'(H_k) = k - 1$. If a proper $(k-1)$-edge-coloring of H_k is given using the colors $1, 2, \ldots k-1$, and every edge incident with the vertex of G_k of degree $n-1$ is assigned the color k, then the edges incident with each vertex of degree k are properly colored with k colors. For any two vertices u and v of G_k, at least one of u and v has degree k in G_k, say $\deg_{G_k} u = k$. Then the set of edges incident with u is a $u - v$ rainbow cut in H. Since this is a rainbow disconnection k-coloring of G, it follows that $\mathrm{rd}(G_k) \leq k$. It is reasonable to conjecture that $\mathrm{rd}(G_k) = k$.

We would still be left with the question of whether every graph H of order n and size $\frac{(k+1)(n-1)}{2} + 1$ must have $\mathrm{rd}(H) > k$. Certainly, every such graph H must contain at least two vertices whose degrees exceed k.

Exercises for Chapter 13

1. Verify the equality in (13.1), that is, $\text{diam}(G) \le \text{rc}(G) \le \text{src}(G) \le m$ for every nontrivial connected graph G of size m.

2. Determine $\text{rc}(G)$ and $\text{src}(G)$ for the graph G in Figure 13.14.

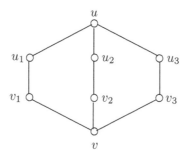

Figure 13.14: The graph G in Exercise 2

3. (a) Show that if G is a graph of diameter 2, then $\text{src}(G + K_1) \le \text{src}(G)$.

    (b) Give an example of a graph G of diameter 2 for which $\text{src}(G + K_1) < \text{src}(G)$.

    (c) Give an example of a graph H of diameter 3 for which $\text{rc}(H + K_1) < \text{rc}(H)$.

4. The rainbow connectivity of the 3-cube Q_3 is known to be 7. Show that $\kappa_r(Q_3) \le 7$.

5. We have seen that if $G = K_3 \,\square\, K_2$, then $\kappa_r(G) = 6$.

    (a) Determine $\text{rc}(G)$ and $\text{src}(G)$.

    (b) Determine the minimum positive integer k for which there exists a k-edge coloring of G such that every two vertices u and v of G are connected by *two* internally disjoint $u - v$ rainbow paths.

6. (a) Prove that $\text{hrc}(P_n \vee \overline{K}_2) = n + 1$ for each integer $n \ge 3$.

    (b) Prove that if G is a graph of order $n \ge 3$ containing a Hamiltonian path, then $\text{hrc}(G \vee \overline{K}_2) = n + 1$.

7. It is known that $\text{hrc}(K_3 \,\square\, K_2) = 7$. Show that $\text{hrc}(K_3 \,\square\, K_2) \le 7$.

8. Let H be a Hamiltonian-connected graph of order $n \ge 4$. Prove that $\text{hrc}(H \,\square\, K_2) \le 2\,\text{hrc}(H) + 2$.

9. (a) Determine $\text{hrc}(C_n^2)$ for $3 \le n \le 6$.

    (b) Prove that $\text{hrc}(C_n^2) \le n$ for each integer $n \ge 7$.

(c) Prove that if G is a Hamiltonian graph of order n, then $\mathrm{hrc}(G^2) \le n$.

(d) Determine a class of Hamiltonian graphs of order n such that
$\mathrm{hrc}(G^2) = n - 1$.

10. Let G be a connected graph of order n and size m. Prove the following:

(a) $\mathrm{pc}(G) = 1$ if and only if $G = K_n$,

(b) $\mathrm{pc}(G) = m$ if and only if $G = K_{1,m}$.

11. Let G be a nontrivial connected graph of order n and size m. Prove that

(a) $\mathrm{spc}(G) = 1$ if and only if $G = K_n$ and

(b) $\mathrm{spc}(G) = m$ if and only if $G = K_{1,m}$.

12. Prove that if G is a tree, then $\mathrm{spc}(G) = \mathrm{pc}(G)$.

13. Prove that if G is a connected graph with $\mathrm{diam}(G) = 2$, then $\mathrm{spc}(G) = \mathrm{src}(G)$.

14. Let G be a nontrivial connected graph that contains bridges. Prove that if b is the maximum number of bridges incident with a vertex in G, then $\mathrm{pc}(G) \ge b$ and $\mathrm{spc}(G) \ge b$.

15. Determine $\mathrm{spc}(C_n)$ for each integer $n \ge 4$.

16. (a) Prove that if G is a connected graph having girth 5 or more, then $\mathrm{spc}(G) = \chi'(G)$.

(b) Determine $\mathrm{spc}(P)$ for the Petersen graph P.

17. Let G be a connected graph of size m_G and H a proper connected subgraph of size m_H in G. Prove that

(a) $\mathrm{pc}(G) \le m_G - m_H + \mathrm{pc}(H)$ and

(b) $\mathrm{spc}(G) \le m_G - m_H + \mathrm{spc}(H)$.

18. For integers $a, b \ge 2$, let $S_{a,b}$ be the double star of order $a + b$ whose central vertices have degrees a and b.

(a) Prove that $\mathrm{pc}(S_{a,b}) = \max\{a, b\}$.

(b) Let G be a connected graph of size $m \ge 3$. Prove that $\mathrm{pc}(G) = m - 1$ if and only if $G = S_{2,m-1}$.

19. For an integer $m \ge 4$, prove that $\mathrm{pc}(K_{1,m-1} + e) = m - 2$ if $m = 4, 5$ and $\mathrm{pc}(K_{1,m-1} + e) = m - 3$ if $m \ge 6$.

20. Let G be a connected graph of size $m \ge 3$. Prove that $\mathrm{spc}(G) = m - 1$ if and only if $G = S_{2,m-1}$.

21. For an integer $m \ge 4$, prove that $\mathrm{spc}(K_{1,m-1} + e) = m - 2$.

22. Prove that if H is a Hamiltonian-connected spanning subgraph of a graph G, then $\text{hpc}(G) \leq \text{hpc}(H)$.

23. For each odd integer $n \geq 3$, prove that $\text{hpc}(C_n \,\square\, K_2) = 3$.

24. Let G be a nontrivial connected graph. Prove that $\text{rc}(G) = 2$ if and only if $\text{rc}_3(G) = 2$.

25. Prove that if G is a connected graph with $\text{diam}(G) \geq 3$, then $\text{rc}_3(G) \geq 3$.

26. (a) Determine $\text{rc}_3(P_n)$ each integer $n \geq 4$.

    (b) Prove that if a graph G contains a Hamiltonian path, then $\text{rc}_3(G) \leq 3$.

27. Let G be a nontrivial connected graph. Prove that $\text{rd}(G \,\square\, K_2) = \Delta(G) + 1$.

28. For integers m and n where $2 \leq m \leq n$, let $G_{m,n} = P_m \,\square\, P_n$. Prove the following.

    (a) for all $n \geq 3$, $\text{rd}(G_{2,n}) = 3$;

    (b) for all $n \geq 4$, $\text{rd}(G_{3,n}) = 3$;

    (c) for all $4 \leq m \leq n$, $\text{rd}(G_{m,n}) = 4$.

Chapter 14

Distance and Colorings

We have seen that there are many ways to color a graph, usually coloring the vertices or the edges of a graph. The three most common types of colorings are proper colorings, rainbow colorings, and monochromatic colorings, where proper colorings are by far the best known and most studied. As we have seen, positive integers are typically used for colors, namely the elements of a set $[k] = \{1, 2, \ldots, k\}$ for some positive integer k. The only requirement for a proper vertex coloring is that every pair of adjacent vertices must be assigned different colors. For nonadjacent vertices, there is no requirement of any kind. This is equivalent to saying that in a connected graph G, a vertex coloring c of G is proper if for every two vertices u and v of G, whenever $d(u, v) = 1$, we must have $c(u) \neq c(v)$; otherwise, there is no condition on $c(u)$ and $c(v)$. In this chapter, we will see that there are many types of colorings $c : V(G) \rightarrow [k]$ of a connected graph G for which $c(u) \neq c(v)$ where u and v are adjacent vertices of G. Necessarily, all of these colorings are proper colorings. Usually, for a connected graph, we are interested in (1) the minimum number of colors used in a proper coloring as well as (2) the minimum positive integer k for a proper coloring. In a standard proper coloring, these two numbers are the same.

A number of graph colorings have their roots in a communications problem known as the *Channel Assignment Problem*. In this problem, there are transmitters, say v_1, v_2, $\ldots$, v_n, located in some geographic region. It is not at all unusual for some pairs of transmitters to interfere with each other. There can be various reasons for this such as their proximity to each other, the time of day, the time of year, the terrain on which the transmitters are constructed, the power of the transmitters, and the existence of power lines in the vicinity. This situation can be modeled by a graph G whose vertices are the transmitters, that is $V(G) = \{v_1, v_2, \ldots, v_n\}$, and such that $v_i v_j \in E(G)$ if v_i and v_j interfere with each other. The goal is then to assign frequencies or channels to the transmitters in a manner that permits clear reception of the transmitted signals. The **Channel Assignment Problem** is the problem of assigning channels to the transmitters in some optimal manner. This problem, with variations, has been studied by the Federal Communications Commission (FCC), AT&T Bell Labs, the National Telecommunications and Information Administration, and the Department of Defense. Interpreting channels as colors (or labels) gives rise to graph coloring (or graph labeling) problems. The idea of studying channel assignment with the aid of graphs is due to B. H. Metzger [150], J. A. Zoellner and C. L. Beall [205], and William K. Hale [110].

14.1   T-Colorings

Suppose that in some geographic region there are n transmitters v_1, v_2, ..., v_n, some pairs of which interfere with each other. A graph G can be constructed that models this situation, namely $V(G) = \{v_1, v_2, \ldots, v_n\}$ and $v_i v_j \in E(G)$ if v_i and v_j interfere with each other. The goal is to assign channels to the transmitters in such a way that the channels of each pair of interfering transmitters differ by a suitable amount, thereby permitting clear reception of the transmitted signals. Suppose that T is a finite set of nonnegative integers containing 0 that represents the disallowed separations between channels assigned to interfering transmitters. Thus, to each vertex (transmitter) v of G, we assign a channel $c(v)$ in such a manner that if $uw \in E(G)$ (and so u and w are interfering transmitters), then $|c(u) - c(w)| \notin T$. The fact that T contains 0 requires interfering transmitters to be assigned distinct channels. Any function $c : V(G) \to \mathbb{N}$ that satisfies these conditions is necessarily a proper coloring of G, called a T-coloring.

More formally then, for a graph G and a given finite set T of nonnegative integers containing 0, a T-**coloring** c of G is a proper vertex coloring of G with positive integers such that if $uw \in E(G)$, then $|c(u) - c(w)| \notin T$. The c-**cap** $\mu_{T,c}(G)$ of a T-coloring c of G is the largest color assigned to a vertex of G by c. The coloring $\bar{c}$ of G defined by $\bar{c}(v) = \mu_{T,c}(G) + 1 - c(v)$ for each vertex v of G, which is also a T-coloring of G, called the **complementary coloring** of c. Thus, $\mu_{T,\bar{c}}(G) = \mu_{T,c}(G)$. The concept of T-colorings is due to William K. Hale [110].

For the set $T = \{0\}$, the only requirement of a T-coloring of a graph G is that colors assigned to every two adjacent vertices of G must differ and so a T-coloring of G in this case is a standard proper coloring of G. In the case where $T = \{0, 1\}$, the colors assigned to every two adjacent vertices of G must differ by at least 2. What this T-coloring requires then is that every two interfering transmitters must be assigned channels that are not only distinct but are not consecutive either. For $T = \{0, 1, 4\}$, Figure 14.1 shows two T-colorings c_1 and c_2 of a graph G. Here, $\mu_{T,c_1}(G) = 8$ and $\mu_{T,c_2}(G) = 6$.

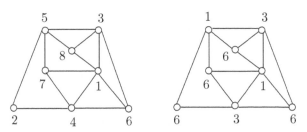

Figure 14.1: Two T-colorings of a graph G for $T = \{0, 1, 4\}$

For a graph G with $V(G) = \{v_1, v_2, \ldots, v_n\}$ and a set T of nonnegative integers containing 0, it is quite easy to construct a channel assignment that is a T-coloring. For example, suppose that r is the largest element of T. The channel assignment (coloring) c defined by

$$c(v_i) = (i - 1)(r + 1) + 1$$

for $1 \le i \le n$ is a T-coloring of G, which implies that

$$\mu_{T,c}(G) = (n - 1)(r + 1) + 1$$

for such a set T and this T-coloring c of G. There are two primary concepts associated with T-colorings of graphs. The **T-chromatic number** $\chi_T(G)$ is the minimum number of colors that can be used in a T-coloring of G. Let T be a finite set of nonnegative integers containing 0. For a graph G, the **T-cap** $\mu_T(G)$ of G is defined as

$$\mu_T(G) = \min\{\mu_{T,c}(G) : c \text{ is a } T\text{-coloring of } G\}.$$

Therefore, if $\mu_T(G) = k$, then there is a T-coloring of G using colors from the set $[k]$ but there is no T-coloring of G using the colors from the set $[p]$ for any positive integer p where $p < k$. The following theorem of Margaret B. Cozzens and Fred S. Roberts [61] explains why the T-cap of a graph is of more interest than the T-chromatic number.

Theorem 14.1 *Let G be a graph. For each finite set T of nonnegative integers containing 0,*

$$\chi_T(G) = \chi(G).$$

Proof. Since every T-coloring of G is also a proper coloring of G, it follows that $\chi(G) \le \chi_T(G)$. Suppose that $\chi(G) = k$ and that r is the largest integer in T. Let there be given a proper k-coloring c of G using the colors $1, 2, \ldots, k$. Define a function $c' : V(G) \to \mathbb{N}$ by

$$c'(v) = (r + 1)c(v)$$

for each vertex v of G. For every two adjacent vertices u and w of G,

$$\begin{aligned}
|c'(u) - c'(w)| &= |(r + 1)c(u) - (r + 1)c(w)| \\
&= (r + 1)|c(u) - c(w)| \ge r + 1
\end{aligned}$$

and so $|c'(u) - c'(w)| \notin T$. Hence, c' is a T-coloring of G. Since k colors are used by the T-coloring c', it follows that $\chi_T(G) \le k = \chi(G)$. Hence, $\chi_T(G) = \chi(G)$. ∎

Since the T-chromatic number of a graph G and the chromatic number of G are the same parameter, the concept $\chi_T(G)$ is not new. Thus, we turn our attention to the T-cap of G. For each T-coloring of a graph G, we may assume that some vertex of G is assigned the color 1. If, for example, c' is a T-coloring of a graph G in which $a > 1$ is the smallest color assigned to any vertex of G, then the coloring c of G defined by

$$c(v) = c'(v) - (a - 1) \text{ for each } v \in V(G)$$

is a T-coloring of G in which some vertex of G is assigned the color 1 by c and in which $\mu_{T,c'}(G)$ is reduced by $a - 1$. It is evident therefore that if $\mu_T(G) = k$, then there is a T-coloring $c : V(G) \to [k]$ of G in which at least one vertex of G is colored 1 and at least one vertex is colored k. Furthermore,

$$\chi_T(G) \le \mu_T(G) \tag{14.1}$$

for every graph G. If G is a graph with $\chi(G) = k$ and T is a set of nonnegative integers containing 0 such that $|T| \ge 2$ and whose smallest positive integer is at least k, then $\chi_T(G) = k = \chi(G)$ as there is a T-coloring of G using all colors in the set $[k]$.

Suppose that we are given a k-chromatic graph G and a finite set T of nonnegative integers containing 0 such that the largest element of T is r. For a proper k-coloring c of G (using the colors $1, 2, \ldots, k$), the coloring c' defined by

$$c'(v) = (c(v) - 1)(r + 1) + 1$$

for every vertex v of G is both a proper k-coloring and a T-coloring with c'-cap $(\chi(G) - 1)(r + 1) + 1$. Hence, we have the following.

Theorem 14.2 *For every graph G and every finite set T of nonnegative integers containing 0 whose largest element is r,*

$$\mu_T(G) \le (\chi(G) - 1)(r + 1) + 1.$$

The following theorem is a consequence of a result due to Cozzens and Roberts [62].

Theorem 14.3 *Let T be a finite set of nonnegative integers containing 0. If G is a k-chromatic graph with clique number ω, then*

$$\mu_T(K_\omega) \le \mu_T(G) \le \mu_T(K_k).$$

Proof. Let c be a T-coloring of G such that $\mu_T(G) = \mu_{T,c}(G)$. Since $\omega(G) = \omega$, it follows that G contains a complete subgraph H of order ω. Hence,

$$\mu_T(K_\omega) \le \max_{u \in V(H)} c(u) \le \max_{u \in V(G)} c(u) = \mu_T(G)$$

and so $\mu_T(K_\omega) \le \mu_T(G)$.

We now establish the second inequality. Let c be a T-coloring of K_k using the colors $1 = r_1, r_2, \ldots, r_k = \mu_T(K_k)$ such that $r_1 < r_2 < \cdots < r_k$. Since $\chi(G) = k$, there also exists a proper k-coloring c' of G using the colors $r_1, r_2, \ldots, r_k$. Because c is a T-coloring of K_k, it follows that $|r_i - r_j| \notin T$ for each pair i, j of integers with $1 \le i, j \le k$ and $i \ne j$. Consequently, c' is also a T-coloring of G and so $\mu_T(G) \le \mu_T(K_k)$.   ∎

For the set $T = \{0, 2, 3\}$, we now determine the T-chromatic number and T-cap for the cycles C_3, C_4, and C_5. By Theorem 14.1,

$$\chi_T(C_3) = \chi_T(C_5) = 3 \text{ and } \chi_T(C_4) = 2.$$

Since the smallest positive integer in T is $2 = \chi(C_4)$, it follows that $\mu_T(C_4) = 2$. For the T-cap of the cycles C_3 and C_5, it turns out that $\mu_T(C_3) = 6$ and $\mu_T(C_5) = 5$. The corresponding T-colorings of these three cycles are shown in Figure 14.2. In particular, this shows that $\mu_T(C_3) \le 6$ and $\mu_T(C_5) \le 5$. We verify for $T = \{0, 2, 3\}$ that the T-cap of C_5 is, in fact, 5. Suppose, to the contrary, that $\mu_T(C_5) = a$ for some integer $a \le 4$. Since $\chi_T(C_5) = \chi(C_5) = 3$, it follows by (14.1) that $a = 3$ or $a = 4$ and so there is a T-coloring c of C_5 in which the largest color used is $a \le 4$. Regardless of whether $a = 3$ or $a = 4$, there are two nonadjacent vertices u and w that are colored the same by c, say $c(u) = c(w) = b$, where $1 \le b \le 4$. We may assume that $b = 1$ or $b = 2$, for otherwise, we could consider the complementary T-coloring $\bar{c}$. There are two adjacent vertices x and y on C_5, neither of which is u or w. Since $c(x) \ne c(y)$, both $c(x)$ and $c(y)$ are different from each other as well as different from b. Also, since $|c(x) - c(y)| \notin \{2, 3\}$, it follows that no such T-coloring of C_5 is possible. Therefore, $\mu_T(C_5) = 5$, as claimed.

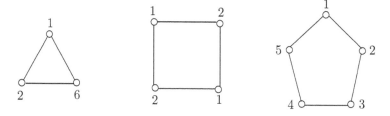

Figure 14.2: T-colorings of C_3, C_4, and C_5

For $T = \{0, 2, 3\}$, a T-coloring c of C_5 with $\chi_T(C_5) = 3$ colors and whose largest color is minimum is shown in Figure 14.3. Thus, in order to assign each vertex of C_5 a color smaller than 6, more than three colors must be used. That is, if c is a T-coloring of a graph G that uses $\chi_T(G)$ colors, this does not imply that $\mu_{T,c}(G) = \mu_T(G)$.

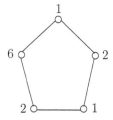

Figure 14.3: A T-coloring of C_5

The following upper bound for the T-cap of a graph is the consequence of a result due to Cozzens and Roberts [61] and is an improvement over that given in Theorem 14.2.

Theorem 14.4 *If G is a k-chromatic graph and T is a finite set of t nonnegative integers containing 0, then*

$$\mu_T(G) \le t(k-1) + 1.$$

Proof.   For each positive integer r, let G_r be the graph with $V(G_r) = \{v_1, v_2, \ldots, v_r\}$ such that $v_i v_j \in E(G_r)$ if $i \ne j$ and $|i - j| \in T$. First, we show that $\chi(G_r) \le t$ for every positive integer r. For a given positive integer r, let H be an induced subgraph of G_r. Suppose that v_i is a vertex of G belonging to H. Since 0 is one of the elements of T, it follows that v_i is adjacent to at most $t - 1$ vertices of H and so $\delta(H) \le \deg_H v_i \le t - 1$. By Theorem 7.8,

$$\chi(G_r) \le 1 + \max\{\delta(H)\},$$

where the maximum is taken over all subgraphs H of G_r. Thus,

$$\chi(G_r) \le 1 + (t - 1) = t,$$

as claimed.

Among all positive integers r for which $\chi(G_r)$ is maximum, let s be the minimum integer. Then

$$\chi(G_s) \le t. \tag{14.2}$$

Let $p = \chi(G_s)(k - 1) + 1$. Thus, $\chi(G_p) \le \chi(G_s)$. By Theorem 6.10,

$$\alpha(G_p) \;\ge\; \frac{p}{\chi(G_p)} \ge \frac{\chi(G_s)(k-1)+1}{\chi(G_s)} > k - 1$$

and so $\alpha(G_p) \ge k$. Let $S = \{v_{j_1}, v_{j_2}, \ldots, v_{j_k}\}$ be an independent set of k vertices of G_p such that $j_1 < j_2 < \cdots < j_k$.

Now let there be given a k-coloring of G, using the colors $1, 2, \ldots, k$. We now replace each color i $(1 \le i \le k)$ by j_i, arriving at a new k-coloring c of G. Hence, if x and y are two adjacent vertices of G, then x and y are assigned distinct colors j_r and j_s, where $1 \le r, s \le k$ and $r \ne s$. Since $v_{j_r}, v_{j_s} \in S$, it follows that $v_{j_r} v_{j_s} \notin E(G_p)$ and so $|j_r - j_s| \notin T$. Hence, c is a T-coloring of G. Since the largest color used in c is j_k and

$$j_k \le p = \chi(G_s)(k - 1) + 1,$$

it follows by (14.2) that

$$\mu_T(G) \le j_k \le \chi(G_s)(k - 1) + 1 \le t(k - 1) + 1,$$

giving the desired result.   ∎

We now show that the upper bound given in Theorem 14.4 for the T-cap of a graph is attainable. Suppose first that $T = \{0, 2, 4\}$ and consider the graphs C_3 and C_5. Then $\chi(C_3) = \chi(C_5) = 3$ and $|T| = 3$. By Theorem 14.4, $\mu_T(C_3) \le 7$ and $\mu_T(C_5) \le 7$. Figure 14.4 shows T-colorings for these graphs with T-cap 7. We show for $T = \{0, 2, 4\}$ that the T-cap of C_3 is, in fact, 7. Assume, to the contrary, that $\mu_T(C_3) = a$, where $a \le 6$. Then there exists a T-coloring of C_3, where some vertex

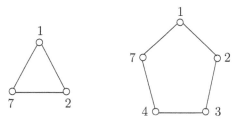

Figure 14.4: T-colorings of C_3 and C_5

u of C_3 is colored 1 and the largest color assigned to a vertex v of C_3 is $a \leq 6$. Since $T = \{0, 2, 4\}$, either $a = 3$ or $a = 5$, and the color of the remaining vertex w of C_3 is of the same parity as either $c(u)$ or $c(v)$, which is impossible since $T = \{0, 2, 4\}$.

An infinite class of graphs verifying the sharpness of the upper bound for $\mu_T(G)$ stated in Theorem 14.4 consists of the complete graphs K_n with $T = \{0, 1, \dots, t-1\}$, where $t \in \mathbb{N}$. By Theorem 14.4,

$$\mu_T(K_n) \leq t(n-1) + 1.$$

Let $V(K_n) = \{v_1, v_2, \dots, v_n\}$. Assigning the color $t(i-1)+1$ to v_i for $i = 1, 2, \dots, n$ gives a T-coloring of K_n. If $\mu_T(K_n) \leq t(n-1)$, then there is a T-coloring of K_n where the difference in colors of two vertices is less than t. This, however, is impossible and so $\mu_T(K_n) = t(n-1) + 1$.

For the graph G shown in Figure 14.1 and the set $T = \{0, 1, 2\}$, the T-coloring c with $\mu_{T,c}(G) = 8$ is not only a proper coloring, it is a rainbow coloring. Since G has order 8, there is no rainbow T-coloring c_1 of G with $\mu_{T,c}(G) < 8$. For the set $T = \{0, 2, 4\}$, Figure 14.4 shows a T-coloring c_2 of C_5 where $\mu_{T,c_2}(C_5) = 7$. In fact, we showed that $\mu_T(C_5) = 7$. This too is a rainbow coloring. This brings up the following problem.

Problem 14.5 *Let G be a graph. Determine for various finite sets T containing 0, the number $\mu_T(G)$ among all rainbow T-colorings of G.*

14.2 L(2, 1)-Colorings

One of the early types of colorings inspired by the Channel Assignment Problem occurred as a result of a communication to Jerrold Griggs by Fred Roberts, who proposed using nonnegative integers to represent radio channels in order to study the problem of optimally assigning radio channels to transmitters at certain locations. As a result of this, Roger Yeh [203] in 1990 and then Griggs and Yeh [102] in 1992 introduced a coloring in which colors (nonnegative integers) assigned to the vertices of a graph depend not only on whether two vertices are adjacent but also on whether two vertices are at distance 2.

For nonnegative integers h and k, an $L(h, k)$-**coloring** c of a graph G is an assignment of colors (nonnegative integers) to the vertices of G such that if u and w are adjacent vertices of G, then $|c(u) - c(w)| \geq h$ while if $d(u, w) = 2$, then

$|c(u) - c(w)| \geq k$. No condition is placed on colors assigned to u and v if $d(u, w) \geq 3$. Hence, an $L(1, 0)$-coloring of a graph G is a proper coloring of G. As with T-colorings, the major problems of interest with $L(h, k)$-colorings concern caps on the colorings. For given nonnegative integers h and k and an $L(h, k)$-coloring c of a graph G, the c-**cap** of G is the largest color assigned to a vertex of G by c, denoted by $\lambda_L(c)$. For given nonnegative integers h and k, the L-**cap** of G is

$$\lambda_L(G) = \min\{\lambda_L(c)\}$$

where the minimum is taken over all $L(h, k)$-colorings c of G. Most of the interest in $L(h, k)$-colorings has been in the case where $h = 2$ and $k = 1$. Therefore, an $L(2, 1)$-**coloring** of a graph G (also called an $L(2, 1)$-**labeling** by some) is an assignment of colors (nonnegative integers) to the vertices of G such that

(1) colors assigned to adjacent vertices must differ by at least 2,

(2) colors assigned to vertices at distance 2 must differ, and

(3) no restriction is placed on colors assigned to vertices at distance 3 or more.

To review, for an $L(2, 1)$-coloring c of a graph G then, the c-**cap** of G is

$$\lambda_L(c) = \max\{c(u) :\ u \in V(G)\}.$$

For simplicity, the c-cap $\lambda_L(c)$ of G is also denoted by $\lambda(c)$ when we are dealing with $L(2, 1)$-colorings. The L-**cap** $\lambda_L(G)$ of G is therefore

$$\lambda_L(G) = \min\{\lambda(c)\},$$

where the minimum is taken over all $L(2, 1)$-colorings c of G. Here too, we write $\lambda(G)$ rather than $\lambda_L(G)$. (Since $\lambda(G)$ is common notation for the edge-connectivity of a graph G, it is essential to know the context in which this symbol is being used.) Therefore, in this context, $\lambda(G)$ is the smallest positive integer k for which there exists an $L(2, 1)$-coloring $c : V(G) \rightarrow \{0, 1, \ldots, k\}$. Since we may always take 0 as the smallest color used in an $L(2, 1)$-coloring of a graph G, it follows that $\lambda(G)$ is the smallest maximum color that can occur in an $L(2, 1)$-coloring of G.

As an example, we determine $\lambda(G)$ for the graph G of Figure 14.5(a). The coloring c of G in Figure 14.5(b) is an $L(2, 1)$-coloring and so $\lambda(c) = 5$. Hence, $\lambda(G) \leq 5$. We claim that $\lambda(G) = 5$. Suppose that $\lambda(G) < 5$. Let c' be an $L(2, 1)$-coloring such that $\lambda(c') = \lambda(G)$. We may assume that c' uses some or all of the colors $0, 1, 2, 3, 4$. Since the vertices u, v, and w are mutually adjacent, these three vertices must be colored $0, 2$, and 4, say $c'(u) = 0$, $c'(v) = 2$, and $c'(w) = 4$. Since $c'(y)$ must differ from $c'(v)$ by at least 2, it follows that $c'(y) = 0$ or $c'(y) = 4$. However, u and w are at distance 2 from y, implying that $c'(y) \neq 0$ and $c'(y) \neq 4$. This is a contradiction. Thus, as claimed, $\lambda(G) = 5$.

For a familiar family of graphs, the L-cap is easy to determine.

Theorem 14.6 *For every positive integer t, the L-cap of a star $K_{1,t}$ is $\lambda(K_{1,t}) = t + 1$.*

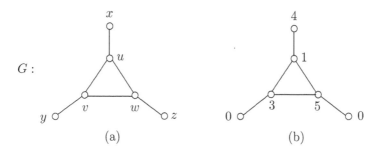

Figure 14.5: A graph G with $\lambda(G) = 5$

Proof. Since the result is immediate if $t = 1$, we may assume that $t \geq 2$. The coloring of $K_{1,t}$ that assigns $0, 1, \ldots, t-1$ to the t end-vertices of $K_{1,t}$ and $t+1$ to the central vertex of $K_{1,t}$ is an $L(2,1)$-coloring of $K_{1,t}$. Thus, $\lambda(K_{1,t}) \leq t+1$.

Suppose that there is an $L(2,1)$-coloring of $K_{1,t}$ using colors in the set $S = \{0, 1, \ldots, t\}$. Since the order of $K_{1,t}$ is $t+1$ and $\text{diam}(K_{1,t}) = 2$, it follows that for each $i \in S$, exactly one vertex of $K_{1,t}$ is assigned the color i. In particular, the central vertex of $K_{1,t}$ is assigned a color $j \in S$. Because some end-vertex of $K_{1,t}$ must be colored $j - 1$ or $j + 1$, this coloring cannot be an $L(2,1)$-coloring of $K_{1,t}$. Hence, we have a contradiction and so $\lambda(K_{1,t}) = t+1$. ∎

The L-cap of a tree with maximum degree Δ can only be one of two values.

Theorem 14.7 *If T is a tree with $\Delta(T) = \Delta \geq 1$, then either*

$$\lambda(T) = \Delta + 1 \text{ or } \lambda(T) = \Delta + 2.$$

Proof. Suppose that the order of T is n. Because $K_{1,\Delta}$ is a subgraph of T and $\lambda(K_{1,\Delta}) = \Delta + 1$ by Theorem 14.6, it follows that $\lambda(T) \geq \Delta + 1$. We now show that there exists an $L(2,1)$-coloring of T with colors from the set

$$S = \{0, 1, \ldots, \Delta + 2\}$$

of $\Delta + 3$ colors. Denote T by T_n and let v_n be an end-vertex of T_n. Let $T_{n-1} = T_n - v_n$ and let v_{n-1} be an end-vertex of T_{n-1}. We continue in this manner until we arrive at a trivial tree T_1 consisting of the single vertex v_1. Consider the sequence $v_1, v_2, \ldots, v_n$. We now give a greedy $L(2,1)$-coloring of the vertices of T with colors from the set S. Assign the color 0 to v_1 and the color 2 to v_2. Suppose now that an $L(2,1)$-coloring of the subtree T_i of T induced by $\{v_1, v_2, \ldots, v_i\}$ has been given, where $2 \leq i < n$. We assign v_{i+1} the smallest color from the set S so that an $L(2,1)$-coloring of the subtree T_{i+1} of T induced by $\{v_1, v_2, \ldots, v_{i+1}\}$ results. From the manner in which the sequence $v_1, v_2, \ldots, v_n$ was constructed, v_{i+1} is an end-vertex of T_{i+1} and so v_{i+1} is adjacent to exactly one vertex v_j with $1 \leq j \leq i$. The vertex v_j is adjacent to at most $\Delta - 1$ vertices in the subtree T_i. There are at most $\Delta - 1$ colors assigned to the neighbors of v_j in T_i. Also, there are at most 3 colors that are either assigned to v_j or are within 1 of the color assigned to v_j. Therefore, there are at most $(\Delta - 1) + 3 = \Delta + 2$ colors that must be avoided when selecting a color for v_{i+1}. Thus, there is at least one available color in S to color v_{i+1} and so $\lambda(T) \leq \Delta + 2$. ∎

We now show that there are trees T for which $\lambda(T) = \Delta(T) + 1$ and trees T for which $\lambda(T) = \Delta(T) + 2$. By Theorem 14.6, $\lambda(K_{1,t}) = \Delta(K_{1,t}) + 1$ for every positive integer t. Thus, $\lambda(P_2) = \Delta(P_2) + 1$ and $\lambda(P_3) = \Delta(P_3) + 1$. The coloring c of $P_4 = (v_1, v_2, v_3, v_4)$ with $c(v_1) = 1$, $c(v_2) = 3$, $c(v_3) = 0$, $c(v_4) = 2$ is an $L(2,1)$-coloring of P_4 and so $\lambda(P_4) = \Delta(P_4) + 1$. For $n \geq 5$, however, the situation is different.

Proposition 14.8 *For $n \geq 5$, $\lambda(P_n) = \Delta(P_n) + 2 = 4$.*

Proof. Let $P_n = (v_1, v_2, \ldots, v_n)$. Consider the subgraph of P_n induced by the vertices v_i $(1 \leq i \leq 5)$, namely $P_5 = (v_1, v_2, v_3, v_4, v_5)$. The $L(2,1)$-coloring of P_5 given in Figure 14.6 shows that $\lambda(P_5) \leq 4$.

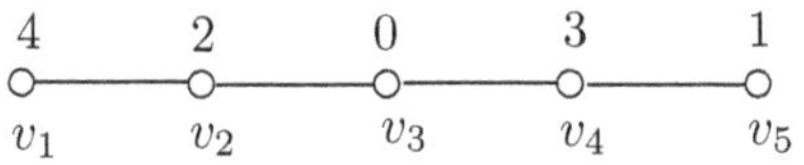

$$
\begin{array}{ccccc}
4 & 2 & 0 & 3 & 1 \\
\circ\!\!-\!\!\!-\!\!\!-\!\!\!-\!\!\circ\!\!-\!\!\!-\!\!\!-\!\!\!-\!\!\circ\!\!-\!\!\!-\!\!\!-\!\!\!-\!\!\circ\!\!-\!\!\!-\!\!\!-\!\!\!-\!\!\circ \\
v_1 & v_2 & v_3 & v_4 & v_5
\end{array}
$$

Figure 14.6: An $L(2,1)$-coloring of P_5

Since $\lambda(P_4) = 3$, it follows that $\lambda(P_5) \geq 3$. Suppose that $\lambda(P_5) = 3$. Then there is an $L(2,1)$-coloring c of P_5 using the colors 0, 1, 2, 3. Either c or $\bar{c}$ assigns the color 0 or 1 to v_3. Suppose that c assigns 0 or 1 to v_3. If $c(v_3) = 0$, then one of v_2 and v_4 is colored 2 and the other is colored 3. We may assume that $c(v_2) = 2$ and $c(v_4) = 3$. Then $c(v_1) = 0$, which is impossible. Hence, $c(v_3) = 1$. However then, at most one of v_1 and v_4 is colored 3, which is impossible. Therefore, $\lambda(P_5) = 4$, which implies by Theorem 14.7 that $\lambda(P_n) = 4$ for $n \geq 5$. ∎

By Theorem 14.7, $\Delta + 1 \leq \lambda(T) \leq \Delta + 2$ for every tree T with maximum degree Δ. If T has order n, then $\Delta \leq n - 1$ and so $\lambda(T) \leq (n-1) + 2 = n + 1$ for every tree T of order n. However, if $\Delta = n - 1$, then T is a star and $\lambda(T) = \Delta + 1 \leq n$. Therefore, for every tree T of order n, $\lambda(T) \leq n$. In fact, $\lambda(G) \leq n$ for every bipartite graph G of order n, which follows from a more general upper bound of Griggs and Yeh [102] for the L-cap of a graph.

Theorem 14.9 *If G is a graph of order n, then*

$$\lambda(G) \leq n + \chi(G) - 2.$$

Proof. Suppose that $\chi(G) = k$. Then $V(G)$ can be partitioned into k independent sets $V_1, V_2, \ldots, V_k$, where $|V_i| = n_i$ for $1 \leq i \leq k$. Assign the colors $0, 1, 2, \ldots, n_1 - 1$ to the vertices of V_1 and for $2 \leq i \leq k$, assign the colors

$$n_1 + n_2 + \cdots + n_{i-1} + (i - 1),$$
$$n_1 + n_2 + \cdots + n_{i-1} + i,$$
$$\vdots$$
$$n_1 + n_2 + \cdots + n_i + (i - 2),$$

to the vertices of V_i. Since this is an $L(2,1)$-coloring of G, it follows that

$$\lambda(G) \le n + k - 2,$$

as desired. ∎

An immediate consequence of Theorem 14.9 is the following.

Corollary 14.10 *If G is a complete k-partite graph of order n, where $k \ge 2$, then*

$$\lambda(G) = n + k - 2.$$

Proof. Let $V_1, V_2, \ldots, V_k$ be the partite sets of G. By Theorem 14.9, $\lambda(G) \le n + k - 2$. Let c be an $L(2,1)$-coloring of G with c-cap $\lambda(G)$ using colors from the set $S = \{0, 1, \ldots, \lambda(G)\}$ and let a_i be the largest color assigned to a vertex of V_i ($1 \le i \le k$). Since every two distinct vertices of G are either adjacent or at distance 2, it follows that c must assign distinct colors to all n vertices of G. Furthermore, since every two vertices of G belonging to different partite sets are adjacent, it follows that no vertex of G can be colored $a_i + 1$ for any i ($1 \le i \le k$). Hence, there are $k - 1$ colors of S that cannot be assigned to any vertex of G, which implies that the largest color that c can assign to a vertex of G is at least $(n - 1) + (k - 1) = n + k - 2$ and so $\lambda(G) \ge n + k - 2$. Therefore, $\lambda(G) = n + k - 2$. ∎

While we have already noted that $\lambda(G) \ge \Delta + 1$ for every graph G with maximum degree Δ, many of the upper bounds for $\lambda(G)$ have also been expressed in terms of Δ. For example, Griggs and Yeh [102] obtained the following.

Theorem 14.11 *If G is a graph with maximum degree Δ, then*

$$\lambda(G) \le \Delta^2 + 2\Delta.$$

Proof. For a given sequence $v_1, v_2, \ldots, v_n$ of the vertices of G, we now conduct a greedy $L(2,1)$-coloring c of G. We begin by defining $c(v_1) = 0$. For each vertex v_i ($2 \le i \le n$), at most Δ vertices of G are adjacent to v_i and at most $\Delta^2 - \Delta$ vertices of G are at distance 2 from v_i. Hence, when assigning a color to v_i, if a vertex v_j adjacent to v_i precedes v_i in the sequence, then we must avoid assigning v_i any of the three colors $c(v_j) - 1$, $c(v_j)$, $c(v_j) + 1$; while if a vertex v_j is at distance 2 from v_i and precedes v_i in the sequence, then we must avoid assigning v_i the color $c(v_j)$. Therefore, there are at most $3\Delta + (\Delta^2 - \Delta) = \Delta^2 + 2\Delta$ colors to be avoided when coloring any vertex v_i ($2 \le i \le n$). Hence, at least one of the $\Delta^2 + 2\Delta + 1$ colors $0, 1, 2, \ldots, \Delta^2 + 2\Delta$ is available for v_i and so $\lambda(G) \le \Delta^2 + 2\Delta$. ∎

Griggs and Yeh [102] also showed that if a graph G has diameter 2, then the bound $\Delta^2 + 2\Delta$ for $\lambda(G)$ in Theorem 14.11 can be improved.

Theorem 14.12 *If G is a connected graph of diameter 2 with $\Delta(G) = \Delta$, then*

$$\lambda(G) \le \Delta^2.$$

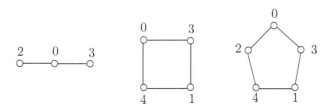

Figure 14.7: $L(2, 1)$-colorings of the three graphs G
with $\Delta(G) = \mathrm{diam}(G) = 2$

Proof. If $\Delta = 2$, then G is either P_3, C_4, or C_5. The $L(2,1)$-colorings of these three graphs in Figure 14.7 show that $\lambda(G) \leq 4$ for each such graph G. Hence, we can now assume that $\Delta \geq 3$. Suppose that the order of G is n. We consider two cases for Δ, according to whether Δ is large or small in comparison with n.

Case 1. $\Delta \geq (n - 1)/2$. Since G is neither a cycle nor a complete graph, it follows from Brooks's theorem (Theorem 7.15) that $\chi(G) \leq \Delta$. By Theorem 14.9,

$$\begin{aligned} \lambda(G) &\leq& n + \chi(G) - 2 \leq (2\Delta + 1) + \Delta - 2 \\ &=& 3\Delta - 1 < \Delta^2, \end{aligned}$$

where the final inequality follows because $\Delta \geq 3$.

Case 2. $\Delta \leq (n - 2)/2$. Therefore, $\delta(\overline{G}) \geq n/2$. By Corollary 3.8, $\overline{G}$ is Hamiltonian and so contains a Hamiltonian path $P = (v_1, v_2, \ldots, v_n)$. Define a coloring c on G by $c(v_i) = i - 1$ for $1 \leq i \leq n$. Since every two vertices of G with consecutive colors are adjacent in $\overline{G}$, these vertices are not adjacent in G. Thus, c is an $L(2,1)$-coloring of G and the c-cap is $n - 1$, which implies that $\lambda(G) \leq n - 1$.

Now, for each vertex v of G, at most Δ vertices are adjacent to v and at most $\Delta^2 - \Delta$ vertices are at distance 2 from v. Since the diameter of G is 2, all vertices of G are within distance 2 of v and so

$$n \leq 1 + \Delta + (\Delta^2 - \Delta) = \Delta^2 + 1.$$

Therefore, $\lambda(G) \leq n - 1 \leq \Delta^2.$ ∎

In the proof of Theorem 14.12 it was shown that if $\Delta \geq 3$ and $\Delta \geq (n - 1)/2$, then $\lambda(G) \leq \Delta^2$. This particular argument did not make use of the assumption that G has diameter 2. This led Griggs and Yeh [102] to make the following conjecture.

Conjecture 14.13 *If G is a graph with $\Delta(G) = \Delta \geq 2$, then $\lambda(G) \leq \Delta^2$.*

In 2008 Frédéric Havet, Bruce Reed, and Jean-Sébastien Sereni [118] established the following.

Theorem 14.14 *There exists a positive integer N such that for every graph G of maximum degree $\Delta \geq N$,*

$$\lambda(G) \leq \Delta^2.$$

A consequence of this theorem is the following.

Corollary 14.15 *There exists a positive integer constant C such that for every positive integer Δ and for every graph G with maximum degree Δ,*

$$\lambda(G) \leq \Delta^2 + C.$$

14.3 Radio Colorings

The concept of $L(h, k)$-colorings has been generalized in a natural way. For non-negative integers $d_1, d_2, \ldots, d_k$, where $k \geq 2$, an $L(d_1, d_2, \ldots, d_k)$-**coloring** c of a graph G is an assignment c of colors (nonnegative integers in this case) to the vertices of G such that $|c(u) - c(w)| \geq d_i$ whenever $d(u, w) = i$ for $1 \leq i \leq k$. The $L(d_1, d_2, \ldots, d_k)$-colorings in which $d_i = k + 1 - i$ for each i ($1 \leq i \leq k$) have proved to be of special interest.

By the Four Color Theorem, the regions of every map, regardless of how many regions there may be, can be colored with four or fewer colors so that every two adjacent regions (regions sharing a common boundary) are assigned distinct colors. However, if a map M contains a large number of regions, then it may be more appealing to use several colors to color the regions rather than trying to minimize the number of colors. One possible difficulty with using many colors is that it becomes more likely that some pairs of colors may be sufficiently similar that the colors are indistinguishable at a casual glance. As a result, it may be difficult to distinguish adjacent regions if they are assigned similar colors. One solution to this problem is to permit regions to be assigned the same or similar colors only when these regions are sufficiently far apart. For two regions R and R', we define the **distance** $d(R, R')$ between R and R' as the smallest nonnegative integer k for which there exists a sequence

$$R = R_0, R_1, \ldots, R_k = R'$$

of regions in M such that R_i and R_{i+1} are adjacent for $0 \leq i \leq k - 1$. Suppose that we have decided to color each region of M with one of 12 colors, namely:

1. White	4. Yellow	7. Orange	10. Purple
2. Silver	5. Gold	8. Red	11. Royal Blue
3. Light Grey	6. Brown	9. Burgundy	12. Black

These colors are listed in an order that may cause two colors with consecutive numbers to be mistaken as the same color if they are assigned to regions that are located close to each other. Indeed, we can use the integers $1, 2, \ldots, 12$ as the colors. We can then assign colors i and j with $1 \leq i, j \leq 12$ to distinct regions R and R', depending on the value of $d(R, R')$. In particular, we could agree to assign colors i and j to R and R' only if $d(R, R') + |i - j| \geq 1 + k$ for some prescribed positive integer k. This gives rise to a coloring of the regions of the map M called a *radio coloring*, a term coined by Frank Harary.

We have seen that with each map M there is associated a dual planar graph G whose vertices are the regions of M and where two vertices of G are adjacent if the corresponding regions of M are adjacent. A **radio coloring** of G is an assignment of colors to the vertices of G such that two colors i and j can be assigned to two distinct vertices u and v only if

$$d(u, v) + |i - j| \geq 1 + k$$

for some fixed positive integer k.

The term "radio coloring" emanates from its connection with the Channel Assignment Problem. In the United States, one of the responsibilities of the Federal Communications Commission (FCC) concerns the regulation of FM radio stations. Each station is characterized by its transmission frequency, effective radiated power, and antenna height. Each FM station is assigned a station class, which depends on a number of factors, including its effective radiated power and antenna height. The FCC requires that FM radio stations located within a certain proximity to one another must be assigned distinct channels and that the nearer two stations are to each other, the greater the difference in their assigned channels must be (see [207]). For example, two stations that share the same channel must be separated by at least 115 kilometers; however, the actual required separation depends on the classes of the two stations. Two channels are considered to be first-adjacent (or simply adjacent) if their frequencies differ by 200 kHz, that is, if they are consecutive on the FM dial. For example, the channels 105.7 MHz and 105.9 MHz are adjacent. The distance between two radio stations on adjacent channels must be at least 72 kilometers. Again, the actual restriction depends on the classes of the stations. The distance between two radio stations whose channels differ by 400 or 600 kHz (second- or third-adjacent channels) must be at least 31 kilometers. Once again, the actual required separation depends on the classes of the stations.

As we have noted, the problem of obtaining an optimal assignment of channels for a specified set of radio stations according to some prescribed restrictions on the distances between stations as well as other factors is referred to as the Channel Assignment Problem. We have also mentioned that the use of graph theory to study the Channel Assignment Problem and related problems dates back at least to 1970 (see Metzger [150]). In 1980, William Hale [110] modeled the Channel Assignment Problem as both a frequency-distance constrained and frequency constrained optimization problem and discussed applications to important real-world problems. Since then, a number of different models of the Channel Assignment Problem have been developed, including T-colorings and $L(2, 1)$-colorings of graphs described in Sections 14.1 and 14.2. For both T-colorings and $L(2, 1)$-colorings of a graph, the major concept of interest has been a parameter called the cap, namely, the T-cap for T-colorings and L-cap for $L(2, 1)$-colorings. For a coloring c of a graph G that is either a T-coloring or an $L(2, 1)$-coloring, the c-cap is the maximum color assigned by c to a vertex of G. The T-cap is then the minimum c-cap over all T-colorings c of G. The L-cap is defined similarly. Since the colors used in T-colorings are typically positive integers (where it can always be assumed that one of the colors used is 1) and the colors used in $L(2, 1)$-colorings are the less frequently used nonnegative integers (where it can always be assumed that one of the colors used is 0), the problem of computing the T-cap and L-cap is essentially that of minimizing the largest color used among all T-colorings or among all $L(2,1)$-colorings. Having made these remarks, we now turn to the primary topic of this section: radio colorings.

Both proper vertex colorings and $L(2,1)$-colorings were extended in 2001 by Chartrand, Erwin, Harary, and Zhang [39]. For a connected graph G of diameter d and an integer k with $1 \le k \le d$, a k-**radio coloring** c of G (sometimes called a **radio k-coloring**) is an assignment of colors (positive integers) to the vertices of G such that

$$d(u,v) + |c(u) - c(v)| \ge 1 + k \qquad (14.3)$$

for every two distinct vertices u and v of G. Thus, a 1-radio coloring of G is simply a proper coloring of G, while a 2-radio coloring is an $L(2,1)$-coloring. Note that a k-radio coloring c does not imply that c is a k-coloring of the vertices of G (a vertex coloring using k colors), it only implies that c is a vertex coloring that satisfies condition (14.3) for some prescribed positive integer k with $1 \le k \le d = \operatorname{diam}(G)$. The **value** $\operatorname{rc}_k(c)$ of a k-radio coloring c of G is defined as the maximum color assigned to a vertex of G by c (where, again, we may assume that some vertex of G is assigned the color 1). The coloring $\overline{c}$ of G defined by

$$\overline{c}(v) = \operatorname{rc}_k(c) + 1 - c(v)$$

for every vertex v of G is also a k-radio coloring of G, referred to as the **complementary coloring** of c. Because it is assumed that some vertex of G has been colored 1 by c, it follows that $\operatorname{rc}_k(\overline{c}) = \operatorname{rc}_k(c)$.

For a connected graph G with diameter d and an integer k with $1 \le k \le d$, the k-**radio chromatic number** (or simply the k-**radio number**) $\operatorname{rc}_k(G)$ is defined as

$$\operatorname{rc}_k(G) = \min\{\operatorname{rc}_k(c)\},$$

where the minimum is taken over all k-radio colorings c of G. (Note that $\operatorname{rc}_k(G)$ is the same notation used for the k-rainbow connection number of a graph G. Since both notations are common, we use the same notation for each. However, these two parameters are never considered at the same time. So, there is no confusion.) Since a 1-radio coloring of G is a proper coloring, it follows that $\operatorname{rc}_1(G) = \chi(G)$. On the other hand, a 2-radio coloring of G is an $L(2,1)$-coloring of G, all of whose colors are positive integers. Thus,

$$\operatorname{rc}_2(G) = 1 + \lambda(G).$$

Since the diameter of the 6-cycle C_6 is 3, $\operatorname{rc}_k(C_6)$ is defined for $k = 1, 2, 3$. Because C_6 is bipartite, $\operatorname{rc}_1(C_6) = \chi(C_6) = 2$. On the other hand, $\operatorname{rc}_2(C_6) = 1 + \lambda(C_6) = 5$ and $\operatorname{rc}_3(C_6) = 8$. The 2-radio coloring of C_6 in Figure 14.8(a) shows that $\operatorname{rc}_2(C_6) \le 5$, while the 3-radio coloring of C_6 in Figure 14.8(b) shows that $\operatorname{rc}_3(C_6) \le 8$. We verify that $\operatorname{rc}_2(C_6) = 5$. Assume, to the contrary, that there is a 2-radio coloring of C_6 with value 4. Since no 2-radio coloring of C_6 can use any color more than twice, either the color 2 or the color 3 is used at least once. We may assume that the color 2 is used to color a vertex of C_6. (If the color 2 is not used in a 2-radio coloring c of C_6, then it is used in the complementary coloring of C_6.)

However, if a vertex u of C_6 is assigned the color 2, then its two neighbors must both be colored 4, which is impossible because the distance between these vertices is 1. Thus, as claimed, $rc_2(C_6) = 5$. (See Exercise 17.)

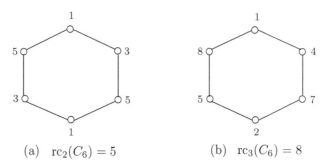

(a) $rc_2(C_6) = 5$ (b) $rc_3(C_6) = 8$

Figure 14.8: k-Radio colorings of C_6 for $k = 2, 3$

Observe that the 2-radio coloring of C_6 shown in Figure 14.8(a) uses each of the colors 1, 3, and 5 twice. Hence, even though the value of this 2-radio coloring of C_6 is 5, the number of colors used is 3. In fact, the minimum number of colors that can be used in a 2-radio coloring of C_6 is 3. (See Exercise 18.)

If $rc_1(G) = \chi(G) = t$ for some graph G, then in any t-coloring of G using the colors $1, 2, \ldots, t$, not only is t the *largest color* used, t is the *number of colors* used. This is not true in general for k-radio colorings of G for $k \geq 2$, as we just observed.

Once $rc_k(G)$ is determined for a connected graph G of order n and diameter d, a simple upper bound exists for $rc_\ell(G)$, where $1 \leq k < \ell \leq d$. The following result is due to Riadh Khennoufa and Olivier Togni [134].

Proposition 14.16 *For a connected graph G of order n having diameter d and for integers k and ℓ with $1 \leq k < \ell \leq d$,*

$$rc_\ell(G) \leq rc_k(G) + (n-1)(\ell - k).$$

Proof. Let c be a k-radio coloring of G such that $rc_k(c) = rc_k(G)$. Let $V(G) = \{v_1, v_2, \ldots, v_n\}$ such that $c(v_i) \leq c(v_{i+1})$ for $1 \leq i \leq n-1$. We define a coloring c' of G by

$$c'(v_i) = c(v_i) + (i-1)(\ell - k).$$

For integers i and j with $1 \leq i < j \leq n$, we therefore have

$$|c'(v_i) - c'(v_j)| = |c(v_i) - c(v_j)| + (j-i)(\ell - k).$$

Since c is a k-radio coloring of G, it follows that

$$|c(v_i) - c(v_j)| \geq 1 + k - d(v_i, v_j).$$

Consequently,

$$
\begin{aligned}
|c'(v_i) - c'(v_j)| &\geq 1 + k + (j-i)(\ell - k) - d(v_i, v_j) \\
&\geq 1 + \ell - d(v_i, v_j).
\end{aligned}
$$

Thus, c' is an ℓ-radio coloring of G with

$$\mathrm{rc}_\ell(c') = \mathrm{rc}_k(G) + (n-1)(\ell - k)$$

and so $\mathrm{rc}_\ell(G) \le \mathrm{rc}_k(G) + (n-1)(\ell - k)$. ∎

Even though k-radio colorings of a connected graph with diameter d are defined for every integer k with $1 \le k \le d$, it is the two smallest and two largest values of k that have received the most attention. For a connected graph G with diameter d, a d-radio coloring c of a connected graph G with diameter d requires that

$$d(u, v) + |c(u) - c(v)| \ge 1 + d$$

for every two distinct vertices u and v of G. In this case, colors assigned to adjacent vertices of G must differ by at least d, colors assigned to two vertices at distance 2 must differ by at least $d - 1$, and so on, up to two vertices at distance d (that is, antipodal vertices), whose colors are only required to differ. A d-radio coloring is sometimes called a **radio labeling** and the d-radio chromatic number (or d-**radio number**) is sometimes called simply the **radio number** $\mathrm{rn}(G)$ of G.

The only connected graph of order n having diameter 1 is K_n and

$$\mathrm{rn}(K_n) = \chi(K_n) = n.$$

For connected graphs G of diameter 2,

$$\mathrm{rn}(G) = \lambda(G) + 1,$$

which was discussed in Section 14.2. The simplest example of a graph G of diameter 3 is P_4. The radio labeling of P_4 in Figure 14.9 shows that $\mathrm{rn}(P_4) \le 6$. We show in fact that $\mathrm{rn}(P_4) = 6$. Let c be a radio labeling of P_4 having the value $\mathrm{rn}(P_4)$. Necessarily either $c(v) \le 3$ or $c(w) \le 3$, for otherwise the complementary coloring $\overline{c}$ has the property that $\overline{c}(v) \le 3$ or $\overline{c}(w) \le 3$. Suppose that $c(v) = a$, where $1 \le a \le 3$. Then $c(u) \ge a + 3$ and $c(w) \ge a + 3$. Since $|c(u) - c(w)| \ge 2$, it follows that either $c(u) \ge a + 5 \ge 6$ or $c(w) \ge a + 5 \ge 6$. Thus, $\mathrm{rn}(P_4) \ge 6$ and so $\mathrm{rn}(P_4) = 6$. The radio labelings of P_3 and P_5 shown in Figure 14.9 also illustrate that $\mathrm{rn}(P_3) = 4$ and $\mathrm{rn}(P_5) = 11$.

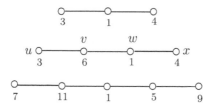

Figure 14.9: Radio labelings of P_n ($3 \le n \le 5$)

Since any radio labeling of a connected graph G of order n and diameter d must assign distinct colors to the vertices of G, it follows that $\mathrm{rn}(G) \ge n$. Furthermore, if $V(G) = \{v_1, v_2, \cdots, v_n\}$, then the coloring c defined by $c(v_i) = 1 + (i-1)d$ for each i ($1 \le i \le n$) is a radio labeling of G with $\mathrm{rn}(c) = 1 + (n-1)d$. These observations are summarized below.

Proposition 14.17 *If G is a connected graph of order n and diameter d, then*

$$n \le \mathrm{rn}(G) \le 1 + (n-1)d. \tag{14.4}$$

We have noted that if $d = 1$ and so $G = K_n$, then $\mathrm{rn}(K_n) = n$. The graph C_5 and the Petersen graph P both have diameter 2 and their radio numbers also attain the lower bound in (14.4), namely $\mathrm{rn}(C_5) = 5$ and $\mathrm{rn}(P) = 10$. Furthermore, for each integer $k \ge 2$, the graph $K_k \,\square\, K_2$ has order $n = 2k$, diameter 2, and $\mathrm{rn}(K_k \,\square\, K_2) = n$. The graph $C_3 \,\square\, C_5$ has order $n = 15$, diameter 3, and radio number 15 (see Figure 14.10).

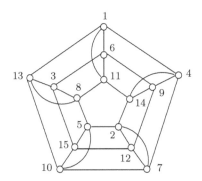

Figure 14.10: A radio labeling of $C_3 \,\square\, C_5$

For a connected graph G of order n and diameter d, the upper bound for $\mathrm{rn}(G)$ given in Theorem 14.17 can often be improved.

Proposition 14.18 *If G is a connected graph of order n and diameter d containing an induced subgraph H of order p and diameter d such that $d_H(u, v) = d_G(u, v)$ for every two vertices u and v of H, then*

$$\mathrm{rn}(H) \le \mathrm{rn}(G) \le \mathrm{rn}(H) + (n-p)d.$$

A special case of Proposition 14.18 is when H is a path.

Corollary 14.19 *If G is a connected graph of order n and diameter d, then*

$$\mathrm{rn}(P_{d+1}) \le \mathrm{rn}(G) \le \mathrm{rn}(P_{d+1}) + (n - d - 1)d.$$

Corollary 14.19 illustrates the value of knowing the radio numbers of paths. The following result was obtained by Daphne Liu and Xuding Zhu [142].

Theorem 14.20 *For every integer $n \ge 3$,*

$$\mathrm{rn}(P_n) = \begin{cases} 2r^2 + 3 & \text{if } n = 2r + 1 \\ 2r^2 - 2r + 2 & \text{if } n = 2r. \end{cases}$$

Combining Corollary 14.19 and Theorem 14.20, we have the following.

Corollary 14.21 *Let G be a connected graph of order n and diameter d.*

(a) *If $d = 2$, then $4 \leq \text{rn}(G) \leq 2n - 2$.*

(b) *If $d = 3$, then $6 \leq \text{rn}(G) \leq 3n - 6$.*

(c) *If $d = 4$, then $11 \leq \text{rn}(G) \leq 4n - 9$.*

While the paths P_{d+1} show the sharpness of the lower bounds in Corollary 14.21, the sharpness of the upper bounds are less obvious. It is not difficult to show that for every integer $n \geq 3$, there exists a connected graph G of diameter 2 with $\text{rn}(G) = 2n - 2$ (see Exercise 21). The graph H of Figure 14.11(a) has order $n = 6$, $\text{diam}(H) = 3$, and $\text{rn}(H) = 12 = 3n - 6$. The graph F of Figure 14.11(b) has order $n = 6$, $\text{diam}(F) = 4$, and $\text{rn}(F) = 14 = 4n - 10$. The number $4n - 10$ does not attain the upper bound for the radio number of a graph of diameter 4 given in Corollary 14.21(c). In fact, it may be that the appropriate upper bound for this case is $4n - 10$ rather than $4n - 9$.

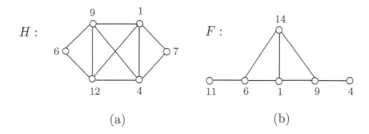

Figure 14.11: Radio numbers of graphs having diameters 3 and 4

For a connected graph G of diameter d, a $(d - 1)$-radio coloring c requires that

$$d(u, v) + |c(u) - c(v)| \geq d$$

for every two distinct vertices u and v of G. A $(d - 1)$-radio coloring c is also referred to as a **radio antipodal coloring** (or simply an **antipodal coloring**) of G since $c(u) = c(v)$ only if u and v are antipodal vertices of G. The **radio antipodal number** or, more simply, the **antipodal number** $\text{an}(c)$ of c is the largest color assigned to a vertex of G by c. The **antipodal chromatic number** or the **antipodal number** $\text{an}(G)$ of G is

$$\text{an}(G) = \min\{\text{an}(c)\},$$

where the minimum is taken over all radio antipodal colorings c of G. If c is a radio antipodal coloring of a graph G such that $\text{an}(c) = \ell$, then the complementary coloring $\bar{c}$ of G defined by

$$\bar{c}(v) = \ell + 1 - c(v)$$

for every vertex v of G is also a radio antipodal coloring of G.

A radio antipodal coloring of the graph H in Figure 14.12 is given with antipodal number 5. Thus, $\mathrm{an}(H) \leq 5$. Let c be a radio antipodal coloring of H with $\mathrm{an}(c) = \mathrm{an}(H) \leq 5$. Since $\mathrm{diam}(H) = 3$, the colors of every two adjacent vertices of H must differ by at least 2 and the colors of two vertices at distance 2 must differ. We may assume that $c(v) \in \{1,2\}$, for otherwise, $\overline{c}(v) \in \{1,2\}$ for the complementary radio antipodal coloring $\overline{c}$ of H. Suppose that $c(v) = a \leq 2$. Then at least one of the vertices u, w, and y must have color at least $a+2$, one must have color at least $a + 3$, and the other must have color at least $a + 4$. Since $a + 4 \geq 5$, it follows that $\mathrm{an}(c) \geq 5$ and so $\mathrm{an}(H) \geq 5$. Hence, $\mathrm{an}(H) = 5$.

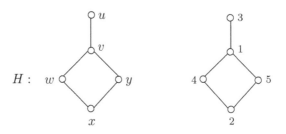

Figure 14.12: A graph with antipodal number 5

Figure 14.13 gives radio antipodal colorings of the paths P_n with $3 \leq n \leq 6$ that give $\mathrm{an}(P_n)$ for these graphs. The antipodal numbers of all paths were determined by Khennoufa and Togni [134].

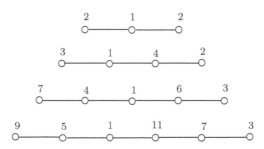

Figure 14.13: Radio antipodal colorings of P_n $(3 \leq n \leq 6)$

Theorem 14.22 *For every integer $n \geq 5$,*

$$\mathrm{an}(P_n) = \begin{cases} 2r^2 - 2r + 3 & \text{if } n = 2r + 1 \\ 2r^2 - 4r + 5 & \text{if } n = 2r. \end{cases}$$

14.4   Hamiltonian Colorings

We saw in Section 14.3 that in a $(d-1)$-radio coloring of a connected graph G of diameter d, the colors assigned to adjacent vertices must differ by at least $d - 1$, the colors assigned to two vertices whose distance is 2 must differ by at least $d - 2$, and so on up to antipodal vertices, whose colors are permitted to be the same. For this reason, $(d - 1)$-radio colorings are also referred to as *antipodal colorings*.

In the case of an antipodal coloring of the path P_n of order $n \geq 2$, only the two end-vertices are permitted to be colored the same. If u and v are distinct vertices of P_n and $d(u, v) = i$, then $|c(u) - c(v)| \geq n - 1 - i$. Since P_n is a tree, not only is i the length of a shortest $u - v$ path in P_n, it is the length of the *only* $u - v$ path in P_n. In particular, i is the length of a longest $u - v$ path.

In Section 1.3 the detour distance $D(u, v)$ between two vertices u and v in a connected graph G is defined as the length of a *longest* $u - v$ path in G. Hence, the length of a longest $u - v$ path in P_n is $D(u, v) = d(u, v)$. Therefore, in the case of the path P_n, an antipodal coloring of P_n can also be defined as a vertex coloring c that satisfies

$$D(u, v) + |c(u) - c(v)| \geq n - 1 \tag{14.5}$$

for every two distinct vertices u and v of P_n.

Vertex colorings c that satisfy (14.5) were extended from paths of order n to arbitrary connected graphs of order n by Chartrand, Nebeský, and Zhang [50]. A **Hamiltonian coloring** of a connected graph G of order n is a vertex coloring c such that

$$D(u, v) + |c(u) - c(v)| \geq n - 1$$

for every two distinct vertices u and v of G. The largest color assigned to a vertex of G by c is called the **value** of c and is denoted by $\mathrm{hc}(c)$. The **Hamiltonian chromatic number** $\mathrm{hc}(G)$ is the smallest value among all Hamiltonian colorings of G.

Figure 14.14(a) shows a graph H of order 5. A vertex coloring c of H is shown in Figure 14.14(b). Since $D(u, v) + |c(u) - c(v)| \geq 4$ for every two distinct vertices u and v of H, it follows that c is a Hamiltonian coloring and so $\mathrm{hc}(c) = 4$. Hence, $\mathrm{hc}(H) \leq 4$. Because no two of the vertices t, w, x, and y are connected by a Hamiltonian path, these vertices must be assigned distinct colors and so $\mathrm{hc}(H) \geq 4$. Thus, $\mathrm{hc}(H) = 4$.

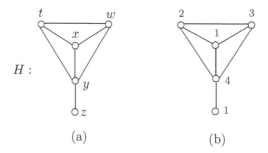

Figure 14.14: A graph with Hamiltonian chromatic number 4

If a connected graph G of order n has Hamiltonian chromatic number 1, then $D(u, v) = n - 1$ for every two distinct vertices u and v of G and consequently G is Hamiltonian-connected, that is, every two vertices of G are connected by a

Hamiltonian path. Indeed, $\mathrm{hc}(G) = 1$ *if and only if G is Hamiltonian-connected.* Therefore, the Hamiltonian chromatic number of a connected graph G can be considered as a measure of how close G is to being Hamiltonian-connected, that is, the closer $\mathrm{hc}(G)$ is to 1, the closer G is to being Hamiltonian-connected. The three graphs H_1, H_2, and H_3 shown in Figure 14.15 are all close (in this sense) to being Hamiltonian-connected since $\mathrm{hc}(H_i) = 2$ for $i = 1, 2, 3$.

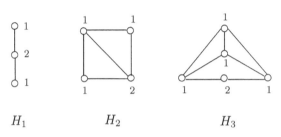

$$H_1 \qquad\qquad H_2 \qquad\qquad H_3$$

Figure 14.15: Three graphs with Hamiltonian chromatic number 2

While there are many graphs of large order with Hamiltonian chromatic number 1, graphs of large order can also have a large Hamiltonian chromatic number.

Theorem 14.23   *For every integer $n \geq 3$,*

$$\mathrm{hc}(K_{1,n-1}) = (n-2)^2 + 1.$$

Proof.   Since $\mathrm{hc}(K_{1,2}) = 2$ (see H_1 in Figure 14.15), we may assume that $n \geq 4$. Let $G = K_{1,n-1}$, where $V(G) = \{v_1, v_2, \cdots, v_n\}$ and v_n is the central vertex. Define the coloring c of G by $c(v_n) = 1$ and

$$c(v_i) = (n-1) + (i-1)(n-3) \text{ for } 1 \leq i \leq n-1.$$

Then c is a Hamiltonian coloring of G and

$$\mathrm{hc}(G) \leq \mathrm{hc}(c) = c(v_{n-1}) = (n-1) + (n-2)(n-3) = (n-2)^2 + 1.$$

It remains to show that $\mathrm{hc}(G) \geq (n-2)^2 + 1$.

Let c be a Hamiltonian coloring of G such that $\mathrm{hc}(c) = \mathrm{hc}(G)$. Because G contains no Hamiltonian path, c assigns distinct colors to the vertices of G. We may assume that

$$c(v_1) < c(v_2) < \cdots < c(v_{n-1}).$$

We now consider three cases, depending on the color assigned to the central vertex v_n.

   Case 1. $c(v_n) = 1$. Since

$$D(v_1, v_n) = 1 \text{ and } D(v_i, v_{i+1}) = 2 \text{ for } 1 \leq i \leq n-2,$$

it follows that

$$c(v_{n-1}) \geq 1 + (n-2) + (n-2)(n-3) = (n-2)^2 + 1$$

and so $hc(G) = hc(c) = c(v_{n-1}) \geq (n-2)^2 + 1$.

Case 2. $c(v_n) = hc(c)$. Thus, in this case,

$$1 = c(v_1) < c(v_2) < \cdots < c(v_{n-1}) < c(v_n).$$

Hence,

$$c(v_n) \geq 1 + (n-2)(n-3) + (n-2) = (n-2)^2 + 1$$

and so $hc(G) = hc(c) = c(v_n) \geq (n-2)^2 + 1$.

Case 3. $c(v_j) < c(v_n) < c(v_{j+1})$ *for some integer* j *with* $1 \leq j \leq n-2$. Thus, $c(v_1) = 1$ and $c(v_{n-1}) = hc(c)$. In this case,

$$
\begin{aligned}
c(v_j) &\geq 1 + (j-1)(n-3), \\
c(v_n) &\geq c(v_j) + (n-2), \\
c(v_{j+1}) &\geq c(v_n) + (n-2), \text{ and} \\
c(v_{n-1}) &\geq c(v_{j+1}) + [(n-1) - (j+1)](n-3).
\end{aligned}
$$

Therefore,

$$
\begin{aligned}
c(v_{n-1}) &\geq 1 + (j-1)(n-3) + 2(n-2) + (n-j-2)(n-3) \\
&= (2n-3) + (n-3)^2 = (n-2)^2 + 2 > (n-2)^2 + 1
\end{aligned}
$$

and so $hc(G) = hc(c) = c(v_{n-1}) > (n-2)^2 + 1$.

Hence, in any case, $hc(G) \geq (n-2)^2 + 1$ and so $hc(G) = (n-2)^2 + 1$. ∎

It is useful to know the Hamiltonian chromatic number of cycles. It is not difficult to see that $hc(C_3) = 1$, $hc(C_4) = 2$, and $hc(C_5) = 3$. Hamiltonian colorings for these three cycles are shown in Figure 14.16. The Hamiltonian chromatic numbers of the cycles C_n $(3 \leq n \leq 5)$ illustrate the following general formula for $hc(C_n)$.

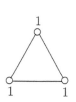 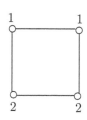 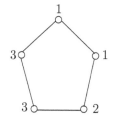

Figure 14.16: Hamiltonian colorings of C_3, C_4, and C_5

Theorem 14.24 *For every integer* $n \geq 3$,

$$hc(C_n) = n - 2.$$

Proof.   Since we noted that $\mathrm{hc}(C_n) = n - 2$ for $n = 3, 4, 5$, we may assume that $n \geq 6$. Let $C_n = (v_1, v_2, \cdots, v_n, v_1)$. Because the vertex coloring c of C_n defined by $c(v_1) = c(v_2) = 1$, $c(v_{n-1}) = c(v_n) = n - 2$, and $c(v_i) = i - 1$ for $3 \leq i \leq n - 2$ is a Hamiltonian coloring, it follows that $\mathrm{hc}(C_n) \leq n - 2$.

Assume, to the contrary, that $\mathrm{hc}(C_n) < n-2$ for some integer $n \geq 6$. Then there exists a Hamiltonian $(n - 3)$-coloring c of C_n. We consider two cases, according to whether n is odd or n is even.

Case 1. n is odd. Then $n = 2k + 1$ for some integer $k \geq 3$. Hence, there exists a Hamiltonian $(2k - 2)$-coloring c of C_n. Let

$$A = \{1, 2, \cdots, k - 1\} \text{ and } B = \{k, k + 1, \cdots, 2k - 2\}.$$

For every vertex u of C_n, there are two vertices v of C_n such that $D(u, v)$ is minimum (and $d(u, v)$ is maximum), namely $D(u, v) = d(u, v) + 1 = k + 1$. For $u = v_i$, these two vertices v are v_{i+k} and v_{i+k+1} (where the addition in $i + k$ and $i + k + 1$ is performed modulo n).

Since c is a Hamiltonian coloring, $D(u, v) + |c(u) - c(v)| \geq n - 1 = 2k$. Because $D(u, v) = k+1$, it follows that $|c(u) - c(v)| \geq k - 1$. Therefore, if $c(u) \in A$, then the colors of these two vertices v with this property must belong to B. In particular, if $c(v_i) \in A$, then $c(v_{i+k}) \in B$. Suppose that there are a vertices of C_n whose colors belong to A and b vertices of C_n whose colors belong to B. Then $b \geq a$. However, if $c(v_i) \in B$, then $c(v_{i+k}) \in A$, implying that $a \geq b$ and so $a = b$. Since $a + b = n$ and n is odd, this is impossible.

Case 2. n is even. Then $n = 2k$ for some integer $k \geq 3$. Hence, there exists a Hamiltonian $(2k - 3)$-coloring c of C_n. For every vertex u of C_n, there is a unique vertex v of C_n for which $D(u, v)$ is minimum (and $d(u, v)$ is maximum), namely $D(u, v) = d(u, v) = k$. For $u = v_i$, this vertex v is v_{i+k} (where the addition in $i + k$ is performed modulo n).

Since c is a Hamiltonian coloring, $D(u, v) + |c(u) - c(v)| \geq n - 1 = 2k - 1$. Because $D(u, v) = k$, it follows that $|c(u) - c(v)| \geq k - 1$. This implies, however, that if $c(u) = k - 1$, then there is no color that can be assigned to v to satisfy this requirement. Hence, no vertex of C_n can be assigned the color $k - 1$ by c. Let

$$A = \{1, 2, \cdots, k - 2\} \text{ and } B = \{k, k + 1, \cdots, 2k - 3\}.$$

Thus, $|A| = |B| = k - 2$. If $c(v_i) \in A$, then $c(v_{i+k}) \in B$. Also, if $c(v_i) \in B$, then $c(v_{i+k}) \in A$. Hence, there are k vertices of C_n assigned colors from A and k vertices of C_n assigned colors from B.

Consider two adjacent vertices of C_n, one of which is assigned a color from A and the other is assigned a color from B. We may assume that $c(v_1) \in A$ and $c(v_2) \in B$. Then $c(v_{k+1}) \in B$. Since $D(v_2, v_{k+1}) = k+1$, it follows that $|c(v_2)-c(v_{k+1})| \geq k-2$. Because $c(v_2), c(v_{k+1}) \in B$, this implies that one of $c(v_2)$ and $c(v_{k+1})$ is at least $2k - 2$. This is a contradiction.   ∎

We now consider some upper bounds for the Hamiltonian chromatic number of a connected graph, beginning with a rather obvious one.

Proposition 14.25 *If H is a spanning connected subgraph of a graph G, then*

$$\mathrm{hc}(G) \leq \mathrm{hc}(H).$$

Proof. Suppose that the order of H is n. Let c be a Hamiltonian coloring of H such that $\mathrm{hc}(c) = \mathrm{hc}(H)$. Then $D_H(u,v) + |c(u) - c(v)| \geq n - 1$ for every two distinct vertices u and v of H. Since $D_G(u,v) \geq D_H(u,v)$ for every two distinct vertices u and v of H (and of G), it follows that $D_G(u,v) + |c(u) - c(v)| \geq n - 1$ and so c is a Hamiltonian coloring of G as well. Hence, $\mathrm{hc}(G) \leq \mathrm{hc}(c) = \mathrm{hc}(H)$. ∎

Combining Theorem 14.24 and Proposition 14.25, we have the following corollary.

Corollary 14.26 *If G is a Hamiltonian graph of order $n \geq 3$, then $\mathrm{hc}(G) \leq n-2$.*

The following result gives the Hamiltonian chromatic number of a related class of graphs.

Proposition 14.27 *Let H be a Hamiltonian graph of order $n - 1 \geq 3$. If G is a graph obtained by adding a pendant edge to H, then $\mathrm{hc}(G) = n - 1$.*

Proof. Suppose that $C = (v_1, v_2, \ldots, v_{n-1}, v_1)$ is a Hamiltonian cycle of H and $v_1 v_n$ is the pendant edge of G. Let c be a Hamiltonian coloring of G. Since $D_G(u,v) \leq n - 2$ for every two distinct vertices u and v of C, no two vertices of C can be assigned the same color by c. Consequently, $\mathrm{hc}(c) \geq n - 1$ and so $\mathrm{hc}(G) \geq n - 1$.

Now define a coloring c' of G by

$$c'(v_i) \;=\; \begin{cases} i & \text{if } 1 \leq i \leq n-1 \\ n-1 & \text{if } i = n. \end{cases}$$

We claim that c' is a Hamiltonian coloring of G. First let v_j and v_k be two vertices of C where $1 \leq j < k \leq n - 1$. Then $|c'(v_j) - c'(v_k)| = k - j$ and

$$D(v_j, v_k) = \max\{k - j, \ (n - 1) - (k - j)\}.$$

In either case, $D(v_j, v_k) \geq n - 1 + j - k$ and so

$$D(v_j, v_k) + |c'(v_j) - c'(v_k)| \geq n - 1.$$

For $1 \leq j \leq n - 1$, $|c'(v_j) - c'(v_n)| = n - 1 - j$, while

$$D(v_j, v_n) \geq \max\{j, \ n - j + 1\}$$

and so $D(v_j, v_n) \geq j$. Therefore,

$$D(v_j, v_n) + |c'(v_j) - c'(v_n)| \geq n - 1.$$

Hence, as claimed, c' is a Hamiltonian coloring of G and so $\mathrm{hc}(G) \leq \mathrm{hc}(c') = c'(v_n) = n - 1$. ∎

In Exercise 12 of Chapter 3, it was stated that if T is a tree of order 4 or more that is not a star, then $\overline{T}$ contains a Hamiltonian path. With the aid of this, an upper bound for the Hamiltonian chromatic number of a graph can be given in terms of its order.

Theorem 14.28 *For every connected graph G of order $n \geq 2$,*

$$\mathrm{hc}(G) \leq (n-2)^2 + 1.$$

Proof. First, if G contains a vertex of degree $n - 1$, then G contains the star $K_{1,n-1}$ as a spanning subgraph. Since $\mathrm{hc}(K_{1,n-1}) = (n-2)^2 + 1$, it follows by Proposition 14.25 that $\mathrm{hc}(G) \leq (n-2)^2 + 1$. Hence, we may assume that G contains a spanning tree T that is not a star and so its complement $\overline{T}$ contains a Hamiltonian path $P = (v_1, v_2, \cdots, v_n)$. Thus, $v_i v_{i+1} \notin E(T)$ for $1 \leq i \leq n - 1$ and so $D_T(v_i, v_{i+1}) \geq 2$. Define a vertex coloring c of T by

$$c(v_i) = (n-2) + (i-2)(n-3) \text{ for } 1 \leq i \leq n.$$

Hence,

$$\mathrm{hc}(c) = c(v_n) = (n-2) + (n-2)(n-3) = (n-2)^2.$$

Therefore, for integers i and j with $1 \leq i < j \leq n$,

$$|c(v_i) - c(v_j)| = (j-i)(n-3).$$

If $j = i + 1$, then

$$D(v_i, v_j) + |c(v_i) - c(v_j)| \geq 2 + (n-3) = n - 1;$$

while if $j \geq i + 2$, then

$$D(v_i, v_j) + |c(v_i) - c(v_j)| \geq 1 + 2(n-3) = 2n - 5 \geq n - 1.$$

Thus, c is a Hamiltonian coloring of T. Therefore,

$$\mathrm{hc}(G) \leq \mathrm{hc}(T) \leq \mathrm{hc}(c) = c(v_n) = (n-2)^2 < (n-2)^2 + 1,$$

which completes the proof.   ∎

Theorem 14.28 shows how large the Hamiltonian chromatic number of a graph G of order n can be. If G is Hamiltonian however, then by Corollary 14.26 its Hamiltonian chromatic number cannot exceed $n - 2$. Moreover, if the Hamiltonian chromatic number is small relative to n, then G must contain cycles of relatively large length. Recall that the circumference $\mathrm{cir}(G)$ a graph G with cycles is the length of a longest cycle in G.

Theorem 14.29 *For every connected graph G of order $n \geq 4$ such that $2 \leq \mathrm{hc}(G) \leq n - 1$,*

$$\mathrm{hc}(G) + \mathrm{cir}(G) \geq n + 2.$$

Proof. Let $\mathrm{hc}(G) = k$. We show that

$$\mathrm{cir}(G) \geq n - k + 2.$$

Let a Hamiltonian k-coloring of G be given using the colors $1, 2, \ldots, k$. For $1 \le i \le k$, let V_i be the color class consisting of the vertices of G that are colored i. Certainly, $V_1 \ne \emptyset$ and $V_k \ne \emptyset$. Since every two vertices of G with the same color are connected by a Hamiltonian path, it follows that if G contains adjacent vertices that are colored the same, then G is Hamiltonian and the result follows. Thus, we may assume that no adjacent vertices are colored the same.

First, suppose that $V_j = \emptyset$ for all j with $1 < j < k$. Thus, $V(G) = V_1 \cup V_k$ and G is a bipartite graph. Since every two vertices in V_1 are connected by a Hamiltonian path, it follows that $|V_1| = |V_k| + 1$. Also, since every two vertices in V_k are connected by a Hamiltonian path, it follows that $|V_k| = |V_1| + 1$. This is impossible. Therefore, $V_j \ne \emptyset$ for some j with $1 < j < k$. Since G is connected, G contains two adjacent vertices $u \in V_j$ and $v \in V_\ell$ with $j \ne \ell$ and

$$|c(u) - c(v)| \le k - 2.$$

Since c is a Hamiltonian coloring, G contains a $u - v$ path of length at least $(n - 1) - (k - 2) = n - k + 1$. Therefore, G contains a cycle of length at least $n - k + 2$ and so $\text{cir}(G) \ge n - k + 2$. ∎

Since there exist Hamiltonian graphs with Hamiltonian chromatic number 2, equality can be attained in Theorem 14.29 when $h(G) = 2$. Equality can also be attained in Theorem 14.29 when $\text{hc}(G) = 3$ since the Petersen graph has Hamiltonian chromatic number 3 and is not Hamiltonian, but does have cycles of length 9 (see Figure 14.17). Equality cannot be attained when $h(G) = 4$ however (see Exercises 25 and 26).

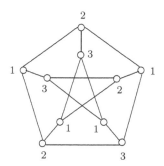

Figure 14.17: A Hamiltonian 3-coloring of the Petersen graph

Every graph G of order $n \ge 3$ that is not Hamiltonian is also not Hamiltonian-connected since no two adjacent vertices of G are connected by a Hamiltonian $u - v$ path. On the other hand, if u and v are nonadjacent vertices of G, then G may contain a Hamiltonian $u - v$ path. Consequently, if u and v are vertices of a graph G of order $n \ge 3$ that is not Hamiltonian, then $D(u, v) \le n - 2$ if u and v are adjacent and $D(u, v) \le n - 1$ if u and v are not adjacent.

In fact, the Petersen graph P (which has order 10) has the property that every two nonadjacent vertices of P are connected by a Hamiltonian path (of length 9)

while no two adjacent vertices of P are connected by a path of length 9 but are connected by a path of length 8.

A connected graph G of order $n \geq 3$ is called **semi-Hamiltonian-connected** if

$$D(u, v) = \begin{cases} n - 2 & \text{if } uv \in E(G) \\ n - 1 & \text{if } uv \notin E(G). \end{cases}$$

The Petersen graph is therefore a semi-Hamiltonian-connected graph, as is the path P_3. Moreover, for semi-Hamiltonian-connected graphs, a vertex coloring c is a Hamiltonian coloring if and only if c is a proper coloring.

Proposition 14.30 *If G is a semi-Hamiltonian-connected graph of order $n \geq 3$, then*

$$hc(G) = \chi(G).$$

Exercises for Chapter 14

1. For $T = \{0, 1, 4, 5\}$, find $\chi_T(K_3)$ and $\mu_T(K_3)$.

2. For $T = \{0, 1, 4, 5\}$, find $\chi_T(C_5)$ and $\mu_T(C_5)$.

3. For $T = \{0, 2, 4\}$, show that $\mu_T(C_5) = 7$.

4. What does Theorem 14.3 say about the T-cap of a perfect graph?

5. For $T = \{0, 1, 4\}$, Figure 14.1 shows a graph G of order 8 and a T-coloring of G such that each vertex of G is colored with exactly one of the colors $1, 2, \ldots, 8$. Give an example of a finite set T of nonnegative integers containing 0 such that $T \neq \{0, 1, 4\}$ and a graph H of order n for which there is a T-coloring of H using each of the colors $1, 2, \ldots, n$ exactly once.

6. (a) Let T be the (infinite) set of nonnegative even integers. Prove that a nontrivial connected graph G is T-colorable if and only if G is bipartite.

   (b) Let T' be the (infinite) set of positive odd integers and let $T = T' \cup \{0\}$. Prove that every graph is T-colorable.

7. Let G be a connected graph of order n. If the vertices of G are assigned distinct colors from the set $\{1, 2, \cdots, n\}$, then the resulting coloring c is necessarily a proper coloring. Equivalently, c is a T-coloring, where $T = \{0\}$. Show that the 3-cube Q_3 has an 8-coloring using the colors $1, 2, \cdots, 8$ that is also a T-coloring when $T = \{0, 1\}$ but no 8-coloring that is also a T-coloring using the colors $1, 2, \ldots, 8$ when $T = \{0, 1, 2\}$.

8. Determine $\mu_T(W_5)$ when
   (a) $T = \{0\}$, (b) $T = \{0, 1\}$, (c) $T = \{0, 1, 2\}$, (d) $T = \{0, 2\}$.

9. Let G be a graph of order n. For a nonnegative integer k, let $f(k)$ denote the smallest positive integer such that for $T = \{0, 1, 2, \ldots, k\}$, the graph G has a T-coloring using distinct elements of the set $\{1, 2, \ldots, f(k)\}$. Determine $f(k)$ for

   (a) $G = C_5$,

   (b) G is the graph obtained from C_4 by adding a pendant edge,

   (c) $G = K_{2,3}$.

10. Determine the L-cap $\lambda(T)$ for the tree T in Figure 14.18.

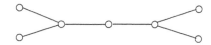

Figure 14.18: The tree T in Exercise 10

11. Determine the L-cap of all double stars.

12. Suppose that G is a 3-connected graph with $\Delta(G) = \Delta$. Prove that

$$\lambda(G) \le \Delta^2 + 2\Delta - 3.$$

   [Hint: Employ the proof technique used for Brooks's theorem.]

13. For the Petersen graph P, where $\Delta(P) = \Delta = 3$, show that $\lambda(P) = \Delta^2$.

14. (a) Draw the incidence graph G of the projective plane of order 3.

   (b) Find an $L(2, 1)$-coloring of the graph G in (a) using the colors $0, 1, \ldots, 12$.

15. Show that $rc_2(C_n) = 4$ for each integer $n \ge 3$.

16. (a) Determine $rc_2(P_n)$ for $2 \le n \le 4$.

   (b) Show that $rc_2(P_n) = 4$ for each integer $n \ge 5$.

17. Show that $rc_3(C_6) = 8$.

18. Prove that the minimum number of colors that can be used in a 2-radio coloring of C_6 is 3.

19. Does there exist a 2-radio coloring of C_6 such that for some positive integer $k \le 6$, every color in the set $\{1, 2, \ldots, k\}$ is assigned to at least one vertex of C_6? If so, find the smallest such k.

20. Does there exist a nontrivial connected graph G and a 3-radio coloring of G that uses distinct consecutive integers $a, a+1, \ldots, b$ for some positive integers a and b with $a < b$ as its colors?

21. Show for each integer $n \ge 3$ that there exists a connected graph G of order n and diameter 2 such that $rn(G) = 2n - 2$.

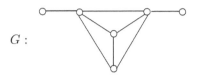

$G:$

Figure 14.19: The graph G in Exercise 22

22. Determine the Hamiltonian chromatic number of the graph G of Figure 14.19.

23. (a) For an integer $r \geq 2$, determine $D(u, v)$ for adjacent vertices u and v of $K_{r,r}$ and for nonadjacent vertices u and v of $K_{r,r}$.

    (b) Show that in every Hamiltonian coloring of $K_{r,r}$, nonadjacent vertices must be colored differently.

    (c) What is the relationship between $\text{hc}(K_{r,r})$ and $\chi(\overline{K}_{r,r})$?

24. Prove that if G is a connected graph of order $n \geq 4$ having circumference $\text{cir}(G) = n - 1$, then $\text{hc}(G) \leq n - 1$.

25. By Theorem 14.29, if G is a connected graph of order $n \geq 4$ such that $2 \leq \text{hc}(G) \leq n-1$, then $\text{hc}(G)+\text{cir}(G) \geq n+2$. We have seen Hamiltonian graphs with Hamiltonian chromatic number 2. Also, the Petersen graph has order $n = 10$, Hamiltonian chromatic number 3, and circumference $n-1 = 9$. Show that there is no graph of order n and circumference $n-2$ having Hamiltonian chromatic number 4. [Hint: Use the argument in the proof of Theorem 14.29.]

26. Prove for every connected graph G of order $n \geq 4$ with $2 \leq \text{hc}(G) \leq n - 2$ that
$$\text{cir}(G) \geq n + 1 - \left\lceil \frac{\text{hc}(G)}{2} \right\rceil.$$

Chapter 15

Domination and Colorings

In Chapter 12, we saw that many of the results and problems in Ramsey theory dealt with coloring the edges of a graph with two colors (typically red and blue) where the coloring is not proper, that is, adjacent edges may be colored the same. In this chapter, the emphasis is once again on coloring a graph with two colors (once again red and blue), but here it is vertex colorings in which we are interested and the subject involved is domination theory.

15.1 Domination Parameters

In recent decades, an area of graph theory that has received increased attention is that of *domination*. The area of domination in graph theory evidently began with Claude Berge in 1958 [14] and Oystein Ore [157] in 1962. It was Ore, however, who actually coined the term *domination*. To many, domination did not become an active area of study until 1977, following an article by Ernest Cockayne and Stephen Hedetniemi [56]. Two books, one written by Teresa Haynes, Stephen Hedetniemi, and Peter Slater [120] and another, by Haynes, Hedetniemi, and Michael Henning [119], are devoted to the subject of domination.

A vertex v in a graph G is said to **dominate** both itself and its neighbors, that is, v dominates every vertex in its closed neighborhood $N[v]$. Therefore, v dominates $\deg_G v + 1$ vertices of G. A set S of vertices in G is called a **dominating set** for (or of) G if every vertex of G is dominated by some vertex in S. Equivalently, S is a dominating set for G if every vertex of G either belongs to S or is adjacent to some vertex in S. The **domination number** $\gamma(G)$ of G is the minimum cardinality of a dominating set for G. If S is a dominating set for G with $|S| = \gamma(G)$, then S is called a **minimum dominating set** or a γ-**set**. (Historically, $\gamma(G)$ is the notation that has been used both for the domination number of G and the genus of G, discussed in Section 5.4.)

Both $S_1 = \{r, u, v, x\}$ and $S_2 = \{t, w, z\}$ are dominating sets for the graph H of order 9 shown in Figures 15.1(a) and 15.1(b). Because S_2 is a dominating set with three vertices, $\gamma(H) \leq 3$. The set S_2 is, in fact, a minimum dominating set. To see why this is the case, suppose that S is a dominating set of H. Since only r, s, and w dominate r, at least one of these three vertices must belong to S. Since only v and z dominate v, at least one of v and z must belong to S. Since the largest degree among the five vertices r, s, w, v, and z is 3, any two of these vertices can dominate at most eight vertices of H. Thus, $|S| \geq 3$ and so $\gamma(H) \geq 3$. Therefore, $\gamma(H) = 3$.

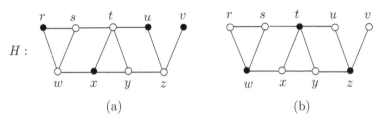

Figure 15.1: Dominating sets in a graph

A dominating set S for a graph G is called a **minimal dominating set** if no proper subset of S is a dominating set of G. Certainly, every minimum dominating set is minimal, but the converse is not true. For example, the dominating set $S_1 = \{r, u, v, x\}$ for the graph H in Figure 15.1(a) is a minimal dominating set that is not a minimum dominating set. The following fundamental property of minimal dominating sets is due to Ore [157].

Theorem 15.1 *If S is a minimal dominating set of a graph G without isolated vertices, then $V(G) - S$ is a dominating set of G.*

Proof. To show that $V(G) - S$ is a dominating set of G, we need to show that every vertex that is not in $V(G) - S$ is dominated by a vertex in $V(G) - S$. Thus, let $v \in V(G)$ such that $v \notin V(G) - S$. Therefore, $v \in S$. Assume, to the contrary, that v is not dominated by any vertex in $V(G) - S$. Then v is adjacent to no vertex in $V(G) - S$. Since S is a dominating set of G, it follows that each vertex in $V(G) - S$ is dominated by some vertex in S other than v, that is, each vertex of $V(G) - S$ is dominated by a vertex in $S - \{v\}$. Since v is not an isolated vertex of G and is adjacent to no vertex in $V(G) - S$, it follows that v is adjacent to some vertex in $S - \{v\}$. Hence, $S - \{v\}$ is a dominating set of G, which contradicts the assumption that S is a minimal dominating set of G.   ∎

One consequence of this theorem is the following.

Corollary 15.2 *If G is a graph of order n without isolated vertices, then*

$$\gamma(G) \leq \frac{n}{2}.$$

Proof. Let S be a minimum dominating set for G. Thus, $|S| = \gamma(G)$. By Theorem 15.1, $V(G) - S$ is a dominating set for G and so

$$\gamma(G) = |S| \leq |V(G) - S| = n - \gamma(G).$$

Therefore, $2\gamma(G) \leq n$ and so $\gamma(G) \leq n/2$. ∎

Recall that the corona $cor(G)$ of a graph G of order k is that graph obtained by adding a new vertex v' to G for each vertex v of G together with the edge vv'. Then the order of $cor(G)$ is $n = 2k$. Since $V(G)$ is a dominating set for $cor(G)$, it follows that $\gamma(cor(G)) \leq k = n/2$. Furthermore, every dominating set for $cor(G)$ must contain either v or v' for each vertex v of G. This implies that $\gamma(cor(G)) = n/2$ and so the bound for the domination number of a graph in Corollary 15.2 is sharp.

Over the years, many variations and generalizations of domination have been introduced. For a positive integer k, a set S of vertices in a graph G is a **k-dominating set** if every vertex not in S is adjacent to at least k vertices in S. Consequently, if S is a k-dominating set in a graph G, then every vertex of S is dominated by itself, while every vertex of G not in S is dominated by at least k vertices of S. The minimum cardinality of a k-dominating set for G is the **k-domination number** $\gamma_k(G)$. Clearly, $\gamma_k(G) \geq k$ and $\gamma_1(G) = \gamma(G)$. For a positive integer k, a set S of vertices in a graph G is a **k-step dominating set** if for every vertex u of G not in S, there exists a $u - v$ path of length k in G for some vertex $v \in S$. The minimum cardinality of a k-step dominating set for G is the **k-step domination number** $\gamma^{(k)}(G)$. Thus, $\gamma^{(1)}(G) = \gamma(G)$.

A dominating set of a graph G that is independent is an **independent dominating set** for G. The minimum cardinality of an independent dominating set for G is the **independent domination number** $i(G)$. A set S of vertices in a graph G containing no isolated vertices is a **total dominating set** for G if every vertex of G is adjacent to some vertex of S. The minimum cardinality of a total dominating set for G is the **total domination number** $\gamma_t(G)$. A total dominating set of cardinality $\gamma_t(G)$ is called a **minimum total dominating set** or a **γ_t-set** for G.

A set S of vertices in a graph G is a **restrained dominating set** if every vertex of G not in S is adjacent to both a vertex in S and a vertex not in S. The minimum cardinality of a restrained dominating set for G is the **restrained domination number** $\gamma_r(G)$.

For the double star G having two vertices of degree 3, the values of the six domination parameters discussed above are shown in Figure 15.2. The solid vertices in each case indicate the members of an appropriate minimum dominating set.

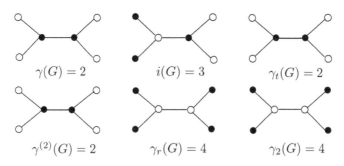

$\gamma(G) = 2$ $\qquad$ $i(G) = 3$ $\qquad$ $\gamma_t(G) = 2$

$\gamma^{(2)}(G) = 2$ $\qquad$ $\gamma_r(G) = 4$ $\qquad$ $\gamma_2(G) = 4$

Figure 15.2: Domination parameters for a double star

15.2   Stratified Domination

We now turn our attention to colorings once again – in fact, to vertex colorings that are not proper vertex colorings in general. In this situation, a k-coloring of G (using the colors $1, 2, \ldots, k$) results in a partition of $V(G)$ into k subsets $V_1, V_2, \ldots, V_k$, where V_i is the set of vertices colored i for $1 \le i \le k$. Since the k-coloring is not necessarily proper, the sets V_i are ordinarily not independent. A graph whose vertex set has such a partition has been referred to as a k-**stratified graph**, a concept introduced by Naveed Sherwani in 1992 and first studied by Reza Rashidi [162].

Most of the interest in this subject has been centered around the case $k = 2$, that is, 2-stratified graphs. In a 2-stratified graph G, $\{V_1, V_2\}$ is a partition of $V(G)$. Here it is common to consider the vertices of V_1 as colored red and the vertices of V_2 colored blue. In a 2-stratified graph then, there is at least one red vertex and at least one blue vertex. When drawing a 2-stratified graph, the red vertices are typically represented by solid vertices and the blue vertices by open vertices. Thus, the graph G of Figure 15.3 represents a 2-stratified graph whose red vertices are $u, w, x,$ and z and whose blue vertices are v and y.

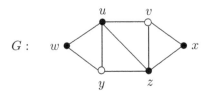

Figure 15.3: A 2-stratified graph

In the current context, a **red-blue coloring** of a graph G is an assignment of the color red or blue to each vertex of G, where all vertices of G may be assigned the same color. With each 2-stratified graph F, there are certain red-blue colorings of a graph G that will be of special interest to us.

Let F be a 2-stratified graph. Therefore, F has at least one red vertex and at least one blue vertex. Some blue vertex of F is designated as the **root** of F and is labeled v. Thus, F is a 2-stratified graph **rooted** at a blue vertex v. Now let G be a graph. By an F-**coloring** of G is meant a red-blue coloring of G such that every blue vertex v of G belongs to a copy of F rooted at v. The F-**domination number** $\gamma_F(G)$ of G is the minimum number of red vertices in any F-coloring of G. The set of red vertices in an F-coloring of a graph is called an F-**dominating set**. An F-coloring of G such that $\gamma_F(G)$ vertices are colored red is called a γ_F-**coloring**. This concept and the next two results are due to Chartrand, Haynes, Henning, and Zhang [45].

For a 2-stratified graph F and a graph G of order n containing no copies of F, the only F-coloring of G is the one in which every vertex of G is assigned the color red. Hence, in this case, $\gamma_F(G) = n$. The simplest example of a 2-stratified graph is $F = K_2$, where one vertex of F is colored red and the other vertex of F is colored blue (necessarily the root v of F). This 2-stratified graph F, a graph G, and two F-colorings of G are shown in Figure 15.4.

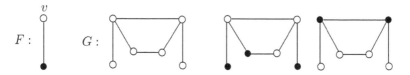

Figure 15.4: F-colorings of a graph

For $F = K_2$, the red vertices of a graph G in any F-coloring of G form a dominating set of G, which implies that $\gamma(G) \leq \gamma_F(G)$. On the other hand, suppose that we were to color the vertices in a minimum dominating set of G red and all remaining vertices of G blue. Then this red-blue coloring of the vertices of G has the property that every blue vertex of G is adjacent to a red vertex of G; that is, this is an F-coloring of G. Hence, $\gamma_F(G) \leq \gamma(G)$ and so $\gamma_F(G) = \gamma(G)$. Consequently, domination can be considered as a certain type of red-blue coloring of G in which the red vertices form a dominating set.

What this shows is that a certain type of domination and domination parameter are associated with each connected 2-stratified graph F. We now look at the various 2-stratified graphs that result from the path P_3 of order 3. In this case, there are five different 2-stratified graphs, denoted by F_i $(1 \leq i \leq 5)$, all of which are shown in Figure 15.5.

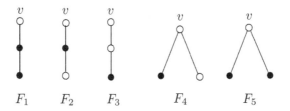

Figure 15.5: The five 2-stratified graphs obtained from P_3

The values of the five domination parameters γ_{F_i} $(1 \leq i \leq 5)$ are shown in Figure 15.6 for the graph $G = \text{cor}(P_4)$.

We saw that if $F = K_2$, then γ_F is the familiar domination number. Indeed, if $F = F_1$, then the domination parameter γ_F is familiar as well.

Theorem 15.3 *If G is a graph without isolated vertices, then*

$$\gamma_{F_1}(G) = \gamma_t(G).$$

Proof. Because G has no isolated vertices, $\gamma_t(G)$ is defined. Let S be a γ_t-set for G. By coloring each vertex of S red and each vertex of $V(G) - S$ blue, an F_1-coloring of G results. Therefore, $\gamma_{F_1}(G) \leq \gamma_t(G)$. It remains therefore to show that $\gamma_t(G) \leq \gamma_{F_1}(G)$.

Among all γ_{F_1}-colorings of G, let c be one that minimizes the number of isolated vertices in the subgraph induced by its red vertices. Since each blue vertex in G

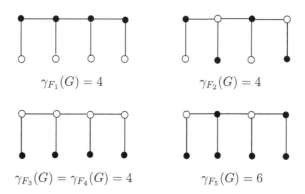

$$\gamma_{F_1}(G) = 4 \qquad\qquad \gamma_{F_2}(G) = 4$$

$$\gamma_{F_3}(G) = \gamma_{F_4}(G) = 4 \qquad\qquad \gamma_{F_5}(G) = 6$$

Figure 15.6: Domination parameters γ_{F_i} ($1 \leq i \leq 5$) of the graph $G = \mathrm{cor}(P_4)$

is adjacent to a red vertex, the red vertices constitute a dominating set S in G. We claim that every red vertex is adjacent to another red vertex, for assume, to the contrary, that there is a red vertex u adjacent only to blue vertices. Let v be a neighbor of u. Since v belongs to a copy of F_1 rooted at v, it follows that v must be adjacent to a red vertex w which itself is adjacent to some other red vertex, which implies that $u \neq w$. Interchanging the colors of u and v produces a new γ_{F_1}-coloring of G having fewer isolated vertices in the subgraph induced by its red vertices, contradicting the choice of c. Hence, as claimed, every red vertex is adjacent to some other red vertex. Therefore, S is a total dominating set of G. This implies that $\gamma_t(G) \leq \gamma_{F_1}(G)$. Consequently, $\gamma_{F_1}(G) = \gamma_t(G)$.   ∎

While the total domination number is only defined for graphs without isolated vertices, the F_1-domination number is defined for all graphs and, in this sense, is more general. Therefore, if G is a graph without isolated vertices, then $\gamma_{F_1}(G) = \gamma_t(G)$; while if G is a nonempty graph with $k \geq 1$ isolated vertices and $G = H + kK_1$, then $\gamma_{F_1}(G) = \gamma_t(H) + k$.

For a connected graph G of order 3 or more and $F = F_2$, the number γ_F is even more familiar.

Theorem 15.4 *If G is a connected graph of order 3 or more, then*

$$\gamma_{F_2}(G) = \gamma(G).$$

Proof. Since the red vertices in any F_2-coloring of G form a dominating set of G, it follows that $\gamma(G) \leq \gamma_{F_2}(G)$. It remains therefore to show that $\gamma_{F_2}(G) \leq \gamma(G)$. Among all γ-sets of G, let S be one such that the corresponding red-blue coloring has the maximum number of blue vertices v belonging to a copy of F_2 rooted at v. We claim that this red-blue coloring is, in fact, an F_2-coloring of G. Suppose that this is not the case. Then there is a blue vertex v that does not belong to a copy of F_2 rooted at v. Since S is a dominating set of G, the blue vertex v is adjacent to a red vertex w. By assumption w is not adjacent to any blue vertex other than v. If v should be adjacent to some blue vertex u, then interchanging the colors of v and w produces a γ-set whose associated red-blue coloring contains more blue vertices v'

that belong to a copy of F_2 rooted at v' than does the associated coloring of S, which is impossible. Hence, v is adjacent to no blue vertex in G. If v is adjacent to a red vertex x different from w, then, by assumption, x is not adjacent to any blue vertex other than v. This, however, implies that $(S - \{w, x\}) \cup \{v\}$ is a dominating set of G of cardinality $\gamma(G) - 1$, which is impossible. Thus, v is an end-vertex of G.

Since the order of G is at least 3, the vertex w is adjacent to some other red vertex, say y. The defining property of S implies that y must be adjacent to a blue vertex z. By interchanging the colors of v and w, a γ-set is produced whose associated red-blue coloring contains more blue vertices v' that belong to a copy of F_2 rooted at v' than does the associated coloring of S, which is a contradiction. Hence, every blue vertex v must belong to a copy of F_2 rooted at v. This implies that the red-blue coloring associated with S is an F_2-coloring of G and so $\gamma_{F_2}(G) \le \gamma(G)$. Therefore, $\gamma_{F_2}(G) = \gamma(G)$. ∎

Before discussing the domination parameter γ_{F_3}, we first turn to γ_{F_4} and γ_{F_5}, which are parameters we've met before. In fact, the parameter γ_{F_4} is the restrained domination number γ_r, while γ_{F_5} is the 2-domination number γ_2 (see Exercises 5 and 6). The table below summarizes the observations we have made concerning the stratified domination parameters γ_{F_i} for $i \in \{1, 2, 4, 5\}$ and other well-known domination parameters.

i	1	2	4	5
γ_{F_i}	γ_t	γ	γ_r	γ_2

While we have seen that γ_{F_1}, γ_{F_2}, γ_{F_4}, and γ_{F_5} are all well-known domination parameters, γ_{F_3} is not. The parameter γ_{F_3} may seem to be the 2-step domination parameter $\gamma^{(2)}$, but it is not. In an F_3-coloring of a graph G, for every blue vertex v of G there must exist a red vertex u such that G contains a $u - v$ path of length 2 whose interior vertex is blue. For example, for $n \ge 3$, $\gamma_{F_3}(K_{1,n-1}) = n$ while $\gamma^{(2)}(K_{1,n-1}) = 2$. The following three results are due to Henning and Maritz [124].

Theorem 15.5 *For each positive integer n,*

$$\gamma_{F_3}(P_n) = \left\lfloor \frac{n+7}{3} \right\rfloor + \left\lfloor \frac{n}{3} \right\rfloor - \left\lceil \frac{n}{3} \right\rceil.$$

Proof. We proceed by the Strong Form of Induction on the order n of P_n. It is straightforward to see that the formula for $\gamma_{F_3}(P_n)$ holds for $n = 1, 2, 3, 4, 5$. Assume for an integer $n \ge 6$ that $\gamma_{F_3}(P_i) = \left\lfloor \frac{i+7}{3} \right\rfloor + \left\lfloor \frac{i}{3} \right\rfloor - \left\lceil \frac{i}{3} \right\rceil$ for every integer i with $1 \le i < n$. We show that $\gamma_{F_3}(P_n) = \left\lfloor \frac{n+7}{3} \right\rfloor + \left\lfloor \frac{n}{3} \right\rfloor - \left\lceil \frac{n}{3} \right\rceil$. Let $P = P_n = (v_1, v_2, \ldots, v_n)$.

We claim that there is a γ_{F_3}-coloring of P in which v_1 and v_4 are colored red and v_2 and v_3 are colored blue. Let there be given a γ_{F_3}-coloring of P. If, in this coloring, v_1 is blue, then v_2 must be blue as well, while v_3 is red. Here, however, there is no copy of F_3 rooted at v_2, which is impossible. Thus, v_1 must be colored red. Similarly, v_n must be colored red.

If v_2 is colored blue, then v_3 is blue and v_4 is red, as desired. Suppose, however, that v_2 is colored red. If v_3 is blue, then v_4 is blue and v_5 is red. However, if we were to interchange the colors of v_2 and v_4, a new γ_{F_3}-coloring of P is produced in which v_1 and v_4 are red and v_2 and v_3 are colored blue, as desired. On the other hand, if v_3 is red, then v_4 must be blue; for if this were not the case, then v_4 is red and the vertices v_2 and v_3 could be recolored blue to produce an F_3-coloring of P having two fewer red vertices, which is impossible. Therefore, v_5 is blue and v_6 is red. However, recoloring v_2 and v_3 blue and recoloring v_4 and v_5 red produces a new γ_{F_3}-coloring of P in which v_1 and v_4 are red and v_2 and v_3 are blue, as desired. This verifies our claim.

Now, let there be given a γ_{F_3}-coloring of P in which v_1 and v_4 are red and v_2 and v_3 are colored blue. Let $P' = (v_4, v_5, \ldots, v_n)$ be the subpath of P order $n - 3$, where the colors of the vertices in P' are those in P. This coloring of P' is an F_3-coloring of P' with $\gamma_{F_3}(P) - 1$ red vertices. Hence, $\gamma_{F_3}(P') \le \gamma_{F_3}(P) - 1$. On the other hand, any γ_{F_3}-coloring of P' colors its end-vertices v_4 and v_n red and can therefore be extended to an F_3-coloring of P by coloring v_1 red and v_2 and v_3 blue. Therefore, $\gamma_{F_3}(P) \le \gamma_{F_3}(P') + 1$ and so $\gamma_{F_3}(P) = \gamma_{F_3}(P') + 1$. By the induction hypothesis,

$$\gamma_{F_3}(P') = \left\lfloor \frac{(n-3)+7}{3} \right\rfloor + \left\lfloor \frac{n-3}{3} \right\rfloor - \left\lceil \frac{n-3}{3} \right\rceil.$$

Thus,

$$\begin{aligned}
\gamma_{F_3}(P_n) &= \left(\left\lfloor \frac{n+4}{3} \right\rfloor + 1 \right) + \left(\left\lfloor \frac{n-3}{3} \right\rfloor + 1 \right) - \left(\left\lceil \frac{n-3}{3} \right\rceil + 1 \right) \\
&= \left\lfloor \frac{n+7}{3} \right\rfloor + \left\lfloor \frac{n}{3} \right\rfloor - \left\lceil \frac{n}{3} \right\rceil.
\end{aligned}$$

Therefore, the formula for $\gamma_{F_3}(P_n)$ holds for each positive integer n. ∎

Let $H_1 = P_6$ and for $k \ge 2$, let H_k be the tree obtained from the disjoint union of the star $K_{1,k+1}$ and a subdivided star $S(K_{1,k})$ by joining a leaf of the star to the central vertex of the subdivided star. Figure 15.7 shows the trees H_1, H_2, and H_3. Let $\mathcal{F} = \{H_k : k \ge 1\}$.

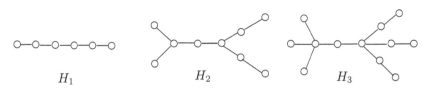

H_1 H_2 H_3

Figure 15.7: The trees H_1, H_2, and H_3

Theorem 15.6 *If T is a tree of order n with* $\operatorname{diam}(T) \ge 3$, *then*

$$\gamma_{F_3}(T) \le \frac{2n}{3}$$

with equality if and only if $T \in \mathcal{F}$.

For the tree H_3 of order 12 in $\mathcal{F}$, it follows by Theorem 15.6 that $\gamma_{F_3}(H_3) = 8 = 2n/3$. A γ_{F_3}-coloring of H_3 is shown in Figure 15.8.

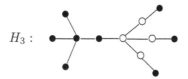

H_3 :

Figure 15.8: $\gamma_{F_3}(H_3) = 8 = 2n/3$

Corollary 15.7 *If T is a tree of order n with* $\mathrm{diam}(T) \geq 6$, *then*

$$\gamma_{F_3}(T) < \frac{2n}{3}$$

and this bound is asymptotically best possible.

Proof. By Theorem 15.5, $\gamma_{F_3}(T) < 2n/3$. It remains therefore to show that the upper bound $2n/3$ is asymptotically best possible. Let $\ell \geq 3$ be a fixed integer and let k be a positive integer. Let T be the tree obtained from H_k by attaching a path of length ℓ at the central vertex w of the subdivided star in H_k. (The tree T is shown in Figure 15.9 for the case where $\ell = k = 3$.) Let P be the resulting path of order $\ell+1$ emanating from w. Then $\mathrm{diam}(T) = \ell+3 \geq 6$ and $n = |V(T)| = 3k+\ell+3$.

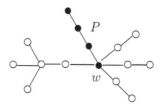

Figure 15.9: The tree T in the proof of Corollary when $\ell = k = 3$

If there is an F_3-coloring of T in which w is red, then all vertices of the subdivided star must be red as well. Furthermore, at least one end-vertex of T is the star must be red and at least $\ell/3$ vertices of P must be red (in addition to w). If there is an F_3-coloring of T in which w is blue, then all end-vertices of T in both the star and the subdivided star must be red, as well as the center of the star must be red. In addition, at least $(\ell+2)/3$ additional vertices (including at least one neighbor of w) are red. Consequently, there are at least $2k+(\ell+5)/3$ red vertices in T. Therefore,

$$\lim_{k\to\infty} \frac{\gamma_{F_3}(G)}{n} = \lim_{k\to\infty} \frac{2k+(\ell+5)/3}{3k+\ell+3} = \lim_{k\to\infty} \frac{6k+\ell+5}{9k+3\ell+9} = \frac{2}{3}$$

and so $\lim_{k\to\infty} \gamma_{F_3}(G) = \frac{2n}{3}$. ∎

15.3 Domination Based on K_3-Colorings

The only other connected 2-stratified graphs of order 3 are those obtained from the graph K_3. There are two such 2-stratified graphs in this case, both shown in Figure 15.10, which are denoted by F_6 and F_7.

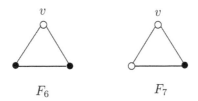

$$F_6 \qquad\qquad F_7$$

Figure 15.10: The two 2-stratified K_3

In any F_6-coloring or F_7-coloring of a graph G, every vertex of G not belonging to a triangle must be colored red. Since F_7 contains one red vertex and F_6 contains two red vertices, it may be expected that $\gamma_{F_7}(G) \leq \gamma_{F_6}(G)$ for every graph G. Trivially, this is the case when $G = K_3$ as $\gamma_{F_7}(K_3) = 1$ while $\gamma_{F_6}(K_3) = 2$. However, this is not true in general (see Exercise 8). In an F_6-coloring of a graph G in which every vertex belongs to a triangle, every blue vertex belongs to a triangle in which the other two vertices are red. In fact, in an F_6-coloring of G, no more than two-thirds of the vertices of G are red.

Theorem 15.8 *If G is a graph of order n in which every vertex is in a triangle, then*

$$\gamma_{F_6}(G) \leq \frac{2n}{3}.$$

Proof. Suppose that the theorem is false. Then there exists a graph G of order n in which every vertex belongs to a triangle but $\gamma_{F_6}(G) > 2n/3$. We may assume that every edge of G belongs to a triangle, for if G contains edges belonging to no triangle, then the graph G' obtained by deleting these edges from G has the property that $\gamma_{F_6}(G') = \gamma_{F_6}(G)$.

Among all γ_{F_6}-colorings of G, we select one that maximizes the number of red triangles. For this coloring, let $B = \{b_1, b_2, \ldots, b_k\}$ denote the set of blue vertices and R the set of red vertices. Then

$$\gamma_{F_6}(G) = |R| = n - k > 2n/3$$

and so $n > 3k$. Thus, $n - k = |R| > 2k$.

We now construct a partition of R into two sets R_1 and R_2 as follows. For $1 \leq i \leq k$, let T_i be a triangle in which b_i is rooted at a copy of F_6. Thus, for each such integer i, T_i contains two red vertices in addition to the blue vertex b_i. Now define

$$R_1 = \left(\cup_{i=1}^{k} V(T_i)\right) - B.$$

Then $|R_1| = 2k - \ell$ for some integer ℓ with $0 \leq \ell \leq 2(k-1)$. We then define $R_2 = R - R_1$. Since $|R_1| + |R_2| = |R| > 2k$, it follows that $|R_2| = \ell + r$ for some positive integer r.

We claim that every triangle of G containing a vertex of R_2 has two blue vertices, but suppose that this is not the case. Then either some vertex of R_2 belongs to a red triangle or to a triangle with exactly one blue vertex. If a vertex of R_2 is in a red triangle, then this vertex may be recolored blue to produce an F_6-coloring of G having fewer than $\gamma_{F_6}(G)$ red vertices, which is impossible. If a vertex x of R_2 is in a triangle with exactly one blue vertex, say b_i, then we can interchange the colors of x and b_i (see Figure 15.11) to produce a new γ_{F_6}-coloring of G that contains more red triangles than our original γ_{F_6}-coloring, which is also impossible. Therefore, as claimed, every triangle with a vertex of R_2 contains two blue vertices.

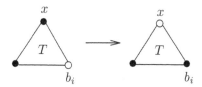

Figure 15.11: A step in the proof of Theorem 15.8

Now observe that $|B| \geq |R_2| + 1$, for if $|B| \leq |R_2|$, then we could interchange the colors of the vertices in $B \cup R_2$ to produce an F_6-coloring of G with at most $\gamma_{F_6}(G)$ red vertices and that contains more red triangles than in the original γ_{F_6}-coloring, which would be a contradiction.

For each $i = 1, 2, \ldots, k$, let e_i be the edge in the triangle T_i that joins the two red vertices in T_i, and let

$$E_B = \{e_i : 1 \leq i \leq k\}.$$

Furthermore, let H be the subgraph induced by the edge set E_B. Then $V(H) = R_1$ and $E(H) = E_B$. Let R_1' be a γ-set of H. Since H has no isolated vertices, it follows by Corollary 15.2 that

$$|R_1'| \leq |V(H)|/2 = k - \ell/2.$$

We now interchange the colors of the vertices in $B \cup (R - R_1')$. We claim that this new red-blue coloring of G is an F_6-coloring of G. Suppose that v is a blue vertex of G. Then $v \in R - R_1'$ and so either $v \in R_2$ or $v \in R_1 - R_1'$. If $v \in R_2$, then in the original red-blue coloring of G, v belongs to a triangle with two blue vertices. Thus, after the color interchange, v belongs to a copy of F_6 rooted at v. On the other hand, if $v \in R_1 - R_1'$, then v is adjacent to a vertex $u \in R_1'$. Thus, $uv = e_i$ for some i $(1 \leq i \leq k)$. After the color interchange, u and b_i are red and so v belongs to a copy of F_6 rooted at v. Therefore, as claimed, this new red-blue coloring of G is an F_6-coloring of G. Since

$$|R_1'| + |B| \leq 2k - \ell/2 < 2k + r = |R| = \gamma_{F_6}(G),$$

the number of red vertices in this F_6-coloring is less than $\gamma_{F_6}(G)$, which is impossible. Therefore, $\gamma_{F_6}(G) \leq 2n/3$.   ∎

The upper bound for $\gamma_{F_6}(G)$ given in Theorem 15.8 is sharp. For example, for the graph G of order $n = 12$ shown in Figure 15.12, $\gamma_{F_6}(G) = 8 = 2n/3$.

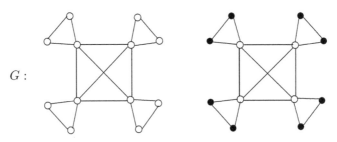

Figure 15.12: A graph G of order n with $\gamma_{F_6}(G) = 2n/3$

We now turn our attention to the domination parameter γ_{F_7}. We noted earlier that for every graph, its independent domination number is at least as large as its domination number. The F_7-domination number, however, always lies between these two numbers.

Theorem 15.9 *If G is a graph in which every edge lies on a triangle, then*

$$\gamma(G) \leq \gamma_{F_7}(G) \leq i(G).$$

Proof. Let G be a graph in which every edge lies on a triangle. In every F_7-coloring of G, every blue vertex is adjacent to a red vertex. Thus, the set of red vertices of G is a dominating set and so $\gamma(G) \leq \gamma_{F_7}(G)$. It therefore remains to show that $\gamma_{F_7}(G) \leq i(G)$. Let S be a minimum independent dominating set of G. Then $|S| = i(G)$. If we color each vertex of S red and all remaining vertices blue, then every blue vertex is adjacent to a red vertex. Since every edge is on a triangle and S is an independent set, it follows that each blue vertex is rooted in a copy of F_7. Hence, this red-blue coloring associated with S is an F_7-coloring of G, and so $\gamma_{F_7}(G) \leq i(G)$.   ∎

The two inequalities in the statements of Theorem 15.9 can be strict. To see this, we consider the graph H of Figure 15.13, where seven vertices of H are labeled as u, v, w, w_1, w_2, w_3, x. Since $\{u, w, x\}$ is a γ-set of H, it follows that $\gamma(H) = 3$. The red-blue coloring whose set of red vertices is $\{u, v, w, x\}$ is a γ_{F_7}-coloring of H. Hence, $\gamma_{F_7}(H) = 4$. Furthermore, $\{u, w_1, w_2, w_3, x\}$ is an i-set and so $i(H) = 5$. Therefore, $\gamma(H) < \gamma_{F_7}(H) < i(H)$ for the graph H. This is illustrated in Figure 15.13.

While Figure 15.13 shows a graph H with $\gamma(H) = 3$, $\gamma_{F_7}(H) = 4$, and $i(H) = 5$, there are no restrictions on the possible values of these three parameters for a graph G other than those given in Theorem 15.9 and that $\gamma(G) \geq 2$ (see [45]).

Theorem 15.10 *For every three integers a, b, c with $2 \leq a \leq b \leq c$, there is a connected graph G in which every edge lies on a triangle such that $\gamma(G) = a$, $\gamma_{F_7}(G) = b$, and $i(G) = c$.*

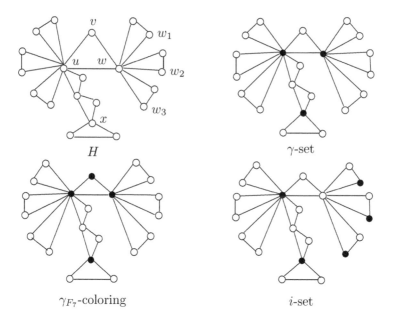

Figure 15.13: A graph H with $\gamma(H) = 3$, $\gamma_{F_7}(H) = 4$, and $i(H) = 5$

In the statement of Theorem 15.9, it is required that every edge of G lie on a triangle of G. If, however, we require only that every vertex of G lie on a triangle of G, then the conclusion does not follow. For example, for the graph G of Figure 15.14, every vertex is on a triangle of G but yet $\gamma(G) = i(G) = 3$ while $\gamma_{F_7}(G) = 4$.

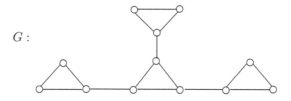

Figure 15.14: A graph G with $\gamma(G) = i(G) = 3$ and $\gamma_{F_7}(G) = 4$

If G is a graph in which every vertex lies on a triangle, then there is a sharp upper bound for $\gamma_{F_7}(G)$ in terms of its order (see [45]).

Theorem 15.11 *If G is a graph of order n in which every vertex lies on a triangle, then*

$$\gamma_{F_7}(G) < \frac{n}{2}.$$

15.4   Stratified Domination by Multiple Graphs

Stratified domination has been extended in a number of ways. For example, let $H_1, H_2, \ldots, H_k$, $k \geq 2$, be 2-stratified graphs, each rooted at a blue vertex and let $\mathcal{H} = \{H_1, H_2, \ldots, H_k\}$. By an $\mathcal{H}$-*coloring* of a graph G is meant a red-blue coloring of the vertices of G such that every blue vertex of G is rooted at a copy of H_i for every integer i with $1 \leq i \leq k$. The $\mathcal{H}$-**domination number** $\gamma_{\mathcal{H}}(G)$ is the minimum number of red vertices in an $\mathcal{H}$-coloring of G.

As above, let $F_1, F_2, \ldots, F_5$ be the five 2-stratified graphs of the path P_3 of order 3 shown in Figure 15.5 and let F_6, F_7 be the two 2-stratified graphs of the triangle K_3 shown in Figure 15.10. If $\mathcal{F} = \{F_1, F_2, \ldots, F_5\}$, then the red-blue coloring of G_1 in Figure 15.15 is an $\mathcal{F}$-coloring of G_1; while if $\mathcal{F}' = \{F_1, F_2, \ldots, F_7\}$, then the red-blue coloring of G_2 in Figure 15.15 is an $\mathcal{F}'$-coloring of G_2.

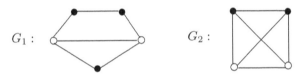

Figure 15.15: Examples of $\mathcal{H}$-colorings

When $\mathcal{F} = \{F_1, F_4\}$, $\mathcal{F}$-colorings of graphs have been investigated in [125, 126]. In this case, every blue vertex is rooted at both a copy of F_1 and a copy of F_4. Here, an $\mathcal{F}$-coloring of a graph G is the same as an F_8-coloring of G, where F_8 is the 2-stratified P_4 rooted at the vertex v, as shown in Figure 15.16.

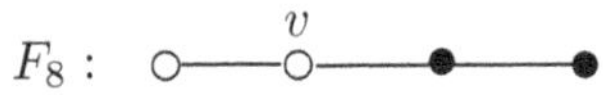

$$F_8 : \quad \circ \!\!-\!\!-\!\! \overset{v}{\circ} \!\!-\!\!-\!\! \bullet \!\!-\!\!-\!\! \bullet$$

Figure 15.16: The 2-stratified P_4

In order to describe a result obtained on the $\mathcal{F}$-domination number of graphs for this set $\mathcal{F}$ of two 2-stratified graphs P_3, we define another additional term. A set S of vertices in a graph G is a **total restrained dominating set** of G if S is both a total dominating set and a restrained dominating set. The **total restrained domination number** of G is the minimum cardinality of a total restrained dominating set for G and is denoted by $\gamma_{tr}(G)$. The following results are due to Henning and Maritz [125].

Theorem 15.12 Let $\mathcal{F} = \{F_1, F_4\}$. If G is a graph without isolated vertices, then

$$\max\{\gamma_t(G), \gamma_r(G)\} \leq \gamma_{\mathcal{F}}(G) \leq \gamma_{tr}(G).$$

Proof.   By coloring the vertices of a minimum total restrained dominating set of a graph G red and the remaining vertices of G blue, an $\mathcal{F}$-coloring of G is achieved, which implies that $\gamma_{\mathcal{F}}(G) \leq \gamma_{tr}(G)$. To verify the lower bound for $\gamma_{\mathcal{F}}(G)$, we first observe that the set of red vertices in an $\mathcal{F}$-coloring of a graph G is a restrained dominating set of G and so $\gamma_r(G) \leq \gamma_{\mathcal{F}}(G)$. Also, observe that the set of red vertices in an $\mathcal{F}$-coloring of a graph G without isolated vertices need not be a total dominating set of G as there could be isolated vertices in the subgraph of G induced by the red vertices. However, every $\mathcal{F}$-coloring of G is also an F_1-coloring of G. Therefore, there exists an F_1-coloring of G with $\gamma_{\mathcal{F}}(G)$ red vertices.

Among all F_1-colorings of G with $\gamma_{\mathcal{F}}(G)$ red vertices, we select one having the minimum number of isolated vertices in the subgraph of G induced by its red vertices. Then, as shown in the proof of Theorem 15.3, every red vertex in such an F_1-coloring is adjacent to another red vertex. Thus, the red vertices form a total dominating set of G, which implies that $\gamma_t(G) \leq \gamma_{\mathcal{F}}(G)$. Therefore, $\max\{\gamma_t(G), \gamma_r(G)\} \leq \gamma_{\mathcal{F}}(G)$. ∎

Even though F_2 and F_4 are 2-stratified trees of order 3, the only $\mathcal{F}$-coloring of a tree, where $\mathcal{F} = \{F_2, F_4\}$, is a vacuous one.

Theorem 15.13 *Let $\mathcal{F} = \{F_2, F_4\}$. A red-blue coloring of a tree T is an $\mathcal{F}$-coloring if and only if all vertices of T are colored red. Consequently $\gamma_{\mathcal{F}}(T) = n$ for every tree T of order n.*

Proof. First, if all vertices of T are colored red, then this coloring is an $\mathcal{F}$-coloring of T. For the converse, assume, to the contrary, that there is a tree T having an $\mathcal{F}$-coloring c in which at least one vertex of T is colored blue. Let v be a blue vertex of T. We consider T as a tree rooted at v. Thus, v is the only vertex at level 0, denoted by L_0. Suppose that the eccentricity of v is k. Then for $i = 0, 1, \ldots, k$, the level L_i consists of all vertices at distance i from v. Since v belongs to a copy of F_4 rooted at v, it follows that v is adjacent to at least one red vertex and at least one blue vertex. Hence, level L_1 contains at least one red vertex and one blue vertex. Let j be the maximum positive integer such that L_j contains a blue vertex w. Since w belongs to a copy of F_4 rooted at w, it follows that w must be adjacent to a blue vertex x in L_{j-1}. On the other hand, because w belongs to a copy of F_2 rooted at w, it follows that w is adjacent to at least one red vertex and all red neighbors of w belong to L_{j+1}. Also, because w belongs to a copy of F_2 rooted at w, it follows that w has a red neighbor $y \in L_{j+1}$ that has a blue neighbor z different from w. Thus, $z \in L_{j+2}$, which is impossible. ∎

We now describe a less restrictive stratified domination defined in terms of more than one 2-stratified graph. Once again, for $k \geq 2$, let $H_1, H_2, \ldots, H_k$ be 2-stratified graphs, each rooted at a blue vertex and let $\mathcal{H} = \{H_1, H_2, \ldots, H_k\}$. By an $\tilde{\mathcal{H}}$-**coloring** of a graph G is meant a red-blue coloring of the vertices of G such that every blue vertex of G is rooted at a copy of *exactly* one H_i, $1 \leq i \leq k$, and for each 2-stratified graph H_i, there is at least one blue vertex v of G such that there is a copy of H_i rooted at v. The $\tilde{\mathcal{H}}$-**domination number** $\gamma_{\tilde{\mathcal{H}}}(G)$ is the minimum number of red vertices in an $\tilde{\mathcal{H}}$-coloring of G.

For example, let $F_1, F_2, \ldots, F_5$ be the five 2-stratified graphs of the path P_3 of order 3 shown in Figure 15.5. In Figure 15.17, an $\tilde{\mathcal{F}}$-coloring of a graph is shown for each of the ten different choices of $\mathcal{F} = \{F_i, F_j\}$ where $1 \leq i < j \leq 5$ as well as an $\tilde{\mathcal{F}}$-coloring of a graph for $\mathcal{F} = \{F_1, F_2, F_3\}$. In each red-blue coloring of a graph in Figure 15.17, we label a blue vertex v by an integer $i \in [5]$ to indicate that v is rooted at a copy of F_i.

1. The coloring of G_1 is an $\tilde{\mathcal{F}}$-coloring where $\mathcal{F} = \{F_1, F_2\}$.

2. The coloring of G_2 is an $\tilde{\mathcal{F}}$-coloring where $\mathcal{F}$ is $\{F_1, F_3\}$ or $\{F_1, F_4\}$.

3. The coloring of G_3 is an $\tilde{\mathcal{F}}$-coloring where $\mathcal{F} = \{F_1, F_5\}$.

4. The coloring of G_4 is an $\tilde{\mathcal{F}}$-coloring where $\mathcal{F} = \{F_2, F_3\}$.

5. The coloring of G_5 is an $\tilde{\mathcal{F}}$-coloring where $\mathcal{F} = \{F_2, F_4\}$.

6. The coloring of G_6 is an $\tilde{\mathcal{F}}$-coloring where $\mathcal{F} = \{F_2, F_5\}$.

7. The coloring of G_7 is an $\tilde{\mathcal{F}}$-coloring where $\mathcal{F} = \{F_3, F_4\}$.

8. The coloring of G_8 is an $\tilde{\mathcal{F}}$-coloring where $\mathcal{F} = \{F_3, F_5\}$.

9. The coloring of G_9 is an $\tilde{\mathcal{F}}$-coloring where $\mathcal{F} = \{F_4, F_5\}$.

10. The coloring of G_{10} is an $\tilde{\mathcal{F}}$-coloring where $\mathcal{F} = \{F_1, F_2, F_3\}$.

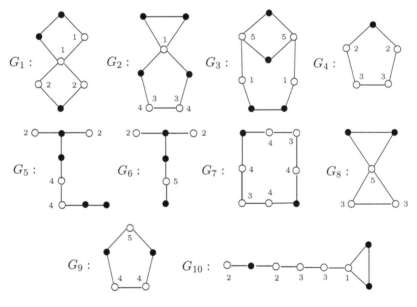

Figure 15.17: Examples of $\tilde{\mathcal{F}}$-colorings

Exercises for Chapter 15

1. We have seen that if G is a connected graph and $H = \text{cor}(G)$ has order n, then $\gamma(H) = n/2$, thereby establishing the sharpness of the bounds in Corollary 15.2. Only one connected graph F of some order n has the property that $\gamma(F) = n/2$ that is not a corona of any graph. What is F?

2. We have seen that if F is a 2-stratified graph and G is a graph of order n containing no copies of F, then $\gamma_F(G) = n$. Is the converse true?

3. We have seen that for $n \geq 3$, $\gamma_{F_3}(K_{1,n-1}) = n$ and $\gamma^{(2)}(K_{1,n-1}) = 2$. Thus, there exists a family of graphs G such that $\gamma_{F_3}(G) - \gamma^{(2)}(G)$ is arbitrarily large. Show that there exists a family of graphs G such that $\gamma_{F_3}(G) - \gamma(G)$ is arbitrarily large.

4. (a) Prove that $\gamma_{F_3}(G) \leq \gamma_{F_4}(G)$ for every graph G.

   (b) Give an example of a graph H such that $\gamma_{F_4}(H) > \gamma_{F_3}(H)$.

5. Prove that if F is the 2-stratified graph F_4 in Figure 15.5, then $\gamma_F(G) = \gamma_r(G)$ for all graphs G.

6. Prove that if F is the 2-stratified graph F_5 in Figure 15.5, then $\gamma_F(G) = \gamma_2(G)$ for all graphs G.

7. For the two 2-stratified graphs F_6 and F_7 shown in Figure 15.10 and for every graph G of order n, show that $\gamma_{F_6}(G) = \gamma_{F_7}(G) = n$ if and only if G is triangle-free.

8. For the graphs H_i $(1 \leq i \leq 3)$ shown in Figure 15.18, determine $\gamma_{F_6}(H_i)$ and $\gamma_{F_7}(H_i)$.

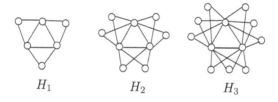

$$H_1 \qquad H_2 \qquad H_3$$

Figure 15.18: The graphs H_i $(1 \leq i \leq 3)$ in Exercise 8

9. Show that for every positive integer k, there is a connected graph G in which every vertex lies on a triangle of G and $\gamma_{F_7}(G) = i(G) + k$.

10. For the graph G of Figure 15.19, determine $\gamma(G)$, $\gamma_{F_7}(G)$, and $i(G)$, where F_7 is the rooted 2-stratified graph shown in Figure 15.10.

11. Let G be a connected graph of order 3 or more. By Theorem 5.1, if S is a minimal dominating set of G, then $V(G) - S$ is a dominating set of G. Furthermore, $\gamma_{F_1}(G) = \gamma_t(G)$ and $\gamma_{F_2}(G) = \gamma(G)$ by Propositions 15.3 and 15.4, respectively. For $i = 1, 2$, show that if S is a minimum F_i-dominating set of G, then $V(G) - S$ need not be an F_i-dominating set of G.

12. Let $\mathcal{F} = \{F_1, F_3\}$.

   (a) Prove that $\gamma_{\mathcal{F}}(C_n) \geq n/2$ for each integer $n \geq 3$.

   (b) Determine all integers $n \geq 3$ for which the bound in (a) is sharp.

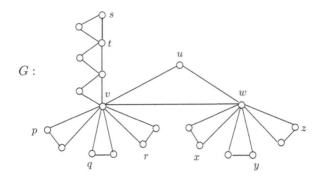

Figure 15.19: The graph G in Exercise 10

(c) Do there exist cycles C_n for which $\gamma_{\tilde{\mathcal{F}}}(C_n) < n$ and C_n has an $\tilde{\mathcal{F}}$-coloring?

13. For $\mathcal{F} = \{F_2, F_3\}$, give an example of an $\tilde{\mathcal{F}}$-coloring of the path P_8 that is neither an F_2-coloring nor an F_3-coloring. What is $\gamma_{\tilde{\mathcal{F}}}(P_8)$?

14. Let $\mathcal{F} = \{F_3, F_4\}$.

   (a) Show for every connected cubic graph G of order n that $\gamma_{\mathcal{F}}(G) \leq n - 2$.
   (b) Determine $\gamma_{\mathcal{F}}(K_{3,3})$ and $\gamma_{\mathcal{F}}(K_2 \,\square\, K_3)$.
   (c) Is the bound in (a) sharp?
   (d) Determine $\gamma_{\mathcal{F}}(P)$ for the Petersen graph P.

15. Let $\mathcal{F} = \{F_3, F_4\}$. Then $\gamma_{\mathcal{F}}(K_n) = 1$ for $n \geq 3$. Also, for each vertex v of K_n, the set $\{v\}$ is an $\mathcal{F}$-dominating set of K_n. Does there exist a connected graph G such that $\gamma_{\mathcal{F}}(G) = k \geq 2$ and every k-element subset of $V(G)$ is an $\mathcal{F}$-dominating set of G?

16. For an integer $k \geq 2$, let $H_1, H_2, \ldots, H_k$ be 2-stratified graphs, each rooted at a blue vertex and let $\mathcal{H} = \{H_1, H_2, \ldots, H_k\}$. By an $\hat{\mathcal{H}}$-**coloring** of a graph G is meant a red-blue coloring of the vertices of G such that every blue vertex v of G is rooted at a copy of H_i for *at least one integer i* with $1 \leq i \leq k$. The $\hat{\mathcal{H}}$-**domination number** $\gamma_{\hat{\mathcal{H}}}(G)$ is the minimum number of red vertices in an $\hat{\mathcal{H}}$-coloring of G. Since every $\mathcal{H}$-coloring is a $\hat{\mathcal{H}}$-coloring and every $\tilde{\mathcal{H}}$-coloring is a $\hat{\mathcal{H}}$-coloring, it follows that $\gamma_{\hat{\mathcal{H}}}(G) \leq \gamma_{\mathcal{H}}(G)$ and $\gamma_{\hat{\mathcal{H}}}(G) \leq \gamma_{\tilde{\mathcal{H}}}(G)$ for every graph G. For $\mathcal{F} = \{F_1, F_3, F_5\}$, give an example of a graph G and an $\hat{\mathcal{F}}$-coloring of G that is neither an $\mathcal{F}$-coloring nor an $\tilde{\mathcal{F}}$-coloring of G.

Chapter 16

Induced Colorings

During the past several decades, there has been increased interest in (typically unrestricted) edge colorings of graphs that give rise to vertex colorings that are often defined in some arithmetic or set theoretic manner from a given edge coloring. In this chapter, we describe several such edge colorings.

16.1 Majestic Colorings

For a positive integer k, let $[k]$ denote $\{1, 2, \ldots, k\}$ and let $\mathcal{P}^*([k])$ denote the set consisting of the $2^k - 1$ nonempty subsets of $[k]$. In 2008, Ervin Győri, Mirko Horňák, Cory Palmer, and Mariusz Woźniak [105] defined an unrestricted edge coloring $c : E(G) \to [k]$ of a nonempty graph G to be a **neighbor-distinguishing coloring** if the vertex coloring $c' : V(G) \to \mathcal{P}^*([k])$, defined so that $c'(v)$ is the set of colors of the edges incident with v, is a proper vertex coloring of G. The minimum positive integer k for which G has a **neighbor-distinguishing k-edge-coloring** was referred to as the **general neighbor-distinguishing index** of G. These concepts were studied further in [19] using the terminology **majestic coloring** and **majestic index** of a graph G, the latter denoted by $\mathrm{maj}(G)$. Since then, this is the terminology that has become commonplace.

A vertex coloring of a nontrivial graph G is **rainbow** or **vertex-distinguishing** if distinct vertices of G are assigned distinct colors. In 1985, Frank Harary and Michael Plantholt [116] considered unrestricted edge colorings that result in vertex-distinguishing vertex colorings defined in the same manner as above, called **set irregular edge colorings**. The minimum positive integer k for which a nonempty graph G has a set irregular k-edge coloring is called the **set irregular chromatic index** of G (also referred to as the **point-distinguishing chromatic index** by Harary and Plantholt). These concepts have also been referred to as **strong majestic colorings** and the **strong majestic index** $\mathrm{smaj}(G)$ of a graph G, which is the terminology we use here. Thus, in either a majestic or a strong majestic k-edge coloring c of a graph G, the color $c(e)$ of an edge e of G is an element of $[k]$ (that is, one of the integers $1, 2, \ldots, k$), while the resulting vertex color $c'(v)$ of a vertex v

of G is a nonempty subset of $[k]$, that is, an element of $\mathcal{P}^*([k])$.

For example, consider the graph G of Figure 16.1. Since G has order 7 and there are only three nonempty subsets of $\{1, 2\}$, it follows that $\mathrm{smaj}(G) \geq 3$. The strong majestic 3-edge coloring of G in Figure 16.1 shows that $\mathrm{smaj}(G) = 3$. For simplicity, we often write the set $\{a\}$ as a, $\{a, b\}$ as ab, $\{a, b, c\}$ as abc, and so on.

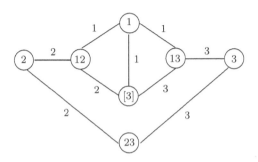

Figure 16.1: A strong majestic 3-edge coloring of a graph

Harary and Plantholt [116] determined the strong majestic indexes (or indices) of all complete graphs.

Theorem 16.1 *For every integer $n \geq 3$, $\mathrm{smaj}(K_n) = \lceil \log_2 n \rceil + 1$.*

Proof. First, suppose that $\mathrm{smaj}(K_n) = k$. Then there exists a strong majestic k-edge coloring of G (using the colors in the set $A = [k] = \{1, 2, \ldots, k\}$). Hence, for distinct vertices u and v in K_n, the vertex colors $c'(u)$ and $c'(v)$ are distinct. However, since $c'(u)$ and $c'(v)$ both contain the color (an element of $[k]$) assigned to uv, it follows that $c'(u) \cap c'(v) \neq \emptyset$. Suppose that $c'(u)$ contains a color that $c'(v)$ does not. Then $\overline{c'(v)} = A - c'(v) \neq \emptyset$. No set $c'(x)$ can be a subset of $\overline{c'(v)}$ for this would imply that $c'(x) \cap c'(v) = \emptyset$, which is impossible. For each color $i \in \overline{c'(v)}$ and for each vertex x, it follows that $c'(x) \subseteq A - \{i\}$. Hence, there are at most 2^{k-1} choices for the set $c'(x)$. Therefore, $n \leq 2^{k-1}$ and so $\log_2 n \leq k - 1$. Thus, $\mathrm{smaj}(K_n) \geq \lceil \log_2 n \rceil + 1$.

It remains to show that $\mathrm{smaj}(K_n) \leq \lceil \log_2 n \rceil + 1$. Let $k = \lceil \log_2 n \rceil + 1$. Thus, $n \leq 2^{k-1}$. We show that there exists a strong majestic k-edge coloring of K_n. Since this is true for $n = 3$, as shown in Figure 16.2, we may assume that $n \geq 4$ and so $n \geq k + 1$.

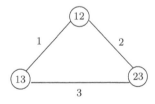

Figure 16.2: A strong majestic 3-edge coloring of K_3

Denote the vertices of K_n by $v_1, v_2, \ldots, v_k, \ldots, v_n$. Assign the set $S_1 = \{1\}$ to v_1, the set $S_i = \{1, i\}$, $2 \le i \le k$, to v_i and the set $S_{k+1} = [k]$ to v_{k+1}. Since $n \le 2^{k-1}$, we can assign distinct subsets S_i $(k+2 \le i \le n)$ of the set $[k]$ that contain 1 and that are distinct from $S_1, S_2, \ldots, S_{k+1}$ to the vertices v_i. Hence, $1 \in S_i$ for all i $(1 \le i \le n)$. For each edge $v_i v_j$ of K_n, assign the largest color belonging to $S_i \cap S_j$ to $v_i v_j$. This gives a k-edge coloring of K_n. (See Figure 16.3 for such a 4-edge coloring of K_6.)

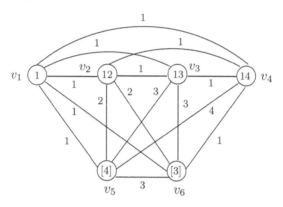

Figure 16.3: A strong majestic 4-edge coloring of K_6

For $1 \le i \le n$, let $c'(v_i)$ be the vertex color which is the set of colors of the incident edges of v_i. Hence, $c'(v_i) \subseteq S_i$ for each i with $1 \le i \le n$. Since $S_1 = \{1\}$, it follows that $c'(v_1) = \{1\}$. For $2 \le i \le k$, we have $i \in S_i$. Since $i \in S_{k+1}$, it follows that $i \in c'(v_i)$ and so $c'(v_i) = S_i$ for all i with $1 \le i \le k$. Consider S_i where $k + 1 \le i \le n$ and suppose that $\ell \in S_i$, where $\ell \ne 1$. Since ℓ is the largest color in S_ℓ, it follows that ℓ is the color of the edge $v_\ell v_i$. Thus, $\ell \in c'(v_i)$ and so $c'(v_i) = S_i$ for all i $(1 \le i \le n)$. Hence, $\mathrm{smaj}(K_n) \le \lceil \log_2 n \rceil + 1$, completing the proof. ∎

The strong majestic indexes of paths, cycles, and n-cubes were determined in [116].

Theorem 16.2 *For each integer $n \ge 3$,*

$$\mathrm{smaj}(P_n) = \min \left\{ 2 \left\lceil \sqrt{\frac{n-1}{2}} \right\rceil, \ 2 \left\lceil \frac{1 + \sqrt{8n - 9}}{4} \right\rceil - 1 \right\}$$

$$\mathrm{smaj}(C_n) = \min \left\{ 2 \left\lceil \sqrt{\frac{n}{2}} \right\rceil, \ 2 \left\lceil \frac{1 + \sqrt{8n + 1}}{4} \right\rceil - 1 \right\}.$$

Theorem 16.3 *For each integer $n \ge 2$, $\mathrm{smaj}(Q_n) = n + 1$.*

While there is no strong majestic coloring of K_2, the strong majestic index of every connected graph of size 2 or more always exists.

Proposition 16.4 *If G is a connected graph of size $m \ge 2$, then $\mathrm{smaj}(G)$ exists and*

$$2 \le \mathrm{maj}(G) \le \mathrm{smaj}(G) \le m.$$

Proof.   Let G be a connected graph with $E(G) = \{e_1, e_2, \ldots, e_m\}$ where $m \geq 2$. We define the edge coloring $c : E(G) \to [m]$ by $c(e_i) = i$ for $1 \leq i \leq m$. Since no two vertices of G have the same set of incident edges, it follows that $c'(x) \neq c'(y)$ for every two distinct vertices x and y of G and so c is a strong majestic m-coloring of G. Therefore, $\operatorname{smaj}(G) \leq m$.   ∎

The lower bound 2 and upper bound m for $\operatorname{maj}(G)$ and $\operatorname{smaj}(G)$ of a connected graph G of size $m \geq 2$ in Proposition 16.4 are both sharp since for the star $G = K_{1,m}$ of size m, it follows that $\operatorname{maj}(G) = 2$ and $\operatorname{smaj}(G) = m$.

For complete graphs, an unrestricted edge coloring is majestic if and only if it is vertex-distinguishing. Therefore, every majestic coloring of a complete graph is a strong majestic coloring. From this observation, the result below follows from Theorem 16.1.

Corollary 16.5   *For every integer $n \geq 3$,*

$$\operatorname{maj}(K_n) = \operatorname{smaj}(K_n) = \lceil \log_2 n \rceil + 1.$$

There is a lower bound for the majestic index of a graph in terms of its chromatic number due to Mirko Horňák and Roman Soták [128].

Theorem 16.6   *If G is a connected graph of order at least 3, then*

$$\operatorname{maj}(G) \geq \lceil \log_2 \chi(G) \rceil + 1.$$

Proof.   Suppose that $\operatorname{maj}(G) = k$. Let there be given a majestic k-edge coloring $c : E(G) \to [k]$ of G. For $A \in \mathcal{P}^*([k])$, let $\overline{A} = [k] - A$ and let V_A denote the set of vertices x of G for which $c'(x) = A$. Therefore, if $V_A \neq \emptyset$, then V_A is an independent set of vertices of G. Suppose that x and y are two vertices of G such that $c'(x) = A$ and $c'(y) = \overline{A}$. Consequently, $c'(x) \cap c'(y) = A \cap \overline{A} = \emptyset$, which implies that $xy \notin E(G)$. Therefore, $\{V_A \cup V_{\overline{A}} : A \in \mathcal{P}^*([k])\}$ forms a partition of $V(G)$ into empty or independent sets of vertices of G and so G has a proper vertex coloring using at most $\frac{1}{2}(2^k) = 2^{k-1}$ colors. Hence, $\chi(G) \leq 2^{k-1}$ and so $k - 1 \geq \log_2 \chi(G)$. Therefore, $\operatorname{maj}(G) = k \geq \lceil \log_2 \chi(G) \rceil + 1$.   ∎

Since $\chi(G) \geq \omega(G)$ for every graph G, Theorem 16.6 leads to a lower bound for the majestic index of G in terms of its clique number.

Corollary 16.7   *If G is a nontrivial connected graph, then*

$$\operatorname{maj}(G) \geq \lceil \log_2 \omega(G) \rceil + 1.$$

With the aid of Theorem 16.6, we have the following result.

Theorem 16.8   *If G is an ℓ-chromatic graph for some integer $\ell \geq 3$ for which there is a partition $\{V_1, V_2, \ldots, V_\ell\}$ of $V(G)$ into color classes such that every vertex v of G is adjacent to at least one vertex in every color class not containing v, then*

$$\operatorname{maj}(G) = \lceil \log_2 \ell \rceil + 1.$$

Proof. By Theorem 16.6, it follows that $\text{maj}(G) \geq \lceil \log_2 \ell \rceil + 1$. It therefore remains to show that $\text{maj}(G) \leq \lceil \log_2 \ell \rceil + 1$. Let K_ℓ be the complete graph of order ℓ with vertex set $\{v_1, v_2, \ldots, v_\ell\}$. Since $\text{maj}(K_\ell) = \lceil \log_2 \ell \rceil + 1$ by Corollary 16.5, it follows that K_ℓ has a majestic coloring c_0 using colors from the set $\{1, 2, \ldots, \lceil \log_2 \ell \rceil + 1\}$. An edge coloring $c : E(G) \rightarrow [\lceil \log_2 \ell \rceil + 1]$ of G is defined by $c(e) = c_0(v_i v_j)$ if $e \in [V_i, V_j]$ for $1 \leq i < j \leq \ell$. Thus, for each integer i with $1 \leq i \leq \ell$ and each vertex $v \in V_i$, it follows that $c'(v) = c_0'(v_i)$. Therefore, c is a majestic coloring of G and so $\text{maj}(G) \leq \lceil \log_2 \ell \rceil + 1$. ∎

The following result is therefore an immediate consequence of Theorem 16.8, which shows that the lower bound in Theorem 16.6 cannot be improved in general.

Corollary 16.9 *If G is a complete ℓ-partite graph where $\ell \geq 3$, then*

$$\text{maj}(G) = \lceil \log_2 \ell \rceil + 1.$$

A special case of Theorem 16.6 is when G is a connected graph with $\chi(G) = 3$. If G is a connected graph with $\chi(G) = 3$, then $\text{maj}(G) \geq 3$. In fact, not only is $\text{maj}(G) \geq 3$ if G is a connected graph with $\chi(G) = 3$, but $\text{maj}(G) = 3$ if $\chi(G) = 3$. We saw in Theorem 16.6 that if G is a connected graph of order at least 3, then $\text{maj}(G) \geq \lceil \log_2 \chi(G) \rceil + 1$. Horňák and Soták [128] showed that if G is a connected graph with $\chi(G) \geq 3$, then $\text{maj}(G)$ cannot exceed this lower bound by much, if at all.

Theorem 16.10 *If G is a connected graph with $\chi(G) \geq 3$, then*

$$\text{maj}(G) \leq \lfloor \log_2 \chi(G) \rfloor + 2.$$

As a consequence of Theorems 16.6 and 16.10, we have the following.

Corollary 16.11 *If G is a connected graph such that $2^{k-1} < \chi(G) < 2^k$ for some integer $k \geq 2$, then $\text{maj}(G) = k + 1$. In particular, if $\chi(G) = 3$, then $\text{maj}(G) = 3$.*

By Corollary 16.5, if $n = 2^k$ for some integer $k \geq 2$, then $\text{maj}(K_n) = k+1$. That is, if G is a complete graph with $\chi(G) = 2^k$ where $k \geq 2$, then $\text{maj}(G) = k + 1$. However, it is not known if there is any connected graph H with $\chi(H) = 2^k$ such that $\text{maj}(H) = k + 2$. This problem was also posed by Horňák and Soták [128].

The preceding four results presented are for connected graphs with chromatic number 3 or more. For chromatic number 2, we are dealing with bipartite graphs of course. First, we consider the case when the graph is a cycle.

Proposition 16.12 *For an integer $n \geq 3$,*

$$\text{maj}(C_n) = \begin{cases} 2 & \text{if } n \equiv 0 \pmod 4 \\ 3 & \text{if } n \not\equiv 0 \pmod 4. \end{cases}$$

Proof. If n is odd, then $\chi(C_n) = 3$ and so $\text{maj}(C_n) = 3$ by Corollary 16.11. Hence we need only consider the case when n is even. It is immediate that $\text{maj}(C_4) = 2$. So, we may assume that $n \geq 6$. Let $C_n = (v_1, v_2, \ldots, v_n, v_{n+1} = v_1)$. In any majestic coloring c of C_n, there are two adjacent edges of C_n that are colored differently, say $c(v_n v_1) = 1$ and $c(v_1 v_2) = 2$.

First, suppose that $n \equiv 0 \pmod 4$. If c is a majestic 2-edge coloring of C_n, then we must have $c(v_2 v_3) = 2$, $c(v_3 v_4) = 1$, and $c(v_4 v_5) = 1$. More generally, $c(v_i v_{i+1}) = 2$ when $i \equiv 1, 2 \pmod 4$ and $c(v_i v_{i+1}) = 1$ when $i \equiv 3, 0 \pmod 4$. Since $n \equiv 0 \pmod 4$, it follows that $c'(v_i) = \{1, 2\}$ if i is odd, while $c'(v_i) = \{1\}$ if $i \equiv 0 \pmod 4$ and $c'(v_i) = \{2\}$ if $i \equiv 2 \pmod 4$. Hence, c is a majestic 2-edge coloring of C_n and so $\mathrm{maj}(C_n) = 2$.

Next, suppose that $n \equiv 2 \pmod 4$. For the majestic 2-edge coloring of C_n defined above, $c'(v_{n-1}) = c'(v_n) = c'(v_1) = \{1, 2\}$ and so c is not a majestic 2-edge coloring of C_n. Therefore, $\mathrm{maj}(C_n) \geq 3$. In this case, changing the colors of both $v_{n-1} v_n$ and $v_n v_1$ to 3 results in $c'(v_{n-1}) = \{1, 3\}$, $c'(v_n) = \{3\}$, and $c'(v_2) = \{2, 3\}$. Since this is a majestic 3-edge coloring, it follows that $\mathrm{maj}(C_n) = 3$ if $n \equiv 2 \pmod 4$.   ∎

All complete bipartite graphs have the same majestic index.

Proposition 16.13 *For positive integers r and s with $r \leq s$ and $s \geq 2$,*

$$\mathrm{maj}(K_{r,s}) = 2.$$

Proof. It suffices to show that $K_{r,s}$ has a majestic 2-edge coloring. Let U and W be the partite sets of $K_{r,s}$, where $|U| = r$ and $W = \{w_1, w_2, \ldots, w_s\}$. Assign the color 1 to each edge incident with w_i for $1 \leq i \leq s - 1$ and the color 2 to each edge incident with w_s. Then $c'(w_i) = \{1\}$ for $1 \leq i \leq s - 1$, $c'(w_s) = \{2\}$, and $c'(u) = \{1, 2\}$ for each $u \in U$. Thus, c is a majestic 2-edge coloring of $K_{r,s}$ and so $\mathrm{maj}(K_{r,s}) = 2$.   ∎

It is well known that if $H \subseteq G$, then $\chi(H) \leq \chi(G)$ and $\chi'(H) \leq \chi'(G)$. This, however, is not the case for the majestic index. For example, $C_6 \subseteq K_{3,3}$; yet, $\mathrm{maj}(C_6) = 3$ and $\mathrm{maj}(K_{3,3}) = 2$ by Propositions 16.12 and 16.13. (In fact, this is also not true even when H is an induced subgraph of G as we will soon see.) Moreover, for a connected graph G of order at least 3, any of the following could occur:

$$\chi(G) < \mathrm{maj}(G), \quad \chi(G) = \mathrm{maj}(G), \quad \chi(G) > \mathrm{maj}(G).$$

For example, if $n \equiv 2 \pmod 4$ and $n \geq 6$, then $\chi(C_n) = 2$ and $\mathrm{maj}(C_n) = 3$ by Proposition 16.12; while if $n \equiv 0 \pmod 4$ and $n \geq 4$, then $\chi(C_n) = \mathrm{maj}(C_n) = 2$. Furthermore, if $k \geq 4$, then $\chi(K_k) = k$ and $\mathrm{maj}(K_k) = \lceil \log_2 k \rceil + 1$ by Corollary 16.5.

We saw in Proposition 16.13 that the majestic index of every complete bipartite graph of order at least 3 is 2 and in Proposition 16.12 that the majestic index of every even cycle is 2 or 3. In fact, this is true for all connected bipartite graphs of order 3 or more (see [19, 105]). To present a proof of this fact, the following concept will be useful. Let u be a vertex in a nontrivial connected graph G. A vertex v distinct from u is called a **boundary vertex** of u if $d(u, v) = k$ for some positive integer k and no $u-w$ geodesic of length greater than k contains v. Equivalently, v is a boundary vertex of u if for every neighbor w of v, it follows that $d(u, w) \leq d(u, v)$. In particular, for every vertex u of G, every end-vertex of G different from u is a boundary vertex of u.

Theorem 16.14 *If G is a connected bipartite graph of order 3 or more, then*

$$\text{maj}(G) \leq 3.$$

Proof. Let U and W be the partite sets of G, where U contains at least two vertices. For a vertex u of U, let

$$\begin{aligned}
U_1 &= \{v \in V(G) : d(u, v) \equiv 0 \pmod 4\} \text{ and} \\
U_2 &= \{v \in V(G) : d(u, v) \equiv 2 \pmod 4\}.
\end{aligned}$$

Thus, $U = U_1 \cup U_2$ and $W = \{v \in V(G) : d(u, v) \text{ is odd}\}$. Assign the color 1 to each edge of G incident with a vertex of U_1 and the color 2 to each edge of G incident with a vertex of U_2. Denote this edge coloring by c and the resulting vertex coloring by c'. If no vertex of W is a boundary vertex of u, then every vertex of W has the color $\{1, 2\}$. Since each vertex of U has the color $\{1\}$ or $\{2\}$, the coloring c is a majestic 2-edge coloring of G and so $\text{maj}(G) = 2$. Suppose, on the other hand, that $w \in W$ is a boundary vertex of u. Then $c'(w) = \{1\}$ or $c'(w) = \{2\}$, say the former. For each neighbor x of w on a $u-w$ geodesic, change the color of xw from 1 to 3. Then $c'(w) = \{3\}$ and $c'(x) = \{1, 3\}$. In this case, the color of every vertex of U is $\{1\}$, $\{2\}$, $\{1, 3\}$, or $\{2, 3\}$, while the color of every vertex of W is $\{1, 2\}$ or $\{3\}$. This new edge coloring is a majestic 3-edge coloring of G and so $\text{maj}(G) \leq 3$. ∎

The following result describes those bipartite graphs having majestic index 2.

Theorem 16.15 *Let G be a connected bipartite graph of order 3 or more. Then $\text{maj}(G) = 2$ if and only if there exists a partition $\{U_1, U_2, W\}$ of $V(G)$ such that $U = U_1 \cup U_2$ and W are the partite sets of G and each vertex $w \in W$ has a neighbor in both U_1 and U_2.*

Proof. First, suppose that U and W are the partite sets of G such that U can be partitioned into two sets U_1 and U_2 for which every vertex in W has a neighbor in each of U_1 and U_2. Define the edge coloring $c : E(G) \to \{1, 2\}$ by $c(e) = i$ if e is incident with a vertex in U_i for $i = 1, 2$. Then $c'(u) = \{i\}$ if $u \in U_i$ for $i = 1, 2$ and $c'(w) = \{1, 2\}$ for each $w \in W$. Hence, c is a majestic 2-edge coloring of G and so $\text{maj}(G) = 2$.

For the converse, suppose that G is a connected bipartite graph of order 3 or more such that $\text{maj}(G) = 2$. Let U and W be the unique partite sets of G and let $c : E(G) \to \{1, 2\}$ be a majestic 2-edge coloring of G. Then U is divided into three sets U_1, U_2, and $U_{1,2}$, where U_i is the set of vertices u with $c'(u) = \{i\}$ for $i = 1, 2$ and $U_{1,2}$ is the set of vertices u with $c'(u) = \{1, 2\}$. Similarly, W is divided into three sets W_1, W_2, and $W_{1,2}$. Observe that the vertices in $U_1 \cup U_2$ can only be adjacent to vertices in $W_{1,2}$ and the vertices in $W_1 \cup W_2$ can only be adjacent to vertices in $U_{1,2}$. Since G is connected and no vertex in $U_{1,2}$ can be adjacent to any vertex in $W_{1,2}$, it follows that either $U_1 \cup U_2 \cup W_{1,2} = \emptyset$ or $W_1 \cup W_2 \cup U_{1,2} = \emptyset$, say $W_1 \cup W_2 \cup U_{1,2} = \emptyset$. Then $\{U_1, U_2, W_{1,2}\}$ is the desired partition of the vertex set of G since each vertex $w \in W_{1,2} = W$ must be adjacent to some vertex in U_1 and some vertex in U_2.   ∎

The following is a consequence of the proof of Theorem 16.14.

Corollary 16.16 *Let G be a connected bipartite graph. If G contains a vertex u such that all boundary vertices of u belong to the same partite set as u, then $\mathrm{maj}(G) = 2$.*

The converse of Corollary 16.16 is not true, however. For example, Figure 16.4 shows a majestic 2-edge coloring of the 3-cube Q_3, where each solid edge is colored 1 and each dashed edge is colored 2. (In Figure 16.4, $\{a\}$ is denoted by a where $a \in \{1, 2\}$ and $\{1, 2\}$ is denoted by 12.) Thus, $\mathrm{maj}(Q_3) = 2$. For each vertex u of Q_3, there is a unique boundary vertex v of u such that $d(u, v) = 3$. Thus, u and v do not belong to the same partite set. Therefore, there is no vertex u in Q_3 all of whose boundary vertices belong to the same partite set as u. In fact, if u and v are boundary vertices of each other, then u and v belong to different partite sets of Q_3. Next, we consider the non-regular bipartite graph G of Figure 16.4. The majestic 2-edge coloring of G in Figure 16.4 (where a solid edge is colored 1 and a dashed edge is colored 2) shows that $\mathrm{maj}(G) = 2$. On the other hand, for $i = 1, 2$, the vertex v_i is a boundary vertex of u_i; while $d(u_i, v_i) = 3$ and so v_i and u_i do not belong to the same partite set. Hence, by symmetry, G has no vertex u all of whose boundary vertices belong to the same partite set as u. For trees, the converse of Corollary 16.16 is true, however, as we show next.

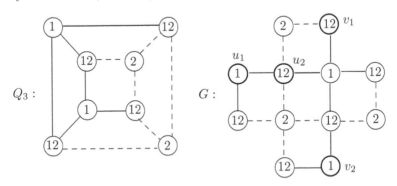

Figure 16.4: A bipartite graph G with $\mathrm{maj}(G) = 2$

Theorem 16.17 *Let T be a tree of order 3 or more. Then $\mathrm{maj}(T) = 2$ if and only if the distance between every two end-vertices is even. Equivalently, $\mathrm{maj}(T) = 2$ if and only if all end-vertices of T belong to the same partite set of T.*

Proof. Suppose that T is a tree of order 3 or more such that the distance between every two end-vertices is even. Let uv be a pendant edge of T, where u is an end-vertex of T. First, the edge uv is colored 1. Let w be any vertex of T such that

$d(u, w)$ is even. If $d(u, w) \equiv 2 \pmod 4$, then all edges incident with w are colored 2; while if $d(u, w) \equiv 0 \pmod 4$, then all edges incident with w are colored 1. Since this is a majestic 2-edge coloring, $\text{maj}(T) = 2$. Note that for any majestic 2-edge coloring of T, the color of every vertex in one partite set of T is $\{1, 2\}$, while the color of a vertex in the other partite set is $\{1\}$ or $\{2\}$.

Next, we verify the converse. Assume to the contrary, that there exists a tree T of order 3 or more such that $\text{maj}(T) = 2$ but T contains a pair u, v of end-vertices for which $d(u, v)$ is odd, say $d(u, v) = 2k + 1$ for some positive integer k. Let $P = (u = u_1, u_2, \ldots, u_{2k+2} = v)$ be the $u - v$ path in T and let c be a majestic 2-edge coloring of T. Since u and v are end-vertices, each of $c'(u)$ and $c'(v)$ is either $\{1\}$ or $\{2\}$. Because u_2 is adjacent to u_1, the color of u_1 must be a proper subset of the color of u_2. So, the color of u_2 is $\{1, 2\}$. Since u_3 is adjacent to u_2, the color of u_3 must be a singleton. Continuing this process, we see that $c'(u_{2i}) = \{1, 2\}$ and $c'(u_{2i-1})$ is either $\{1\}$ or $\{2\}$. In particular, $c'(u_{2k+2}) = c'(v) = \{1, 2\}$, which is a contradiction. ∎

By Theorem 16.17, it follows that for each integer $n \geq 3$,

$$\text{maj}(P_n) = \begin{cases} 2 & \text{if } n \text{ is odd} \\ 3 & \text{if } n \text{ is even.} \end{cases} \tag{16.1}$$

We have seen that there are connected bipartite graphs G with $\text{maj}(G) = 3$ such that $\delta(G) = 1$ or $\delta(G) = 2$. However, there are also connected bipartite graphs G with $\text{maj}(G) = 3$ such that $\delta(G) \geq 3$ as well. We present such an example now. The **Heawood graph** (the unique 6-cage) shown in Figure 16.5 is a 3-regular bipartite graph of order 14.

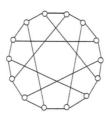

Figure 16.5: The Heawood graph: the unique 6-cage

Proposition 16.18 *The majestic index of the Heawood graph is 3.*

Proof. Assume, to the contrary, that the majestic index of the Heawood graph H is 2. Thus, there exists a majestic 2-edge coloring $c : E(H) \to \{1, 2\}$ of H such that the induced vertex coloring c' of G is proper. Since each edge of H is colored 1 or 2, each vertex of H is colored $\{1\}$, $\{2\}$, or $\{1, 2\}$. Since no two adjacent vertices can be colored by a singleton, every vertex in one partite set of H is colored $\{1, 2\}$ and every vertex in the other partite set of H is colored by a singleton. Let $U = \{u_1, u_2, \ldots, u_7\}$ be the set of vertices colored $\{1, 2\}$ by c', let X be the set of vertices

colored $\{1\}$, and let Y be the set of vertices colored $\{2\}$. Thus, U and $X \cup Y$ are
the partite sets of H. Since the girth of H is 6 and so H has no 4-cycles,

$$|N(v) \cap N(w)| \leq 1 \text{ for every two distinct vertices } v \text{ and } w \text{ in } H. \qquad (16.2)$$

Since (i) each vertex in U is adjacent to at least one vertex in X and at least
one vertex in Y and (ii) H is 3-regular, it follows by (16.2) that $|X| \geq 3$ and
$|Y| \geq 3$. We may assume that $|X| = 3$ and $|Y| = 4$. Let $X = \{x_1, x_2, x_3\}$ and let
$Y = \{y_1, y_2, y_3, y_4\}$. By (16.2), $|N(x_1) \cap N(x_2)| \leq 1$. Hence, we may assume that
$N(x_1) = \{u_1, u_2, u_3\}$ and $\{u_4, u_5\} \subset N(x_2)$. We consider the following two cases.

Case 1. $N(x_1) \cap N(x_2) = \emptyset$. In this case, we may assume $N(x_2) = \{u_4, u_5, u_6\}$
and $N(x_3) = \{u_1, u_4, u_7\}$. Since H has no 4-cycles, no two vertices in $\{u_1, u_2, u_3\}$
have a common neighbor in Y. In addition, exactly one edge incident with u_1 is
colored $\{2\}$ and exactly two edges incident with each of u_2 and u_3 are colored $\{2\}$,
say $u_1 y_1 \in E(G)$. However then, each of u_2 and u_3 is adjacent to two vertices in
$\{y_2, y_3, y_4\}$ and so u_2 and u_3 have a common neighbor in Y, which is impossible.

Case 2. $|N(x_1) \cap N(x_2)| = 1$, say $N(x_1) \cap N(x_2) = \{u_1\}$. Thus, $N(x_2) =$
$\{u_1, u_4, u_5\}$. Since each of u_6 and u_7 is adjacent to exactly one of x_1 and x_2, it
follows that u_6 and u_7 are both adjacent to x_3. Since $\deg x_3 = 3$ and $c'(u_1) =$
$\{1, 2\}$, it follows that x_3 is adjacent to a vertex in $\{u_2, u_3, u_4, u_5\}$. Thus, either
$N(x_1) \cap N(x_3) = \emptyset$ or $N(x_2) \cap N(x_3) = \emptyset$. Whichever of these situations occurs,
we can apply Case 1 to produce a contradiction.   ∎

As we saw in Theorem 16.14, for every connected graph G of order 3 or more
with $\chi(G) = 2$, either $\text{maj}(G) = 2$ or $\text{maj}(G) = 3$. On the other hand, if $\chi(G) \geq 3$,
then it is impossible that $\text{maj}(G) = 2$ by Theorem 16.6.

16.2   Royal and Regal Colorings

In a majestic coloring of a graph G, the colors assigned to the edges of G are elements
of some set $[k]$ for a positive integer k, which results in a proper vertex coloring of
G where the color of a vertex v is the set of colors of the edges incident with v.
In a strong majestic coloring, the vertex coloring is vertex-distinguishing. Here, we
consider two types of edge colorings, one type called royal colorings and strong royal
colorings and the other type called regal colorings and strong regal colorings, where
in each type the colors assigned to the edges of a graph are nonempty subsets of a
set $[k]$ rather than elements of $[k]$.

For a connected graph G of order 3 or more, let $c : E(G) \to \mathcal{P}^*([k])$ be an
unrestricted edge coloring of G for some positive integer k. The edge coloring c
produces the vertex coloring $c' : V(G) \to \mathcal{P}^*([k])$ defined by

$$c'(v) = \bigcup_{e \in E_v} c(e),$$

where E_v is the set of edges of G incident with v. That is, each edge e of G is
assigned a nonempty subset of $[k]$. This is the color $c(e)$ of e. Then each vertex v

of G is assigned the union of the colors of the edges incident with v. This is the color $c'(v)$ of v. Hence, the colors of the edges and vertices of G are both nonempty subsets of $[k]$.

If c' is a proper vertex coloring of G, then c is called a **royal k-edge coloring** of G. An edge coloring c is a **royal coloring** of G if c is a royal k-edge coloring for some positive integer k. The minimum positive integer k for which a graph G has a royal k-edge coloring is the **royal index** roy(G) of G. If c' is vertex-distinguishing, then c is a **strong royal k-edge coloring** of G. An edge coloring c is a **strong royal coloring** of G if c is a strong royal k-edge coloring for some positive integer k. The minimum positive integer k for which a graph G has a strong royal k-coloring is the **strong royal index** sroy(G) of G. Assigning a color $a \in [k]$ to an edge e in a majestic or strong majestic coloring of a graph G is equivalent to assigning the singleton set $\{a\} \in \mathcal{P}^*([k])$ to e in a royal or strong royal coloring of G. That is, royal and strong royal colorings are generalizations of majestic and strong majestic colorings, respectively. While there are k possible colors for the edges of a graph G in a majestic or strong majestic k-edge coloring of G, there are $2^k - 1$ possible colors for the edges of G in a royal or strong royal k-edge coloring of G. Since $2^k - 1 > k$ when $k \geq 2$, there are, in general, many more choices to color the edges of a graph so that the resulting vertex coloring is either proper or vertex-distinguishing. Furthermore, since, by Proposition 16.4, smaj(G) exists if G is a connected graph of order at least 3, we have the following observation.

Proposition 16.19 *If G is a connected graph of order 3 or more, then G has a strong royal coloring and therefore a royal coloring. Furthermore,*

$$\text{roy}(G) \leq \text{maj}(G) \text{ and } \text{sroy}(G) \leq \text{smaj}(G).$$

For the graph K_3, we have roy$(K_3) = $ maj$(K_3) = $ sroy$(K_3) = $ smaj$(K_3) = 3$. The only other connected graph of order 3 is P_3. In this case, roy$(P_3) = $ maj$(P_3) = $ sroy$(P_3) = $ smaj$(P_3) = 2$. Since $|\mathcal{P}^*([2])| = 3$, it follows that sroy$(G) \geq 3$ for every connected graph G of order 4 or more. Thus, P_3 is the only connected graph having strong royal index 2. Figure 16.6 shows a royal 2-edge coloring, a strong royal 3-edge coloring, and a strong majestic 4-edge coloring of the star $G = K_{1,4}$ of size 4, where, as before, we write a for the set $\{a\}$, ab for $\{a, b\}$, and abc for $\{a, b, c\}$. In fact, roy$(G) = $ maj$(G) = 2$, sroy$(G) = 3$, and smaj$(G) = 4$ for this graph G. Thus, the values of the three parameters roy(G), sroy(G), and smaj(G) can be different for a graph G.

For the graph $G = K_{1,4}$ in Figure 16.6, we observed that roy$(G) = $ maj(G). Because roy$(G) \leq $ maj(G) for every connected graph G of order 3 or more and because in many instances there are so many more choices for edge colors in a royal k-edge coloring of a graph than in a majestic k-edge coloring of the graph, it appears there should be examples of connected graphs G for which roy$(G) < $ maj(G) – in fact examples where these two numbers are quite different. We discuss this now. An argument similar to the proof of Theorem 16.6 gives the following result.

Theorem 16.20 *If G is a connected graph of order at least 3, then*

$$\text{roy}(G) \geq \lceil \log_2 \chi(G) \rceil + 1.$$

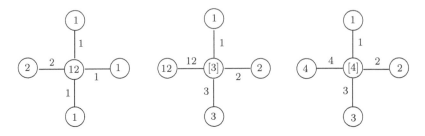

Figure 16.6: A graph G with $\text{roy}(G) = 2, \text{sroy}(G) = 3$, and $\text{smaj}(G) = 4$

We saw in Theorem 16.10 that if G is a connected graph with $\chi(G) \geq 3$, then $\text{maj}(G) \leq \lfloor \log_2 \chi(G) \rfloor + 2$. Furthermore, by Proposition 16.19, $\text{roy}(G) \leq \text{maj}(G)$ for every connected graph G of order 3 or more. Thus, the following is a consequence of Theorem 16.10, Proposition 16.19, and Theorem 16.20.

Theorem 16.21 *If G is a connected graph with $\chi(G) \geq 3$, then*

$$\lceil \log_2 \chi(G) \rceil + 1 \leq \text{roy}(G) \leq \text{maj}(G) \leq \lfloor \log_2 \chi(G) \rfloor + 2.$$

Consequently, if $\chi(G)$ is not a power of 2, then $\text{roy}(G) = \text{maj}(G) = \lceil \log_2 \chi(G) \rceil + 1$.

In the case of complete graphs, the royal index does not depend on its chromatic number (that is, its order) not being a power of 2.

Proposition 16.22 *For every integer $n \geq 3$,*

$$\text{roy}(K_n) = \text{maj}(K_n) = \lceil \log_2 n \rceil + 1.$$

Proof.   First, $\text{roy}(K_n) \geq \lceil \log_2 n \rceil + 1$ by Theorem 16.21. Furthermore, $\text{roy}(K_n) \leq \text{maj}(K_n)$ by Proposition 16.19. Thus, $\text{roy}(K_n) = \text{maj}(K_n) = \lceil \log_2 n \rceil + 1$ by Corollary 16.5. ∎

Since $\text{sroy}(K_n) = \text{roy}(K_n)$ and $\text{smaj}(K_n) = \text{maj}(K_n)$, it follows by Proposition 16.22 that $\text{sroy}(K_n) = \text{smaj}(K_n)$. There are many other connected graphs H for which $\text{sroy}(H) = \text{smaj}(H)$. While there are also many connected graphs G for which $\text{sroy}(G) \neq \text{smaj}(G)$, we know of no graph G for which $\text{roy}(G) \neq \text{maj}(G)$. By Theorem 16.21, if G is a connected graph with $\chi(G) \geq 3$, then $\text{maj}(G) - \text{roy}(G) \leq \lfloor \log_2 \chi(G) \rfloor - \lceil \log_2 \chi(G) \rceil + 1$. That is, as observed, if $\chi(G)$ is not a power of 2, then $\text{roy}(G) = \text{maj}(G)$. Therefore, the only possibility for $\text{maj}(G) \neq \text{roy}(G)$ is when $\chi(G) \geq 4$ is a power of 2, in which case it may occur that $\text{maj}(G) = \text{roy}(G) + 1$. Consequently, our emphasis here is on the strong royal indexes of graphs. We now present a lower bound for the strong royal index of any connected graph of order 4 or more in terms of its order.

Proposition 16.23 *If G is a connected graph of order $n \geq 4$, then*

$$\text{sroy}(G) \geq \lceil \log_2(n+1) \rceil = \lfloor \log_2 n \rfloor + 1.$$

Proof. Suppose that $\text{sroy}(G) = k$ and $c : E(G) \to \mathcal{P}^*([k])$ is a strong royal k-edge coloring of G. Then the vertex coloring $c' : V(G) \to \mathcal{P}^*([k])$ is vertex-distinguishing. Since $c'(v) \neq \emptyset$ for each vertex v of G and $|\mathcal{P}^*([k])| = 2^k - 1$, it follows that $n \leq 2^k - 1$ and so $\text{sroy}(G) = k \geq \lceil \log_2(n+1) \rceil = \lfloor \log_2 n \rfloor + 1$. ∎

Next, we present an upper bound for the strong royal index of a connected graph G of order 4 or more in terms of the strong royal indexes of the connected spanning subgraphs of G.

Proposition 16.24 *If G is a connected graph of order 4 or more, then*

$$\text{sroy}(G) \leq 1 + \min\{\text{sroy}(H) : H \text{ is a connected spanning subgraph of } G\}.$$

In particular,

$$\text{sroy}(G) \leq 1 + \min\{\text{sroy}(T) : T \text{ is a spanning tree of } G\}. \tag{16.3}$$

Proof. Among all connected spanning subgraphs of G, let H be one having the minimum strong royal index, say $\text{sroy}(H) = k$. Let $c_H : E(H) \to \mathcal{P}^*([k])$ be a strong royal k-edge coloring of H. Then $c'_H(x) \neq c'_H(y)$ for every two vertices x and y of H. We extend c_H to an edge coloring $c_G : E(G) \to \mathcal{P}^*([k+1])$ of G by defining

$$c_G(e) = \begin{cases} c_H(e) & \text{if } e \in E(H) \\ \{k+1\} & \text{if } e \in E(G) - E(H). \end{cases}$$

Since c'_H is vertex-distinguishing and either $c'_G(x) = c'_H(x) \subseteq [k]$ or $c'_G(x) = c'_H(x) \cup \{k+1\}$ for each $x \in V(G)$, it follows that c'_G is vertex-distinguishing. Therefore, c_G is a strong royal $(k+1)$-edge coloring of G and so $\text{sroy}(G) \leq k + 1 = \text{sroy}(H) + 1$. The inequality (16.3) then follows immediately. ∎

By Proposition 16.24, if we know the strong royal indexes of all spanning trees of a connected graph G, then we have an upper bound for $\text{sroy}(G)$. This shows the value of knowing the strong royal indexes of trees. By Proposition 16.23, if T is a tree of order $n \geq 4$, then $\text{sroy}(T) \geq \lceil \log_2(n+1) \rceil$. There is equality for this bound when T is a star or a path.

Proposition 16.25 *For every integer $n \geq 4$, $\text{sroy}(K_{1,n-1}) = \lceil \log_2(n+1) \rceil$.*

Proof. Let $k = \lceil \log_2(n+1) \rceil \geq 3$ and let $G = K_{1,n-1}$ be a star of order n, where $V(G) = \{v, v_1, v_2, \ldots, v_{n-1}\}$ and $\deg_G v = n - 1$. By Proposition 16.23, it suffices to show that G has a strong royal k-edge coloring. Since $k = \lceil \log_2(n+1) \rceil \geq 3$, it follows that

$$2^{k-1} - 1 \leq n - 1 \leq 2^k - 2.$$

Let $S_1, S_2, \ldots, S_{2^k-2}$ be the $2^k - 2$ distinct nonempty proper subsets of $[k]$, where $S_i = \{i\}$ for $1 \leq i \leq k$. Define the coloring $c : E(G) \to \mathcal{P}^*([k])$ by $c(vv_i) = S_i$ for $1 \leq i \leq n-1$. Thus, $c'(v_i) = S_i$ for $1 \leq i \leq n-1$. Since $2^k - 1 \geq k$, it follows that $c'(v) = [k]$ and so c' is vertex-distinguishing. Therefore, c is a strong royal k-edge coloring of G and so $\mathrm{sroy}(G) = \lceil \log_2(n+1) \rceil$. ∎

Theorem 16.26 *For every integer $n \geq 4$, $\mathrm{sroy}(P_n) = \lceil \log_2(n+1) \rceil$.*

Proof. Let $k = \lceil \log_2(n+1) \rceil \geq 3$. Then $2^{k-1} \leq n \leq 2^k - 1$. By Proposition 16.23, it suffices to show that G has a strong royal k-edge coloring. For $4 \leq n \leq 7$, $\lceil \log_2(n+1) \rceil = 3$. A strong royal 3-edge coloring of P_n for each such integer n is shown in Figure 16.7 and so $\mathrm{sroy}(P_n) = 3$. For $8 \leq n \leq 15$, $\lceil \log_2(n+1) \rceil = 4$. A strong royal 4-edge coloring of P_n for each such integer n is also shown in Figure 16.7 and so $\mathrm{sroy}(P_n) = 4$.

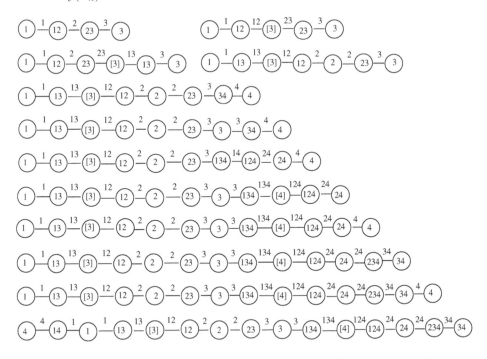

Figure 16.7: Showing that $\mathrm{sroy}(P_n) = \lceil \log_2(n+1) \rceil$ for $4 \leq n \leq 15$

Suppose that $\mathrm{sroy}(P_n) = k$ for an integer $n \geq 8$ such that $2^{k-1} \leq n \leq 2^k - 1$. Let $c_n : E(P_n) \to \mathcal{P}^*([k])$ be a strong royal k-edge coloring of P_n. Since $2^{k-1} \leq n \leq 2^k - 1$, it follows that $2^k \leq 2n < 2^{k+1} - 1$ and $2^k < 2n + 1 \leq 2^{k+1} - 1$. Hence, $\lceil \log_2(2n+1) \rceil = \lceil \log_2(2n+2) \rceil = k + 1$. We construct strong royal $(k+1)$-edge colorings of P_{2n} and P_{2n+1} from the strong royal k-edge coloring c_n of P_n as follows. Let P_{2n} be constructed from two copies $(u_1, u_2, \ldots, u_n)$ and $(v_1, v_2, \ldots, v_n)$ of P_n by adding the edge $u_n v_n$ and let P_{2n+1} be obtained from P_{2n} by adding a new vertex v_0 and the new edge $v_0 v_1$. Define the edge coloring $c_{2n} : E(P_{2n}) \to \mathcal{P}^*([k+1])$

of P_{2n} as follows:

$$c_{2n}(e) = \begin{cases} c_n(e) & \text{if } e = u_iu_{i+1} \text{ for } 1 \leq i \leq n-1 \\ c_n(u_{n-1}u_n) & \text{if } e = u_nv_n \\ c_n(u_iu_{i+1}) \cup \{k+1\} & \text{if } e = v_iv_{i+1} \text{ for } 1 \leq i \leq n-1. \end{cases}$$

Since $c'_{2n}(u_i) = c'_n(u_i)$ and $c'_{2n}(v_i) = c'_n(u_i) \cup \{k+1\}$ for $1 \leq i \leq n$, it follows that c'_{2n} is vertex-distinguishing. Thus, c_{2n} is a strong royal $(k+1)$-edge coloring of P_{2n}. Next, we extend this strong royal $(k+1)$-edge coloring c_{2n} of P_{2n} to a strong royal $(k+1)$-edge coloring c_{2n+1} of P_{2n+1} by assigning the color $\{k+1\}$ to the edge v_0v_1. Since $c'_{2n+1}(v_0) = \{k+1\}$ and $c'_{2n+1}(x) = c'_{2n}(x) \neq \{k+1\}$ if $x \neq v_0$, it follows that c'_{2n+1} is vertex-distinguishing. Hence, c_{2n+1} is a strong royal $(k+1)$-edge colorings of P_{2n+1}. This is illustrated in Figure 16.8 for $n = 8$ and $k = 4$, where a strong royal 5-edge coloring of P_{17} is constructed from a strong royal 4-edge coloring of P_8. Deleting the vertex labeled 5 from P_{17} results in a strong royal 5-edge coloring of P_{16}.

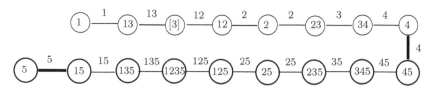

Figure 16.8: Constructing strong royal 5-edge colorings of P_{16} and P_{17}

Therefore, $\text{sroy}(P_n) = \lceil \log_2(n+1) \rceil$ for each integer $n \geq 4$. ∎

Unlike the situation for paths, there are two exceptional situations for the strong royal index of cycles.

Theorem 16.27 *For every integer $n \geq 3$,*

$$\text{sroy}(C_n) = \begin{cases} \lceil \log_2(n+1) \rceil & \text{if } n \neq 3, 7 \\ 1 + \lceil \log_2(n+1) \rceil & \text{if } n = 3, 7. \end{cases}$$

That is, if C_n is a cycle of length $n \geq 3$ where $2^{k-1} \leq n \leq 2^k - 1$ for some integer k, then $\text{sroy}(C_n) = k$ unless $n = 3$ or $n = 7$, in which case, $\text{sroy}(C_3) = 3$ and $\text{sroy}(C_7) = 4$.

We discuss the two exceptional situations of this statement, namely, $\text{sroy}(C_3) = 3$ and $\text{sroy}(C_7) = 4$. Figure 16.9 shows a strong royal 3-edge coloring of C_3 and a strong royal 4-edge coloring of C_7, which shows that $\text{sroy}(C_3) \leq 3$ and $\text{sroy}(C_7) \leq 4$. If there were to exist a strong royal 2-edge coloring of C_3, then one vertex of C_3 must be colored 1 and another colored 2, implying that two edges of C_3 are colored with each of these two colors, which is impossible since the size of C_3 is 3. Similarly, if there exists a strong royal 3-edge coloring of C_7, then there are vertices of C_7 colored 1, 2, and 3, implying that two edges of C_7 are colored with each of these three colors. Regardless of how the seventh edge of C_7 is colored, however, the resulting set of vertex colors is not $\mathcal{P}^*([3])$. Consequently, $\text{sroy}(C_3) = 3$ and $\text{sroy}(C_7) = 4$. Figure 16.9 also shows a strong royal 3-edge coloring for each of C_4, C_5, and C_6 and so $\text{sroy}(C_n) = 3$ for $n = 4, 5, 6$ by Proposition 16.23.

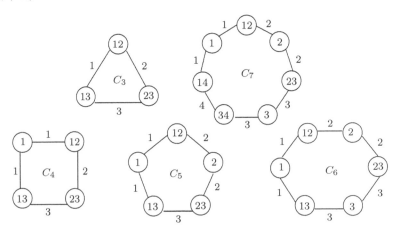

Figure 16.9: Strong royal colorings of C_n where $3 \leq n \leq 7$

Not only is the strong royal index of every star and every path of order $n \geq 4$ equal to $\lceil \log_2(n+1) \rceil$, many other trees have been shown to have the same strong royal index (see [3, 42]). This suggests the following conjecture.

Conjecture 16.28 *If T is a tree of order $n \geq 4$, then $\text{sroy}(T) = \lceil \log_2(n+1) \rceil$.*

If Conjecture 16.28 is true, then by Propositions 16.23 and 16.24 there are only two possible values for the strong royal index of any connected graph of order $n \geq 4$, which is stated in the conjecture below.

Conjecture 16.29 *If G is a connected graph of order $n \geq 4$, then either*

$$\text{sroy}(G) = \lceil \log_2(n+1) \rceil \ \ or \ \ \text{sroy}(G) = \lceil \log_2(n+1) \rceil + 1.$$

Since no connected graph of order $n \geq 4$ has a strong royal index less than $\lceil \log_2(n+1) \rceil$, Conjecture 16.29 is equivalent to the following conjecture.

Conjecture 16.30 *If G is a connected graph of order $n \geq 4$ where $2^{k-1} \leq n \leq 2^k - 1$ for an integer k, then there exists a strong royal $(k+1)$-edge coloring of G.*

It was shown in [27] that Conjecture 16.30 is true if $(k+1)$-edge coloring is replaced by $(k+2)$-edge coloring. However, no graph G of order $n \geq 4$ where $2^{k-1} \leq n \leq 2^k - 1$ with $\text{sroy}(G) = k+2$ has ever been found. Conjecture 16.29 suggests the study of two classes of connected graphs of order 4 or more. A connected graph G of order $n \geq 4$ is a **royal-zero graph** if $\text{sroy}(G) = \lceil \log_2(n+1) \rceil$ and is a **royal-one graph** if $\text{sroy}(G) = \lceil \log_2(n+1) \rceil + 1$. Conjectures 16.29 and 16.30, if true, state that there is no other class of graphs.

Conjecture 16.31 *Every connected graph of order at least 4 is royal-zero or royal-one.*

By Theorem 16.27, the cycles C_3 and C_7 are royal-one while all other cycles are royal-zero. Even though C_7 is royal-one (that is, sroy$(C_7) = 4$), sroy$(C_7 + e) = 3$ for every chord e of C_7, which is shown in Figure 16.10. That is, while one might suspect that sroy$(G + uv) \geq$ sroy(G) for every connected graph G and every pair u, v of nonadjacent vertices of G, this is not the case.

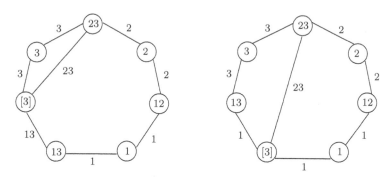

Figure 16.10: Showing that sroy$(C_7 + e) = 3$ for each $e \notin E(C_7)$

We have seen that every tree has either been shown or conjectured to be royal-zero and, with two exceptions (namely C_3 and C_7), every cycle has been shown to be royal-zero. If $n \geq 4$ is a power of 2, then the complete graph K_n is a royal-zero graph; however, all other complete graphs are royal-one graphs. No connected graph whose order is a power of 2 is known to be a royal-one graph.

A common and much-studied class of graphs is that of the prisms, each of which is the Cartesian product of a graph and K_2. The strong royal index of the Cartesian product of a connected graph G and K_2 never exceeds the strong royal index of G by more than 1.

Proposition 16.32 *If G is a connected graph of order $n \geq 4$, then*

$$\text{sroy}(G \,\square\, K_2) \leq \text{sroy}(G) + 1.$$

Proof. Let G be a connected graph of order $n \geq 4$ where sroy$(G) = k$ for some positive integer k and let $H = G \,\square\, K_2$ where G_1 and G_2 are the two copies of G in the construction of H. Suppose that $V(G_1) = \{u_1, u_2, \ldots u_n\}$ where u_i is labeled v_i in G_2. Thus, $V(G_2) = \{v_1, v_2, \ldots, v_n\}$ and

$$E(H) = E(G_1) \cup E(G_2) \cup \{u_i v_i : 1 \leq i \leq n\}.$$

Since sroy$(G) = k$, there is a strong royal k-edge coloring

$$c_{G_1} : E(G_1) \to \mathcal{P}^*([k]) \text{ of } G_1.$$

Define an edge coloring $c_H : E(H) \to \mathcal{P}^*([k+1])$ by

$$c_H(e) = \begin{cases} c_{G_1}(e) & \text{if } e \in E(G_1) \\ c_{G_1}(u_i u_j) \cup \{k+1\} & \text{if } e = v_i v_j \in E(G_2) \text{ for } 1 \le i, j \le n \text{ and } i \ne j \\ c'_{G_1}(u_i) & \text{if } e = u_i v_i \text{ for } 1 \le i \le n. \end{cases}$$

The vertex coloring $c'_H : V(H) \to \mathcal{P}^*([k+1])$ is then given by

$$c'_H(u_i) = c'_{G_1}(u_i) \text{ and } c'_H(v_i) = c'_{G_1}(u_i) \cup \{k+1\}$$

for $1 \le i \le n$. Since c'_H is vertex-distinguishing, it follows that c'_H is a strong royal $(k+1)$-edge coloring of H. Thus,

$$\text{sroy}(H) \le k + 1 = \text{sroy}(G) + 1,$$

as desired.   ∎

If G is a royal-zero graph of order $n \ge 4$, then $\text{sroy}(G) = k$ where $2^{k-1} \le n \le 2^k - 1$. Since the order of $G \,\square\, K_2$ is $2n$ and $2^k \le 2n < 2^{k+1} - 1$ and $\text{sroy}(G \,\square\, K_2) \ge k + 1$ by Proposition 16.23, it follows by Corollary 16.33 that $\text{sroy}(G \,\square\, K_2) = k + 1$, resulting in the following corollary.

Corollary 16.33 *If G is a royal-zero graph, then so too is $G \,\square\, K_2$.*

Since C_4 is a royal-zero graph by Theorem 16.27, it follows by Corollary 16.33 that $C_4 \,\square\, K_2 = Q_3$ is a royal-zero graph. By repeated application of Corollary 16.33, we have the following corollary.

Corollary 16.34 *For each integer $n \ge 2$, the hypercube Q_n is a royal-zero graph.*

As stated in Corollary 16.33, if G is a royal-zero graph, then $G \,\square\, K_2$ is also a royal-zero graph. On the other hand, it is possible that G is a royal-one graph and $G \,\square\, K_2$ is a royal-zero graph. To see an example of this, we return to the 7-cycle C_7, which we saw (in Theorem 16.27) is a royal-one graph. Figure 16.11 shows a strong royal 4-edge coloring of $C_7 \,\square\, K_2$ and so $\text{sroy}(C_7) = \text{sroy}(C_7 \,\square\, K_2) = 4$. Thus, C_7 is royal-one, while $C_7 \,\square\, K_2$ is royal-zero.

As noted in Proposition 16.22, the complete graph K_7 is a royal-one graph. However, $H = K_7 \,\square\, K_2$ is royal-one as well. That there is a strong royal 5-edge coloring of H is straightforward. To show that $\text{sroy}(K_7 \,\square\, K_2) = 5$, however, it is necessary to show that there is no strong royal 4-edge coloring of H, for assume that such an edge coloring c of H exists. Since the order of H is 14, the induced vertex colors of H must consist of 14 elements of $\mathcal{P}^*([4])$. In particular, at least three of the four singleton subsets of $[4]$ must be vertex colors of H. Suppose that H_1 and H_2 are the two copies of K_7 in the construction of H. Therefore, at least one of H_1 and H_2 has at least two singleton subsets as its vertex colors, say $c'(u_1) = \{1\}$ and $c'(u_2) = \{2\}$ where $u_1, u_2 \in V(H_1)$, which is impossible since u_1 and u_2 are adjacent and $c'(u_1)$ and $c'(u_2)$ are disjoint. Hence, $\text{sroy}(K_7 \,\square\, K_2) = 5$.

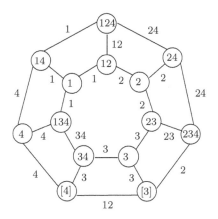

Figure 16.11: A strong royal 4-edge coloring of $C_7 \,\square\, K_2$

We mentioned that every tree has either been shown to be royal-zero or conjectured to be royal-zero. Consequently, there appears to be an abundance of royal-zero graphs. On the other hand, if G is a connected graph whose size is large with respect to its order, then it appears likely that G is not a royal-zero graph. In order to present a sufficient condition for a connected graph not to be royal-zero, we first describe a graph G_k, $k \geq 3$, of order $2^k - 1$. The vertices of G_k are labeled with the $2^k - 1$ distinct elements of $\mathcal{P}^*([k])$. For each vertex v of G_k, let $\ell(v)$ denote its label. Thus, $\{\ell(v) : v \in V(G_k)\} = \mathcal{P}^*([k])$. Two vertices u and v of G_k are adjacent in G_k if and only if $\ell(u) \cap \ell(v) \neq \emptyset$. The graph G_k is therefore the royal-zero graph of largest size having order $2^k - 1$. The vertex set $V(G_k)$ is partitioned into k subsets $V_1, V_2, \ldots, V_k$ where $V_i = \{v \in V(G_k) : |\ell(v)| = i\}$ for $1 \leq i \leq k$. Therefore, $G_k[V_k] = K_1$ and $G_k[V_1] = \overline{K}_k$ is empty. If $k = 2p + 1$ is odd, then $G_k[V_{p+1} \cup V_{p+2} \cup \cdots \cup V_k] = K_{2^{k-1}}$. If $k = 2p$ is even, then let V'_p be the subset consisting of those elements S in V_p for which $1 \in S$. Then $|V'_p| = \frac{1}{2}\binom{k}{p}$ and $G_k[V'_p \cup V_{p+1} \cup V_{p+2} \cup \cdots \cup V_k] = K_{2^{k-1}}$. The size of G_k is denoted by m_k. The graph G_3 of order $7 = 2^3 - 1$ and size $m_3 = 15$ is shown in Figure 16.12.

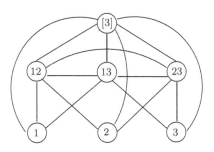

Figure 16.12: The graph G_3 of order $7 = 2^3 - 1$

There is an obvious condition under which a connected graph cannot be a royal-zero graph.

Proposition 16.35 *Let G be a connected graph of order $n \geq 4$ and size m where $2^{k-1} \leq n \leq 2^k - 1$ for an integer k. If G is not a subgraph of the graph G_k, then $\mathrm{sroy}(G) \geq k+1$. Consequently, if $m > m_k$, then $\mathrm{sroy}(G) \geq k+1$.*

In order for Proposition 16.35 to be useful, it is of value to obtain an appropriate expression for m_k.

Proposition 16.36 . *For each integer $k \geq 3$, the size of the graph G_k is*

$$m_k = \frac{1}{2}\left(4^k - 3^k - 2^k + 1\right).$$

Proof.   Recall that the $2^k - 1$ vertices of G_k are labeled with the distinct elements of $\mathcal{P}^*([k])$. If we denote the label of each vertex v of G_k is $\ell(v)$, then

$$\{\ell(v): \ v \in V(G_k)\} = \mathcal{P}^*([k]).$$

Let $\{V_1, V_2, \ldots, V_k\}$ be the partition of of $V(G_k)$ described earlier, where then

$$V_i = \{v \in V(G_k) : |\ell(v)| = i\}$$

for $1 \leq i \leq k$. Let $v \in V_i$ for some integer i with $1 \leq i \leq k$. Then $\ell(v) = S$ is some i-element subset of $[k]$. There are $2^i - 1$ nonempty subsets of S and 2^{k-i} subsets of $[k] - S$. For each nonempty subset S' of S and each subset T of $[k] - S$, the vertex v is adjacent to that vertex w of G_k for which $\ell(w) = S' \cup T$. Since v is not adjacent to itself, however, it follows that

$$\deg_{G_k} v = (2^i - 1)2^{k-i} - 1.$$

Furthermore, there are $\binom{k}{i}$ vertices in V_i for $1 \leq i \leq k$. Therefore,

$$
\begin{aligned}
m_k &= \frac{1}{2}\sum_{i=1}^{k}\binom{k}{i}\left[(2^i - 1)2^{k-i} - 1\right] \\
&= \frac{1}{2}\sum_{i=1}^{k}\binom{k}{i}(2^k - 2^{k-i} - 1) \\
&= \frac{1}{2}\left[\sum_{i=1}^{k}\binom{k}{i}2^k - \sum_{i=1}^{k}\binom{k}{i}2^{k-i} - \sum_{i=1}^{k}\binom{k}{i}\right] \\
&= \frac{1}{2}\left[2^k\sum_{i=1}^{k}\binom{k}{i} - 2^k\sum_{i=1}^{k}\binom{k}{i}\left(\frac{1}{2}\right)^i - \sum_{i=1}^{k}\binom{k}{i}\right] \\
&= \frac{1}{2}\left\{2^k(2^k - 1) - 2^k\left[\left(1 + \frac{1}{2}\right)^k - 1\right] - (2^k - 1)\right\} \\
&= \frac{1}{2}(4^k - 3^k - 2^k + 1),
\end{aligned}
$$

as desired. ∎

In particular, if $k = 3$, then the size of G_3 is $m_3 = 15$, as we saw in Figure 16.12. Propositions 16.35 and 16.36 then give us the following information.

Proposition 16.37 *Let G be a graph of order $n \geq 4$ and size m where $2^{k-1} \leq n \leq 2^k - 1$ for some integer $k \geq 3$. If*

$$m > \frac{1}{2}(4^k - 3^k - 2^k + 1),$$

then either $\mathrm{sroy}(G) = k+1$ *or* $\mathrm{sroy}(G) = k+2$, *and so G is not a royal-zero graph.*

For each integer $k \geq 3$, the minimum degree $\delta(G_k)$ of the graph G_k is $2^{k-1} - 1$. Consequently, if G is a graph of order $n \geq 4$ and size m where $2^{k-1} \leq n \leq 2^k - 1$ for which $\delta(G) \geq 2^{k-1}$, then it may occur that $m < m_k$ but yet G is not a subgraph of G_k, and so (by Proposition 16.35) $\mathrm{sroy}(G) \geq k+1$. However, in this case, more can be said. It is useful to recall that every path P_n for $n \geq 4$ is royal-zero (see [27, 42]).

Proposition 16.38 *Let G be a connected graph of order $n \geq 4$ where $2^{k-1} \leq n \leq 2^k - 1$ for some integer $k \geq 2$. If $\delta(G) \geq 2^{k-1}$, then $\mathrm{sroy}(G) = k+1$ and so G is a royal-one graph.*

Proof. We have already observed that $\mathrm{sroy}(G) \geq k + 1$ for such a graph. Since $\delta(G) \geq 2^{k-1}$ and $n \leq 2^k - 1$, it follows that $\delta(G) \geq (n + 1)/2$ and therefore G has a Hamiltonian path (in fact, a Hamiltonian cycle). Since $\mathrm{sroy}(P_n) = k$ for every path P_n of order n by Theorem 16.26, it follows by Proposition 16.24 that $\mathrm{sroy}(G) \leq k + 1$ and so $\mathrm{sroy}(G) = k + 1$. ∎

In particular, both 4-regular graphs of order $n = 7$ have size 14 and are royal-one graphs because they contain a Hamiltonian path (see Figure 16.13). However, in this case, $k = 3$ and

$$\frac{1}{2}(4^k - 3^k - 2^k + 1) = 15 = m_3.$$

So, even though $m < m_3$, we can nevertheless conclude that both graphs are royal-one.

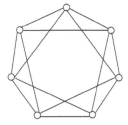 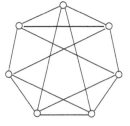

Figure 16.13: Two royal-one graphs of order 7

In a royal coloring of a connected graph G, each edge of G is colored with a nonempty subset of $[k]$ for some positive integer k and the color of each vertex v of G is that nonempty subset of the set $[k]$ consisting of all those elements of $[k]$ belonging to at least one color of an edge incident with v. There is a coloring of G related to a royal coloring. A **regal coloring** of a connected graph G is an edge coloring of G in which each edge is colored with a nonempty subset of the set $[k]$ for some positive integer k and the color of each vertex v is that *nonempty* subset of $[k]$ consisting of all those elements of $[k]$ belonging to the color of *every* edge incident with v. More formally then, for a connected graph G, let $c : E(G) \to \mathcal{P}^*([k])$ be an unrestricted edge coloring of G for some positive integer k with the vertex coloring $c' : V(G) \to \mathcal{P}^*([k])$ defined by

$$c'(v) = \bigcap_{e \in E_v} c(e),$$

where E_v is the set of edges incident with a vertex v of G. This definition then requires the edge coloring c to have the property that $c'(v) \neq \emptyset$ for each vertex v of G. If c' is a proper vertex coloring of G, then c is called a **regal k-edge coloring** of G. The minimum positive integer k for which a graph G has a regal k-edge coloring is called the **regal index** $\mathrm{reg}(G)$ of G. If c' is vertex-distinguishing, then c is called a **strong regal k-edge coloring** of G. The minimum positive integer k for which a graph G has a strong regal k-edge coloring is called the **strong regal index** $\mathrm{sreg}(G)$ of G. A **regal coloring** of a graph G is a regal k-edge coloring of G for some positive integer k and a **strong regal coloring** is a strong regal k-edge coloring for some positive integer k. These concepts were introduced in [44]. Since every strong regal coloring is also a regal coloring, it follows that $\mathrm{reg}(G) \leq \mathrm{sreg}(G)$ for every connected graph G of order at least 3. For example, Figure 16.14 shows a regal 3-edge coloring and a strong regal 4-edge coloring of the path P_8 of order 8. (As before, we write the set $\{a\}$ as a, $\{a, b\}$ as ab, and $\{a, b, c\}$ as abc.) In fact, $\mathrm{reg}(P_8) = 3$ and $\mathrm{sreg}(P_8) = 4$, as we will see later.

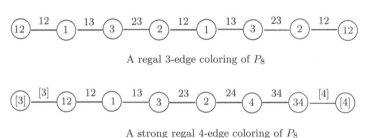

A regal 3-edge coloring of P_8

A strong regal 4-edge coloring of P_8

Figure 16.14: A regal 3-edge coloring and a strong regal 4-edge coloring of P_8

Every connected graph of order 3 or more has a strong regal coloring (and therefore a regal coloring as well). To see why this is the case, we first make the following observation.

Proposition 16.39 *Let G be a connected graph of order 3 or more and let H be a connected spanning subgraph of G. If H has a strong regal k-edge coloring for some positive integer k, then so does G. Consequently, $\operatorname{sreg}(G) \leq \operatorname{sreg}(H)$.*

Proof. Suppose that H has a strong regal coloring and that $\operatorname{sreg}(H) = k$. Let $c_H : E(H) \to \mathcal{P}^*([k])$ be a strong regal k-edge coloring of H. Then $c'_H(x) \neq c'_H(y)$ for every two vertices x and y of H. Now, we extend c_H to an edge coloring $c_G : E(G) \to \mathcal{P}^*([k])$ of G by defining

$$c_G(e) = \begin{cases} c_H(e) & \text{if } e \in E(H) \\ [k] & \text{if } e \in E(G) - E(H). \end{cases}$$

Since $c'_G(x) = c'_H(x)$ for each vertex x of G and c'_H is vertex-distinguishing, it follows that c'_G is vertex-distinguishing as well. Therefore, c_G is a strong regal k-edge coloring of G and so $\operatorname{sreg}(G) \leq k = \operatorname{sreg}(H)$. ∎

Theorem 16.40 *Every connected graph of order 3 or more has a strong regal coloring and therefore a regal coloring.*

Proof. By Proposition 16.39, it suffices to show that every tree of order 3 or more has a strong regal coloring. We proceed by induction on the order $n \geq 3$ of a tree T to show that there exists a strong regal coloring $c : E(T) \to \mathcal{P}^*([n])$. Assigning the colors $\{1,2\}$ and $\{1,3\}$ to the two edges of P_3 produces a strong regal 3-edge coloring of P_3 and so $\operatorname{sreg}(P_3) \leq 3$ (in fact, $\operatorname{sreg}(P_3) = 3$), establishing the base step of the induction. Suppose that every tree of order $n - 1 \geq 3$ has a strong regal coloring whose edges are colored with elements of $\mathcal{P}^*([n-1])$. Let T be a tree of order n where v is an end-vertex of T and let $T_0 = T - v$. By the induction hypothesis, T_0 has a strong regal $(n-1)$-coloring $c_0 : E(T_0) \to \mathcal{P}^*([n-1])$. Let u be the vertex adjacent to v. Suppose that $c'_0(u) = S \subseteq [n-1]$. Then $c'_0(u) \neq c'_0(x)$ for all $x \in V(T_0) - \{u\}$. Define an edge coloring $c : E(T) \to \mathcal{P}^*([n])$ by

$$c(e) = \begin{cases} c_0(e) & \text{if } e \in E(T_0) \\ [n] & \text{if } e = uv. \end{cases}$$

Thus, $c'(x) = c'_0(x) \subseteq [n-1]$ for all $x \in V(T_0)$ and $c'(v) = [n]$. Since $c'(v) \neq c'(x)$ for all vertices $x \in V(T_0)$ and $c'(x) \neq c'(y)$ for every two vertices x and y of T, it follows that c' is vertex-distinguishing and so c is a strong regal n-coloring of T. Therefore, $\operatorname{sreg}(G)$ exists and so does $\operatorname{reg}(G)$. ∎

The proof of Theorem 16.40 not only establishes the fact that every connected graph G of order $n \geq 3$ has a strong regal index but that $\operatorname{sreg}(G) \leq n$. As we will see later, this upper bound appears to be far from sharp for connected graphs of sufficiently large order.

If c is a regal coloring of a connected graph G of order at least 3 whose colors are nonempty subsets of some set $[k]$, then $c(uv)$ cannot be a singleton subset of $[k]$ for any edge uv of G, for in this case $c'(u) = c'(v)$, which is impossible. We state this observation next.

Proposition 16.41 *If c is a regal coloring of a connected graph G of order at least 3, then $|c(e)| \geq 2$ for each $e \in E(G)$ and so $\operatorname{reg}(G) \geq 2$.*

Even Proposition 16.41 can be improved, however. The following result gives a lower bound for the regal index (and the strong regal index) of a graph in terms of its chromatic number.

Theorem 16.42 *If G is a connected graph of order 3 or more, then*

$$\max\{3, \lceil \log_2(\chi(G)+1) \rceil\} \leq \operatorname{reg}(G) \leq \operatorname{sreg}(G).$$

Proof. Suppose that $\operatorname{reg}(G) = k$. Then $k \geq 2$ by Proposition 16.41. However, if there exists a regal 2-edge coloring of G using the colors in $\mathcal{P}^*([2])$, then each edge e of G must be colored $\{1,2\}$ by Proposition 16.41, but then the induced vertex coloring assigns $\{1,2\}$ to every vertex of G, which is impossible. Thus, $k \geq 3$. Next, let $c : E(G) \to \mathcal{P}^*([k])$ be a regal k-edge coloring of G where $k \geq 3$. Since $c' : V(G) \to \mathcal{P}^*([k])$ is a proper vertex coloring of G, it follows that $\chi(G) \leq |\mathcal{P}^*([k])| = 2^k - 1$. Therefore, $k \geq \log_2(\chi(G)+1)$ and so $k \geq \lceil \log_2(\chi(G)+1) \rceil$. Thus, $\operatorname{reg}(G) \geq \max\{3, \lceil \log_2(\chi(G)+1) \rceil\}$.   ∎

By Theorem 16.42, if G is a connected graph of order at least 3, then $\operatorname{reg}(G) \geq 3$. This lower bound is sharp. The only connected graphs of order 3 are P_3 and K_3. In each case, the regal index and strong regal index is 3. Strong regal 3-edge colorings of P_3 and K_3 are shown in Figure 16.15. Indeed, there are numerous graphs of order 3 or more with regal index 3. In fact, every connected bipartite graph of order 3 or more has regal index 3.

Figure 16.15: Strong regal 3-edge colorings of P_3 and K_3

Theorem 16.43 *Every connected bipartite graph of order at least 3 has regal index 3.*

Proof. Let G be a connected bipartite graph of order at least 3. By Theorem 16.42, it suffices to show that there exists a regal 3-edge coloring $c : E(G) \to \mathcal{P}^*([3])$ of G. Let v be a peripheral vertex of G. Thus, either v is an end-vertex of G or $\deg v \geq 2$ and v is not adjacent to any end-vertex of G. For each integer i with $0 \leq i \leq d = \operatorname{diam}(G) \geq 2$, let

$$V_i = \{u \in V(G) : d_G(u,v) = i\}.$$

Thus, $V_0 = \{v\}$ and $V_1 = N_G(v)$ is the neighborhood of v. Since G is bipartite, each set V_i $(0 \le i \le d)$ is an independent set of vertices of G. For $0 \le i \le d - 1$ and for each edge $v_i v_{i+1} \in [V_i, V_{i+1}]$, where $v_i \in V_i$ and $v_{i+1} \in V_{i+1}$, define

$$c(v_i v_{i+1}) = \begin{cases} 12 & \text{if } i \equiv 0 \pmod 3 \\ 13 & \text{if } i \equiv 1 \pmod 3 \\ 23 & \text{if } i \equiv 2 \pmod 3. \end{cases}$$

Thus, $c'(v) = 12$. For $1 \le i \le d - 1$, each vertex $v_i \in V_i$ is adjacent to a vertex in V_{i-1}. If v_i is also adjacent to a vertex in V_{i+1}, then

$$c'(v_i) = \begin{cases} 1 & \text{if } i \equiv 1 \pmod 3 \\ 3 & \text{if } i \equiv 2 \pmod 3 \\ 2 & \text{if } i \equiv 0 \pmod 3; \end{cases}$$

while if v_i is adjacent only to vertices in V_{i-1}, then

$$c'(v_i) = \begin{cases} 12 & \text{if } i \equiv 1 \pmod 3 \\ 13 & \text{if } i \equiv 2 \pmod 3 \\ 23 & \text{if } i \equiv 0 \pmod 3. \end{cases}$$

Consequently, $|c'(u)| \le 2$ for each vertex u of G. This coloring is illustrated Figure 16.16. Since $c'(u) \ne c'(w)$ for every two adjacent vertices u and w of G, it follows that c' is a proper vertex coloring of G and so c is a regal 3-edge coloring of G. Therefore, $\text{reg}(G) = 3$. ∎

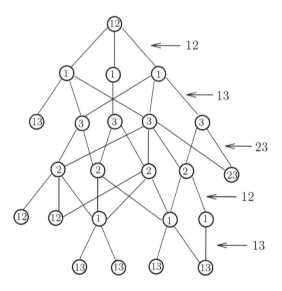

Figure 16.16: A regal 3-edge coloring of a bipartite graph

16.3   Rainbow Mean Colorings

We now consider edge colorings of a graph that induce vertex colorings defined in an arithmetic manner. In this instance, the edge colors and vertex colors are both positive integers. Let G be a connected graph of order $n \geq 3$ with an edge coloring $c : E(G) \to \mathbb{N}$. If the edge coloring c gives rise to a vertex coloring cm defined by

$$\mathrm{cm}(v) = \frac{\sum_{e \in E_v} c(e)}{\deg v}$$

for each vertex v of G such that $\mathrm{cm}(v)$ is a positive integer, then c is called a **mean coloring**. That is, $\mathrm{cm}(v)$ is the mean (or average) of the colors of the edges incident with v. The positive integer $\mathrm{cm}(v)$ is called the **chromatic mean** of v. If v is a vertex in a graph G possessing a mean coloring, then

$$\mathrm{cs}(v) = \sum_{e \in E_v} c(e)$$

is called the **chromatic sum** of v. Therefore, $\mathrm{cs}(v) = \deg v \cdot \mathrm{cm}(v)$. If the chromatic sums of the vertices in a connected graph G with a mean coloring are summed, then the color of each edge of G is counted twice.

Proposition 16.44 *If c is a mean coloring of a connected graph G, then*

$$\sum_{v \in V(G)} \mathrm{cs}(v) = 2 \sum_{e \in E(G)} c(e).$$

A vertex v in a connected graph G with a mean coloring c is **chromatically odd** if $\mathrm{cs}(v)$ is odd and is **chromatically even** otherwise. The following is an immediate consequence of Proposition 16.44.

Proposition 16.45 *Let G be a connected graph of order 3 or more with a mean coloring. Then G has an even number of chromatically odd vertices.*

If distinct vertices of a connected graph G of order 3 or more with a mean coloring c have distinct chromatic means, then c is called a **rainbow mean coloring** of G. Every connected graph of order 3 or more has such a coloring.

Theorem 16.46 *Every connected graph of order 3 or more has a rainbow mean coloring.*

Proof. Suppose that G is a connected graph of size m with $E(G) = \{e_1, e_2, \ldots, e_m\}$. Thus, $\Delta(G) = \Delta \geq 2$. Let $k = 2\Delta$ and $t = \Delta! k^m$. Define the edge coloring $c : E(G) \to [t]$ by $c(e_i) = \Delta! k^i$ for $1 \leq i \leq m$. We show that the coloring c has the desired property. Since $\mathrm{cm}(v)$ is an integer for each vertex v of G, it follows that c is a mean coloring of G. It remains only to show that c is a rainbow mean coloring. Assume, to the contrary, that there are two distinct vertices u and v of G such that $\mathrm{cm}(u) = \mathrm{cm}(v)$. Let $\deg u = r$ and $\deg v = s$, where $r \leq s$ say, and let

$E_u = \{e_{i_1}, e_{i_2}, \dots, e_{i_r}\}$ and $E_v = \{e_{j_1}, e_{j_2}, \dots, e_{j_s}\}$ where $1 \le i_1 < i_2 < \cdots < i_r \le m$ and $1 \le j_1 < j_2 < \cdots < j_s \le m$. If $uv \notin E(G)$, then $E_u \cap E_v = \emptyset$; while if $uv \in E(G)$, then $E_u \cap E_v = \{uv\}$. Consequently,

$$
\begin{aligned}
\mathrm{cm}(u) &= \frac{\Delta!}{r}\left(k^{i_1} + k^{i_2} + \cdots + k^{i_r}\right) \\
\mathrm{cm}(v) &= \frac{\Delta!}{s}\left(k^{j_1} + k^{j_2} + \cdots + k^{j_s}\right).
\end{aligned}
$$

We consider two cases, according to whether $r = s$ or $r < s$.

Case 1. $r = s$. Then $k^{i_1} + k^{i_2} + \cdots + k^{i_r} = k^{j_1} + k^{j_2} + \cdots + k^{j_r}$.

- First, suppose that $i_r \ne j_r$. We may assume that $i_r < j_r$. Let $p = j_r \ge 2$. Since $k = 2\Delta \ge 4$, it follows that $1 > \frac{1}{k^{p-1}} + \frac{1}{k^{p-2}} + \cdots + \frac{1}{k}$ and so $k^p > k + k^2 + \ldots + k^{p-1}$. However then,

$$
k^{j_1} + k^{j_2} + \cdots + k^{j_r} \ge k^{j_r} = k^p > k + k^2 + \ldots + k^{p-1} \ge k^{i_1} + k^{i_2} + \cdots + k^{i_r},
$$

which is a contradiction.

- Next, suppose that $i_r = j_r$. Then $k^{i_1} + k^{i_2} + \cdots + k^{i_{r-1}} = k^{j_1} + k^{j_2} + \cdots + k^{j_{r-1}}$. If $i_{r-1} \ne j_{r-1}$, then we can apply the argument above to produce a contradiction. If $i_{r-1} = j_{r-1}$ then we can proceed until we arrive at $i_q \ne j_q$ for some integer q with $1 \le q \le r - 2$.

Case 2. $r < s$. Then $s\left[k^{i_1} + k^{i_2} + \cdots + k^{i_r}\right] = r\left[k^{j_1} + k^{j_2} + \cdots + k^{j_s}\right]$.

- First, suppose that $i_r < j_s$. Let $p = j_s \ge 2$. Since $1 > \frac{1}{k^{p-1}} + \frac{1}{k^{p-2}} + \cdots + \frac{1}{k}$, it follows that

$$
2 > \frac{1}{k^{p-1}} + \frac{1}{k^{p-2}} + \cdots + \frac{1}{k} + 1 > \frac{1}{k^{p-2}} + \frac{1}{k^{p-3}} + \cdots + \frac{1}{k} + 1.
$$

Hence, $k = 2\Delta > \Delta\left(\frac{1}{k^{p-2}} + \frac{1}{k^{p-3}} + \cdots + 1\right)$. Because $\Delta \ge s/r$, it follows that

$$
\begin{aligned}
k^{j_1} + k^{j_2} + \cdots + k^{j_s} &\ge k^{j_s} = k^p = k(k^{p-1}) \\
&> \Delta\left(\frac{1}{k^{p-2}} + \frac{1}{k^{p-3}} + \cdots + 1\right)k^{p-1} \\
&= \Delta(k + k^2 + \cdots + k^{p-1}) \\
&\ge \frac{s}{r}(k + k^2 + \cdots + k^{p-1}) \\
&\ge \frac{s}{r}\left[k^{i_1} + k^{i_2} + \cdots + k^{i_r}\right],
\end{aligned}
$$

which is a contradiction.

- Next, suppose that $i_r \ge j_s$. The argument in Case 1 shows that

$$
k^{i_1} + k^{i_2} + \cdots + k^{i_r} > k^{j_1} + k^{j_2} + \cdots + k^{j_s}.
$$

Since $r < s$, it follows that $1 > r/s$ and so

$$k^{i_1} + k^{i_2} + \cdots + k^{i_r} > k^{j_1} + k^{j_2} + \cdots + k^{j_s} > \tfrac{r}{s}\left[k^{j_1} + k^{j_2} + \cdots + k^{j_s}\right],$$

which is a contradiction.   ∎

For a rainbow mean coloring c of a connected graph G, the maximum vertex color is the **rainbow chromatic mean index** (or simply, the **rainbow mean index**) $\mathrm{rm}(c)$ of c. That is,

$$\mathrm{rm}(c) = \max\{\mathrm{cm}(v) : v \in V(G)\}.$$

The **rainbow chromatic mean index** (or the **rainbow mean index**) $\mathrm{rm}(G)$ of the graph G itself is defined as

$$\mathrm{rm}(G) = \min\{\mathrm{rm}(c) : c \text{ is a rainbow mean coloring of } G\}.$$

By Theorem 16.46, the rainbow mean index is defined for every connected graph of order 3 or more. We now present some useful observations.

Proposition 16.47 *If G is a connected graph of order $n \geq 3$, then $\mathrm{rm}(G) \geq n$.*

Proposition 16.48 *If G is a connected graph of order $n \geq 3$ with a rainbow mean coloring c with $\mathrm{rm}(c) = n$, then*

$$\sum_{v \in V(G)} \mathrm{cm}(v) = \binom{n+1}{2}.$$

To illustrate these concepts, we determine the rainbow mean index of the path P_4.

Proposition 16.49 $\mathrm{rm}(P_4) = 5$.

Proof. By Proposition 16.47, $\mathrm{rm}(P_4) \geq 4$. Since the rainbow mean coloring in Figure 16.17 has rainbow mean index 5, it follows that $\mathrm{rm}(P_4) \leq 5$. Next, we show that $\mathrm{rm}(P_4) \neq 4$. Assume, to the contrary, that there is a rainbow mean coloring c of P_4 such that $\mathrm{rm}(c) = 4$. Let $P_4 = (v_1, v_2, v_3, v_4)$. Since $\{\mathrm{cm}(v_i) : 1 \leq i \leq 4\} = [4]$, no two edges can be colored the same. Consequently, since only one vertex is colored 1, this implies that $\mathrm{cm}(v_1) = 1$ or $\mathrm{cm}(v_4) = 1$, say the former. Therefore, $c(v_1 v_2) = 1$. Hence, the edges of P_4 are colored with distinct odd integers. If some edge of P_4 is colored 7 or more, then some vertex of P_4 is colored 5 or more, which is impossible. Thus, $\{c(v_i v_{i+1}) : i = 1, 2, 3\} = \{1, 3, 5\}$ and so $\{c(v_2 v_3), c(v_3 v_4)\} = \{3, 5\}$. Whether $c(v_2 v_3) = 3$ or $c(v_2 v_3) = 5$, it follows that $\{\mathrm{cm}(v_i) : 1 \leq i \leq 4\} \neq [4]$, a contradiction. Thus, $\mathrm{rm}(P_4) \neq 4$ and so $\mathrm{rm}(P_4) = 5$. ∎

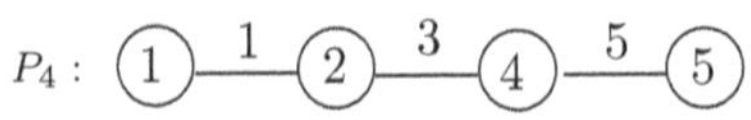

$$P_4 : \;\; \textcircled{1} \!\overset{1}{\rule{1.2cm}{0.4pt}}\! \textcircled{2} \!\overset{3}{\rule{1.2cm}{0.4pt}}\! \textcircled{4} \!\overset{5}{\rule{1.2cm}{0.4pt}}\! \textcircled{5}$$

Figure 16.17: A rainbow mean coloring of P_4

The following result obtained in [43] shows that $n = 4$ is an exceptional situation for the rainbow mean index of paths.

Theorem 16.50 *For each integer $n \geq 3$ and $n \neq 4$, $\mathrm{rm}(P_n) = n$.*

In the case of cycles C_n, $n \geq 3$, half have rainbow mean index n, while the other half have rainbow mean index $n + 1$ (see [43]).

Theorem 16.51 *For each integer $n \geq 3$,*

$$\mathrm{rm}(C_n) = \begin{cases} n & \text{if } n \equiv 0, 1 \pmod 4 \\ n + 1 & \text{if } n \equiv 2, 3 \pmod 4. \end{cases}$$

We have seen many connected graphs of order $n \geq 3$ whose rainbow mean index is n as well as some whose rainbow mean index is $n+1$. There are certain connected graphs of order $n \geq 6$ where it is impossible that $\mathrm{rm}(G) = n$.

Proposition 16.52 *If G is a connected graph of order $n \geq 6$ with $n \equiv 2 \pmod 4$ all of whose vertices are odd, then $\mathrm{rm}(G) \geq n + 1$.*

Proof. Assume, to the contrary, that $\mathrm{rm}(G) = n$. Consequently, there exists a rainbow mean coloring $c : E(G) \to \mathbb{N}$ of G such that $\{cm(v) : v \in V(G)\} = [n]$. Since $n \equiv 2 \pmod 4$, it follows that $n = 4k+2$ for some positive integer k. Thus, the set $[n]$ contains $2k + 1$ odd integers, say $a_i = 2i - 1$ for $i = 1, 2, \ldots, 2k + 1$. Suppose that $u_1, u_2, \ldots, u_{2k+1}$ are the vertices of G such that $cm(u_i) = a_i$ for $1 \leq i \leq 2k+1$. Since every vertex of G has odd degree, the vertices $u_1, u_2, \ldots, u_{2k+1}$ are the only chromatically odd vertices. This contradicts Corollary 16.45. ∎

To illustrate Proposition 16.52, we state the following result from [43] and prove one case.

Theorem 16.53 *For each integer $n \geq 3$,*

$$\mathrm{rm}(K_n) = \begin{cases} n & \text{if } n \geq 4 \text{ and } n \equiv 0, 1, 3 \pmod 4 \\ n + 1 & \text{if } n = 3 \text{ or } n \equiv 2 \pmod 4. \end{cases}$$

Proof. We only prove the case when $n \geq 6$ and $n \equiv 2 \pmod 4$. Thus, $n = 4k + 2$ for some positive integer k. By Proposition 16.52, $\mathrm{rm}(K_n) \geq n + 1 = 4k + 3$.

It therefore remains to show that there is a rainbow mean coloring c_n of K_n with $\mathrm{rm}(c_n) = n+1$. In order to do this, we construct an $n \times n$ symmetric matrix M_n by constructing a sequence $A_1, A_2, \ldots, A_k$ of symmetric matrices, where A_1 is a 6×6 matrix and A_i is a 4×4 matrix for $2 \leq i \leq k$. For $a = n - 1$, let

$$B = \begin{bmatrix} 0 & a & a & 2a \\ a & 0 & 2a & a \\ a & 2a & 0 & a \\ 2a & a & a & 0 \end{bmatrix}.$$

Define

$$
A_1 = \begin{bmatrix}
0 & 1 & 1 & 1 & 1 & 1 \\
1 & 0 & a+1 & 1 & 1 & 1 \\
1 & a+1 & 0 & 1 & 1 & a+1 \\
1 & 1 & 1 & 0 & a+1 & 2a+1 \\
1 & 1 & 1 & a+1 & 0 & 3a+1 \\
1 & 1 & a+1 & 2a+1 & 3a+1 & 0
\end{bmatrix} \text{ and}
$$

$$
A_2 = \begin{bmatrix}
0 & a+1 & 3a+1 & 3a+1 \\
a+1 & 0 & 3a+1 & 4a+1 \\
3a+1 & 3a+1 & 0 & 3a+1 \\
3a+1 & 4a+1 & 3a+1 & 0
\end{bmatrix}.
$$

For $3 \le i \le k$, define

$$
A_i = A_{i-1} + B = A_2 + (i-2)B.
$$

To describe the $n \times n$ matrix M_n, we begin with a $k \times k$ matrix $A = [a_{i,j}]$ and then replace the entry $a_{i,i}$ on the main diagonal of A by the matrix A_i for $1 \le i \le k$ and each entry off the main diagonal of A by the matrix J, each of whose entries is 1. Thus, $a_{1,1}$ is replaced by the 6×6 matrix A_1 and $a_{i,i}$ for $2 \le i \le k$ is replaced by the 4×4 matrix A_i. That is, $M_n = [M_{i,j}]$ is an $n \times n$ matrix where

$$
M_{i,j} = \begin{cases}
A_i & \text{if } 1 \le i = j \le k \\
J & \text{if } 1 \le i \ne j \le k.
\end{cases}
$$

Thus, $M_6 = A_1$, $M_{10} = \begin{bmatrix} A_1 & J \\ J & A_2 \end{bmatrix}$ and $M_{14} = \begin{bmatrix} A_1 & J & J \\ J & A_2 & J \\ J & J & A_3 \end{bmatrix}$. In particular,

$$
M_6 = \begin{bmatrix}
0 & 1 & 1 & 1 & 1 & 1 \\
1 & 0 & 6 & 1 & 1 & 1 \\
1 & 6 & 0 & 1 & 1 & 6 \\
1 & 1 & 1 & 0 & 6 & 11 \\
1 & 1 & 1 & 6 & 0 & 16 \\
1 & 1 & 6 & 11 & 16 & 0
\end{bmatrix}.
$$

We now define a rainbow mean coloring $c : E(K_n) \to \mathbb{N}$ by $c(v_i v_j) = m_{i,j}$ for each pair i, j of integers with $1 \le i \le j \le n$ and $i \ne j$. For example, the matrix M_6 gives rise to the rainbow mean coloring of K_6 as shown in Figure 16.18, where again each edge drawn in a thin line is colored by 1. Since $\mathrm{rm}(c) = n+1$, it follows that $\mathrm{rm}(K_n) = n+1$ for each integer $n \ge 6$ with $n \equiv 2 \pmod{4}$. ∎

The rainbow mean index has also been determined for complete bipartite graphs of order 3 or more. The following observation shows that there are values of s and t for which $\mathrm{rm}(K_{s,t}) > s + t$.

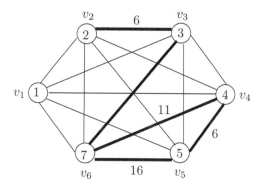

Figure 16.18: A rainbow mean coloring of K_6

Proposition 16.54 *If s and t are odd integers with $s, t \geq 3$, then*
$$\mathrm{rm}(K_{s,t}) \geq s + t + 1.$$

Proof. Let $G = K_{s,t}$ where $s, t \geq 3$. Since s and t are both odd, it follows that $s = 2a + 1$ and $t = 2b + 1$ for some nonnegative integers a and b with $a + b \geq 2$. If $s \equiv t \pmod{4}$, then the statement follows by Proposition 16.52. Nevertheless, we verify the statement without this assumption. Assume, to the contrary, that there is a rainbow mean coloring $c : E(G) \to \mathbb{N}$ of G with $\mathrm{rm}(c) = s + t$. Thus,

$$\sum_{v \in V(G)} cm(v) = \binom{s+t+1}{2} = \frac{(s+t+1)(s+t)}{2} = (2a + 2b + 3)(a + b + 1).$$

Let $\{X, Y\}$ be a partition of the set $[s + t]$ where $|X| = t$ and $|Y| = s$ such that the sum of elements in X is x and the sum of elements in Y is y. Since $x + y = (2a + 2b + 3)(a + b + 1)$ and $sx = ty$, it follows that

$$sx = ty = t[(2a + 2b + 3)(a + b + 1) - x] = t(2a + 2b + 3)(a + b + 1) - xt.$$

Thus, $sx + tx = x(s + t) = t(2a + 2b + 3)(a + b + 1)$ or $(2a + 2b + 2)x = t(2a + 2b + 3)(a + b + 1)$. However then, $2x = t(2a + 2b + 3)$, which is an odd integer. This is a contradiction. Therefore, $\mathrm{rm}(G) \geq s + t + 1$. ∎

The following result was obtained in [113].

Theorem 16.55 *If s and t are positive integers with $\min\{s, t\} \geq 2$, then*

$$\mathrm{rm}(K_{s,t}) = \begin{cases} s + t & \text{if at least one of } s \text{ and } t \text{ is even} \\ s + t + 1 & \text{if both } s \text{ and } t \text{ are odd.} \end{cases}$$

We describe a proof of Theorem 16.55 in the case when s and t are both even. Let $G = K_{s,t}$ with partite sets $U = \{u_1, u_2, \ldots, u_s\}$ and $W = \{w_1, w_2, \ldots, w_t\}$, where s and t are both even. We show that $\mathrm{rm}(G) = s + t$. By Proposition 16.47, it suffices to show that there is a rainbow mean coloring of $K_{s,t}$ with rainbow mean index $s + t$. We proceed with the following three steps:

(1) The set $[s + t]$ is partitioned into the two subsets $X = \{x_1, x_2, \ldots, x_t\}$ and $Y = \{y_1, y_2, \ldots, y_s\}$ where $x_1 < x_2 < \cdots < x_t$ and $y_1 < y_2 < \cdots < y_s$ such that $s \sum_{i=1}^{t} x_i = t \sum_{j=1}^{s} y_j$.

(2) A $t \times s$ matrix $M = [a_{ij}]$ is constructed such that sx_i is the sum of the entries in row i for $1 \le i \le t$ and ty_j is the sum of the entries in column j for $1 \le j \le s$.

(3) The matrix $M = [a_{ij}]$ is used to construct a rainbow mean coloring c of $K_{s,t}$. For each vertex u_j of $K_{s,t}$ where $1 \le j \le s$, we define a t-vector $\text{code}(u_j) = (a_{1j}, a_{2j}, \ldots, a_{tj})$ to be column j in M. This in turn gives rise to the corresponding s-vectors $\text{code}(w_i) = (a_{i1}, a_{i2}, \ldots, a_{is})$ to be row i in M for each vertex w_i where $1 \le i \le t$. The edge coloring $c : E(K_{s,t}) \to \mathbb{N}$ is defined by $c(w_i u_j) = a_{ij}$ for each pair i, j of integers with $1 \le i \le t$ and $1 \le j \le s$. Since the chromatic means of the vertices of $K_{s,t}$ are given by $\text{cm}(u_j) = x_j$ for $1 \le j \le s$ and $\text{cm}(w_i) = y_i$ for $1 \le i \le t$, it follows that $\{\text{cm}(v) : v \in V(K_{s,t})\} = [s + t]$ and so $\text{rm}(c) = s + t$.

We may assume that $s \le t$. Since s and t are positive even integers, it follows that $s = 2a$ and $t = 2b$ for some integers a and b with $1 \le a \le b$. First, we partition the $(2a + 2b)$-element set $[2a + 2b]$ into the two subsets

$$X = [b] \cup [b + 2a + 1, 2a + 2b] \text{ and } Y = [b + 1, b + 2a],$$

where then $|X| = 2b$ and $|Y| = 2a$. Let $X = \{x_1, x_2, \ldots, x_{2b}\}$ where $x_1 < x_2 < \cdots < x_{2b}$ and $Y = \{y_1, y_2, \ldots, y_{2a}\}$ where $y_1 < y_2 < \cdots < y_{2a}$. Since

$$x = \sum_{i=1}^{2b} x_i = 2b^2 + 2ab + b \text{ and } y = \sum_{i=1}^{2a} y_i = 2a^2 + 2ab + a,$$

it follows that $2ax = 2by = 4a^2b + 4ab^2 + 2ab$.

Next, we define a $(2b) \times (2a)$ matrix $M = [a_{i,j}]$ such that

- $2ax_i$ is the sum of the entries in row i for $1 \le i \le 2b$ and

- $2by_j$ is the sum of the entries in column j for $1 \le j \le 2a$.

For the first b rows in M, we define each entry in row i to be i where $1 \le i \le b$. That is,

$$M = \begin{bmatrix} 1 & 1 & 1 & \cdots & \cdots & \cdots & 1 \\ 2 & 2 & 2 & \cdots & \cdots & \cdots & 2 \\ \vdots & \vdots & \vdots & \vdots & \vdots & \vdots & \vdots \\ b & b & b & \cdots & \cdots & \cdots & b \\ \vdots & \vdots & \vdots & \vdots & \vdots & \vdots & \vdots \end{bmatrix}. \qquad (16.4)$$

Thus, the sum of the entries in row i is $2ax_i = 2ai$ for $1 \le i \le b$. Next, we determine the remaining entries in the last $b - 1$ rows of M. Since we want the sum of the entries in column 1 to be $2by_1 = 2b(b + 1)$ and $\sum_{i=1}^{b} a_{i1} = \frac{b(b+1)}{2}$, it follows that

$$\sum_{i=b+1}^{2b} a_{i1} = 2b(b + 1) - \frac{b(b+1)}{2} = \frac{b(3b+3)}{2}.$$

We now choose each of the remaining b entries $a_{b+1,1}, a_{b+2,1}, \ldots, a_{2b,1}$ in column 1 as either

$$\left\lfloor \frac{b(3b+3)}{2b} \right\rfloor = \lfloor \tfrac{3b+3}{2} \rfloor \text{ or } \left\lceil \frac{b(3b+3)}{2b} \right\rceil = \lceil \tfrac{3b+3}{2} \rceil$$

so that the sequence $a_{b+1,1}, a_{b+2,1}, \ldots, a_{2b,1}$ is nondecreasing and the column sum is $2by_1 = 2b(b+1)$. Furthermore, since we want the sum of the entrees in row $b+1$ to be $2ax_{b+1} = 2a(b+2a+1)$, we choose each of the entries $a_{b+1,2}, a_{b+1,3}, \ldots, a_{b+1,2a}$ as either

$$\left\lfloor \frac{2a(b+2a+1)-a_{b+1,1}}{2a-1} \right\rfloor \text{ or } \left\lceil \frac{2a(b+2a+1)-a_{b+1,1}}{2a-1} \right\rceil$$

so that the sequence $a_{b+1,2}, a_{b+1,3}, \ldots, a_{b+1,2a}$ is nondecreasing and the row sum is $2ax_{b+1} = 2a(b+2a+1)$. We now proceed in this manner to determine the remaining entries in M.

Finally, we define the edge coloring $c : E(G) \to \mathbb{N}$ by $c(w_i u_j) = a_{ij}$ for every pair i, j of integers where $1 \le j \le 2a$ and $1 \le i \le 2b$. By the defining property of the matrix M, it follows that $\mathrm{cm}(u_j) = x_j$ for $1 \le j \le 2a$ and $\mathrm{cm}(w_i) = y_i$ for $1 \le i \le 2b$. Thus, $\{\mathrm{cm}(v) : v \in V(G)\} = X \cup Y = [2a + 2b]$ and so $\mathrm{rm}(c) = 2a + 2b = s + t$. As an illustration, we construct a rainbow mean coloring c of $K_{4,6}$ with $\mathrm{rm}(c) = 10$. In this case, $a = 2$ and $b = 3$. We partition the set $[10]$ into the two sets $X = [3] \cup [8, 10] = \{1, 2, 3, 8, 9, 10\}$ and $Y = \{4, 5, 6, 7\}$. Thus, $x = 33$ and $y = 22$ and so $4x = 6y = 132$. Using the technique described above, we obtain the

$$6 \times 4 \text{ matrix } M = \begin{bmatrix} 1 & 1 & 1 & 1 \\ 2 & 2 & 2 & 2 \\ 3 & 3 & 3 & 3 \\ 6 & 8 & 9 & 9 \\ 6 & 8 & 11 & 11 \\ 6 & 8 & 10 & 16 \end{bmatrix}. \text{ The resulting rainbow mean coloring } c \text{ of}$$

$K_{4,6}$ with $\mathrm{rm}(c) = 10$ is shown in Figure 16.19, where the four vertices u_1, u_2, u_3, u_4 of the partite set U of $K_{4,6}$ are drawn in bold.

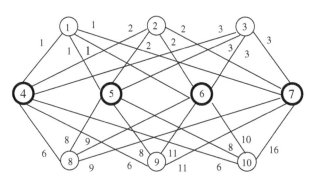

Figure 16.19: Constructing a rainbow mean coloring of $K_{4,6}$

There is one class of complete bipartite graphs not considered in Theorem 16.55, namely the stars. The following result was obtained in [43].

Theorem 16.56 *If G is a star of order $n \geq 3$, then*

$$\mathrm{rm}(G) = \begin{cases} n & \text{if } n \text{ is odd} \\ n+2 & \text{if } n \text{ is even.} \end{cases}$$

Consequently there are connected graphs G of order $n \geq 3$ for which $\mathrm{rm}(G)$ is n, $n+1$, or $n+2$. In fact, the following conjecture appeared in [43].

Conjecture 16.57 *For every connected graph G of order $n \geq 3$,*

$$n \leq \mathrm{rm}(G) \leq n+2.$$

Exercises for Chapter 16

1. Let k denote the smallest positive integer n for which $\mathrm{smaj}(P_n) \neq \mathrm{maj}(P_n)$.

   (a) What is k?
   (b) If $n > k$, is $\mathrm{smaj}(P_n) \neq \mathrm{maj}(P_n)$?

2. For a connected graph G of order 3 or more with $\mathrm{maj}(G) = k$, the **majestic chromatic number** $\chi_{\mathrm{maj}}(G)$ of G is the minimum number of vertex colors in a majestic k-edge coloring of G.

   (a) Prove that $\chi_{\mathrm{maj}}(G) = 3$ for every connected graph G having $\mathrm{maj}(G) = 2$.
   (b) Give an example of two connected graphs G_1 and G_2 such that $\mathrm{maj}(G_i) = 3$ for $i = 1, 2$ but $\chi_{\mathrm{maj}}(G_1) \neq \chi_{\mathrm{maj}}(G_2)$.
   (c) Do there exist three non-isomorphic connected graphs H_1, H_2, and H_3 such that $\mathrm{maj}(H_i) = 3$ for $i = 1, 2, 3$ but no two of which have the same majestic chromatic number?

3. Determine the strong majestic index of the 3-regular graphs of order 6.

4. Let G be a connected graph of order 3 or more and let c be a strong majestic coloring of G. If c is also a proper edge coloring of G, then c is called a **strong edge coloring** of G. The minimum positive integer k for which G has a strong k-edge coloring is called the **strong chromatic index** $\chi'_s(G)$ of G.

   (a) Show that $\chi'_s(G) \geq \max\{\chi'(G), \mathrm{smaj}(G)\}$ for every connected graph G of order 3.
   (b) Find the smallest positive integer k for which there is a graph G with $\chi'_s(G) = k$ and $\mathrm{smaj}(G) < k$.

5. Let G_0 be a connected graph of order 3 or more with edge set $E(G) = \{e_1, e_2, \ldots, e_m\}$. For $1 \leq i \leq m$, the graph G_i is obtained from G_0 by subdividing the edge e_i exactly once. If $\max\limits_{0 \leq i \leq m} \{\mathrm{maj}(G_i \vee K_2)\} = 3$, then what can be said about G_0?

6. Give an example of a connected graph G of order at least 3 such that $\text{maj}(G) \neq \text{maj}(H)$, where H is the graph obtained by subdividing each edge of G exactly once.

7. The graph G of Figure 16.20 has $\chi(G) = 3$ and therefore, by Corollary 16.13, $\text{maj}(G) = 3$. Let $\{V_1, V_2, V_3\}$ be the partition of $V(G)$ into the three color classes $V_1 = \{u_1, u_2, u_3, u_4\}$, $V_2 = \{v_1, v_2, v_3, v_4\}$, and $V_3 = \{w_1, w_2, w_3, w_4\}$. Complete a majestic 3-edge coloring $c : E(G) \rightarrow [3]$ where (i) $c(e) = 1$ if $e = uv$ where $u \in V_1$ and $v \in V_2$, (ii) $c(e) = 2$ if $e = uw$ where $u \in V_1$, $w \in V_3$, and (iii) $c(e) = 3$ if $e = vw$ where $v \in V_2$ and $w \in V_3$ and v and w are adjacent to a vertex (not necessarily the same vertex) in V_1.

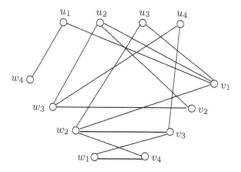

Figure 16.20: The graph in Exercise 7

8. Prove Theorem 16.20: *If G is a connected graph of order at least 3, then* $\text{roy}(G) \geq \lceil \log_2 \chi(G) \rceil + 1$.

9. Show that if G is a connected graph of order 4, then $\text{sroy}(G) = 3$.

10. For a graph G, recall that the corona $\text{cor}(G)$ of G is that graph obtained from G by adding a pendant edge at each vertex of G.

    (a) Prove that if G is a connected graph of order 4 or more, then

    $$\text{sroy}(\text{cor}(G)) \leq \text{sroy}(G) + 1.$$

    (b) Use (a) to prove that if G is a royal-zero graph, then so is $\text{cor}(G)$.

    (c) For a royal-one graph G, must $\text{cor}(G)$ be royal-one?

11. Let P denote the Petersen graph.

    (a) Show that $\text{roy}(P) = 3$.

    (b) Let S denote the set of 2-element and 3-element subsets of $[4]$. Thus, $|S| = 10$. Verify that P is a royal-zero graph by showing that there exists a strong royal 4-edge coloring c of P such that $\{c'(v) : v \in V(P)\} = S$.

(c) Let T denote the set of 3-element and 4-element subsets of $[4]$. Thus, $|T| = 5$. Show that there exists a strong royal 4-edge coloring c of P such that for each $A \in T$, there are two (nonadjacent) vertices u and v such that $c'(u) = c'(v) = A$.

12. For a connected graph G of order 3 or more with $\mathrm{roy}(G) = k$, the **royal chromatic number** $\chi_{\mathrm{roy}}(G)$ of G is the minimum number of vertex colors in a royal k-edge coloring of G. Let $\chi_{\mathrm{maj}}(G)$ be the majestic chromatic number of G as defined in Exercise 2.

   (a) Recall that we know of no connected graphs G for which $\mathrm{maj}(G) \neq \mathrm{roy}(G)$. Determine $\chi_{\mathrm{maj}}(C_{10})$ and $\chi_{\mathrm{roy}}(C_{10})$.

   (b) What does the problem in (a) tell us?

13. We mentioned that we know of no connected graph G for which $\mathrm{maj}(G) \neq \mathrm{roy}(G)$. However, there are many connected graphs G for which $\mathrm{smaj}(G) \neq \mathrm{sroy}(G)$. Indeed, these two numbers can differ by a great deal. Determine a positive integer r such that $\mathrm{smaj}(K_{1,r})$ and $\mathrm{sroy}(K_{1,r})$ differ by exactly 100.

14. We have seen that C_7 is a royal-one graph but $C_7 + e$ is a royal-zero graph for every chord e of C_7. Show that the same thing occurs in the case of regal colorings, that is, for the regal-one graph C_7, show that $C_7 + e$ is a regal-zero graph for every chord e of C_7.

15. We have seen that $P_7 = (v_1, v_2, \ldots, v_7)$ is a royal-one graph, as is the 7-cycle $P_7 + v_1 v_7$.

   (a) Show that there is an edge e in $\overline{P}_7$ different from $v_1 v_7$ such that $P_7 + e$ is a regal-one graph.

   (b) Show that there is an edge f in $\overline{P}_7$ such that $P_7 + f$ is a regal-zero graph.

16. For a positive integer k, $\mathcal{P}^*([k])$ is the collection of all nonempty subsets of $[k]$. Furthermore, let $\mathcal{P}^\circ([k])$ be the collection of all proper subsets of $[k]$. Thus, $|\mathcal{P}^*([k])| = |\mathcal{P}^\circ([k])| = 2^k - 1$.

   (a) Let G be a connected graph of order 4 or more with $\mathrm{sroy}(G) = k$. Show that there exists a k-edge coloring $f : E(G) \to \mathcal{P}^\circ([k])$ of G such that the vertex coloring $f' : V(G) \to \mathcal{P}^*([k])$ defined by $f'(v) = \bigcap_{e \in E_v} f(e)$ for each $v \in V(G)$ is vertex-distinguishing.

   (b) Show that if $\mathrm{sroy}(G) = k$ for the graph G in (a), then there is no $(k-1)$-edge coloring $g : E(G) \to \mathcal{P}^\circ([k-1])$ of G such that the vertex coloring $g' : V(G) \to \mathcal{P}^*([k-1])$ defined by $g'(v) = \bigcap_{e \in E_v} g(e)$ for each $v \in V(G)$ is vertex-distinguishing.

17. We have seen that $\mathrm{sroy}(P_7) = 3$ and $\mathrm{sreg}(P_7) = 4$. Therefore, there exists a strong royal 3-edge coloring of P_7, but there is no strong regal 3-edge coloring of P_7. (For a positive integer k, $\mathcal{P}^\circ([k])$ is defined in Exercise 16.)

(a) Show, however, that there exists a 3-edge coloring $f : E(P_7) \to \mathcal{P}^\circ([3])$ of P_7 such that the vertex coloring $f' : V(P_7) \to \mathcal{P}^\circ([3])$ defined by $f'(v) = \bigcap_{e \in E_v} f(e)$ for each $v \in V(P_7)$ is vertex-distinguishing.

(b) Does there exist a 3-edge coloring $g : E(P_7) \to \mathcal{P}^\circ([3])$ of P_7 such that the vertex coloring $g' : V(G) \to \mathcal{P}^\circ([3])$ defined by $g'(v) = \bigcup_{e \in E_v} g(e)$ for each $v \in V(P_7)$ is vertex-distinguishing?

18. We have seen that $\mathrm{sroy}(K_7) = 4$ and $\mathrm{sreg}(K_7) = 3$. Therefore, there is a strong regal 3-edge coloring of K_7, but there is no strong royal 3-edge coloring of K_7. Thus, there exists an edge coloring $c : E(K_7) \to \mathcal{P}^*([3])$ of K_7 such that the vertex coloring $c' : V(K_7) \to \mathcal{P}^*([3])$ defined by $c'(v) = \bigcap_{e \in E_v} c(e)$ for each $v \in V(K_7)$ is vertex-distinguishing. Does such an edge coloring of K_7 exist if $\mathcal{P}^*([3])$ is replaced by $\mathcal{P}^\circ([3])$?

19. For which sets $S \in \mathcal{P}([3])$, does there exist an edge coloring $c : E(P_7) \to \mathcal{P}^*([3]) - \{S\}$ of P_7 such that the vertex coloring $c' : V(P_7) \to \mathcal{P}^*([3]) - \{S\}$ defined by $c'(v) = \bigcap_{e \in E_v} c(e)$ for each $v \in V(P_7)$ is vertex-distinguishing?

20. (a) Let G be a connected graph with chromatic number 3. Prove that if there exists a partition $\{V_1, V_2, V_3\}$ of $V(G)$ into three independent sets such that the subgraph $H = G[V_1 \cup V_2]$ induced by the set $V_1 \cup V_2$ is a connected graph of order 3 or more, then $\mathrm{reg}(G) = 3$.

(b) Prove or disprove: If G is a connected graph of order 4 or more with $\chi(G) = k \geq 3$, then there exists an independent set X of vertices of G such that $H = G - X$ is a connected graph of order 3 or more and $\chi(H) = k - 1$.

21. Show for each integer $n \geq 4$ that there is a regal 3-edge coloring c of C_n such that $c'(v) \neq [3]$ for every vertex v of C_n. Consequently, $\mathrm{reg}(C_n) = 3$ for every integer $n \geq 3$.

22. Let $W_n = C_n \vee K_1$ be the wheel of order $n + 1 \geq 5$. Prove that $\mathrm{reg}(W_n) = 3$.

23. Let G be a connected graph of order 3 or more with $\mathrm{reg}(G) = k \geq 3$. The **regal chromatic number** $\chi_{\mathrm{reg}}(G)$ of G is the minimum number of vertex colors in a regal k-edge coloring of G.

(a) Prove that if $\mathrm{reg}(G) = 3$, then $\chi_{\mathrm{reg}}(G) \geq 3$.

(b) Find a class of connected graphs G with $\mathrm{reg}(G) = \chi_{\mathrm{reg}}(G) = 3$.

24. Show that the First Theorem of Graph Theory (Theorem 1.1) is a consequence of Proposition 16.44 and the fact that every graph has an even number of odd vertices is a consequence of Proposition 16.45.

25. If G is a connected graph of order $n \geq 3$ such that $\mathrm{rm}(G) = n + 1$, then what statement corresponds to Proposition 16.48 in this case?

26. Show that $\mathrm{rm}(C_3) = \mathrm{rm}(C_4) = 4$, $\mathrm{rm}(C_5) = 5$, and $\mathrm{rm}(C_6) = 6$.

27. A **proper mean coloring** of a connected graph G of order 3 or more is a mean coloring c of G such that $cm(u) \neq cm(v)$ for every pair u, v of adjacent vertices of G. The maximum value of a vertex color in a proper mean coloring c of G is the **proper mean index** $\mu(c)$ of c. The minimum proper mean index among all proper mean colorings of G is the (proper) **mean chromatic number** $\mu(G)$ of G.

    (a) Show that $\mu(G)$ exists for every connected graph G of order 3 or more.

    (b) Show that $\mu(G) \geq \chi(G)$ for every connected graph G of order 3 or more.

    (c) Give an example of a graph G for which $\mu(G) = \chi(G)$.

28. Determine $\mu(K_{s,t})$ for all complete bipartite graphs $K_{s,t}$ where $s+t \geq 3$. (See Exercise 27 for the definition of $\mu(G)$.)

Chapter 17

The Four Color Theorem Revisited

We began in Chapter 0 by describing the origin of graph colorings. As we noted, this subject grew out of attempts to solve the famous Four Color Problem. From this study, a variety of colorings of planar graphs resulted, namely, proper region colorings of planar graphs, proper vertex colorings of planar graphs, and proper edge colorings of planar graphs (especially cubic maps). This study then led to proper vertex colorings and proper edge colorings of graphs in general. Eventually, more general colorings of graphs resulted. In this final chapter, we once again return to proper colorings of planar graphs and describe a vertex labeling of planar graphs that has a connection with the Four Color Problem.

17.1 Zonal Labelings of Planar Graphs

The trees constitute one of the best-known classes of connected planar graphs. Figure 17.1 shows all trees of order n for $2 \le n \le 5$. For each of these trees, every vertex is labeled with one of the two nonzero elements of $\mathbb{Z}_3$, namely 1 or 2, and the sum (in $\mathbb{Z}_3$) of the labels of the vertices in each tree is 0. These are all examples of a special type of vertex labeling possessed by some connected planar graphs.

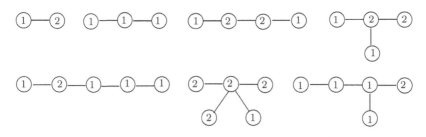

Figure 17.1: The trees of order $2, 3, 4,$ and 5

We now consider connected planar graphs embedded in the plane in general, namely connected plane graphs. For a connected plane graph G, let $f : V(G) \to \mathbb{Z}_3^* = \{1, 2\}$ be a vertex labeling of G with the nonzero elements of $\mathbb{Z}_3$. The **label of a region** R of G is defined as the sum in $\mathbb{Z}_3$ of the labels of the vertices on the boundary of R. A region of a plane graph G is also called a **zone** of G. If the label of every zone of G is 0, then the labeling f is called a **zonal labeling** of G. If a connected plane graph G has a zonal labeling, then G is a **zonal graph**. The topic of zonal labelings of plane graphs was introduced by Cooroo Egan in 2014 and discussed in [38]. Since a tree has only one zone, every tree in Figure 17.1 is a zonal graph. In fact, we have the following.

Theorem 17.1 *Every nontrivial tree is zonal.*

Proof. Let T be a tree of order $n \geq 2$. If n is even, then we can assign the label 1 to half of the vertices of T and the label 2 to the other half, giving us a sum of 0 in $\mathbb{Z}_3$. Suppose, then, that $n \geq 3$ is odd. Thus, $n = 2k + 1$ for some positive integer k. Then $n = 2k + 1 = (k - 1) + (k + 2)$. If we assign the label 1 to $k + 2$ vertices of T and assign the label 2 to the other $k - 1$ vertices of T and add these labels, we have $1 \cdot (k + 2) + 2(k - 1) = (k + 2) + 2k - 2 = 3k = 0$ in $\mathbb{Z}_3$.   ■

Another familiar class of planar graphs are the cycles. While each cycle has two zones, the boundary of each zone is the same, namely the cycle itself. Using the same proof as in Theorem 17.1, we have the following.

Theorem 17.2 *Every cycle is zonal.*

There are two zonal labelings of the 3-cycle C_3, namely the labeling in which all three vertices of C_3 are labeled 1 and the labeling where all three vertices are labeled 2. In a zonal labeling of the 4-cycle C_4, two vertices must be labeled 1 and two vertices must be labeled 2. The labels of the vertices of C_4 could alternate 1 and 2, or two vertices of the same label can be adjacent. For a zonal labeling of the 5-cycle C_5, one vertex must be labeled 1 or 2 with the other four vertices receiving the remaining label. These zonal labelings of C_3, C_4, and C_5 are shown in Figure 17.2.

Let f be a labeling of the vertices of a plane graph G with the elements of $\mathbb{Z}_3^*$. The labeling $\overline{f}$ of the vertices of G with the elements of $\mathbb{Z}_3^*$ defined by $\overline{f}(v) = 3 - f(v)$ for each vertex v of G is called the **complementary labeling** of G. Our interest in complementary labelings lies with the following theorem.

Theorem 17.3 *If f is a zonal labeling of a connected plane graph G, then the complementary labeling $\overline{f}$ is also a zonal labeling of G.*

While all nontrivial trees have a single zone and all cycles have two zones, the plane graphs $K_4 - e$ and K_4 in Figure 17.3 have three and four zones, respectively. A zonal labeling of a triangle must assign the same label, say 1, to all three vertices. This implies that all four vertices of both $K_4 - e$ and K_4 must be labeled 1. While every zone of K_4 is labeled 0 (and therefore K_4 itself is zonal), only the two interior

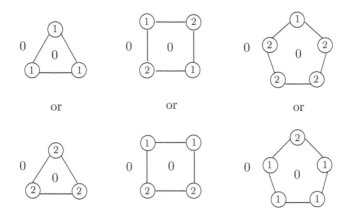

Figure 17.2: Zonal labelings of small cycles

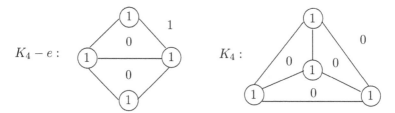

Figure 17.3: The plane graphs $K_4 - e$ and K_4

zones of $K_4 - e$ have the label 0. The label of the exterior zone of $K_4 - e$ is 1 and so $K_4 - e$ is not zonal. The complementary labeling of $K_4 - e$ also results in each interior zone being labeled 0 but with the exterior zone being labeled 2.

To determine whether a given plane graph G is zonal, it is sometimes important to know the manner in which G is embedded in the plane. For example, consider the planar graph H_1 shown in Figure 17.4 together with a zonal labeling of H_1. Thus, H_1 is a zonal plane graph. In the same figure, a different planar embedding of H_1 is shown, resulting in the plane graph H_2, which is not zonal.

In 1932, Whitney [201] proved that if G is a 3-connected planar graph, then there is only one way that G can be embedded in the plane, that is, G is uniquely embeddable in the plane, always resulting in the same regions regardless of how G is embedded in the plane. Since K_4 is 3-connected, it is uniquely embeddable in the plane. On the other hand, $K_4 - e$ is not 3-connected. We state this theorem for future reference.

Theorem 17.4 (Whitney's Theorem) *Every 3-connected planar graph is uniquely embeddable in the plane.*

The planar graph $C_3 \square K_2$ shown in Figure 17.5 is also 3-connected and therefore can be embedded in the plane in only one way. This planar graph has five zones and is zonal. A zonal labeling is also shown in Figure 17.5.

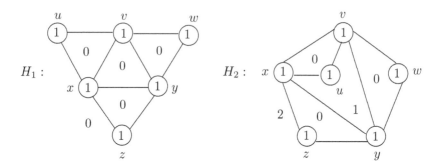

Figure 17.4: Two planar embeddings of a planar graph

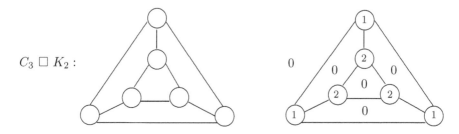

Figure 17.5: The graph $C_3 \square K_2$ and a zonal labeling

Not only is the graph $C_3 \square K_2$ zonal, but $C_n \square K_2$ is zonal for every integer $n \geq 3$.

Theorem 17.5 *The graph $C_n \square K_2$ is zonal for every integer $n \geq 3$*

Proof.   Let $C = (u_1, u_2, \ldots, u_n, u_{n+1} = u_1)$ and $C' = (v_1, v_2, \ldots, v_n, v_{n+1} = v_1)$ be two n-cycles. By adding the edges $u_i v_i$, $1 \leq i \leq n$, to C and C', the graph $C_n \square K_2$ results. Since $C_n \square K_2$ is 3-connected, there is a unique embedding of this graph in the plane, resulting in $n + 2$ zones, one having C as its boundary, one having C' as its boundary, and n having a 4-cycle as their boundary, namely $(u_i, v_i, v_{i+1}, u_{i+1}, u_i)$ for $1 \leq i \leq n$. Let f be a zonal labeling of an n-cycle C_n and let $\overline{f}$ be the complementary labeling of f. Define a zonal labeling g of $C_n \square K_2$ by $g(u_i) = f(u_i)$ and $g(v_i) = \overline{f}(v_i)$ for $1 \leq i \leq n$ This is illustrated in Figure 17.6 for $n = 5$. This is a zonal labeling of $C_n \square K_2$.   ∎

17.2   Zonal Labelings and Edge Colorings

We have now seen that the cubic graphs K_4 and $C_n \square K_2$, $n \geq 3$, are all zonal. These graphs are all members of a special class of plane graphs. A connected bridgeless cubic plane graph is referred to as a **cubic map**. This class of graphs was encountered in Chapter 10 when we stated a theorem of Peter Tait (Theorem 10.16)

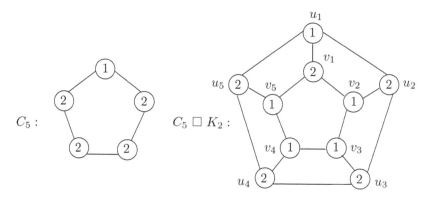

Figure 17.6: A zonal labeling of $C_5 \square K_2$

that dealt with proper edge colorings of cubic maps and the Four Color Problem. First, let's recall this theorem.

Theorem 17.6 (**Tait's Theorem**) *The zones* (regions) *of a cubic map G can be properly colored with four or fewer colors if and only if there exists a proper 3-edge coloring of G.*

As we mentioned in Chapter 10, Tait believed that this theorem would lead to a proof of the Four Color Theorem as he thought it would be quite easy to prove that there is a proper 3-edge coloring of every cubic map. However, it turned out that this problem is equivalent to (and therefore as difficult as) the Four Color Problem. After the Four Color Theorem was verified, we now know that there is a proper 3-edge coloring of every cubic map.

If a planar graph G is 2-connected, then the boundary of every zone in a planar embedding of G is a cycle. Therefore, to determine whether 2-connected planar graphs are zonal, it is useful to obtain additional information on zonal labelings of cycles. At the same time, we will have a special interest in proper edge colorings of cycles. While proper edge colorings of odd cycles require three colors and proper edge colorings of even cycles require only two colors, we will consider proper edge colorings of all cycles using three colors, namely colors from the set $\{1, 2, 3\}$. Therefore, a proper edge coloring of an odd cycle must use all three of these colors and a proper edge coloring of an even cycle may use all three colors. We represent an n-cycle C, $n \geq 3$, as follows: $C = (v_1, v_2, \ldots, v_{n-1}, v_n, v_{n+1} = v_1)$. We assume that C is drawn in the plane as indicated in Figure 17.7.

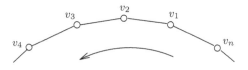

Figure 17.7: Proceeding about a cycle in a counter-clockwise direction

In a proper edge coloring c of the n-cycle C, the color of an edge $v_i v_{i+1}$ ($1 \le i \le n$) of C is then denoted by $c(v_i v_{i+1})$, which, therefore, is one of the colors $1, 2$, or 3. For a proper edge coloring c of a cycle with colors from the set $\{1, 2, 3\}$, the edge coloring $\bar{c}$ of C defined by $\bar{c}(e) = 4 - c(e)$ for each edge e of C is called a **complementary coloring** c. An important characteristic of a complementary coloring of a proper 3-edge coloring of a cycle is the following.

Proposition 17.7 *The complementary coloring of a proper 3-edge coloring of a cycle C with colors from the set $\{1, 2, 3\}$ is also a proper 3-edge coloring of C with colors from $\{1, 2, 3\}$.*

For such a cycle C drawn in the plane, we will assume that we proceed about the interior of C in a counter-clockwise direction. In this case, after an edge $v_{i-1} v_i$ is encountered, the next edge encountered is $v_i v_{i+1}$, as indicated in Figure 17.7. If the color $c(v_i v_{i+1})$ of the edge $v_i v_{i+1}$ immediately follows the color $c(v_{i-1} v_i)$ of the edge $v_{i-1} v_i$ numerically (that is, if 2 follows 1, 3 follows 2, or 1 follows 3), then the vertex v_i on C is said to be of **type** 1. Otherwise, v_i is of **type** 2. Figure 17.8 shows the case where $c(v_{i-1} v_i) = 1$.

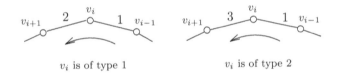

v_i is of type 1 v_i is of type 2

Figure 17.8: The type of a vertex

In addition to proper 3-edge colorings of cycles, we will be interested in labelings of the vertices of cycles where each label is an element of $\mathbb{Z}_3^*$. Therefore, if there is such a labeling of the vertices of a cycle C such that the sum in $\mathbb{Z}_3$ of the labels of the vertices of C is 0, then the labeling is a zonal labeling of C.

An important characteristic of complementary colorings is the following.

Proposition 17.8 *If a vertex v of C is of type i where $i \in \{1, 2\}$ in a proper 3-edge coloring c of C, then v is of type $3 - i$ in the complementary coloring $\bar{c}$ of the edges of C.*

Figure 17.9 shows possible 3-edge colorings of both a 3-cycle and a 4-cycle with colors from the set $\{1, 2, 3\}$, where the vertices of each cycle are labeled with their types. In each case, the resulting vertex labeling is a zonal labeling. We show that this is not a coincidence.

Theorem 17.9 *Each proper 3-edge coloring of an n-cycle C, $n \ge 3$, with colors from the set $\{1, 2, 3\}$ results in a zonal labeling of C.*

Proof. We proceed by induction on the order of a cycle. Other than their complementary colorings, there is only one proper 3-edge coloring of a 3-cycle and

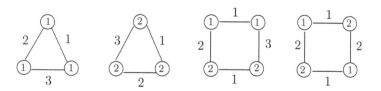

Figure 17.9: Zonal labelings of a 3-cycle and a 4-cycle

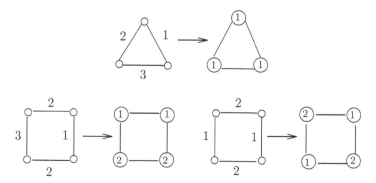

Figure 17.10: The 3-colorings of a 3-cycle and a 4-cycle

essentially two proper 3-colorings of a 4-cycle. (See Figure 17.10.) In each case, the labeling of the vertices with their types is a zonal labeling.

Assume for an $(n-1)$-cycle C', $n \geq 4$, with a proper 3-edge coloring of C' using the colors from the set $\{1, 2, 3\}$ that the labeling f of the vertices of C' with their types is a zonal labeling. Let $C = (v_1, v_2, \ldots, v_n, v_1)$ be an n-cycle having a proper 3-edge coloring c of C with the colors from the set $\{1, 2, 3\}$. If only two of the three colors, say 1 and 2, are used in the coloring, then the colors 1 and 2 alternate, n is even, and the types of the vertices alternate between 1 and 2. Thus, the labeling f of the vertices of C with their types is a zonal labeling. Hence, we may assume that all three colors $1, 2$, and 3 are used in a proper 3-edge coloring of C. Necessarily, there are three consecutive edges of C having distinct colors, say $c(v_1 v_2) = 1$, $c(v_2 v_3) = 2$, and $c(v_3 v_4) = 3$. Then both v_2 and v_3 are of type 1. Let C' be the $(n-1)$-cycle obtained from C by identifying v_2 and v_3 and denoting the resulting vertex by v. Thus, $C' = (v_1, v, v_4, \ldots, v_n, v_1)$ is an $(n-1)$-cycle. Then $c(v_1 v) = 1$ and $c(v v_4) = 3$. This gives rise to a proper 3-edge coloring of C' where v is of type 2. All other vertices of C' have the same type as in C. By the induction hypothesis, the labeling of the vertices of C' with their types is a zonal labeling of C'. Let $s(C')$ denote the sum in $\mathbb{Z}_3$ of the vertex labels of C'. Since

$$s(C) = s(C') + f(v_2) + f(v_3) - f(v) = s(C') + 1 + 1 - 2 = s(C') = 0,$$

it follows that the labeling of the vertices of C with their types is also a zonal labeling. The theorem then follows by the Principle of Mathematical Induction. ∎

The converse of Theorem 17.9 is also true.

Theorem 17.10 *For every zonal labeling f of an n-cycle C_n, $n \geq 3$, there exists a proper 3-edge coloring of C_n such that the type of each vertex v of C_n is $f(v)$.*

Proof. We proceed by the Strong Form of Induction on the order of a cycle. There is only one zonal labeling of C_3 (as well as its complementary zonal labeling) and two zonal labelings of C_4, as shown in Figure 17.10. Each corresponds to a proper 3-edge coloring of cycles (also shown in Figure 17.10).

Assume for each zonal labeling f' of a k-cycle C_k for $3 \leq k < n$, where $n \geq 5$, that there corresponds a proper 3-edge coloring of C_k such that for each vertex v of C_k, the type of v is $f'(v)$. Let $C = (v_1, v_2, \ldots, v_n, v_1)$ be an n-cycle with a zonal labeling f. We show that there exists a proper 3-edge coloring of C such that the type of each vertex v of C is $f(v)$.

If each vertex of C has the same label, say 1, then $n \equiv 0 \,(\mathrm{mod}\ 3)$ and the coloring $1, 2, 3, 1, 2, 3, \ldots$ of the edges of C in a counter-clockwise direction results in each vertex of C having type 1. Hence, we may assume that both labels 1 and 2 are used in the zonal labeling f of C. Therefore, there are adjacent vertices of C with distinct labels, say $f(v_3) = 1$ and $f(v_4) = 2$. Let C' be the $(n-2)$-cycle obtained from C by deleting v_3 and v_4 and joining v_2 to v_5. Then $C' = (v_1, v_2, v_5, v_6, \ldots, v_n, v_1)$ is an $(n-2)$-cycle. Let f' be the labeling of C' where $f'(v) = f(v)$ for each vertex v of C'. Since f is a zonal labeling of C, so is f'. By the induction hypothesis, there is a proper 3-edge coloring of C' such that the type of each vertex v of C' is $f'(v)$. We may assume that $c'(v_5v_6) = 1$. We now consider four cases, depending on the labels $f'(v_2)$ and $f'(v_5)$ in C'.

Case 1. $f'(v_2) = f'(v_5) = 1$. Thus, $c'(v_2v_5) = 3$ and $c'(v_1v_2) = 2$. By defining $c(v_4v_5) = 3$, $c(v_3v_4) = 1$, and $c(v_2v_3) = 3$, we obtain a proper 3-edge coloring of C such that the type of each vertex v of C is $f(v)$.

Case 2. $f'(v_2) = f'(v_5) = 2$. Thus, $c'(v_2v_5) = 2$ and $c'(v_1v_2) = 3$. By defining $c(v_4v_5) = 2$, $c(v_3v_4) = 3$, and $c(v_2v_3) = 2$, we obtain a proper 3-edge coloring of C such that the type of each vertex v of C is $f(v)$.

Case 3. $f'(v_2) = 1$ *and* $f'(v_5) = 2$. Thus, $c'(v_2v_5) = 2$ and $c'(v_1v_2) = 1$. By defining $c(v_4v_5) = 3$, $c(v_3v_4) = 1$, and $c(v_2v_3) = 3$, we obtain a proper 3-edge coloring of C such that the type of each vertex v of C is $f(v)$.

Case 4. $f'(v_2) = 2$ *and* $f'(v_5) = 1$. Thus, $c'(v_2v_5) = 3$ and $c'(v_1v_2) = 1$. By defining $c(v_4v_5) = 3$, $c(v_3v_4) = 1$, and $c(v_2v_3) = 3$, we obtain a proper 3-edge coloring of C such that the type of each vertex v of C is $f(v)$.   ■

Corollary 17.11 *For every zonal labeling f of an n-cycle C_n, $n \geq 2$, and for a color a from the set $\{1, 2, 3\}$ assigned to an arbitrary edge e of C_n, there exists a unique proper 3-edge coloring c of C_n such that $c(e) = a$ and the type of each vertex v of C_n is $f(v)$.*

We have now seen that there is a close connection between proper 3-edge colorings and zonal labelings of cycles. We now return to Tait's theorem (Theorem 17.6) to determine possible connections between proper 3-edge colorings and zonal labelings of cubic maps.

Let G be a cubic map. That is, G is a connected, bridgeless, cubic planar graph embedded in the plane. As we noted earlier, the boundary of every zone of G is a cycle. By the Four Color Theorem, there is a proper 3-edge coloring of G. Let such a coloring of G be given, using the colors $1, 2, 3$. We now define the type of a vertex in G. Let v be a vertex of G. The three edges incident with v are colored 1, 2, 3 in some order. The vertex v is of **type** 1 if the colors are encountered in the order 1-2-3, as we proceed clockwise about v and of **type** 2 if the colors are encountered in the order 1-3-2, as we proceed clockwise about v. This is illustrated in Figure 17.11.

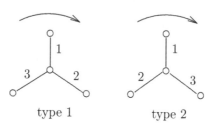

type 1   type 2

Figure 17.11: The vertex types in a proper 3-edge coloring of a cubic map

As an illustration, consider the proper 3-edge coloring of the 3-cube Q_3 shown in Figure 17.12(a) with the colors 1, 2, 3. This coloring results in the vertex types in Figure 17.12(b). The vertex labeling of Q_3 with these types is a zonal labeling since the label of every zone is 0.

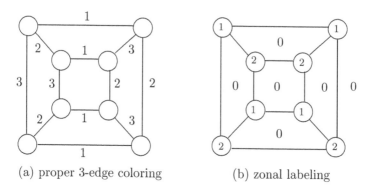

(a) proper 3-edge coloring   (b) zonal labeling

Figure 17.12: A proper 3-edge coloring and a zonal labeling of Q_3

The fact that the proper 3-edge coloring of the cubic map Q_3 produces a zonal labeling with the types of the vertices is, in fact, a characteristic of all cubic maps. Before showing this, we state an observation.

Proposition 17.12 *Let C be the boundary cycle of a zone in a connected cubic plane graph G. A zonal labeling of G restricted to C is a zonal labeling of C.*

Theorem 17.13 *Every proper 3-edge coloring of a cubic map G with the colors $1, 2, 3$ produces a zonal labeling of G.*

Proof.   Let R be a zone of G and let C be the boundary cycle of R. Now, let there be given a proper 3-edge coloring of G with the colors $1, 2, 3$. First, observe that the type of a vertex v on C according to the colors of its three incident edges is the same as the type of v for the coloring of the edges of C. By Theorem 17.9, the labeling of the vertices of C with their types is a zonal labeling of C. Since this is the case for every boundary cycle, this labeling of the vertices of G is a zonal labeling of G. $\blacksquare$

According to Theorem 17.13 then, for each proper 3-edge coloring of a cubic map G with the colors $1, 2, 3$, there is a corresponding zonal labeling of G. In fact, for every zonal labeling of a cubic map G, there is also a corresponding proper 3-edge coloring of G except for the choice of colors used. We now verify this converse of Theorem 17.13.

Theorem 17.14   *Let G be a cubic map on which is defined a zonal labeling f. Then there exists a proper 3-edge coloring of G with the colors $1, 2, 3$ such that the type of each vertex v of G is $f(v)$. In fact, if some edge of G is initially assigned one of the colors $1, 2, 3$, then a proper 3-edge coloring of G is uniquely determined.*

Proof.   Let G be a cubic map with a zonal labeling f and let e be an edge of G, where the color $c(e) \in \{1, 2, 3\}$ is given. Let u be a vertex of G that is incident with e. Thus, $f(u)$ is the label of u in the given zonal labeling of G. Interpreting $f(u)$ as the type of u, we see that the colors of the three edges incident with u are now uniquely determined. Next, we show that a proper 3-edge coloring of G is uniquely determined such that the type of each vertex x of G is $f(x)$. Let h be an arbitrary edge of G. If h is incident with u, then we saw that the color of h is uniquely determined. We may therefore assume that h is not incident with u. Let v be a vertex incident with h. Thus, $v \neq u$.

Since G is 2-connected, it follows by Theorem 2.4 that there are cycles in G containing both u and v. Among all cycles containing u and v, let C be one in which the number of zones interior to C is minimum. If the number of zones interior to C is 1, then C is the boundary cycle of a zone of G. By Proposition 17.12, the zonal labeling f restricted to the vertices of C is a zonal labeling of C. It then follows by Corollary 17.11 that the colors of all edges of C are uniquely determined. Thus, the colors of the two edges of C incident with v are uniquely determined, as is the color of the edge h. We may therefore assume that C encloses more than one zone of G.

Let $C^{(1)}$ be the boundary cycle of a zone $R^{(1)}$ of G lying within C such that u lies on $C^{(1)}$. Thus, v does not lie on $C^{(1)}$. The restriction of the zonal labeling f to the vertices of $C^{(1)}$ is a zonal labeling of $C^{(1)}$ by Proposition 17.12. By Corollary 17.11, the colors of the edges of $C^{(1)}$ are uniquely determined. There exists a sequence $C^{(1)}, C^{(2)}, \ldots, C^{(k)}$, $k \geq 2$, of boundary cycles, where $C^{(i)}$ is the boundary cycle of a zone $R^{(i)}$ lying within C such that $C^{(i)}$ and $C^{(i+1)}$ have an edge in common for $1 \leq i \leq k - 1$ and v is a vertex on $C^{(k)}$. Since the colors of the edges of $C^{(1)}$ are uniquely determined and an edge of $C^{(1)}$ belongs to $C^{(2)}$, the colors of the edges of $C^{(2)}$ are uniquely determined. Continuing in this manner, we see that the colors of the edges of $C^{(k)}$ are uniquely determined and so the colors of the edges on

$C^{(k)}$ incident with v are uniquely determined. Therefore, the color of h is uniquely determined. ∎

As a consequence of Theorem 17.14, it follows that for every cubic map G possessing a zonal labeling, there exists a proper 3-edge coloring of G. This in turn implies the Four Color Theorem. Therefore, if it can be shown that every cubic map has a zonal labeling, then this would provide an alternative proof of the Four Color Theorem. Of course, since the Four Color Theorem is known to be true, we know that there is a proper 3-edge coloring of every cubic map. Hence, the following corollary is a consequence of Theorem 17.13.

Corollary 17.15 *Every cubic map has a zonal labeling.*

Theorem 17.6 (Tait's theorem), Theorem 17.14, and Corollary 17.15 give rise to the following theorem.

Theorem 17.16 *There exists a 4-coloring of the zones of a cubic map G if and only if G has a zonal labeling.*

By Corollary 17.15, we know that every cubic map is zonal. That is, if G is a connected cubic plane graph with no bridges, then G is zonal. Such is not case, however, for any connected cubic plane graph possessing bridges. To show this, it is useful to recall some terminology from Chapter 10. A graph G of odd order n and size m is called *overfull* if $m > \Delta(G)(n-1)/2$. We also recall the following theorem (Corollary 10.10).

Theorem 17.17 *If G is an overfull graph, then G does not possess a proper $\Delta(G)$-edge coloring.*

The following argument is similar to the one used in the proof of Theorem 17.14.

Theorem 17.18 *If G is a connected cubic plane graph with bridges, then G is not zonal.*

Proof. Assume, to the contrary, that there is a zonal connected cubic plane graph G with bridges. Since G is zonal, there exists a zonal labeling f of G. Let $e = uv$ be a bridge of G such that v belongs to an end-block B of G. Since G is cubic, the order of B is at least 5 and v is incident with two edges belonging to B, say vx and vy. Assign the colors $1, 2, 3$ to the three edges incident with v such that the type of v is $f(v)$. Next, we show that the color $c(e)$ of e and the zonal labeling f of G uniquely determines a proper 3-edge coloring c of B such that the type of each vertex v' of B is $f(v')$.

Let h be an arbitrary edge of B where h is incident with a vertex w of B. Since the color of each edge incident with v is uniquely determined by $c(e)$ and $f(v)$, we may assume that h is not incident with v. Since the end-block B is 2-connected, there are cycles in B containing both v and w. Furthermore, each such cycle contains the edges xv and vy. Let C be a cycle containing v and w such that there is a minimum number of zones lying within C.

If C is the boundary cycle of some zone of B, then it follows by Proposition 17.12 that the zonal labeling f restricted to the vertices of C is a zonal labeling of C. Since the colors of the edges vx and vy of C are given, it follows by Corollary 17.11 that the colors of all edges of C are uniquely determined. In particular, the colors of the two edges of C incident with w are uniquely determined, as is the color of the edge h. Hence, we may assume that C is not the boundary cycle of any zone of B. Therefore, C encloses more than one zone of B.

Let $C^{(1)}$ be the boundary cycle of a zone $R^{(1)}$ of B lying within C such that v lies on $C^{(1)}$. The zonal labeling f restricted to the vertices of $C^{(1)}$ is a zonal labeling of $C^{(1)}$. By Proposition 17.12 and Corollary 17.11, the colors of the edges of $C^{(1)}$ are uniquely determined. There exists a sequence $C^{(1)}, C^{(2)}, \ldots, C^{(k)}$ of cycles, where $C^{(i)}$ is the boundary cycle of a zone $R^{(i)}$ lying within C such that $C^{(i)}$ and $C^{(i+1)}$ have an edge in common for $1 \leq i \leq k-1$ and w is a vertex of $C^{(k)}$. Since the colors of the edges of $C^{(1)}$ are uniquely determined and an edge of $C^{(1)}$ belongs to $C^{(2)}$, the colors of the edges of $C^{(2)}$ are uniquely determined. Continuing in this manner, we see that the colors of the edges of $C^{(k)}$ are uniquely determined as well, as is the color of the edge f. Thus, there is a proper 3-edge coloring of B.

Let $p \geq 5$ be the order of B. Therefore, B has $p-1$ vertices of degree 3 and one vertex of degree 2. Thus, p is odd and the size of B is $(3p-1)/2$. Since

$$\frac{3p-1}{2} > \frac{3(p-1)}{2} = \frac{\Delta(B)(p-1)}{2},$$

it follows that B is an overfull graph. By Theorem 17.17, there is no proper 3-edge coloring of B. This is a contradiction.   ∎

By Corollary 17.15 and Theorem 17.18, we then have the following.

Theorem 17.19 *A connected cubic plane graph G is zonal if and only if G is bridgeless.*

Exercises for Chapter 17

1. In a zonal labeling of the 6-cycle C_6, there are r vertices labeled 1 and s vertices labeled 2, where $r + s = 6$. What are the possible values of r and s?

2. In a zonal labeling of the 7-cycle C_7, there are r vertices labeled 1 and s vertices labeled 2, where $r + s = 7$. What are the possible values of r and s?

3. Under what conditions can a zonal labeling of C_n, $n \geq 3$, have r vertices labeled 1 and s vertices labeled 2, where $r + s = n$?

4. Prove Theorem 17.3: *If f is a zonal labeling of a connected plane graph G, then the complementary labeling $\overline{f}$ is also a zonal labeling of G.*

5. Give an example of a connected planar graph G different from the graph of Figure 17.4 where one embedding of G is zonal and another is not zonal.

6. Find a zonal labeling of the planar graph $C_7 \,\square\, K_2$.

7. Give an example of a cubic map G_1 that is not 3-connected with the property that there is a planar embedding G_2 of G_1 and distinct from G_1. Determine if G_1 and G_2 have zonal labelings.

8. For which integers $n \geq 3$, does there exist a proper 3-edge coloring of C_n such that every vertex of C_n has the same type?

9. Prove Proposition 17.12: *Let C be the boundary cycle of a zone in a connected cubic plane graph G. A zonal labeling of G restricted to C is a zonal labeling of C.*

10. Let c be a proper 3-edge coloring of a cubic map G and let C be the boundary cycle of a zone R of G. Let f be the zonal labeling of G obtained from c and let $v \in V(C)$. Prove that if $f(v)$ is the type of v in G determined by c, then $f(v)$ is also the type of v in C determined by c.

11. Determine whether the connected cubic plane graph F shown in Figure 17.13 has a zonal labeling (without using Theorem 17.19).

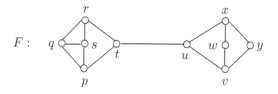

Figure 17.13: The connected cubic plane graph F in Exercise 11

12. Determine the types of the vertices of the properly 3-edge colored cubic map H shown in Figure 17.14.

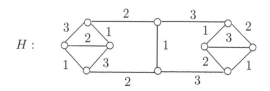

Figure 17.14: The cubic amp H in Exercise 12

13. Figure 17.15 shows a 2-connected plane graph H whose vertices are labeled with elements of $\mathbb{Z}_3^*$.

    (a) Is this labeling a zonal labeling of H?

    (b) Let $C = (u, v, z, x, u)$. Is the sum of the labels of the vertices of C equal to $0 \in \mathbb{Z}_3$?

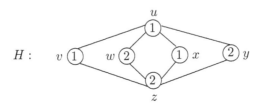

Figure 17.15: The 2-connected plane graph H in Exercise 13

14. We have seen that every nontrivial tree and every cycle are zonal. Which unicyclic graphs are zonal?

Bibliography

[1] M. O. Albertson, You can't paint yourself into a corner. *J. Combin. Theory Ser. B* **73** (1998) 189-194.

[2] M. O. Albertson and E. H Moore. Extending graph colorings. *J. Combin. Theory Ser. B* **77** (1999) 83-95.

[3] A. Ali, G. Chartrand, J. Hallas, and P. Zhang, *Extremal problems in royal colorings of graphs.* Preprint.

[4] N. Alon and M. Tarsi, Colorings and orientations of graphs. *Combinatorica* **12** (1992) 125-134.

[5] E. Andrews, Z. Bi, D. Johnston, C. Lumduanhom, and P. Zhang, On k-Ramsey numbers of stripes. *Util. Math.* **106** (2018) 233-249.

[6] E. Andrews, G. Chartrand, C. Lumduanhom, and P. Zhang, Stars and their k-Ramsey numbers. *Graphs and Combin.* **33** (2017) 257-274.

[7] E. Andrews, F. Fujie, K. Kolasinski, C. Lumduanhom, and A. Yusko, On monochromatic subgraphs of edge-colored complete graphs. *Discuss. Math. Graph Theory.* (2014) 5-22.

[8] E. Andrews, E. Laforge, C. Lumduanhom, and P. Zhang, On proper-path colorings in graphs. *J. Combin. Math. Combin. Comput.* **97** (2016) 189-207.

[9] D. Archdeacon and P. Huneke, A Kuratowski theorem for non-orientable surfaces. *J. Combin. Theory Ser. B* **46** (1989) 172-231.

[10] J. P. Ballantine, A postulational introduction to the four color problem. *Univ. Washington Publ. Math.* **2** (1930) 1-16.

[11] M. Behzad, *Graphs and Their Chromatic Numbers.* Ph.D. Thesis, Michigan State University (1965).

[12] L. W. Beineke and A. J. Schwenk, On a bipartite form of the ramsey problem. In *Proceedings of the Fifth British Combinatorial Conference* (Univ. Aberdeen, Aberdeen) (1975), 17–22.

[13] A. Benjamin, G. Chartrand, and P. Zhang, *The Fascinating World of Graph Theory*. Princeton University Press, Princeton, NJ (2015).

[14] C. Berge, *Theory of Graphs and Its Applications*. Methuen, London (1962).

[15] C. Berge, Perfect graphs. *Six Papers on Graph Theory*. Indian Statistical Institute, Calcutta (1963) 1-21.

[16] Z. Bi, A. Byers, and P. Zhang, Rainbow Hamiltonian-connected graphs. *Bull. Inst. Combin. Appl.* **78** (2016) 8-22.

[17] Z. Bi, A. Byers, and P. Zhang, Proper Hamiltonian-connected graphs. *Bull. Inst. Combin. Appl.* **79** (2017) 31-48.

[18] Z. Bi, G. Chartrand, and P. Zhang, A new view of bipartite Ramsey numbers. *J. Combin. Math. Combin. Comput.* **108** (2019) 193-203.

[19] Z. Bi, S. English, I. Hart, and P. Zhang, Majestic colorings of graphs. *J. Combin. Math. Combin. Comput.* **102** (2017) 123-140.

[20] A. Bialostocki and W. Voxman, Generalizations of some Ramsey-type theorems for matchings. *Discrete Math.* **239** (2001) 101-107.

[21] H. Bielak and M. M. Syslo, Peripheral vertices in graphs. *Studia Sci. Math. Hungar.* **18** (1983) 269-275.

[22] G. D. Birkhoff, A determinant formula for the number of ways of coloring a map. *Ann. Math.* **14** (1912) 42-46.

[23] B. Bollobás, Uniquely colorable graphs. *J. Combin. Theory Ser. B* **25** (1978) 54-61.

[24] B. Bollobás and A. J. Harris, List-colourings of graphs. *Graphs Combin.* **1** (1985) 115-127.

[25] J. A. Bondy and V. Chvátal, A method in graph theory. *Discrete Math.* **15** (1976) 111-136.

[26] V. Borozan, S. Fujita, A. Gerek, C. Magnant, Y. Manoussakis, L. Montero, and Z. Tuza, Proper connection of graphs. *Discrete Math.* **312** (2012) 2550-2560.

[27] N. Bousquet, A. Dailly, E. Duchêne, H. Kheddouci, and A. Parreau, A Vizing-like theorem for union vertex-distinguishing edge coloring. *Discrete Appl. Math.* **232** (2017) 88-98.

[28] R. C. Brigham and R. D. Dutton, A compilation of relations between graph invariants. *Networks* **15** (1985) 73-107.

[29] R. L. Brooks, On coloring the nodes of a network. *Proc. Cambridge Philos. Soc.* **37** (1941) 194-197.

[30] D. Bruce, Gallai-Ramsey number for C_7 with multiple colors, Honors in the Major Theses, University of Central Florida (2017).

[31] F. Buckley, Z. Miller, and P. J. Slater, On graphs containing a given graph as center. *J. Graph Theory* **5** (1981) 427-434.

[32] S. A. Burr and J. A. Roberts, On Ramsey numbers for stars. *Utal. Math.* **4** (1973) 217-220.

[33] D. Cartwright and F. Harary, On colorings of signed graphs. *Elem. Math.* **23** (1968) 85-89.

[34] F. Castagna and G. Prins, Every generalized Petersen graph has a Tait coloring. *Pacific J. Math.* **40** (1972) 53-58.

[35] P. A. Catlin, Hajós' graph-coloring conjecture: Variations and counterexamples. *J. Combin. Theory Ser. B* **26** (1979) 268-274.

[36] G. J. Chang, L-D. Tong, J-H. Yan, and H-G, Yeh, A note on the Gallai-Roy-Vitaver Theorem. *Discrete Math.* **256** (2002) 441-444.

[37] G. Chartrand, S. Devereaux, and P. Zhang, Color-connected graphs and information-transfer paths. *Ars Combin.* **144** (2019) 249-263.

[38] G. Chartrand, C. Egan, and P. Zhang, *How to Label a Graph.* Springer, New York (2019).

[39] G. Chartrand, D. Erwin, F. Harary, and P. Zhang, Radio labelings of graphs. *Bull. Inst. Combin. Appl.* **33** (2001) 77-85.

[40] G. Chartrand and J. B. Frechen, On the chromatic number of permutation graphs. *Proof Techniques in Graph Theory* (F. Harary, ed.), Academic Press, New York (1969) 21-24.

[41] G. Chartrand and D. P. Geller, On uniquely colorable planar graphs. *J. Combin. Theory* **6** (1969) 271-278.

[42] G. Chartrand, J. Hallas, and P. Zhang, Royal colorings of graphs. *Ars Combin.* To appear.

[43] G. Chartrand, J. Hallas, E. Salehi, and P. Zhang, Rainbow mean colorings of graphs. *Discrete Math. Lett.* **2** (2019) 18-25.

[44] G. Chartrand, J. Hallas, and P. Zhang, Color-induced graph colorings. Preprint.

[45] G. Chartrand, T. W. Haynes, M.A. Henning, and P. Zhang, Stratification and domination in graphs. *Discrete Math.* **272** (2003)171-185.

[46] G. Chartrand, T. W. Haynes, H. T. Hedetniemi, and P. Zhang, Rainbow disconnection in graphs. *Discuss. Math. Graph Theory* **38** (2018) 1007-1021.

[47] G. Chartrand, G. L. Johns, K. A. McKeon, and P. Zhang, Rainbow connection in graphs. *Math. Bohem.* **133** (2008) 85-98.

[48] G. Chartrand, G. L. Johns, K. A. McKeon, and P. Zhang, The rainbow connectivity of a graph. *Networks.* **54** (2009) 75-81.

[49] G. Chartrand and J. Mitchem, Graphical theorems of the Nordhaus-Gaddum class. *Recent Trends in Graph Theory.* Springer-Verlag, Berlin (1971) 55-61.

[50] G. Chartrand, L. Nebeský, and P. Zhang, Hamiltonian colorings of graphs. *Discrete Appl. Math.* **146** (2005) 257-272.

[51] A. G. Chetwynd and A. J. W. Hilton, Star multigraphs with three vertices of maximum degree. *Math. Proc. Cambridge Philos. Soc.* **100** (1986) 303-317.

[52] M. Chudnovsky, N. Robertson, P. Seymour, and R. Thomas, The strong perfect graph theorem. *Ann. Math.* **164** (2006) 51-229.

[53] F. R. K. Chung and R. L. Graham, Edge-colored complete graphs with precisely colored subgraphs. *Combinatorica.* **3** (1983) 315-324.

[54] V. Chvátal, Tree-complete ramsey numbers. *J. Graph Theory* **1** (1977) 93.

[55] V. Chvátal and P. Erdős, A note on hamiltonian circuits. *Discrete Math.* **2** (1972) 111-113.

[56] E. J. Cockayne and S. T. Hedetniemi, Towards a theory of domination in graphs. *Networks* (1977) 247-261.

[57] M. Codish M. Frank, A. Itzhakov, and A. Miller, Computing the Ramsey number R(4,3,3) using abstraction and symmetry breaking, *Constraints* **21** (2016) 365–393.

[58] W. C. Coffman, S. L. Hakimi, and E. Schmeichel, Bounds for the chromatic number of graphs with partial information. *Discrete Math.* **263** (2003) 47-59.

[59] S. Cook, The complexity of theorem proving procedures. *Proceedings of the Third Annual ACM Symposium on Theory of Computing* (1971) 151-158.

[60] R. Courant and H. E. Robbins, *What is Mathematics?* Oxford University Press, New York (1941).

[61] M. B. Cozzens and F. S. Roberts, T-colorings of graphs and the channel assignment problem. *Congr. Numer.* **35** (1982) 191-208.

[62] M. B. Cozzens and F. S. Roberts, Greedy algorithms for T-colorings of complete graphs and the meaningfulness of conclusions about them. *J. Combin. Inform. System Sci.* **16** (1991) 286-299.

[63] B. Csaba, D. Kühn, A. Lo, D. Osthus, and A. Treglown, Proof of the 1-factorization and Hamiltonian decomposition conjectures. *Memoirs of the AMS.* (2016).

[64] B. Descartes, A three colour problem. *Eureka* (1947) 21.

[65] S. Devereaux, G. L. Johns, and P. Zhang, Color connection in graphs intermediate to proper and rainbow connection. *J. Combin. Math. Combin. Comput.* **106** (2018) 309-325.

[66] G. A. Dirac, Some theorems on abstract graphs. *Proc. London Math. Soc.* **2** (1952) 69-81.

[67] G. A. Dirac, A property of 4-chromatic graphs and some remarks on critical graphs. *J. London Math. Soc.* **27** (1952) 85-92.

[68] G. A. Dirac, In abstrakten Graphen vorhande vollständigene 4-Graphen und ihre Unterteilungen. *Math. Nachr.* **22** (1960) 61-85.

[69] G. A. Dirac, On rigid circuit graphs. *Abh. Math. Sem. Univ. Hamburg* **25** (1961) 71-76.

[70] P. Elias, A. Feinstein, and C. E. Shannon, A note on the maximum flow through a network. *IRE Trans. on Inform. Theory* **IT 2** (1956) 117-119.

[71] S. English, D. Johnston, and D. Olejniczak, Ramsey type numbers of graphs. Research Report (2015).

[72] S. English, D. Johnston, D. Olejniczak, and P. Zhang, Proper Ramsey numbers of graphs. *J. Combin. Math. Combin. Comput.* **101** (2017) 281-299.

[73] P. Erdös and R. Rado, A combinatorial theorem. *J. London Math. Soc.* **25** (1950) 249-255.

[74] P. Erdős and R. Rado, A partition calculus in set theory. *Bull. Amer. Math. Soc.* **62** (1956) 427-489.

[75] P. Erdős, A. L. Rubin, and H. Taylor, Choosability in graphs. *Proceedings of the West Coast Conference on Combinatorics, Graph Theory and Computing* (Humboldt State University, Arcata, CA, 1979). *Congr. Numer.* **126** (1980) 125-157.

[76] P. Erdös and G. Szekeres, A combinatorial problem in geometry. *Compositio Math.* **2** (1935) 463-470.

[77] P. Erdős and R. J. Wilson, On the chromatic index of almost all graphs. *J. Combin. Theory Ser. B* **23** (1977) 255-257.

[78] A. B. Ericksen, A matter of security, *Graduating Engineer & Computer Careers* (2007) 24-28.

[79] L. Eroh, *Rainbow Ramsey numbers*. Ph.D. Dissertation, Western Michigan University (2000).

[80] L. Euler, Solutio problematis ad geometriam situs pertinentis. *Comment. Academiae Sci. I. Petropolitanae* **8** (1736) 128-140.

[81] R. J. Faudree, R. J. Gould, M. S. Jacobson, and C. Magnant, Ramsey numbers in rainbow triangle free colorings. *Australas. J. Combin.* **46** (2010) 269-284.

[82] R. J. Faudree, S. L. Lawrence, T. D. Parsons, and R. H. Schelp, Path-cycle Ramsey numbers. *Discrete Math.* **10** (1974), 269–277.

[83] R. J. Faudree and R. H. Schelp, All ramsey numbers for cycles in graphs. *Discrete Math.* **8** (1974), 313–329.

[84] H. J. Finck, On the chromatic numbers of a graph and its complement. In: *Theory of Graphs* (Proc. Colloq., Tihany, 1966) Academic Press, New York (1968) 99-113.

[85] C. Flye Sainte-Marie, Solution to problem number 48. *L'Intermédiare des Mathématiciens* **1** (1894) 107-110.

[86] L. R. Ford, Jr. and D. R. Fulkerson, Maximal flow through a network. *Canad. J. Math.* **8** (1956) 399-404.

[87] P. Franklin, The four color problem. *Amer. J. Math.* **44** (1922) 225-236.

[88] O. Frink and P. A. Smith, Irreducible non-planar graphs. *Bull. Amer. Math. Soc.* **36** (1930) 214.

[89] S. Fujita and C. Magnant, Gallai-Ramsey numbers for cycles. *Discrete Math.* **311** (2011) 1247-1254.

[90] S. Fujita, C. Magnant, and K. Ozeki, Rainbow generalizations of Ramsey theory - a dynamic survey. *Theory and Applications of Graphs.* **0** Article 1. (2014)

[91] T. Gallai, Maximum-minimum Sätze über Graphen. *Acta Math. Acad. Sci. Hungar.* **9** (1958) 395-434.

[92] T. Gallai, Transitiv orientierbare Graphen. (German) *Acta Math. Acad. Sci. Hungar* **18** (1967) 25-66.

[93] T. Gallai, On directed paths and circuits. In: *Theory of Graphs; Proceedings of the Colloquium held at Tihany, Hungary*, 1969 (P. Erdős and G. Katona, eds). Academic Press, New York (1969) 115-118.

[94] F. Galvin, The list chromatic index of a bipartite multigraph. *J. Combin. Theory Ser. B* **63** (1995) 153-158.

[95] M. R. Garey and D. S. Johnson, *Computers and Intractability: A Guide to the Theory of NP-Completeness.* W. H. Freeman, New York (1979).

[96] D. P. Geller, Problem 5713. *Amer. Math. Monthly* **77** (1970) 85.

[97] L. Gerencsér and A. Gyárfas, On Ramsey-type problems. *Ann. Univ. Sci. Budapest. Eötvös Sect. Math.* **10** (1967), 167–170.

[98] W. Goddard, M. A. Henning, and O. R. Oellermann, Bipartite Ramsey numbers and Zarankiewicz numbers. *Discrete Math.* **219** (2000), 85–95.

[99] J. Goedgebeur, On minimal triangle-free 6-chromatic graphs, arXiv:1707.07581.

[100] R. E. Greenwood and A. M. Gleason, Combinatorial relations and chromatic graphs. *Canad. J. Math.* **7** (1955) 1-7.

[101] J. Gregory, Gallai-Ramsey number of an 8-cycle, Electronic Theses & Dissertations. 1435 (2016).

[102] J. R. Griggs and R. K. Yeh, Labelling graphs with a condition at distance two. *SIAM J. Discrete Math.* **5** (1992) 586-595.

[103] E. J. Grinberg, Plane homogeneous graphs of degree three without Hamiltonian circuits. *Latvian Math. Yearbook* **4** (1968) 51-58.

[104] R. P. Gupta, The chromatic index and the degree of a graph. *Notices Amer. Math. Soc.* **13** (1966) 719.

[105] E. Györi, M. Horňák, C. Palmer, and M. Woźniak, General neighbour-distinguishing index of a graph. *Discrete Math.* **308** (2008) 827-831.

[106] H. Hadwiger, Über eine Klassifikation der Streckenkomplexe. *Vierteljschr. Naturforsch. ges Zürich* **88** (1943) 133-143.

[107] R. Haggkvist and A. G. Chetwynd, Some upper bounds on the total and list chromatic numbers of multigraphs. *J. Graph Theory* **16** (1992) 503-516.

[108] A. Hajnal and J. Surányi, Über die Auflösung von Graphen in vollständige Teilgraphen. *Ann. Univ. Sci. Budapest, Eötvös. Sect. Math.* **1** (1958) 113-121.

[109] G. Hajós, Über eine Konstruktion nicht *n*-färbbarer Graphen. *Wiss. Z. Martin-Luther-Univ. Halle-Wittenberg. Math.-Nat. Reihe.* **10** (1961) 116-117.

[110] W. K. Hale, Frequency assignment: Theory and applications. *Proc. IEEE* **68** (1980) 1497-1514.

[111] M. Hall, C. Magnant, K. Ozeki, and M. Tsugaki, Improved upper bounds for Gallai-Ramsey numbers of paths and cycles. *J. Graph Theory* **75** (2014) 59-74.

[112] P. Hall, On representation of subsets. *J. London Math. Soc.* **10** (1935) 26-30.

[113] J. Hallas, E. Salehi, and P. Zhang, Rainbow mean colorings of bipartite graphs. Preprint.

[114] F. Harary, Graphs and matrices. *SIAM Review* **9** (1967) 83-90.

[115] F. Harary, *Graph Theory*. Addison-Wesley, Reading, PA (1969).

[116] F. Harary and M. Plantholt, The point-distinguishing chromatic index. In: *Graphs and Applications*. Wiley, New York (1985) 147-162.

[117] J. H. Hattingh and M. A. Henning, Bipartite Ramsey theory. *Util. Math.* **53** (1998) 217-230.

[118] F. Havet, B. Reed, and J-S. Sereni, $L(2, 1)$-Labelling of graphs. *Proceedings of the ACM-SIAM Symposium on Discrete Algorithms*. (2008) 621-630.

[119] T. W. Haynes, S. T. Hedetniemi, and M. A. Henning, *Domination in Graphs*. Springer, New York (2020).

[120] T. W. Haynes, S. T. Hedetniemi, and P. J. Slater, *Fundamentals of Domination in Graphs*. Marcel Dekker, New York (1998).

[121] P. J. Heawood, Map colour theorem. *Quarterly J. Math.* **24** (1890) 332-338.

[122] P. J. Heawood, On the four-colour map theorem. *Quarterly J. Math.* **29** (1898) 270-285.

[123] L. Heffter, Über das Problem der Nachbargebiete. *Math. Ann.* **38** (1891) 477-508.

[124] M. A. Henning and J. E. Maritz, Stratification and domination in graphs II. *Discrete Math.* **286** (2004) 203-211.

[125] M. A. Henning and J. E. Maritz, Simultaneous stratification and domination in graphs with minimum degree two. *Quaestiones Mathematicae* **29** (2006) 1-6.

[126] M. A. Henning and J. E. Maritz, Total restrained domination in graphs with minimum degree two. *Discrete Math.* **308** (2008), 1909–1920.

[127] C. Hierholzer, Über die Möglichkeit, einen Linienzug ohne Wiederholung und ohne Unterbrechung zu umfahren. *Math. Ann.* **6** (1873) 30-32.

[128] M. Horňák and R. Soták, General neighbour-distinguishing index via chromatic number. *Discrete Math.* **310** (2010) 1733-1736.

[129] F. Jaeger, A survey of the cycle double cover conjecture. *Ann. Discrete Math.* **27** (1985) 1-12.

[130] T. Jensen and G. F. Royle, Small graphs with chromatic number 5: A computer search. *J. Graph Theory* **19** (1995) 107-116.

[131] D. Johnston and P. Zhang, A note on the 2-Ramsey numbers of 4-cycles. *J. Combin. Math. Combin. Comput.* **98** (2016) 271-279.

[132] R. M. Karp, Reducibility among combinatorial problems, in *Complexity Computer Computations* (R. E. Miller and J. W. Thatcher, eds.), Plenum, New York (1972) 85-103.

[133] J. Kelly and L. Kelly, Path and circuits in critical graphs. *Amer. J. Math.* **76** (1954) 786-792.

[134] R. Khennoufa and O. Togni, A note on radio antipodal colourings of paths. *Math. Bohem.* **130** (2005) 277-282.

[135] M. Kneser, Aufgabe 300. *Jahresber. Deutsch., Math. Verein.* **58** (1955) 27.

[136] D. König, Über Graphen ihre Anwendung auf Determinantentheorie und Mengenlehre. *Math. Ann.* **77** (1916) 453-465.

[137] D. König, *Theorie der endlichen und unendliehen Graphen.* Teubner, Leipzig (1936).

[138] K. Kuratowski, Sur le problème des courbes gauches en topologie. *Fund. Math.* **15** (1930) 271-283.

[139] S. A. J. Lhuilier, Mémoir sur la polyédrométrie; contenant une démonstration directe du Théorème d'Euler sur les polyédres, et un examen des diverses exceptions auxquelles ce théorème est assujetti. *Annales de Mathématiques Pures et Appliquées* **3** (1812–1813) 169-189.

[140] H. Li, A generalization of the Gallai-Roy Theorem. *Graphs and Combin.* **17** (2004) 681-685.

[141] X. Li and C. Magnant, Properly colored notions of connectivity - a dynamic survey, *Theory and Applications of Graphs*: Vol. 0: Iss. 1, Article 2. (2015).

[142] D. Liu and X. Zhu, Multi-level distance labelings and radio number for paths and cycles. *SIAM J. Discrete Math.* **3** (2005) 610-621.

[143] S. C. Locke, Problem 11086. *Amer. Math. Monthly.* **111** (2004) 440-441.

[144] L. Lovász, A characterization of perfect graphs. *J. Combin. Theory Ser. B* **13** (1972) 95-98.

[145] L. Lovász, Normal hypergraphs and the perfect graph conjecture. *Discrete Math.* **2** (1972) 253-267.

[146] L. Lovász, Kneser's conjecture, chromatic number, and homotopy. *J. Combin. Theory Ser. A* **25** (1978) 319-324.

[147] M. H. Martin, A problem in arrangements. *Bull. Amer. Math. Soc.* **40** (1934) 859-864.

[148] D. W. Matula, The cohesive strength of graphs. In: *The Many Facets of Graph Theory* (G. Chartrand and S. F. Kapoor, eds.) Springer-Verlag, Berlin **110** (1969) 215-221.

[149] K. Menger, Zur allgemeinen Kurventheorie. *Fund. Math.* **10** (1927) 95-115.

[150] B. H. Metzger, Spectrum management technique. Paper presented at 38th National ORSA Meeting, Detroit, MI (1970).

[151] G. J. Minty, A theorem on n-coloring the points of a linear graph. *Amer. Math. Monthly.* **69** (1962) 623-624.

[152] M. Mirzakhani, A small non-4-choosable planar graph. *Bull. Inst. Combin. Appl.* **17** (1996) 15-18.

[153] M. Molloy and B. Reed, A bound on the total chromatic number. *Combinatorica* **18** (1998) 241-280.

[154] J. Mycielski, Sur le coloriage des graphes. *Colloq. Math.* **3** (1955) 161-162.

[155] E. A. Nordhaus and J. W. Gaddum, On complementary graphs. *Amer. Math. Monthly* **63** (1956) 175-177.

[156] O. Ore, Note on Hamilton circuits. *Amer. Math. Monthly* **67** (1960) 55.

[157] O. Ore, *Theory of Graphs*. Math. Soc. Colloq. Pub., Providence, RI (1962).

[158] O. Ore, Hamilton connected graphs. *J. Math. Pures Appl.* **42** (1963) 21-27.

[159] J. Petersen, Die Theorie der regulären Graphen. *Acta Math.* **15** (1891) 193-220.

[160] S. P. Radziszowski, Small Ramsey numbers, *Electronic J. Combin.* Dynamic Surveys (2014).

[161] F. Ramsey, On a problem of formal logic. *Proc. London Math. Soc.* **30** (1930) 264-286.

[162] R. Rashidi, *The Theory and Applications of Stratified Graphs*. Ph.D. Dissertation, Western Michigan University (1994).

[163] R. C. Read, An introduction to chromatic polynomials. *J. Combin. Theory* **4** (1968) 52-71.

[164] B. Reed, ω, Δ, and χ. *J. Graph Theory* **27** (1998) 177-212.

[165] G. Ringel, Problem 25. In: *Theory of Graphs and its Applications* Proc. Symposium Smolenice (1963).

[166] G. Ringel, Das Geschlecht des vollständigen paaren graphen. *Abh. Math. Sem. Univ. Hamburg* **28** (1965) 139-150.

[167] G. Ringel and J. W. T. Youngs, Solution of the Heawood map-coloring problem. *Proc. Nat. Acad. Sci. USA* **60** (1968) 438-445.

[168] N. Robertson and P. D. Seymour, Graph minors. XX Wagner's Conjecture. *J. Combin. Theory Ser. B* **92** (2004) 325-357.

[169] N. Robertson, P. Seymour, and R. Thomas, Hadwiger's conjecture for K_5-free graphs. *Combinatorica* **14** (1993) 279-361.

[170] N. Robertson, P. D. Seymour, and R. Thomas, Tutte's edge-colouring conjecture. *J. Combin. Theory Ser. B* **70** (1997) 166-183.

[171] N. Robertson, P. D. Seymour, and R. Thomas, Cyclically 5-connected cubic graphs. *J. Combin. Theory Ser. B* **125** (2017) 132-167.

[172] N. Robertson, P. D. Seymour, and R. Thomas, Excluded minors in cubic graphs. Preprint.

[173] V. Rosta, On a ramsey-type problem of J. A. Bondy and P. Erdős. *J. Combin. Theory Ser. B* **15** (1973) 94-104.

[174] B. Roy, Nombre chromatique et plus longs chemins d'un graph. *Rev AFIRO* **1** (1967) 127-132.

[175] D. P. Sanders, P. D. Seymour, and R. Thomas, Edge 3-coloring cubic double-cross graphs. Preprint.

[176] D. P. Sanders and R. Thomas, Edge 3-coloring cubic apex graphs. Preprint.

[177] D. P. Sanders, and Y. Zhao, Planar graphs with maximum degree seven are class I. *J. Combin. Theory Ser. B* **83** (2001) 201-212.

[178] P. D. Seymour, Sums of circuits. In: *Graph Theory and Related Topics* (J. A. Bondy and U. S. R. Murty, eds.) Academic Press, New York (1979) 342-355.

[179] P. D. Seymour, Nowhere-zero 6-flows. *J. Combin. Theory Ser. B* **30** (1981) 130-135.

[180] C. E. Shannon, A theorem on coloring the lines of a network. *J. Math. Phys.* **28** (1949) 148-151.

[181] T. Slivnik, Short proof of Galvin's theorem on the list-chromatic index of a bipartite multigraph. *Combin. Probab. Comput.* **5** (1996) 91-94.

[182] B. M. Stewart, On a theorem of Nordhaus and Gaddum. *J. Combin. Theory* **6** (1969) 217-218.

[183] G. Szekeres, Polyhedral decompositions of cubic graphs. *Bull. Austral. Math. Soc.* **8** (1973) 367-387.

[184] G. Szekeres and H. S. Wilf, An inequality for the chromatic number of a graph. *J. Combin. Theory.* **4** (1968) 1-3.

[185] P. G. Tait, Remarks on the colouring of maps *Proc. Royal Soc. Edinburgh* **10** (1880) 501-503, 729.

[186] C. Thomassen, 3-List-coloring planar graphs of girth 5. *J. Combin. Theory Ser. B* **64** (1995) 101-107.

[187] C. Thomassen, Some remarks on Hajós' conjecture. *J. Combin. Theory Ser. B* **93** (2005) 95-105.

[188] W. T. Tutte, On the imbedding of linear graphs in surfaces. *Proc. London Math. Soc.* **51** (1949) 474-483.

[189] W. T. Tutte, A short proof of the factor theorem for finite graphs. *Canad. J. Math.* **6** (1954) 347-352.

[190] W. T. Tutte and C. A. B. Smith, On unicursal paths in a network of degree 4. *Amer. Math. Monthly* **48** (1941) 233-237.

[191] T. van Aardenne-Ehrenfest and N. G. de Bruijn, Circuits and trees in oriented linear graphs. *Simon Stevin* **28** (1951) 203-217.

[192] L. M. Vitaver, Determination of minimal coloring of vertices of a graph by means of Boolean powers of the incidence matrix. (Russian) *Dokl. Akad. Nauk. SSSR* **147** (1962) 758-759.

[193] V. G. Vizing, On an estimate of the chromatic class of a *p*-graph. (Russian) *Diskret. Analiz.* **3** (1964) 25-30.

[194] V. G. Vizing, Critical graphs with given chromatic class. *Metody Diskret. Analiz.* **5** (1965) 9-17.

[195] V. G. Vizing, Coloring the vertices of a graph in prescribed colors (Russian) *Diskret. Analiz* No. 29 *Metody Diskret. Anal. v Teorii Kodov i Shem* (1976) 3-10.

[196] M. Voigt, List colourings of planar graphs. *Discrete Math.* **120** (1993) 215-219.

[197] K. Wagner, Üer eine Eigenschaft der ebene Komplexe. *Math. Ann.* **114** (1937) 570-590.

[198] M. E. Watkins, A theorem of Tait colorings with an application to the generalized Petersen graphs. *J. Combin. Theory* **6** (1969) 152-164.

[199] D. J. A. Welsh and M. B. Powell, An upper bound for the chromatic number of a graph and its application to timetabling problems. *Computer J.* **10** (1967) 85-86.

[200] P. Wernicke, Über den kartographischen Vierfarbensatz. *Math. Ann.* **58** (1904) 413-426.

[201] H. Whitney, Congruent graphs and the connectivity of graphs. *Amer. J. Math.* **54** (1932) 150-168.

[202] H. Whitney, The coloring of graphs. *Ann. Math.* **33** (1932) 688-718.

[203] R. K. Yeh, *Labelling Graphs with a condition at distance two.* Ph.D. Thesis, University of South Carolina (1990).

[204] J. W. T. Youngs, Minimal imbeddings and the genus of a graph. *J. Math. Mech.* **12** (1963) 303-315.

[205] J. A. Zoellner and C. L. Beall, A breakthrough in spectrum conserving frequency assignment technology. *IEEE Trans. Electromag. Comput. EMC* – 19 (1977) 313-319.

[206] A. A. Zykov, On some properties of linear complexes (Russian). *Mat. Sbornik* **24** (1949) 163-188.

[207] Minimum distance separation between stations. *Code of Federal Regulations*, Title 47, sec. 73.207.

Index

Names

Albertson, Michael 242-244, 279
Allaire, Frank 25
Alon, Noga 147, 240
Appel, Kenneth 23-26, 208
Archdeacon, Daniel 139
Archimedes 110

Ballantine, John Perry 242
Baltzer, Richard 13
Beall, C. L. 376
Behzad, Mehdi 282
Beineke, Lowell 297
Berge, Claude 163-164, 169, 405
Bertrand, Louis 135
Bialostocki, Arie 327
Bielak, Halina 36
Birkhoff, George David 18-19, 21, 211
Blanuša, Danilo 268
Bollobás, Béla 229, 279
Bondy, J. Adrian 84
Borozan, Valentin 352, 356
Brigham, Robert C. 188
Brooks, Rowland Leonard 186
Burr, Stefan A. 296

Carr, John W. 23
Carroll, Lewis 12, 268, 286
Cartwright, Dorwin 224
Castagna, Frank 268
Catlin, Paul 209-210
Cayley, Arthur 6, 8, 12-13, 205, 262
Chang, Gerard J. 190
Chartrand, Gary 197, 225, 227, 322, 352,

356-357, 364, 389, 395, 408
Chetwynd, Amanda G. 102, 261
Chudnovsky, Maria 170
Chung, Fan 336
Chvátal, Vašek 84, 99, 294
Cockayne, Ernie 405
Codish, Michael 296
Coffman, William C. 183
Collins, Karen 279
Cook, Stephen 179
Courant, Richard 20, 219
Coxeter, Harold Scott MacDonald 13
Cozzens, Margaret 377-379
Cray, Stephen 21
Csaba, Béla 102

de Bruijn, Nicolaas Govert 76-78
De Morgan, Augustus 2-6, 205
Descartes, Blanche 24, 161
Descartes, René 110
Dinitz, Jeffrey Howard 279
Dirac, Gabriel Andrew 12, 66, 83, 102, 167, 208
Dodgson, Charles Lutwidge 12, 268
Dürre, Karl 21-22
Dutton, Ronald D. 188

Edison, Thomas 5
Egan, Cooroo 462
Ehrenfest, Paul 76
Einstein, Albert 20, 76
Elias, Peter 363
Erdős, Paul 99, 230, 234, 256, 292, 300, 327, 329, 334
Ericksen, Anne Baye 343
Eroh, Linda 329

Mathematical Terms